THE BIRTH OF TERRITORY

THE BIRTH OF TERRITORY

STUART ELDEN

THE UNIVERSITY OF CHICAGO PRESS

CHICAGO AND LONDON

STUART ELDEN is professor of political theory and geography at the University of Warwick.

The University of Chicago Press, Chicago 60637
The University of Chicago Press, Ltd., London
© 2013 by The University of Chicago
All rights reserved. Published 2013.
Printed in the United States of America

22 21 20 19 18 17 16 15 14 13   1 2 3 4 5

ISBN-13: 978-0-226-20256-3 (cloth)
ISBN-13: 978-0-226-20257-0 (paper)
ISBN-13: 978-0-226-04128-5 (e-book)

Library of Congress Cataloging-in-Publication Data
Elden, Stuart, 1971-
    The birth of territory / Stuart Elden.
       pages. cm.
    Includes bibliographical references and index.
    ISBN 978-0-226-20256-3 (cloth : alk. paper)—ISBN 978-0-226-20257-0 (pbk. : alk.
paper)—ISBN 978-0-226-04128-5 (e-book) 1. Political geography. 2. Geography,
Ancient. 3. Geography, Medieval. I. Title.
    JC319.E44 2013
    320.1′2—dc23

                                                          2013005902

CONTENTS

ACKNOWLEDGMENTS

This book has been in gestation for many years, and would have taken many more had it not been for the award of a Leverhulme Major Research Fellowship. I am extremely grateful to the Leverhulme Trust for this wonderful opportunity. Although I have been employed by Durham University for almost all of the time I have been working on this project, research leave and the Leverhulme award have meant that I have spent time elsewhere in visiting posts. I would like to thank the University of Virginia; University of Tasmania; University of British Columbia; University of California, Los Angeles; New York University; National University of Singapore; Queen Mary, University of London; University of Washington, Seattle; and the Australian National University for being excellent hosts. Libraries at all of the above institutions and in addition the British Library, Senate House Library, Warburg Institute, Columbia University, and University of York provided the materials I have used. The Geography Department at Durham University provided some financial assistance to help with the costs involved in this book's publication. Chris Orton, Amy Kuttner, and the Palace Green Library helped with images.

I have greatly benefited from the enthusiasm, support, and advice of a number of people. I would like to particularly mention John Agnew, Suzanne Conklin Akbari, Luiza Bialasiewicz, Neil Brenner, Ed Casey, Jeremy Crampton, Veronica della Dora, Elgin Diaz, Colin Gordon, Laurence Hemming, Barbara Hooper, Emma Hutchinson, Engin Isin, Keith Lilley, David Livingstone, Jeff Malpas, Eduardo Mendieta, Alec Murphy, Nisha Shah, Charlie Withers and Haim Yacobi. At Durham I have a number of excellent colleagues, of whom Louise Amoore, Ben Anderson, Angharad Closs Stephens, Ray Hudson, Joe Painter, and Martin Pratt deserve special mention. Abby Collier at the University of Chicago Press has been an excellent

ix

and supportive editor, and Susan Cohan an exceptionally careful copy-editor. Samuel A. Butler and Andrew Burridge helped with the proofs and compiled the index. I am grateful for the reports from three anonymous referees for suggestions and pushing me to justify my approach. Above all, I am grateful to Susan for her love and support.

I have given talks on this project over many years, including in Australia (University of Tasmania; University of Western Sydney; University of New South Wales; Australian National University); Canada (University of Toronto; University of British Columbia; University of Victoria); Cyprus (University of Cyprus); Finland (University of Turku; Lapland University); Denmark (Danish Institute of International Studies); Germany (Technische Universität Berlin; Friedrich-Alexander Universität Erlangen-Nürnberg); Holland (Radboud University Nijmegen); Israel (Ben Gurion University); Hong Kong, China SAR (Hong Kong Baptist University); Italy (University of Palermo; Scuola Superiore Sant'Anna Pisa); Japan (Kyoto University); Macau, China SAR (University of Macau); Singapore (National University of Singapore); the United States (Association of American Geographers, San Francisco, Boston, and Washington, DC; New School for Social Research; University of Pittsburgh; University of Virginia; University of Washington, Seattle; Ohio State University; University of Oregon; University of California, Los Angeles; University of California, Berkeley; University of Arizona; New York University; Yale University); and the United Kingdom (Heythrop College, University of London; Durham University; Bath Royal Literary and Scientific Institution; Staffordshire University; University of Salford; Swansea University; University of Leicester; University of Lancaster; Royal Holloway, University of London; Queen Mary, University of London; Open University; University College London; University of Cambridge; Queen's University Belfast; Newcastle University; University of Westminster; Aberystwyth University; Royal Geographical Society; University of Birmingham; King's College London; Leeds University).

An earlier and longer version of the introduction was published as "Land, Terrain, Territory," *Progress in Human Geography* 24 (2010): 799–817. Parts of chapter 1 appeared in "Another Sense of *Demos*: Kleisthenes and the Greek Division of the *Polis*," *Democratization* 10 (2003): 135–56; and a section of "The Place of the *Polis*: Political Blindness in Judith Butler's *Antigone's Claim*," *Theory and Event* 8 (2005). One section of chapter 3 draws on the second half of "Place Symbolism and Land Politics in *Beowulf*," *Cultural Geographies* 16 (2009): 447–63. The Leibniz discussion in chapter 9 develops claims first made in "Missing the Point: Globalization, Deterritorialization and the Space of the World," *Transactions of the Institute of British Geographers* 30 (2005): 8–19. I am grateful to Sage, Taylor & Francis, Johns Hopkins University Press, and Wiley-Blackwell for permission to reuse this material.

All primary texts are referenced back to the original language, with English translations noted where available. I have frequently modified existing translations. With primary texts, references are made to book, chapter, and section, where possible, to facilitate reference to different editions.

A t the beginning of the second book of his discourse on inequality, Jean-Jacques Rousseau declares:

> The first man who, having fenced off a plot of land [*enclos un terrain*], thought of saying, *this is mine,* and found people simple enough to believe him, was the real founder of civil society. How many crimes, wars, murders, how many miseries and horrors might the human race had been spared by the one who, upon pulling up the stakes or filling in the ditch, had shouted to his kind: Beware of listening to this impostor; You are lost if you forget the fruits of the earth belong to all and that the Earth [*Terre*] belongs to no one.[1]

Conflict over land, at a variety of spatial scales, is a major factor in human affairs, and, as Rousseau suggests, its effects have been almost entirely negative. Yet his argument here is twofold. First, that this event was the foundation of civil society—which, at the time he was writing, still meant *civilized* society, that is, society with some form of structure and power relations. Civil society was, effectively, a society with some form of government, some form of state. It was opposed to the idea of a "state of nature," rather than civil society and state being contrasted, as they were only after Hegel.[2] Second, that if the consequences of this event were to be prevented, the time to challenge was at that precise moment. It was not something to contest subsequently, lest the challenge be seen as a rival plan for division rather than to see division itself as the problem. To believe the imposter was to mean all was lost. Yet, as Rousseau immediately concedes:

But in all likelihood things had by then reached a point where they could not continue as they were; for this idea of property, depending as it does on many prior ideas which could only arise successively, did not take shape all at once in the human mind: Much progress had to have been made, industry and enlightenment acquired, transmitted, and increased from one age to the next, before this last stage of the state of Nature was reached. Let us therefore take up the thread earlier, and try to fit this slow succession of events and of knowledge together from a single point of view, and in their most natural order.[3]

Several things might be said of this continuation. He recognizes that the question of property in land did not arise all of a sudden, but as a stage in a complicated set of relations that would stretch back in time. As he later notes, "From the cultivation of land [*terres*], its division [*partage*] necessarily followed; and from property, once recognized, the first rules of justice necessarily followed."[4] Similar questions can be asked about a very particular understanding of property and political power over land, that of the relation between the state and its territory.

⋘⋙

Territory continues to matter today in a whole range of registers. Take, for example, the post-1989 territorial changes within central and eastern Europe, where successor states to the Soviet Union, Czechoslovakia, and Yugoslavia emerged and in many instances fought over the delineation of their boundaries. Kosovo, Trans-Dnistra, Chechnya, and the breakaway areas of Georgia show the continuation of these issues. We could also look at the conflict between Ethiopia and Eritrea in east Africa; Somalia's fragmentation into *de facto* but unrecognized states; the independence of South Sudan and the ongoing border tensions; the Arab-Israeli conflict; the territorial dimensions of the "war on terror," environmental disasters, resource ownership, migration, and climate change, especially in terms of melting sea ice in the Arctic and the need to delimit maritime boundaries. Self-determination movements, such as the campaign for an independent Kurdistan, the independence of East Timor, the long-running disputes in Western Sahara, Tibet, East Turkistan, and many other areas show that numerous groups seek control of territory occupied by a state.[5] Yet what are these groups claiming? What is being fought over, divided, mapped, distributed, or transformed? Where did this idea of exclusive ownership of a portion of the earth's surface come from? What kinds of complexities are hidden behind that seemingly straightforward definition? Is the standard story that it emerged with the Peace of Westphalia in 1648 sufficient? What different elements made up the modern notion of "territory," and what roots do they have in different historical lineages? Why is it, as Pascal suggests, that "three degrees of latitude upset the whole of jurisprudence and one meridian determines what is true. . . . It is a funny sort of justice marked by a river! True on this side of the Pyrenees, false on the other."[6]

While there are some excellent and important investigations of particular territorial configurations, disputes, or issues,[7] and some valuable textbooks on the topic,[8] there is little that investigates the term *territory* conceptually or historically. This is, in part, because it is generally assumed that territory is self-evident in meaning, and that its particular manifestations–territorial disputes, the territory of specific countries, etc.—can be studied without theoretical reflection on *territory* itself. Although it is a central term within political theory, geography, and international relations, the *concept* of territory has been underexamined.[9] Where it is defined, territory is either assumed to be a relation that can be understood as an outcome of territoriality, or as a bounded space, in the way that Giddens described the state as a "bordered power-container."[10] In the first, the historical dimension is neglected; in the second, the conditions of possibility of such a configuration are assumed rather than examined. Both take the thing that needs explaining as the explanation.

There is a range of reasons for the comparative neglect of territory. First, there is the turn away from reflection on the state, with a rejection of terms associated with territory, such as "boundedness, identity, integrity, sovereignty and spatial coherence."[11] Second, there is the fear of what John Agnew identified as the "territorial trap,"[12] summed up by his admonition that "the spatiality of power . . . need not be invariably reduced to state territoriality."[13] While he was right to insist that territory is only one kind of spatiality,[14] all too often his warnings have not led to a more careful examination of what territory is, and its intrinsic limits, but rather to an avoidance of the topic altogether. It is through a historical conceptual examination that moving beyond "the territorial trap," rather than simply skirting around it, is possible.[15] Third, there is an unhealthy degree of conceptual imprecision regarding the terms *territory* and *territoriality*. This makes it appear that, because there is a wide-ranging literature on territoriality, there is plenty of discussion of territory.

The first thing to note with regard to territoriality is that unlike, say, "spatiality," which is generally understood as a property or condition of space, something pertaining to it, *territoriality* has today a rather more active connotation. The other, older sense of *territoriality*, as the condition, or status of territory, rather than a mode of operating toward that territory, is generally lost, though it would be good to retrieve it. It is equally important to recognize that there are conflicting traditions in the use of the term, in this more modern sense: the first biological, the second social. These may not actually be distinct, and care should be taken to suggest an implied nature/culture divide, but advocates of territoriality do present them in this way. There is therefore a logic to approaching these works under their own terminological division. Earlier work outlined ways in which territory can be understood through a basis in a fundamental biological drive and as a form of animal association.[16] Their work often covers a great deal of ground, within a broad historical sweep, but they continually blur territory and territoriality together, seeing territoriality as a constant human element, played out in different contexts. What is interesting about their work is that they trade on work in animal ethology—itself taking a term from the analysis of humans—in order to understand human behavior.[17] The problem with this is that while it can tell us something about human behavior in space, it is not at all clear that it can tell us something about "territory." In part this is due to the obvious point that human social organization has changed more rapidly than biological drives.

A rather different approach is offered by Robert Sack in *Human Territoriality*.[18] Despite its title, Sack does not suggest a purely biological, determinist approach. He suggests that territoriality is a geopolitical strategy and not a basic vital instinct. Sack claims that while he sees "territoriality as a basis of power, I do not see it as part of an instinct, nor do I see power as essentially aggressive."[19] Sack labels the area or place delimited and controlled through territoriality a *territory*. This means that he uses the term in a very general and nonspecific way. A place can be a territory at times but not at others; "territories require constant effort to establish and maintain"; and as a corollary of the previous definition, they are "the results of strategies to affect, influence, and control people, phenomena, and relationships."[20] Indeed, in his later *Homo Geographicus*, Sack conceives of the general "role of place as territory," suggesting that "the meaning of place in this current book is then very much like that of territory."[21]

Sack effectively argues that territoriality is a social construct, forged through interaction and struggle, and thoroughly permeated with social relations. While his work has some excellent analyses, none of it really gets to grips with the complexities in the term *territory* itself. The problem with this mode of analysis—a problem it shares with the biological approach—is that it is both historically and geographically imprecise. These kinds of understandings seem to transcend historical periods and uneven geographical development, and also function beyond geographical scale. Territories seem to exist at all times and in all geographical contexts: there is no sense of a history of the *concept*. Perhaps this is only to be expected given that the focus is on "territoriality" instead of territory. Specific territories have histories, and Sack is at his best when he approaches the question of territoriality historically, such as in the passages on Renaissance thought, or on the role of capitalism in shaping understandings of space and time.[22] But this is to reduce the complexity to different historical arrangements of the same questions rather than address the much more challenging question of the very concepts themselves having histories. As Soja notes, "Neither my earlier work nor Sack's however, provide a satisfactory social ontology of territoriality."[23] Soja rightly points to the lack of a fundamental basis to the inquiries that were being pursued. How did the concept of territory emerge?

A related analysis to Sack can be found in some of the writings of the Swiss geographer Claude Raffestin. Like Sack, Raffestin is cautious about assuming too straightforward a relation between animal and human territoriality.[24] Rather, he develops a rich account grounded in a reading of Foucault and Lefebvre together. While this has become more common in recent years, Raffestin was pioneering in reading them together in his 1980 book *Pour une géographie du pouvoir*. Raffestin develops Foucault's theory of power, suggesting that "relational space-time is organised by a combination of energy and information."[25] In a sense, energy can be read alongside power; and information with knowledge, the other two terms of the Foucauldian triad of space, knowledge, and power. For Raffestin, "population, territory and authority" are the three elements of the state, and he suggests that "the entire geography of the state derives from this triad."[26]

Raffestin contends that *space* and *territory* are not equivalent, and that using them indiscriminately has led to a lot of confusion. Space is, for Raffestin, the anterior term, because territory is generated from space, through the actions of an actor, who "territorialises" space.[27] This is the potential danger, in that while Raffestin wishes to make an argument for the conceptual precision of territory, he invokes territoriality as the way into this term. The displacement of territory by territoriality blunts the potential of his analysis.[28] What it means is that space becomes something transformed, rather than something that is itself socially produced, of which territory is a specific, historically limited, form. Yet at times Raffestin offers some very valuable insights, particularly evident in his careful and historical examination of the notion of the frontier.[29]

⌘

In identifying some of the reasons why territory has been neglected as a topic of examination, Painter has suggested that "'territoriality' is often treated as complex and dynamic; 'territory' as more straightforward and not in need of sophisticated analysis."[30] While it is difficult to dispute the complexities surrounding territoriality, its dynamism appears not to be historical. Indeed, given that territoriality is so widespread in animal and human behavior, it can only help us to understand territory if that is a term without a history. Rather, it is territory that is conceptually prior to territoriality, even if existentially second. Linguistically the historical record certainly supports this. Strategies and processes toward territory—of which territoriality is but a fraction—conceptually presuppose the object that they practically produce. It is therefore more fruitful to approach territory as a concept in its own right.

The best general study of *territory* remains Gottmann's *The Significance of Territory*, published in 1973. It trades on his earlier book *La politique des États et leur géographie*, in which he claims that "one cannot conceive a State, a political institution, without its spatial definition, its territory."[31] Nonetheless, both there and in *The Significance of Territory*, he also tends to employ the term in an undifferentiated historical sense, as a concept used throughout history.[32] Thus, while he makes a detailed and valuable analysis, he is still perhaps too willing to see territory existing at a variety of spatial scales and in a variety of historical periods. This tends to create an ahistorical, and potentially ageographical, analysis. Recent works by Saskia Sassen and Jeremy Larkins have recognized that territory has a history.[33] Yet unlike both these books, the current study takes "territory" as a concept to be historically examined rather than simply differently ordered at different times. In examining the relation between *place* and *power*—to use these terms as relatively neutral for the moment—in a wide range of historical settings and texts, I show how the concept of territory *emerged* within Western political thought and practice. The history of the concept provides the basis for the more radical claim that the

term *territory* became the way used to describe a particular and historically limited set of practices and ideas about the relation between place
and power.

Territory therefore requires the same kind of historical, philosophical
analysis that has been undertaken by Edward Casey for another key geographical concept, that of place.[34] This is not to suggest, of course, that
territory is the privileged object of social/spatial theory, but rather that
compared to other dimensions, it has been underexamined. There is simply no study of territory comparable to Casey's for place; it is conceptually
much less examined than network; and other terms, such as *landscape*
and *nature*, have received much more careful historical analysis.[35]

As the following chapters demonstrate, a range of questions need to be
considered in thinking about the emergence of territory. One is that territory is a word, a concept, and a practice, and the relation between these
can only be grasped historically. Bishai has suggested that territory can
be "examined in a similar fashion as sovereignty—through conceptual
history."[36] Conceptual history, *Begriffsgeschichte*, pioneered by Reinhart
Koselleck and his colleagues, offers a valuable emphasis on the use of terminology.[37] As Koselleck suggests, "Through the alternation of semasiological and onamasiological questions, *Begriffsgeschichte* aims ultimately
at *Sachsgeschichte*."[38] Translated, this suggests that the alternation needs
to be between which concepts are implied by words (meaning) and what
words are used to denote specific concepts (designation), and thus conceptual history enables us to speak of material history. Yet this work is weak
on practices, and has not, with partial exceptions, been turned toward the
question of territory explicitly.[39] One of the very few attempts to offer a
conceptual history of territory, aside from Bishai herself, is found in the
work of Paul Alliès. His book *L'invention du territoire* was originally a
thesis supervised by Nicos Poulantzas in 1977, entitled "Le territoire dans
la formation de l'Etat national." Alliès suggests that "territory always
seems linked to possible definitions of the state; it gives it a physical basis
which seems to render it inevitable and eternal."[40] It is precisely in order
to disrupt that inevitability and eternal nature that an interrogation of the
state of territory is necessary.

The work of the Cambridge school of contextualist approaches to the
history of political thought, of which Quentin Skinner and J. G. A. Pocock
are perhaps the most significant figures, offers some guidance on methodological principles, but only tangentially in terms of its focus.[41] It is helpful through its insistence on trying to read texts back into the frames in

which they were forged, and to avoid imposing retrospective concerns on them. As such, there is a great deal of emphasis here on language, and the specific words and formulations used. Equally, attempts are made to render these arguments contextually specific. As Skinner stresses, it is important to understand what purpose was being served by a text, and we need to know why someone was writing, and often whom the person was writing against. Otherwise, he suggests, "We shall find ourselves in a position comparable to that of someone listening to prosecution or the defence in a criminal trial without having heard the other side's case."[42]

Important though such methods are, the approach employed here is closer to a genealogical account, of the type Foucault developed from Nietzsche and Heidegger's work.[43] Foucault makes it clear that though the relation between words and things is important, we should not mistake one for the other. Foucault's insistence on the relation between knowledge and power is crucial, as it enables us to move beyond simply the word-concept relation and bring in practices. That said, most of what Foucault says about territory specifically is at best misleading, as the more thorough treatment here demonstrates.[44] Genealogy, though, understood as a historical interrogation of the conditions of possibility of things being as they are, is helpful for a number of reasons. There is no need to choose exclusively between genealogy and these other accounts.[45] Genealogy, as I practice it here, makes use of the kinds of textual and contextual accounts offered by *Begriffsgeschichte* or the Cambridge school but is critical of notions that the production of meaning is reliant on authorial intent.[46] It makes use of the full range of techniques—including etymology, semantics, philology, and hermeneutics—that should inform the history of ideas but pairs them with an analysis of practices and the workings of power. Such a study cannot simply function as a counterhistory, running up against and challenging the established overview. While that might be possible in some instances, for different concepts where a standard history exists, it would be reductive to what a genealogy is. But such a way of writing is wholly inappropriate for a concept whose substantive history does not exist, such as territory. This history needs to be reconstructed, and in detail, in order to provide the foundation upon which the story I am telling can be situated. There is a fundamental need to return to the texts that reveal the concepts that inform the practices. The approach employed is thus both textual, with all references traced back to their original languages, and contextual, in which texts are resituated in their time and place. And it is avowedly political, undertaking this work as part of a wider project that aspires to be a "history of the present."[47]

Territory should be seen as inherently related to, yet ultimately distinct from, two different concepts: land and terrain. Land is a relation of property, a finite resource that is distributed, allocated, and owned—a political-economic question. Land can be bought, sold, and exchanged; it is a resource over which there is competition. Some of Marx's work recognizes the three-way relation of "land-capital-labor," but his comments are relatively cursory.[48] This theme has been picked up by other writers, perhaps most fundamentally in Perry Anderson's *Passages from Antiquity to Feudalism* and *Lineages of the Absolutist State*, which provide a large-scale analysis of state development from within this broad perspective, concentrating on the material forces and economic conditions for different political formations.[49]

Property is important as an indicator, but as Anderson and other writers recognize, conflict over land is twofold: both over its possession and conducted on its terrain.[50] Land is both the site and stake of struggle. In this it differs from conflict over other resources. Strategic-military reasons thus become significant. These can be understood through a notion of terrain, a relation of power, with a heritage in geology and the military, the control of which allows the establishment and maintenance of order. As a "field," a site of work or battle, it is a political-strategic question. While terrain is seen as land form rather than process—that is, as something that is acted upon rather than itself active—work on military uses has recognized the importance of terrain analysis to military success.

Max Weber's analysis of the historical development of the state, and Michael Mann's study of the changing dynamics of power,[51] where they do discuss territory, could be seen to be operating in a way that sees territory as terrain, a political-strategic relation. In his interview with the geographers of the *Hérodote* journal, Foucault deflects their inquiry about his use of spatial categories, suggesting that they are not primarily geographical but instead shot through with power. As he declares, "*Territory* is no doubt a geographical notion, but it's first of all a juridico-political one: the area controlled by a certain kind of power."[52] As his interviewers respond, "Certain spatial metaphors are equally geographical and strategic, which is only natural since geography grew up in the shadow of the military." They make the explicit linkage between the region of geographers and the commanded region, from *regere*; the conquered territory of a province, from *vincere*; and the field as battlefield. Foucault then notes how "the politico-strategic term is an indication of how the military and administration actually come to inscribe themselves both on a material soil and within forms of discourse."[53]

Land and terrain are obviously important notions, and political-economic and political-strategic understandings of territory have considerable merit. Yet, like the approach through territoriality, they tend to fail the historically specific test. As a political-economic relation, the importance of property in land is clear from as far back as there is recorded human history. Political-strategic concerns about conflict over terrain can be similarly seen in a range of contexts. Territory in distinction, at least in its modern sense, but the case can be made for the term *in itself*, seems to be dependent on a number of techniques and on the law, which are more historically and geographically specific. In taking these dimensions into account, this approach exceeds merely *conceptual* history but begins to fold the analysis of practices into its genealogical narrative. Land, terrain, and territory need to be conceptually distinguished, even if in many instances they are practically intertwined. Of course it would be unusual or reductive to see the political-economic, political-strategic, legal, or technique-based models in strict isolation. Political-economic accounts often indicate a strategic relation; strategic work recognizes the dependence on measure and calculation. Yet it is only in seeing the elements together, and in privileging the legal and the technical, that an understanding of territory can be usefully attained. To concentrate on the political-economic risks reducing territory to land; to emphasize the political-strategic blurs it with a sense of terrain. Recognizing both, and seeing the development made possible by emergent techniques, allows us to understand "territory" as a distinctive mode of social/spatial organization, one that is historically and geographically limited and dependent, rather than a biological drive or social need. "Territory" needs to be thought of in its specificity.

⌘

This book therefore seeks to offer an account of the emergence of the concept of territory in Western political thought. It does so primarily through a contextualized reading of the texts of that tradition with one key question: what is the relation between place and power? It is therefore historical in its execution, philosophical in its interrogation of texts, and political and geographical in its significance. Taking a broad historical period— ancient Greece to the seventeenth century—it traces the relation between politics and place in a range of different texts and contexts. This historical period looks at the key moments that led to the formation of our modern concepts. The account shows in detail how elements from classical, medieval, and Renaissance thought differ from our own time, and yet how they came together, were reread in new situations, and were transformed to give the idea of territory we have today. As such, the majority of the book does not discuss "territory" in a narrow, modern sense. The category is foreign to ancient Greek thought, and even the very rare instances of the Latin word *territorium* do not straightforwardly map onto our modern notion. The point is to look at how place and power were understood in these different texts and contexts, and to trace how the modern concept of territory emerged out of these debates.

Chapter 1 begins with discussion of Greek myths of autochthony, the idea that founders of cities were born from the very soil they are situated upon. It offers readings of a range of historians and poets, including Homer, Euripides, and Aeschylus, but particularly concentrates on what Sophocles's *Antigone* can tell us about the relation between place and the *polis*. The chapter then moves to a detailed discussion of Kleisthenes's urban reforms of Athens, and readings of Plato's *Laws* and Aristotle's *Politics* for their determination of political rule and its geographical basis. While Plato was concerned with outlining a design for the *polis*, Aristotle's intent was much more to adumbrate its manifestations and to derive some more general rules. The chapter concludes with a discussion of how we should understand the *polis* as simultaneously a site and a community, in the Greek sense of a *koinon*, a place and the people who inhabit it.

Chapter 2 offers detailed readings of the writings of Julius Caesar and Cicero, the former treating the question of terrain and the military-geography terms he uses and the latter the *res publica*. These are followed by a discussion of the Latin historians, and the spatial vocabulary they used, with a specific focus on Tacitus. The chapter then proceeds with substantial analyses of two key terms: *imperium* and *limes*. This helps establish the understanding of the political and that of boundaries or frontiers in ancient Rome. The Romans understood spatial relations in a rather different way to contemporary politics, even though modern notions are often read back into the earlier period. The question of how we should translate *territorium* is not straightforward: it means lands surrounding a place, usually a city. The lands so described are outside the city walls, predominantly agricultural lands. Yet, on the other hand, the Romans had plenty of ways to describe lands belonging to people or towns: *terra*, *ager*, or the area within *fines*, boundaries. The discussion of the *limes*, the edges or limits of the empire, raises the question of how Rome saw the rest of the world. The chapter discusses the civil war, practices of land reform, the founding myth of Rome, the names of Octavian/Augustus, and ends with a discussion of practices of land surveying that are outlined in the *Corpus Agrimensorum Romanorum* and of the later historian Ammianus Marcellinus.

Chapter 3 begins with a reading of Saint Augustine's two cities, and reads him, along with Jerome and Paulus Orosius, in the context of the barbarian invasions. It moves to an analysis of the work of Boethius and Isidore of Seville and their attempts to preserve the classical heritage. The political context of the time is the fracturing of the West following the collapse of the Roman Empire. Yet this time is unfairly characterized as the "dark ages." Christianity was in the ascendant, and there was a flowering of national histories of various Germanic tribes, including Gregory of Tours on the Franks, Bede on the English, Isidore on the Goths, and Saxo Grammaticus on the Danes. These texts are not merely accounts of these people but actively shape their sense of identity and consequent political practice. The chapter also provides an analysis of the land politics inherent in the *Beowulf* poem, both in terms of the economics of exchange, gifting, and inheritance, but also a more "geopolitical" sense of conflict over land.

Chapter 4 looks at the establishment of the Carolingian Empire. It begins with a discussion of the Donation of Constantine, which claimed to be a text from the fourth century, was forged in the late eighth century, and finally exposed as such in the fifteenth century by Nicholas of Cusa and Lorenzo Valla. The chapter then moves to a discussion of the crowning of Charlemagne and the practices of political ritual and naming that accompanied it. A range of works are analyzed to show what precisely was being established: a new Roman Empire, a political form of Christendom, or more simply a Frankish kingdom. The position of Europe, particularly in relation to the rise of Islam, is discussed. The chapter moves to a discussion of cartography from Rome to the medieval period. Cartography is a key political practice that both represents and produces political space. Jerusalem is often centrally located on maps of this time, providing a context in which to understand the Crusades undertaken to recapture it. The chapter ends with a discussion of feudalism, stressing the political-economic importance of property in land and practices that went alongside it.

Chapter 5 provides a reading of the organic idea of the body politic in the work of John of Salisbury. It examines the idea of the "two swords," in which the pope claimed both temporal power (over the span of human life on the earth) and spiritual power (over sin, salvation, and people's eternal souls). The pope laid claim to supremacy in the latter by right, and appointed or anointed secular rulers such as kings or emperors to act on his behalf in the former. However, this split, originally proposed by papal theorists, began to articulate a scope and purpose of a separate kind of power, which secular rulers and theorists started to develop. The chapter also discusses in detail the rediscovery of Aristotle's political writings and their translation into Latin, initially through the Arabic. Translation is not simply a textual question, but one of practice, because the availability of these texts changed both the language and the substance of political thought. Here there is a particular focus on the work of Thomas Aquinas and Ptolemy of Lucca, and in particular their guidance on how to act politically.

Chapter 6 begins with a discussion of the dispute between Pope Boniface VIII and King Philip the Fair of France. This was concerned with whether the king could tax clergy within his kingdom and who had jurisdiction if members of the clergy committed a crime. Administrative practices therefore have a direct impact on the shaping of the terms of political discourse. The dispute was also directly productive of some extremely important political theory, notably the writings of Giles of Rome and John of Paris. These took opposing views over the respective competencies of the spiritual and temporal rulers. The chapter then moves to detailed readings of three theorists of temporal power: Dante, Marsilius of Padua, and William of Ockham. Dante, better known as the poet of the *Commedia*, was author of the important *Monarchia*, which argued for a resurgent empire free from papal control. Marsilius offered a defense of the smaller political unit of the city. Ockham, who became a political theorist late in life, was an advocate of the Franciscan vow of poverty and believed that the church should be poor. Yet this was not simply a view about property, but a view that the church should absent itself from all worldly concerns.

Chapter 7 discusses the importance of Roman law, and in particular its compilation and codification under the Byzantine emperor Justinian. These texts were unknown to the Latin West for centuries, and when they were discovered, much academic labor by the so-called glossators was needed to make them intelligible. The focus of the chapter is on the two most important Post-Glossators or commentators: Bartolus of Sassoferrato and Baldus de Ubaldis. Bartolus and Baldus put the law to work in fourteenth-century Italian cities, and crucially made the argument that *territorium* and jurisdiction went together. In establishing a spatial determination of legal power, they took the notion of land, or land belonging to an entity, as the thing to which jurisdiction applies, thus providing the extent of rule. Crucially, *territorium* becomes not simply a property of a ruler but the object of rule itself. This was an inherently practical set of arguments: Bartolus and Baldus both made their living from offering legal opinions on cases presented to them, and indeed Bartolus's work on river boundary law is a combination of legal argument and practical techniques. The final part of the chapter looks at how this work provided a missing basis for assertions of temporal power: in distinction to the universal aspirations of the papacy, temporal power was geographically determined. Within his kingdom, the king had the same power as the emperor in the empire. The legacy of this work is found in the reform of church law of Nicholas of Cusa and in secular legal theorists such as Francisco de Vitoria's writing on colonization and Hugo Grotius's work on the law of the sea and the rights of war and peace. Again, these texts are all interventions in contemporary political issues.

Chapter 8 looks at the relation between the Renaissance and the conquest and mapping of the New World. These political events provide an essential background to the texts from this time. Despite how Machiavelli is often read, and translated, he did not have a concept of territory and did not see political power as preeminently related to land. Instead, we need to make sense of his ambiguous notion of *lo stato*. The second part of the chapter looks at the Reformation, and in particular the political writings of Erasmus, Thomas More, and Martin Luther. The establishment of polities with different confessions to Catholicism produced a political as well as religious fracturing within the Holy Roman Empire. Some of these issues are worked through in the writings of Jean Bodin and Giovanni Botero, the former known for his discussions of sovereignty and the latter for the notion of reason of state. But Bodin's work is complicated by looking at the French and Latin versions of his *Six Books of the Republic*, and Botero's writings on the city and the world also need to be interrogated. The chapter concludes with a reading of the role of property in and struggles over land in Shakespeare's *King Lear*.

Chapter 9 begins with a detailed discussion of some unjustly neglected thinkers of the early seventeenth century whose work was integral to thinking through the political and geographical legacy of the Reformation. These include Richard Hooker, Andreas Knichen, and Johannes Althusius. The next part of the chapter offers a reading of the political implications of the scientific revolution, with special focus on Descartes, Spinoza, and the Newton/Leibniz dispute. Hobbes, Filmer, and Locke are then discussed in terms of the relation between politics and land (or at times territory) in their work. The colonial context is particularly crucial to understanding Locke. But the chapter ends by suggesting that Gottfried

Leibniz is the most important political thinker on territory of this period. Leibniz, like Theodor Reinking, Bogislaw Philipp von Chemnitz, and Samuel Pufendorf, is trying to make sense of the fractured political geographies of the Holy Roman Empire, especially in the wake of the Peace of Westphalia. In distinguishing between the majesty of the emperor and the territorial supremacy of the princes, Leibniz provides a strikingly modern definition.

The coda returns to Rousseau and suggests that he comes conceptually too late. He suggests that the time to challenge the person putting up a fence or ditch was at the very moment it was established. But by the time he was writing, the state of territory was widely assumed: it had become almost the static background behind the action of political struggles. His own writings operate within that context. Subsequent thinkers such as Montesquieu, Hume, and Kant all effectively work within the framework of state-territorial politics. For this reason, the book conceptually ends here. Yet state practices and techniques of cartography, surveying, and statistics all continue to develop, and there are many particular histories of states and their territories. The coda therefore outlines ways in which territory came to be understood and practiced as a political technology. This political technology is one of the means by which we can understand the emergence and development of the modern state. The book's aim is to reinscribe the history of space both in the history of political theory and in the history of the state. In this respect, this book is both a history of space and a spatial history, in which questions of space function as both an object and a tool of analysis.[54] It therefore offers an alternative history of the emergence of the modern state from the perspective of its territory. Taking the story of the birth of territory as a lens allows us to shed new light on the history of political thought.

⟨≈⟩

It is important to stress that this is an approach derived from, and directed toward, Western political thought. The problematic term *West* is of course open to question, but it is intended here to be read in relation to a chronology of thought that can be traced from ancient Greece to Roman appropriations and late medieval Latin rediscoveries, providing the conceptual frame within which the emergence of the modern state and its territory occurred. Other traditions would have very different histories, geographies, and conceptual lineages. The specificity of the analysis begun here militates against generalization and pretensions to universalism.[55] Nonetheless, it is hoped that the historical conceptual approach and its specifics would be useful in other such analyses, even if it would need to be supplemented, developed, and critiqued.

The definition of political thought has been widely debated.[56] There is something of an established canon of great thinkers—Plato, Aristotle, Augustine, Aquinas, Machiavelli, Hobbes, etc.—all of whom receive due attention here, even if some of the most familiar are revealed to be less than central to the development of territory. But there are a number of other important thinkers who are known either for work in other areas (Descartes or Leibniz, for instance) or barely at all (writers such as Bartolus, Baldus, and Knichen). Not all of these would have self-identified as "political theorists," but their work offers valuable insights into political questions. A whole range of other texts and practices—legal documents, constitutions, papal bulls, treatises, histories, and works of literature—are utilized along the way. In this sense, I take a catholic approach to the question of genre. Sophocles's *Antigone* is not just a work of great literature that can be read politically, but a political work of literature. *Beowulf* reveals something of attitudes of the time toward questions of land ownership, transfer, and conflict. Shakespeare's greatest works reveal, comment upon, and engage with the politics of his time, even as they speak beyond them.

Territory contains a mix of political, geographical, legal, technical, practical, and relational questions. These are arranged in a particular way in the modern notion. Where these different elements come from is, however, not straightforward, as they have different lineages, emergences, and descents. How different elements were arranged in other political systems, and how they were labeled is the point of this study. In examining the relation between place and power, this study looks at the history of Western political thought to try to trace the emergence of this political technology.

It is a political technology not because it is merely technical. While advances in geometry, land surveying, navigation, cartography, and statistics play a crucial role in the development of territory, the question of technique is broader than this. As Heidegger argued, the essence of technology is not, in itself, technological. Rather, it is a way of grasping and conceiving of the world. These ways of conceiving, which make possible the narrowly defined technological, are crucial to this study. Yet by techniques it is also meant to imply the broader sense of the Greek *techne*, which Foucault examined in his last decade. These techniques, or arts, of governance have an important bearing on the development being examined here. These techniques include legal systems and arguments; political debates, theories, concepts, and practices; colonization and military excursions; works of literature and dictionaries; historical studies, myths, and— the technical in the narrower sense—geometrical instruments, statistical handbooks, maps, land-surveying instruments, and population controls.

Territory is not simply an object: the outcome of actions conducted toward it or some previously supposedly neutral area. Territory is itself a process, made and remade, shaped and shaping, active and reactive. Just as David Harvey argued we should think of the urban process, so too should we think about territory as process or the territorial process.[57] But this may not be enough. One approach of more recent times that is helpful in beginning to broaden the scope of process is the idea of the urban assemblage.[58] While assemblage is a somewhat misleading translation of Deleuze and Guattari's notion of *agencement*, it seeks to capture the plural, heterogenous, contested, and multiple elements that coalesce only to break apart and re-form in the urban fabric, its continual transformation and contestation. But this work has been neglected for the sometimes absence of the political—not merely the political-economic—from its analyses.[59] The idea of a political technology seeks to capture the processual, multiple, and conflictual nature of the bundle of political techniques—in that expanded sense—that make up and transform the contested and diverse notion of territory. Territory cannot simply be understood as the political-economic notion of land, nor even as a political-strategic sense of terrain, but instead comprises the techniques used to—among other elements— measure land and control and manage terrain. The different elements that make up our modern notion can be found in translations of Greek political thought, compilations and rediscoveries of Roman law, struggles in German political action, and the advances of the scientific revolution, among other practices.

At times, the question of territory, or even the more general and plural notion of the place of power, will seem to disappear from the study. One key example is the discussion of the relation between temporal and spiritual power in the late Middle Ages. Yet this does not mean that the debates here have no bearing on the wider inquiry of this study. Indeed, one of the key arguments of this book is that seemingly unconnected discussions are sometimes recoded in significant ways. The temporal-spiritual, or secular-religious, division of power—a distinction based in part on an understanding of time—has important implications for how later thinkers discussed the understanding of space in relation to politics. Crudely put, and to anticipate a discussion that will be made later in much more detail in subsequent chapters, spiritual power, as the power of the church and the pope, becomes understood as power that knows no earthly limits, whereas temporal power, by its nature plural, is divided, limited, and spatially constrained. That latter form of power will come to be understood as exercised over and limited by territory, and eventually as the idea of territorial sovereignty. But this is to anticipate a very long and involved story, or set of stories.

The Birth of Territory builds on the analysis of this topic developed in *Terror and Territory: The Spatial Extent of Sovereignty*. That book demonstrates why territory continues to matter in global politics today, taking the post–Cold War world generally and the "war on terror" specifically as its focus. This book, in distinction, is a far more historical and conceptual study of this crucial topic. The approach is to try to grasp how political-geographical relations were understood in different times and places rather than to assume that the categories with which people in other times and places thought were the same as our own. The idea of a territory as a bounded space under the control of a group of people, with fixed boundaries, exclusive internal sovereignty, and equal external status is historically produced. This book seeks to understand how and why. There is, of course, a danger of presupposing the thing we are looking for, which we then find. But the intent here is more to examine the relation between what is named *territory* and cognate terms, on the one hand, and how particular politics-power-place-practices are labeled, on the other. These semasiological and onomasiological questions—the relation between meaning and designation, between concepts and practices—allow us to trace the birth of what we now, unproblematically, call *territory*.

PART I

CHAPTER ONE

The *Polis* and the *Khora*

AUTOCHTHONY AND THE MYTH OF ORIGINS

Foucault warns us that genealogists will never confuse themselves with a search for origins.[1] It is for this reason that we cannot simply find *a* birth of territory, a singular moment, which could be outlined and its lineages traced backward. Rather, the approach taken here is to ask questions of the texts in terms of the relations between place and power that they pose, to see how they understood things in different ways and with different vocabularies, in order to try to see where strands emerge, intertwine, run to nothing, are picked up, and transformed. So, where do we begin? With a suspicion, a doubt, a question? The intent and attempt of this project has been outlined in the introduction, but the question of beginnings remains to be resolved. It is not the intention to begin this inquiry into the state of territory with an Ur-state, an *Ursprung*, or a primal political leap. Instead, we join the story some way along the path, at a familiar, though less well-known than might be imagined, point, at the site of the Greek *polis*. Martin Bernal's important and ongoing inquiries should act as caution to see this as the root or fountain of Western culture,[2] and earlier configurations of location and political rule should not be downplayed.[3] But a study has to begin somewhere, and the kind of approach being offered here requires some limits of temporality, scope, and especially linguistic competence.

Greek myth is a notoriously complicated and contentious field. To cite it in support of an argument may seem tantamount to collusion with the unconfirmed. A more verifiable source is tragedy, although this too is debatable in supporting a case. But both myth and tragedy were essential to a living *polis*, and so are potentially valuable in recapturing the use of the term.

The myth discussed here, which is often drawn upon in tragedy, is
that of autochthony, the idea that men sprang up fully formed, born of the
earth. There are many variants and variant interpretations of the myth
of autochthony. Loraux draws a distinction between the Platonic myth
of the *gêgenis*, the idea that people were born (*gen*) of the earth (*gê*); and
the autochthonous Athenian or Theban myths—from *autokhthôn*, born
from the earth (*khthôn*) itself (*autos*) of one's homeland.[4] These three main
areas—the role of *gêgenis* and autochthony in Plato, Athens, and Thebes—
will be the focus here, though, as shall be seen, the distinction is not quite
as clear-cut as Loraux suggests.

An early version of the story, which lies behind many of the others, is
found in Isocrates's *Panegyricus*:

> We did not become dwellers in this land by expelling others, nor by
> finding it uninhabited, nor by coming together here as a motley horde
> of many races. We are a lineage so noble and pure that we have for
> all time continued in possession of the very land which gave us birth,
> since we are autochthonous, and can address our *polis* by the very
> names which apply to our nearest kin; for we alone of the Greeks have
> the right to call it at once fatherland, nurse and mother.[5]

In Plato, there are a number of references to the myth of autochthony.
In the little-known *Menexenus*, Socrates is repeating a speech of Aspasia,
the mistress of Pericles. As the speech is a funeral oration, it is not difficult to detect a level of satire against Thucydides's report of Pericles's own
oration,[6] though the speech referring to autochthony is also a parody of
Isocrates.[7] According to Socrates, Aspasia suggested that Athenians were
descended from men who were:

> not foreigners, nor are these their sons settlers in this land, descended
> from strangers who came to our country from abroad. These men were
> autochthonous, sprung from the land itself, living and dwelling in their
> true fatherland, nurtured by no stepmother, as others are, but by their
> mother the land [*khoras*] where they dwell. And now in death they lie
> in the place proper to them, received back again by the mother who
> bore and nurtured them.[8]

In the *Republic*, autochthony is the basis of the "noble lie" (*pseudos*). It
is suggested that with a single noble lie, the rulers themselves, or at least
the rest of the *polis*, can be indoctrinated.[9] This lie, called a "Phoenician

lie,"[10] which probably refers to the tale of Cadmus the Phoenician, discussed below,[11] will be to suggest that

> all the nurture and education we provided happened in a kind of dreamworld; in actual fact, they were at that time being formed and nurtured deep inside the earth. . . . When they were finished products, the earth, their mother, sent them up above ground; and now in their policy making they must regard the country [*khoras*] they find themselves in as their mother and their nurse, they must defend her against invasion, and they should think of the rest of the inhabitants of the *polis* as their earth-born [*gegenon*] brothers.[12]

The noble lie serves a key purpose: it will enable all people to claim noble origins.[13] It makes explicit the close and organic link between the people, the land (*khora*), and the *polis*. In the *Republic*, Plato recognizes the important political implications this can have, even as, in the *Menexenus*, he satirizes the idea. It may not be true, but if it can be believed, it can have a powerful effect as a founding myth. The notion is also treated in the dialogue known as the *Statesman*. *Statesman* is a limited English equivalent of *Politikos*, which means "the possessor of *politiké tekhné* or the skill of uniting and organising a political community."[14] Here, the visitor relates a story of a past age, in which people were born from the earth rather than from other humans.[15] The earthborn (*gêgenis*) race would reform in the earth after their death and come back to life. This would be in accord with the "reversal undergone by all natural cycles."[16] In time the earthborn race was exhausted, because every soul had fulfilled its quota of incarnations.[17]

What is important about the use of the myth in the *Statesman* is that it refers back to a past age, which precedes the current one; and that all humans at that time were earthborn. The implication is that no one can claim uniqueness in being descended from these earthborn humans, because, at the same time, none and all were. As Lane puts it, "No city can claim its founders in these earthbound humans, lodged firmly in an era without politics and deprived of the sexual intercourse by which the *polis* is perpetuated."[18] However, the treatment in the *Statesman* seems to be the exception, and in the use to which Plato envisages the myth can be put in the *Republic*, there is a reflection of the actual situation in Athens and Thebes.[19] The autochthonous birth of Athenians or Thebans is enough to set them apart.[20] Others might be initially migrant people who settled in a certain area, but the people of Athens and Thebes had a deeply rooted attachment to the soil, to the particular place. They were not just born there, but born from there. As Aristotle notes in the *Rhetoric*, good birth for a nation or *polis* is either autochthonous or at least ancient.[21]

The story of Athens is passed down largely through mythic accounts such as those recounted by Apollodorus and is found in Herodotus's *Histories*, and drawn upon in Euripides's play *Ion*. A standard version of the story is that Erichthonios was a miraculous child born from the earth (*ge*), or Gaia, made fertile by Hephaistos's desire for the virgin Athena.[22] Athena had been born from Zeus, with Hephaistos acting as a kind of midwife, splitting Zeus's head open so she could spring forth. It is unclear whether Hephaistos's desire for her was immediate or consequent; usually the story is that she went to have some armor fashioned by him. Hephaistos tried to rape her, and in so doing, he spilled semen on her leg, which she cleaned off with a piece of wool. Athena dropped the wool to the earth, and Erichthonios was born. Earth gave the child to Athena, who brought him up in her temple. Euripides says that Erichthonios was *gêgenous*, "born of the earth"; that Athena took him up from the earth with "virginal hands."[23] Erichthonios's name derives from this act: *erion* (wool) or *eris* (struggle) joined with *khthon* (earth).[24] He is sometimes fused with his grandson Erechtheus and their stories conflated.[25] This gave Athenians a language for speaking about the origin of the city. For Euripides, they are the "renowned earth-born [*autokhthonas*] inhabitants of Athens";[26] for Aristophanes, "The true-born Attics are the genuine old autochthones, native children of the ground."[27]

Erichthonios is both autochthonous *and* a product of a bisexual transaction. Athenians can thus claim to be the children of earth and gods,[28] and in Homer's *Iliad*, Athena fostered the child born by earth.[29] Loraux suggests that Kekrops, the first king of Athens, is a witness or even arbiter of this divine *eris*, but though he is the first king, it is Erichthonius who is the first Athenian. "Kekrops rules and establishes order in a barely civilised land; Erichthonios, in Herodotus, exercises a power that is already political."[30] It is for this reason that Loraux calls Athens the "most 'political' of all the Greek *poleis*."[31] She notes how this notion of autochthony functions as a civic bond, particularly in the funeral orations, of which Pericles's is only the most famous. She suggests that the funeral oration utilizes the patriotic and civic myth of autochthony in order to promote the unity of the Athenian community, and that it "is a political symbol more than a military theme."[32] Despite the original king Kekrops, or the first Athenian Erichthonius, the loyalty of Athenians is not to either of them, but rather to the idea of autochthony.

Not all tales of founding work in this register. Coming to the site of Pharos, Alexander the Great was struck by the advantages of the location, which was a broad isthmus, between a lagoon and sea ending in a broad harbor. He decided to found a city that would bear his name: Alexandria. He wanted to mark out the outer defenses of this new *polis*, but had no chalk to do so. One of the men suggested that they use the barley meal from the soldiers' packs and spread it out on the ground, following Alexander's footsteps. Plutarch recounts that this was "a semi-circle, which was divided into equal segments by lines radiating from the inner arc to the circumference."[33] In Arrian, the soothsayers suggest that this means that the town would prosper and in particular benefit from the fruits of the earth.[34] Plutarch provides a bit more detail. He says that the king was admiring the design when suddenly flocks of all kinds of birds came from the nearby river and lagoon and ate all the meal. While Alexander was concerned about this, the diviners told him it would mean that the city would not merely have sufficient for itself but also provide for neighboring lands.[35]

It is also in tragedy that traces of the story of the autochthonous birth of Thebes can be found.[36] Cadmus wished to sacrifice a cow to Athena, so sent some of his companions to draw water from the spring of Aves. Most of his men were killed by a dragon or serpent that guarded the spring for Aves. The dragon itself is described as earthborn (*gêgenis*) by Euripides.[37] After killing the dragon, Cadmus sacrificed the cow, and Athena commanded him to sow the dragon's teeth into the ground. Warriors, "a golden-helmeted harvest of sown-men [*spartoi*],"[38] burst forth from the ground and fought one another. The story is either that they fought unprovoked, or that Cadmus threw a stone into their midst, and they fought because they blamed one another. Stories agree, however, that they fought until there were only five survivors—Echion (snake-man), Udaeus (man of the ground), Chthonius (man of the earth), Hyperenor (arrogant), and Pelorus (monster).[39] These five found the noble house of the *polis* Thebes, on the land they were born from. Echion is the father of Pentheus in Euripides's *Bacchae*. The story follows that Cadmus had to atone to Aves for a year for the death of the dragon. In Plato's *Laws*, the Athenian Stranger suggests that the story of "the sowing of the teeth and the birth of armed men from them" shows a potential legislator "that the souls of the young can be persuaded of anything if they try."[40]

Saxonhouse has noted that the theme of autochthony is useful in a number of ways. First, and as noted above, it provides a unity to the *polis*. Second, the boundaries of the *polis* are set by nature rather than human agreements. The *polis* is natural, rather than set in opposition to nature.

Third, the land is seen to belong to the people by right, by birth. There was no need for conquest and forced movement of previous inhabitants. Playing a role similar to that social contract theory would many centuries later, the origins of a *polis* could be assumed to be peaceful. The consequence of this is the existing regime is the original and only one. In other words, it is not a regime that had to overthrow a previous one, but the only possible regime, thereby enhancing its legitimacy and security.[41] However, the myth of autochthonous birth had some less desirable consequences too. One of these negatives was the obvious xenophobia toward those who were not descended in the same way and, as a partner to this, a tendency toward an aristocracy.[42] Another is the attitude to women. The public *polis* is the realm of male warriors sprung from the earth.[43] By excluding women from the birth origins of the city, their position generally tends toward marginalization. Indeed, in Athens's case, it is Athena's legitimate distaste for Hephaistos's advances that leads to birth from the earth rather than a woman. The Athenians, like Athena herself, "can be the children of fathers only."[44] However, it is worth noting the feminine imagery of *mother earth* in Isocrates, as well as in Plato's *Menexenus* and *Republic*.[45]

This theme gives a good sense of some of the issues behind the notion of the *polis*. The site of birth and the community of people within that site are key issues. The interplay of *polis*, *khora*, and community are central. Thebes and tragedy remain the focus as these themes are pursued through a reading of Sophocles's *Antigone*.

ANTIGONE AND THE *POLIS*

As Euben notes, "Greek tragedy was about boundaries of space, time and place, about being inside and outside."[46] This is particularly the case in *Antigone*, in which the questions of burial inside or outside the *polis* and exile play central roles. This is a play that has been read and written about by numerous eminent thinkers, among them Aristotle, Hegel, Nietzsche, Heidegger, Lacan, and, most recently, Butler. Here at least the intent is not to discuss their readings at length but to read *Antigone* without accepting a simplistic translation of *polis* as "state" or "city."[47]

The crucial elements of the story are the following. The principal characters—both alive and dead—are members of the royal household. Oedipus had four children by his mother, Jocasta. The two sons—Eteocles and Polyneices—have been fighting over the *polis* of Thebes: Eteocles defending the *polis*, Polyneices attacking it. They meet at the seventh gate

and die by each other's hand. Their maternal uncle Creon, the king of
Thebes, has decreed that Polyneices should be left unburied, unmourned:
"Whoever disobeys in the least will die, his doom is sealed: murder by
public stoning inside the *polis* walls."[48] The opening scene is a discussion between Antigone and Ismene—sisters to the brothers—about what
should be done. Antigone decides to bury the corpse, alone, for Ismene declares she has no strength to "defy the people of the *polis* [*politon*]."[49] Antigone is caught in the act, and confesses instantly when Creon questions
her. Antigone refuses Ismene's attempt to share the blame, claiming, "I
do not care for a loved one who loves in words alone."[50] She is condemned
to be entombed alive, ostensibly "that the *polis* may avoid defilement,"[51]
but the denial of a death with ritual mourning and burial is a symbolic
punishment for one who valued these rites so highly.[52] Equally, though,
the wish to avoid pollution seems a little inadequate given the pollution
caused by Polyneices's lack of burial and Creon's disregard for it.[53] Rather,
Creon seems to have realized by this time that a public stoning—that is,
not simply a stoning in public, but by the public—as originally proposed[54]
will not have the support of the community. Despite persuasion from his
son, Haemon, who is to marry Antigone, Creon is unmoved, and it is only
when the blind prophet Tiresias suggests the gods' disquiet that Creon re-
lents. He realizes his neglect of sacred duty to Polyneices, whose body by
this time has been ravaged by birds and dogs, and first cleans, then burns,
then buries the body. He then makes for Antigone's tomb. The prophecy of
Tiresias had mentioned the interment of the living in a tomb, and the denial of burial for the dead,[55] and when the chorus had instructed Creon to
follow this prophesy, they too had suggested the opposite order to what he
actually does.[56] That is, Creon is supposed to attend to Antigone first, and
with speed,[57] but his delay means that by the time he arrives at her tomb,
he is too late. Antigone has hung herself,[58] Haemon kills himself in grief,
followed by Creon's wife, Eurydice. Realizing the horrific results of his actions, Creon is led from the stage.

The royal household of Thebes—Laius and Jocasta, Oedipus and Jocasta, Eteocles, Polyneices, Antigone, and Ismene—seems fated. At the
beginning of the play, Antigone asks her sister if she knows of any evil
that Zeus will not bring to pass "to those who stem from Oedipus."[59] It is
Ismene's wish to avoid a similar fate to the rest of her family that leads her
not to join Antigone in the burial.[60] When the chorus confront Antigone,
they wonder if her ordeal is payment of a debt from her father.[61] Antigone
confesses this is her most painful thought, that it is a destiny that attends

her "house," which had been earlier hinted at by the chorus.[62] Creon himself, who as the brother of Jocasta claims authority "by closeness of kinship to the dead,"[63] is hardly immune. Despite not being himself a blood
relative of Laius, it is Haemon and Eurydice of his family who are the last
deaths of the play. But it should not be forgotten that this personal, familial, tragedy is also a "political" tragedy, or rather, a tragedy for the *polis*.
The Greek *polis* was founded on kinship, and for the royal family to suffer
such fates inevitably impacts on the *polis*.[64] As even Creon recognizes, his
creation of disorder within his blood kin will impact on relations outside
his family—to act rightly in his family will be to do his duty in the *polis*.[65]
Equally, when he confronts the sisters, he likens the treachery to a viper
in his house.[66] Braun suggests that Creon is "a political tyrant, probably
he has long been a domestic one." (While the terms *political* and *domestic*
are problematic, the point is still well made.) For modern audiences, Braun
suggests this dichotomy is "more apparent than real"; for Athenians, there
would have been no doubt that they are inseparable.[67]

In *Oedipus at Colonus*, Oedipus asks, "Will they even shroud my body
in Theban soil?"[68] Because of his crimes, this is impossible, and he is buried out of sight, and out of site, by Theseus.[69] As Butler has noted, Antigone
"mimes the act of the strong and true Theseus and buries her brother
out of sight, making sure that Polyneices's shade is composed of Theban
dust." She suggests that this burial "might be understood to be for both,
a burial that once reflects and institutes the equivocation of brother and
father. They are, after all, already interchangeable for her, and yet her act
reinstitutes and reelaborates that interchangeability."[70] Elsewhere, Butler
claims that the invocation of the notion of brother taking precedence over
the decree of Creon is ambiguous, as there is nothing "that can successfully restrict its scope of referentiality to the single person, Polyneices."[71]
It could refer to Oedipus or Eteocles. To make this perfectly explicit, consider two points: Oedipus is both father *and* brother to Antigone, as they
share a mother in Jocasta;[72] and though the burial of Polyneices is "out
of sight," it is not out of *site*, for it is within the bounds of the *polis* of
Thebes.[73]

It is important to note that the play initially seems to suggest a need
for balance, because there is a real dilemma between the two sides. On the
one hand, the religious rites to be accorded to the dead, the law of the gods;
on the other, the danger that this dead man posed to the *polis* and its laws.
As Vernant perceptively notes, the term *nomos* (law or convention) for
Antigone "designates the contrary to what Creon, in his circumstances,

also calls *nomos* . . . and in fact the semantic field of *nomos* is extended enough to cover both of these senses, amongst others."[74] It seems that loyalty to the wider community must take priority over a blood relation, even though it is royal blood. This loyalty is most obviously embodied in the chorus, the old citizens of Thebes. They describe the borders of the *polis* as "our borders," its walls as "our walls."[75] When Creon notes that "whoever proves his loyalty to the *polis*—I'll prize that man in death as well as life," the Leader gives cautious assent.[76] But it is not loyalty to the king, because when Creon's actions are revealed to be threatening to the community by Tiresias, the chorus turns against Creon.[77] This has been coming for some time. They express some doubts when they suggest that the initial burial was the work of the gods,[78] but Creon browbeats them into submission. When Creon suggests to Antigone that she is alone in her view, she perceptively remarks that the chorus share her view, "but they keep their mouths shut for you."[79] As she had noted earlier, "Fear seals their mouths."[80] Haemon later points out that the people of the *polis* support Antigone.[81] The chorus, as representatives of the community, show that the *polis* can be reduced neither to the strictly familial (as late as line 875, they still say authority should not be breached) nor to the rule of the king. The *polis* must be understood in a dual sense—as the site *where* the action takes place and as the people *who* live there.

The *polis* is in need of protection as a site—the defense of its walls, the defense of its integrity—and as a populace. Antigone's loyalty is loyalty both to a particular instance (the polysemantic sense of *brother*) and in a wider sense (the laws of the gods). As noted, on the initial discovery of the burial, the chorus question whether this may be the act of the gods (*theelaton*), but in the famous second choral ode, they judge that the *polis* casts out those who "wed themselves to inhumanity"[82]—that is, to a purpose outside of the human community. Therefore, because Antigone's loyalty seems in conflict with the *polis*, they side with Creon. However, they turn to her when the actions of Creon seem to be more challenging to the *polis*. From the other side, Creon's initial actions seem to be directed first and foremost toward the maintenance of the *polis*, and therefore, the chorus support him. He recognizes that as a new ruler, he will stand or fall on the basis of his actions.[83] It is as his actions become increasingly despotic that he is clearly the greater danger. This is encapsulated in the dialogue he has with his son. Initially, the chorus side with Creon ("You seem to say what you have to say with sense")[84] then recognize that both Creon and Haemon should learn from each other ("You are both speaking wisely").[85]

It is when Creon fails to heed this advice and accuses Antigone of treason that the transition is complete:

> Haemon: The whole community [*homoptolis leos*] of Thebes denies it.
> Creon: And is the *polis* about to tell me how to rule?
> Haemon: Now you see? Who's talking like a child?
> Creon: Am I to rule this land [*khthonos*] for others—or myself?
> Haemon: It's no *polis* at all, owned by one man alone.
> Creon: What? The *polis is* the king's—that's the law!
> Haemon: What a splendid king you'd make of a desert island—you and
> you alone.[86]

This passage is revealing in a number of ways. For Creon, Polyneices's treason is mirrored in Antigone's defiance of the edict prohibiting burial; but the community denies this, which Creon immediately identifies with the *polis* telling him how to rule. The *demos*, at this moment, in this text, can be seen as an embodiment of the *polis*. As Haemon points out, to simply rule the *polis* in the interests of one is to neglect what is distinct about the *polis*. In this sense, Antigone's defiance can be seen as upholding the values of a just *polis*.[87] Creon's position of equating the *polis* with the law of the king is sufficiently extreme for the gods and subsequently the chorus to recognize that he is the greater threat. In other words, the human community is an important part of the *polis*. Antigone's laments toward the end of the play include the lines "O my *polis*, all your fine rich sons! [literally "men," *andres*]," and "Land of Thebes, city of all my fathers [*astu patroion*]."[88] Antigone also invokes the chorus as *patrias politai*, "citizens of my fatherland."[89] The familial and the political entity are joined, as they are explicitly in the conjoined term *mother-polis* (*matropolin*) used by the chorus, and again in the term *patroian* used by Creon.[90] The notion of lineage and a temporal, historical sense of the legacy of previous inhabitants is clearly stressed.

But equally, the *polis* is situated, a site or place. Creon asks, "Am I to rule this land [*khthonos*] for others—or myself?" and elsewhere Creon describes himself as king of the realm (*khoras*).[91] In avoiding the word *polis* here, Sophocles is stressing that the site is anterior to the *polis*, even though it clearly forms its ground.[92] This opposition between a place and a *polis* is also found in *Oedipus at Colonus*.[93] Haemon's final retort here demonstrates this again: a desert island with a single person is no *polis*. It has the site, but no people. While the site is necessary to the *polis*, and a recognition of this is necessary to an understanding, it is not itself sufficient.

THE REFORMS OF KLEISTHENES

If *Antigone* demonstrates that the *polis* is simultaneously a place and the people who inhabit it, it also seems evident that these determinations are understood in a qualitative rather than quantitative way. Issues of number and calculation might appear to be alien to the determination of the Greek *polis*. But later understandings of the *polis* demonstrate a number of issues about the division of land and the *demos*. Rather than turning directly to the work of Plato and Aristotle, the discussion is first of the important reforms of Athens by the *politicus* Kleisthenes.

These reforms indeed hinge around the terms *polis* and, particularly, *demos*. This last term is traditionally understood as "people," sometimes the whole community, particularly when assembled, and sometimes a particular section of the community.[94] However, the term is also translated as the "deme," a location. For example, in Homer's *Iliad*, the term *demos* sometimes means people, sometimes land.[95] *Demos*, translated as "deme," was the name used of the units into which Kleisthenes divided Attica.[96] In the work known as *The Athenian Constitution*, attributed to Aristotle but more likely by one of his pupils, the reform is described in a few short but ambiguous and much-disputed lines, partly based on Herodotus's *Histories*.[97]

There are three key passages that will be looked at here. In the first, the general scope of the reform is outlined. Kleisthenes "divided the land [*khoran*] of Attica by demes [*demous*] into thirty parts—ten parts in the city [*astu*], ten in the coastal region [*paralia*] and ten inland [*mesógeois*]— and he called these parts thirds [*trittyes*], and allotted three to each tribe [*phyle*] in such a way that each tribe should have a share in all the areas [*topon*]."[98] The second passage discusses the membership and naming. "He made the men living in each deme fellow-demesmen of one another, so that they should not use their fathers' names and make it obvious who were the citizens but should be named after their demes. He instituted demarchs with the same responsibilities as the old *naucrari* [an earlier division of Athens]; for he named some of the demes after their areas, and some after their founders (not all were there any longer)."[99] The third passage precedes the first two, and notes that Kleisthenes refused to utilize the existing four tribes: "He refused to divide the Athenians into twelve tribes, to avoid allocating them to the already existing thirds (the four tribes were divided into twelve thirds) as if he had used them he would not have succeeded in mixing up the people."[100] The four previous tribes had been named after the sons of Ion—Geleon, Aegicones, Argades, and

Hoples; the new tribes were named after other heroes.[101] A related point is
made in Aristotle's *Politics*: "The sorts of institutions used by Kleisthenes
at Athens, when he wanted to enlarge the democracy . . . are useful . . .
for one should make more and different tribes, combine private rites into
a few common ones, and use every sophism to mix people up as much as
possible with each other and dissolve previous bonds of familiarity."[102]

A number of reasonably uncontested points can be summarized here:
the division of Attica was by demes; these demes were grouped in *trittyes*, or thirds, of which ten were in the city, ten on the coast, and ten inland; and each newly created tribe had three *trittyes*, one from each area.
There are, however, a number of contested issues. All those of concern
here hinge around the meaning of the term *deme*, and its characteristics.
There is one main issue, that of the territorial nature of these reforms,
that will be the close focus. Several subsidiary issues arise from this. One
that is of passing interest in itself is the number of demes. If Herodotus is
read literally, Kleisthenes gave "ten demes to each tribe,"[103] which would
give 100 demes. However, Whitehead suggests reading this as "the demes
in ten groups to the tribes," which seems more plausible.[104] In his detailed
study, Traill proposes 139 demes initially (12 of which were upper and
lower divisions), and 2 later additions.[105] The best source from antiquity
is Strabo, who gives a figure of 170 or 174,[106] but this dates from sometime
after the reforms and so may reflect an increased later figure. For Traill,
though, the discrepancy is explained by the fact that Strabo was referring
to demes in their sense as villages and not as the political units of Kleisthenes's reforms.[107]

It is therefore clear that the meaning of the term *deme* is essential.
Lévêque and Vidal-Naquet note that before Kleisthenes the deme referred
only to rural areas, but with his reforms it was also applied to the urban
wards of Athens.[108] What the deme meant in Kleisthenes's reforms is, however, debatable. For a long time, the consensus was that it was a geographical term of relatively fixed limits. As Whitehead notes, even before *The
Athenian Constitution* was available, "it was regarded as self-evident that
Kleisthenes' procedure was indeed one of *territorial division*—in essence
a task of cartography, with the fixing of boundaries between demes as the
crucial exercise."[109] In his important book on the coastal demes, Eliot suggests that "a Kleisthenic deme was a fixed area of land with an inhabited
centre from which the deme was administered,"[110] and that "each rural
deme possessed one or more inhabited centres or villages and an area of
land around the settlement or settlements determined at the time of the
Kleisthenic organisation."[111] For Eliot, therefore, a deme is both the village

at the center and surrounding areas, which are demarcated and divided. Kleisthenes, Eliot suggests, established "the geographical extent of each deme."[112] But the term *deme* also meant "village," before *and* after Kleisthenes, so there is an ambiguity.

In their valuable study of these reforms, Lévêque and Vidal-Naquet suggest that the needs of administration required subdivisions of the polity, which they suggest were independent of the hereditary *cadres* of the tribes. They note the forty-eight naucraries, which had existed since the seventh century, but suggest that unfortunately we know almost nothing about them. However, they do argue that with the system of naucraries, Attica possessed a rudimentary spatial division, "for both secular and pragmatic ends, alongside the ancient system of tribes, phratries, *gene*, founded on birth and shot through with religious elements."[113] In other words, Attica had two kinds of division, one religious, tribal, and concerned with the populace,[114] and one secular and concerned with places.

Vidal-Naquet notes in a retrospective of this work that he and Lévêque had had an early suspicion that the figures within Kleisthenes's reforms—three, five, and ten—were also important in the contemporaneous thought of the followers of Pythagoras.[115] These numbers are debatable in their importance to Kleisthenes's reforms. Three is obviously important in the *trittyes*, there were ten tribes, but five is much less certain. Lévêque and Vidal-Naquet also interpret Herodotus to mean that there were ten demes in each tribe, which seems inaccurate.[116] However, the conclusion of the work is that "it is not Pythagoreanism that illuminates Kleisthenes's reforms but Kleisthenes's reforms that allow us to grasp certain aspects of Pythagoreanism," including his politics.[117] "If Kleisthenes constructed the first geometric city, it was not Athens that had the first geometric philosophies"—we should not therefore think that geometry created the city, or the city geometry.[118] However, they suggest that "the intellectual atmosphere at the end of the 6th century was characterised by a certain coincidence between the geometric vision of the world, such as formulated by Anaximander, and the political vision of a rational and homogeneous city, such as realised by Kleisthenes."[119] Regardless of the causality, there is here an initial glimpse of the interplay between political and geometric conceptions of number and space. It is possible, they note, that Pythagoras was a pupil of Thales or Anaximander, but whatever, he had acquired an overall conception of the *cosmos*, founded on both astronomy and geometry.[120] But this relation disappeared in the fifth century. However, whatever the later changes, Lévêque and Vidal-Naquet suggest that the essential feature of Kleisthenes's reforms, "the creation of a political space and time," remained largely intact.[121] A number of issues about their study are potentially contestable, but the most significant is the use of the term *space* to describe the divisions. Indeed, there is an important debate concerning these reforms, as to whether the demes had strict territorial boundaries at all.

As Kain and Baignet suggest, there is plenty of landscape and archaeological evidence of systematic and regular urban/rural divisions in the Greek colonies, but there is no evidence this was mapped.[122] And yet, to have demes that were divided in the way Eliot, Lévêque, and Vidal-Naquet suggest would have required an extensive land survey and mapping. Dilke has explicitly challenged this assumption, suggesting that there is "little evidence of systematic land surveying," and that the land was not generally divided.[123] It is no surprise, therefore, that their claims and the previous consensus have come under sustained scrutiny. The challenge to the consensus stems from the work of Thompson, who suggests that though the demes were local units, it does not follow that they had formal boundaries or that boundaries were important to Kleisthenes. For Thompson, demes were isolated villages rather than blocks of territory, and these villages were places where people might register. He suggests that there were not the resources, skills, or time to conduct a proper cadastral survey.[124] As Lambert notes, demes had personal and territorial characteristics: people registered at the deme center nearest their abode, but the demes were unlikely to have been mapped.[125] This would be an act of self-identification by residents of local communities.[126] It follows from this that the *trittyes* were not units of land, because the demes were not contiguous.[127] A related challenge is found in the work of Lewis. He suggests that Kleisthenes would have found drawing lines on the map difficult, especially in the city.[128] He argues that the emphasis on him as "geometer conceals completely the difficulties of geography on the actual ground."[129] Eliot thinks that the *trittyes* of the south had natural boundaries and were geographical units, but Lewis notes that "Eliot's map of the trittyes is not fully argued," so "it does leave it open to doubt whether geographical divisions were as strong as he thinks."[130] Lewis suggests the key issue is whether Kleisthenes was concerned with land or people; lines on a map or deme-registers; the deme as territory or the people living there.[131]

Although initially this challenge was not well received, as Whitehead notes, "Its attractions have grown more compelling; and the traditional geographical trittyes . . . may have to be modified, even discarded."[132] Indeed, though most recent studies now follow the line of Thompson and Lewis, a counterchallenge has been mounted by Langdon, who suggests

that most demes already had some kind of territorial division, and therefore there was "no need for elaborate surveying or map-making." Langdon
argues that both in those demes that existed before Kleisthenes's reforms
and in those created by them, "the end result was the same: units composed of villages *plus* land within official boundaries." This reform would
have taken months rather than years.[133] In the *astu* within the city walls,
Langdon argues that the majority of scholars, including Thompson, admit
the need for boundaries.[134] For Langdon, the boundaries could be streets,
or the limits of the acropolis or *agora*, neither of which were in demes
themselves.[135] She cites a much-disputed scholium to Aristophanes, *Aves*,
which says, "as is written in the *Horismoi* of the city."[136] She argues that
this implies that written records were kept of the boundaries of the urban
demes, and that "the situation in Athens is likely to reflect that in rural
Attica."[137] However, back in the 1950s, Finley had argued against the reading of *horoi* as "boundary stones" in any simple sense. Through a reading
of inscriptions on stones, Finley argued that geographical considerations
took second place to legal and property aspects.[138] Though he retains
the Greek term *horos* in his text, he suggests "hypothecation stone" or
"stone marking legal encumbrance" may be more accurate than "boundary stone."[139] *Horos* certainly does mean "limit" or "boundary," and came
to mean an object that marked such, but this can be understood in a legal
sense of limitation rather than a geographical delimitation.[140]

In this context, note a passage of Strabo:

> For if there be no accurate boundaries [*akribon horon*] of stone posts
> [*stolon*], for example, or enclosures [*perubolon*]—take the case of Co-
> lytus and Melite [two Attic demes]—we can say only "this is Colytus"
> and "that is Melite" but we should not be able to point out the bound-
> aries [*horous*], and this is the reason why disputes often arise concern-
> ing districts.[141]

This seems to imply that there are no stone posts or enclosures, but
that the locals would know roughly where Colytus and Melite were.
Strabo—writing later than these reforms—seems to be recognizing the
limits of indistinct districts.[142]

A number of issues arise from this debate. The first of these is whether
Kleisthenes distributed the *trittyes* to the tribes by lot. The translation
here of the passage in *The Athenian Constitution* is "allotted three to each
tribe," though a more common translation is "gave three to each tribe by
lot." For Eliot, it is clear that the distribution would have needed a great

deal of time, and was conducted by Kleisthenes rather than randomly.[143] This rests on Eliot's assumption that the *trittyes* had clear natural boundaries and were essentially geographic units. As Lewis notes, were this true, "geographical units will differ in size, and therefore, to a varying extent, in population." It would follow, he suggests, that for Kleisthenes to have produced equal tribes, he must have matched large *trittyes* to small, small to medium, and so forth. Therefore, "Aristotle" cannot be believed.[144] But even if the geographical basis of the demes is disputed, it is still difficult to see how the demes could be allocated to the *trittyes*, and the *trittyes* to the tribes in any simple way. If the distribution were by lot, then it is hard to imagine how the tribes could in any way be equal. There have been various attempts to map the distribution of demes, *trittyes*, and tribes,[145] but as Rhodes suggests, the "regional boundaries are purely schematic."[146] For Andrewes, "the natural assumption that trittyes would be blocks of continuous territory began to crumble some while ago."[147]

The second is in the suggestion that Kleisthenes "named some of the demes after their areas, and some after their founders (not all were there any longer)."[148] Again, the translation preserves the ambiguity. The issue rests on the parenthetical "not all were there any longer." What is the subject of that clause? It could either be the *demoi* or the *ktisantes*, the founders. The phrase has been variously translated as "for not all were still connected with a particular locality," "for there were no longer founders in existence for all the places,"[149] or "not all the founders of the demes were known any longer."[150] Rhodes suggests that the most acceptable reading is one that emphasizes the founders—this certainly fits better with the preceding clause—but even this, he suggests, is not clear.[151]

For Langdon, the implication of this is that either "the people who constituted some demes were no longer living in the places associated with the eponymous founders of their villages, in which case the artificial territorial nature of the demes newly created by Kleisthenes is demonstrated, or else not all places had people who still remembered or honoured their founders, in which case there is nothing opposing the conclusion that the demes had definite territorial identity." She suggests that the latter had better support, but that the sentence is ambiguous enough that it cannot work as evidence for either side.[152] This is of course disingenuous, because her interpretation supports her argument in either a strong or a weak form. But it is this weak form that is most persuasive. It is entirely consistent with either reading to suggest that the primary purpose of Kleisthenes's reforms was to catalog people, and that a rough land division was the easiest way to do this. As Rhodes suggests, a deme for this purpose "could be a village and the land around it (or perhaps better, a village and its inhabitants)."[153] This does not imply that the demes had fixed boundaries, much less ones that were rigidly established and fixed in public records.

Indeed, and this highlights the final consequential point, the geographical nature of these reforms, such that it was, was only temporary. This is because the principle of membership was hereditary. The membership of the previous four tribes had been hereditary, based on kinship.[154] Despite the attempt to redistribute the people among the ten new tribes, it is certain that Kleisthenes made membership of his tribes hereditary, just as membership of the individual demes was.[155] However, the new division may have been geographical in some sense at the moment of the reform, but it also was hereditary following. Because even if a person moved, he still belonged to his father's deme.[156] By making deme membership hereditary, Kleisthenes undid the shift to land, even in a loose sense, and brought it more closely back to kinship.[157] However, the name of the deme, the demotic, started to replace the patronymic.[158] In other words, there was a mix-up of the previous situation, for which location was important as a distributive principle, but the underlying rationale remained largely unchanged. Rather than looking for a fundamental shift in the logic of governing the *polis*, there is some sense in Stanton's suggestion that the replacement of the four Ionian tribes with ten artificial tribes was a partisan reform that benefited Kleisthenes himself.[159]

Disputed though these reforms clearly are, three key points can be noted. First, it is clear that the term *demos*—like that of *polis*—has a meaning of both a particular place and the community within it.[160] Second, the mechanisms of division of the *polis* into *demos* may be related to the conceptions of mathematics current at the time. No direct causal link, but a relation nonetheless. The key point, however, contrary to Lévêque and Vidal-Naquet, is that the quantitative division is more in accord with understandings of arithmetic rather than geometry.[161] Third, and finally, these were actual reforms, political practices, rather than philosophers' schemes, such as those of Plato and Aristotle.

PLATO'S *LAWS*

While Plato's *Republic* is his most often cited political text, for a more concrete analysis, the late dialogue the *Laws* is actually more constructive.[162] Rather than the *Republic*'s utopianism, in the *Laws* there is the plan for the design of an actual *polis*. And unlike the rigor of the *Republic*, but like the *Statesman*, Plato no longer considers political rule to be math-

ematical in design and application.[163] Here too, as in *Antigone*, there is the
entwinement of site and community. The *Laws* are a dialogue between
three old men—Kleinias from Crete, Megillus from Sparta, and an Athenian Stranger, who is effectively Plato himself.[164] In book 3, after some
initial preliminaries, Kleinias reveals to the others that Crete is attempting to found a colony, and he asks the others to help him—initially in
theory—to set out its laws.[165] Note that a colony was usually founded by a
group of settlers from an existing *polis*, who then enjoyed autonomy in the
new *polis*.[166] Though separate, they would have closer links to the original
polis than to others,[167] but naturally they could make no claims for autochthony. For our purpose here, there are a number of interesting and important discussions. This is especially highlighted because contemporary
writers, such as Gottmann, claim that we find a use of the word *territory*
in Plato, and specifically in the *Laws*.

Gottmann cites two passages, one from book 4 and one from book 5.[168]
The first key passage is that in which the Athenian and Kleinias are discussing the place where the new *polis* is to be, and the second is that in
which they are discussing the division of its land.[169] Gottmann suggests
that this shows a notion of territory is implicit in the discussion of the
founding of a state.[170] The danger here is that of reading back contemporary notions into ancient thought. There are four important words for our
purposes in these passages, which are *polis*, *topos*, *khora*, and *ge*, but it
is interesting that translations or commentaries of this work rarely prob-
lematize these notions.

The first three can be found in the discussion of the situation of the
new *polis*. As Strauss argues, "The first serious question, however, concerns the location of the future city or, more generally, the nature of its
territory."[171] The Athenian asks whether it is on the sea or inland, about
its harbor, about the surrounding land, and if there are any neighboring
poleis.[172] The first term, *polis*, itself has already been discussed, but it is
worth noting here that it is usually translated as "state" (Saunders and
Bury), "city" (Strauss and Pangle), or "city-state." As with the reading of
Antigone, the term will be untranslated with a view to letting its meaning
emerge through the course of the discussion. The second is the word *topos*.
This word is usually translated as "locale" or "place." Here Pangle uses
"locale," Bury "locality," Strauss "location," Morrow "site," and Saunders
"actual foundation." These words are generally sufficient, but *locale* and
related words are words rooted in the Latin *locus*, which can be confusing, so *place* will be used instead. However, in the context, *site* or *situation* might actually be better, as the discussion is of the place where the

new *polis* is found. The Athenian asks how far it is from the sea—eighty stades, which is about ten miles;[173] whether it has a good harbor; and so on.[174] Indeed, Morrow suggests that the description Kleinias "gives of the site of the new colony . . . suggests that Plato has in mind a definite place in Southern Crete."[175]

The third word is found in the passage in which the Athenian asks about the land surrounding the *polis*, the *khora*. The word *khora* is one of the most difficult words in the entire Platonic lexicon. It is briefly mentioned in the *Timaeus*, a theme that has been commented upon by a range of thinkers, including Aristotle, Plutarch, Plotinus, Schelling, Hegel, Heidegger, and Derrida.[176] This has a stressed, philosophically significant, sense. For John Sallis, there is a risk of making the sense of *khora* equivalent to "place," when he suggests that the *khora* should be seen as "the other, the outside, of being, that which makes externality possible, that which makes it possible for something outside being nonetheless to be."[177] It is this that determines its possibility as mother, receptacle, womb, a reading that has been advanced by some feminist accounts.[178] Sallis notes that *khora* appears with "incomparably greater frequency than in any other dialogue (including the *Timaeus*)" in the *Laws*, "and yet, for the most part the word seems to have settled back into its prephilosophical senses, e.g., land, territory, country, rather than calling forth the level of the discourse woven around the word in the chorology."[179] But even in the *Laws*, the word is complicated. It is important to note that Plato does not speak about the land *of* the *polis*, but that surrounding it. It is this that Gottmann and Leo Strauss both see as "territory."[180] Now, is this justified? The Athenian is not looking at this in terms of a dominion for the new *polis*, but as resources, more a question of what the area around it is like than a survey of property. Saunders's translation as "surrounding countryside," Bury's as "surrounding country,"' or even Pangle's as "land surrounding" is more faithful. The use of *peri* seems to show that the *khora* is external to the *polis*. Morrow suggests in his commentary that what is at stake is the *terrain*.[181] Picard has suggested that the Greek *khora* was used in Hellenistic Asia Minor, where it "defines the dependent territory of a village," and its Latin equivalents would be *terra*, *ager*, or *pagus*.[182] What is interesting in this instance is that this is not land that is likely to be claimed by another *polis* nearby, for Kleinias informs the Athenian that there is none. The idea of modern territory as a politically and geographically bounded *space* belonging to, or under the control of, a *state* would seem to be alien to the discussion.

The fourth word comes in the discussion of the division and distribu-

tion of the land.[183] Now, here is something very important. Plato's suggestion is that the land be divided into equal portions and distributed by lot.
The land, *ge*, must therefore be measured and divided. The land dividers,
geonomoi, are charged with working out an equitable way of doing this.
In designing a new *polis*, the Athenian Stranger suggests that they are fortunate, because they can avoid vicious and dangerous disputes about land
and cancellation of debts and distribution of property. Older states that
are forced to legislate to solve these problems encounter difficulties, as
both leaving them as they are and reforming them are both equally impossible.[184] The solution is proposed, and essentially combines a sense of
justice and a need of indifference to wealth,[185] because poverty is a matter
of increased greed rather than diminished wealth.[186] Though the new *polis* proposed does not need to solve an already existing problem, it should
adopt this broad policy in its establishment, to avoid such problems later.
There are assumed to be no problems between the people to inhabit this
polis, so a distribution that created ill will would be criminally stupid.[187]

In order to avoid these problems, the number of people ought to be derived from the land available, and then that land distributed equally. The
land obviously needs to be great enough to support the people in modest
comfort, but no more is needed. Equally the number of people should be
sufficient to defend themselves, and to help out neighboring communities. An actual survey of the land is not attempted here, and the Athenian
Stranger assumes a figure of 5,040 adult males and their families. These
men are farmers, and, as Lacey notes, also soldiers,[188] and the number is
chosen because of its large number of divisors—fifty-nine in total, including one to ten. This facilitates division of the number for the various purposes of the military, administration, contracts, and taxes.[189] The division
of the land must also include provision for sacred sites for gods or spirits,
or heroes.[190]

While an ideal society would share everything, Plato considers this
unrealistic, and even suggests that farming in common is beyond the sort
of people these legislators have to deal with.[191] There is therefore a division
of land among the 5,040, but though each man receives this parcel of land,
he is supposed to consider it as the common possession of the *polis* as a
whole. The law of succession will be to the favored son, and the intention
is to keep to the number of 5,040 at all costs.[192] What is important is that
for Plato the *polis* is not a collection of detached citizens, but a "union of
households or families."[193] (It is important to realize here that "citizen," as
a translation of *polites*, is essentially as problematic as "city" is for *polis*.)[194]
The qualification for citizenship is not ownership of land, because many

others might own land, and because women are described as citizens too, though they do not own land.[195] There will be strict punishment for those who trade in this distributed land; property in land will be inalienable.[196] Equally, those who seek to move boundaries—that is, to acquire land unjustly—can expect serious punishment;[197] and there are clear prohibitions of overstepping the boundaries, allowing cattle to graze outside the boundaries, planting trees or burning wood too close to someone else's land, and attempting to attract another man's bees.[198] However, Plato does not think that land should be distributed equally, but rather at four levels or classes depending on how they are initially measured. This is grounded on the argument about indiscriminate equality leading to inequality.[199] It will, however, be possible for people to move through these different classes, as they become richer or poorer,[200] but it is not quite clear how this would work. What is clarified is that the value of the holding alone should be the lower limit of wealth, and four times as much the upper limit. People holding wealth above that level should be required to hand it over to the *polis*, and the *polis* should ensure that no one drops lower than the worth of the holding.[201] As Morrow notes, in the *Republic* there is no private land for the guardians, and he wonders if this is a departure.[202]

The *polis* as a whole should equally be divided. The *polis* itself should be at the center of the *khora*, or as near as convenient if the site is not suited. A central point of the *polis* should be designated the *acropolis* as a sacred place for Hestia, Zeus, and Athena. As Cartledge notes, "Spatially, the civic *agora*, the human 'place of gathering' and the *acropolis*, the 'high city' where the gods typically had their abode, were the twin, symbiotic nodes of ancient Greek political networking."[203] As Loraux adds, the Athenian *agora* also had a temple of the Mother of the gods.[204] The whole area should then be divided into twelve, with the boundaries radiating from the center, which Morrow notes will mean "each division will be a continuous area from the acropolis of the city to the borders of the state, including land within the city proper and the country outside."[205] Each of the twelve divisions will have a village, in which there will be an *agora* and shrines for Athena, Zeus, and Hestia, as well their own patron deity.[206] These divisions should then be subdivided into 5,040 lots, which should be equal in value, with those having poorer soil larger areas and so on. These lots should be further divided into two, which each man having one lot near the center and one toward the periphery. The twelve divisions would be given roughly equal rich and poor men, and separate gods. They will be called the tribes, and comprise 420 citizens (this number too has plenty of divisors, including one to eight, twelve, fifteen, and twenty).[207] Each tribe would be made up of sections (*meron*), which Thompson suggests is the parallel of Athens's *demes*.[208] This fascination with numbers is continued for other aspects of administration. All sorts of measures and divisions of the citizenry can be derived from this.[209]

While the protection of the *polis* itself is a job for centralized authority—the executive, generals, and other military commanders—the protection of the rest of the country (*khora*) is a task for the individual tribes. Each tribe appoints country wardens who choose assistants and take a different section of the *khora* each month so that over their two-year term they will experience all the divisions, and in all seasons.[210] When in place, they have various duties in order to defend the land. They should erect fortifications and excavate ditches; they must enable passage by road and regulate drainage and irrigation. They must provide water for temples, erect gymnasia for the good of the community, and keep a stock of dry wood for fuel, among other duties.[211] The architectural design of the city of a whole is important for Plato, and the Athenian suggests the need for temples to be built around the *agora* and around the perimeter of the *polis*, for both protection and sanitation.[212] However, he is against walls for the *polis* itself, because they encourage complacency in the minds of the inhabitants, and because the fortifications and ditches in the *khora* will be sufficient to prevent invasion. If a wall was to be had, it should be constructed from the private houses, so that their walls formed the *polis* walls.[213]

ARISTOTLE'S *POLITICS*

While Plato was concerned with outlining a design for the *polis*, Aristotle's intent was much more to catalog its manifestations and to derive some more general rules. For this reason, his work on *Politics* is often taken as the classic definition of a *polis*. It should of course be supplemented with the argument of the *Nicomachean Ethics*, with which it forms a continuous inquiry. Aristotle suggests that people join together in associations or communities, the first of which is the family (*oikos*), in order to improve life, and that these associations are the foundation of the larger political community (*politikes*), the *polis*, which too is an association of some kind.[214] It is when the congregation of village-size associations reaches a limit of self-sufficiency (*autarkeia*) that the association can be called a *polis*.[215] All these associations seem to be parts of the political community (*politikes*), and people come together with something useful in mind, to supply something for life. For Aristotle, the political community originally came together for the sake of what is useful, and continues for the same reason.[216] Because the first associations exist by nature, and it is natural for people to congregate in the interests of living well, the *polis* exists by nature. The human is therefore—in the oft-cited phrase—by nature a political animal (*anthropos physei politikon zoon*). However, in more appropriately Aristotelian language, the human is defined as that living being whose nature—i.e., whose highest purpose, or goal, *telos*—is to live in a *polis*. As Aristotle continues, "Anyone who is without a *polis* (*apolis*), not by luck but by nature, is either a poor specimen or else superhuman."[217]

Aristotle notes that ten people do not make a *polis*, nor do a hundred thousand; rather, the right number is somewhere within a certain range.[218] He uses a parallel with a ship to describe the ideal size of a *polis*. A ship that is one span (that is, seven and a half inches) or one that is two stades (that is, twelve hundred feet) will not be a ship at all.[219] The size therefore relates to a possible range. Too small and it will not be self-sufficient; too large and it might be a nation (*ethnos*), but will not easily have a constitution, the multitude will be hard to command, and the herald will find it hard to be heard.[220] His summary is therefore that the ideal *polis* will have "the greatest size of multitude that promotes life's self-sufficiency and that can be easily surveyed as a whole."[221] For Aristotle, similar things hold for the land (*khoras*).[222]

Indeed, at one point Aristotle suggests that in Plato's *Laws*, "it is stated that a legislator should look to just two things in establishing his laws: the land [*khoran*] and the people [*anthropos*]."[223] Aristotle does not hold to this equal valuation, but emphasizes the people over the land. However, he does make some important points about land that are worth discussing here. The land or location of a *polis* must be of sufficient size, but equally not too vast. Like the multitude of people, it should be easy to survey as a whole, because a land that is easy to survey is also easy to defend. Defensive troops should have easy access to all parts of the land. Its layout is, he suggests, not difficult to describe, because it should be difficult for enemies to invade and easy for the citizens to leave. However, on some points the advice of military experts should also be taken. Essentially, for a *polis* to be ideally sited, its location in relation to the sea and the surrounding land should be considered. "The remaining defining principle is that the *polis* should be accessible to transportation, so that crops, timber, and any other such materials the surrounding land [*khoras*] happens to possess can be easily transported to it."[224]

In book 7, chapter 11, Aristotle goes into some more detail about the situation of a *polis*. There are, he says, four factors, though the list can be

read in a number of ways. Health is a necessity, and this includes fresh
air and clean water, and there are political and military requirements—it
should be "easy for the citizens themselves to march out from but difficult
for their enemies to approach and blockade"—and order or beauty.[225] According to Aristotle, it was Hippodamus of Miletus in the fifth century
who invented the division of *poleis* and also laid out the street plan for
Piraeus. The division of the *polis* was both of the citizens (into craftsmen,
farmers, and defenders possessing weapons) and of the land (sacred, public, and private).[226] The Hippodamean mode of laying out the *polis* was to
have private dwellings in straight lines, which Aristotle says is pleasanter
and more useful for general purposes, but for security purposes the ancient
method is better.[227] This is because it makes it harder for foreign troops to
enter and orientate themselves. Aristotle suggests that houses be laid out
like vine clusters—that is, like the arrangement of five spots on a die—or
in an "x" shape. This will mean some parts and areas are in straight rows,
but not the *polis* as a whole. This will serve both security and beauty.[228] A
similar balance is required with walls. They are necessary, and *poleis* that
have walls have the option of acting as if they do or do not have them, but
those that do not have no choice. To suggest that it should not have walls
is akin to saying mountains should be leveled to make the land easier to
invade, or saying walls on houses make the inhabitants cowardly. Military requirements are therefore important, especially with the recent introduction of siege weaponry, and military expertise should be allowed to
determine the kinds of walls necessary. But the walls should enhance the
beauty of the *polis* too.[229]

Aristotle notes that the land "should belong to those who possess
weapons and participate in the constitution" and that he has explained
why the class of farmers should be different from them; and how much
land there should be and of what sort. He therefore thinks a subsidiary
task is to discuss the distribution of the land, who the farmers should be,
and what sort of people they should be. He suggests that he does not agree
"with those who claim that property should be communally owned, but
it should be commonly used, as it is among friends, and no citizen should
be in need of sustenance."[230] Therefore, "the land must be divided into
two parts, one of which is communal and another that belongs to private
individuals. And each of these must again be divided in two: one part of
the communal land should be used to support public services to the gods,
the other to defray the cost of messes. Of the private land one part must
be near the border, the other near the *polis*, so that, with two allotments
assigned to each citizen, all of them may share in both locations [*topon*]."

The reason for this is not simply justice and equality, but because it would be beneficial in the case of war with neighbors. Those who live far from the border may otherwise be unconcerned at the prospect of war, those near overly concerned.[231]

Therefore, in Aristotle, the qualification that the citizens must in the first instance "share their location [*topo*]; for one *polis* occupies one location [*topos*], and citizens share that one *polis*," is central.[232] Aristotle discusses the identity conditions for a *polis* and suggests that "the most superficial way to investigate this problem is by looking to location [*topon*] and people." The people of a *polis* can be split, and "some can live in one place and some in another."[233] Nor is it sufficient to say that people inhabiting the same location should be thought of as a single *polis*.[234] Equally for the constitution of a *polis* it is not sufficient that they share their dwelling place, as others such as foreigners and slaves do too:

> Evidently then, a *polis* is not a sharing of a common location [*topo*], and does not exist for the purpose of preventing mutual wrongdoing and exchanging goods. Rather, while these must be present if indeed there is to be a *polis*, when all of them *are* present there is still not yet a *polis*, but only when households and families live well as an association whose end is a complete and self-sufficient life. But this will not be possible unless they do inhabit one and the same location and practice intermarriage [*khromenon epigamiais*].[235]

Therefore, the essential definition of a *polis* for Aristotle is that it is a "sort of association, an association of citizens [*koinonia politon*] sharing a constitution [*politeias*]."[236] The link between the association of the family and the *polis* is not insignificant. In the *Eudemian Ethics*, Aristotle suggests that the human is not only a political but also a householding animal (*oikonomikon zoon*)—that is, the human is also a being whose nature is to live in a household.[237] As Aristotle continues, "In the household lie the primary origins of friendship, *politeia* and the just."[238] At the beginning of book 3 of the *Politics*, Aristotle recognizes that the first real question concerning constitutions is what a *polis* is. The first question needs to be further divided, because a *polis* is a composite, and the first part of this is the citizens, for "a *polis* is some sort of multitude of citizens." As noted, it is not enough to say that a citizen is such by residing in a place, because foreigners and slaves might share this dwelling place. Rather, a citizen is someone who is eligible to take part in the offices of a *polis*; and a *polis* is therefore a multitude of such people, adequate for self-sufficiency.[239]

Aristotle's understanding of the *polis* can be profitably compared with that of Plato or Kleisthenes. While in both of these earlier plans there was a strong emphasis on the numerical division of the land and inhabitants, in Aristotle there is a contrary emphasis on the need for relation and balance. While Plato provides numerical requirements and chooses numbers precisely because they admit of a large number of dividers, Aristotle is more interested in the range of possible sizes. Just as Aristotle's understanding of geometry is distinct from that of arithmetic, here too his understanding of political place admits of no easy division. Where Plato's understanding of civic land is shot through with a crude quantification—a reduction of geometry to a mode of arithmetic—and Kleisthenes's reforms owe much to mathematical models at the time, Aristotle is providing an understanding based on qualitative measure. As Vilatte puts it, for Aristotle, "all quantitative definition of the city, of men and space, is defective."[240]

The difference between arithmetic and geometry is therefore crucial. In Aristotle, it is summarized by bearing in mind that arithmetic is concerned with *monas*, the unit, geometry with *stigme*, the point. The *monas* is related to *monon*, the unique or the sole, and is indivisible according to quantity. The *stigme* is, like *monas*, indivisible, but unlike *monas*, it has the addition of a *thesis*—a position, an orientation, an order or arrangement. *Monas* is *athetos*, unpositioned; *stigme* is *thetos*, positioned.[241] *Monas* and *stigme* cannot be the same, shows Aristotle, for the mode of their connection is different. Numbers have a sequence, the *ephekses*. On the other hand, everything perceivable has stretch, size, *megethos*, which should be understood as *synekhes*, the *continuum*. This is a succession, not only where the ends meet in one place, but where the ends of one are identical with the next. Points are characterized by *haptesthai*, by touching; indeed, they are *ekhomenon*—an *ephekses* determined by *haptesthai*. But the units have only the *ephekses*. The mode of connection of the geometrical, of points, is characterized by the *synekhes*; the series of numbers—where no touching is necessary—by the *ephekses*.[242]

In other words, geometry is not concerned with division, for this will never get to the heart of the matter. The higher geometrical figures for Aristotle are not simply made up of the lower ones—there is more, for example, to a line than a string of points.[243] Arithmetic is concerned with number, with the possibility of division. Geometry, for Aristotle, is more concerned with place, position. Because everything tends toward its correct place, it is therefore a measure of quality rather than quantity, with ratio, relation, and balance more than division and calculability. On the other hand, in both Kleisthenes and Plato, division of the people and land is understood in a quantitative rather than qualitative way. Issues of number and calculation are crucial to their determination of the Greek *polis*. This does not accord with understanding of geometry current at the time. Geometry, for Plato, was much more an abstract deductive science than the physical land measuring it had been for the Egyptians.[244] In Aristotle, however, where the *polis* and *demos* are concerned with qualitative measures—relation and balance—the program is much closer to his understanding of geometry.[245]

This distinction can be found in a number of places in his work. In the *Politics*, Aristotle suggests that all *poleis* can be measured by either qualitative measures (*poion*) such as freedom, wealth, education, and status, or by quantitative measures (*poson*), by the greater number.[246] This means that in a *polis*, the poor may outnumber the rich, but the rich may outweigh the poor on a qualitative measure. Consequently there are two types of equality, of number (*arithmoi*) and worth (*axian*). Aristotle suggests that he uses "'number' to cover that which is equal and the same in respect of either size or quantity, and 'worth' for that which is equal by ratio [*logoi*]."[247] The problem with democracy, for Aristotle, is that it works on a crude type of equality (arithmetic equality), in which all are treated equally, instead of a more relational or proportionate (*analogian*) type of equality (geometric equality), in which only equals are treated equally.[248] This is why he argues that voting should combine both a numerical weighting and a qualitative balance.[249] Effectively this means that some votes should count more than others. He makes a similar argument about justice in the *Nicomachean Ethics*—justice is in accordance with proportion (*analogon*) rather than crude equality, geometric equality rather than arithmetic equality.[250]

There is, then, an interesting paradox. While Plato's division of the land appears to be a geometric division, it is actually closer to his understanding of arithmetic. The same argument could be made for Kleisthenes, who, despite many of the assumptions in the literature, was more concerned with division of people than of land. Yet in Aristotle, where the importance of land and location is downplayed, it appears that geometry—understood as something concerned with qualitative rather than quantitative measure, with relation, ratio, and balance rather than calculation—is more important than arithmetic.

SITE AND COMMUNITY

Aristotle's definition is usually taken as the classic one, as suggested. But as de Polignac points out, this definition of the *polis* as a multitude of citizens is too narrow. Even in the classical age, it excludes any who do not rule themselves, and is inaccurate for earlier periods when this distinction was not so clear-cut. At that time, de Polignac suggests, it would be a looser social entity—the network of relations between various members of a community based on a particular location,[251] the *polis* would be "the social unit constituted on a territory with a central inhabited area, the site of its political institutions."[252] For Nicole Loraux, a *polis* is a "group of citizens established on its territory." There are therefore two key elements in defining a *polis*: the *khora*, and *andres*, men.[253] "Citizen" obviously implies quite a select group: not foreigners, not slaves, and only men. But she then goes on to clarify this:

> At the centre of the *khôra*, the urban space, the physical place of civic life, punctuated by three summits [*haut lieux*]: the Acropolis, the Agora, the Kermeikos—the hill of power and the sacred, the public square [*place*], the national cemetery. A community of citizens with their wives (who are entitled to the name of Athenians [*Athênaiai*], but not that of citizens), and two categories of noncitizens, Metics and slaves.[254]

Loraux's two-part definition, supported by de Polignac and others, seems useful, and is the best broad interpretation of what a *polis* is. The readings above—of myth, tragedy, Kleisthenes, Plato, and Aristotle—can all be understood within this general context. But, as has been seen, they do not all accord in any simple way.

Reeve suggests that when Aristotle fleshes out his abstract theory, "we see that an Aristotelian *polis* is quite like a modern state in these important respects: it establishes the constitution, designs and enacts the laws, sets foreign and domestic policy, controls the armed forces and police, declares war, enforces the law, and punishes criminals."[255] However, as Heidegger has persuasively argued, to read modern notions of the political back into Greek thought is problematic. Our understanding of the political is derived from the *polis*, but has been developed and changed over time, and so we cannot use a modern notion of the political to understand the *polis*.[256] It can be seen from the preceding discussions that the notion of the *polis* is indeed neither straightforward nor reducible to modern notions of the state. The distinctions between Greek conceptions of the *polis* and modern ones of the state are helpfully outlined by Cartledge.[257] For Cartledge, the contrasts are the direct, unmediated, participatory character of political action in Greece; the lack of any Hegelian-style civil society; the lack of separation of powers; no real sense of sovereignty; and no parties, opposition, police force, or individual rights. "State" therefore seems to be a problematic translation, because the tendency is to think of it in terms of modern states.[258] As Castoriadis notes, "The Greek polis is *not* a 'State' in the modern sense. The very term *State* does not exist in ancient Greek (characteristically, modern Greeks had to invent a word, and they used the ancient κράτος [*kratos*], which means 'sheer force')."[259]

"City" is likewise problematic, because of our modern understanding of what a city is. A *polis* would necessarily have some rural areas because of a need to feed its inhabitants, and though the urban area would generally be the focus, the term *polis* does not simply mean the urban center.[260] As Sealey notes, sometimes larger *poleis* had dependent villages too.[261] De Polignac suggests that when agriculture is a major mode of food production, the need to acquire land permanently is more important, and temporary pillage insufficient.[262] He suggests this led to a shift in how military forces were understood, as instead of champions who could go on short-term, occasional plundering raids, there was a need for systematic defense of territory. Agriculture therefore led to the birth of the territorial community of the *polis*.[263] Etymologically the term *polis* links to *akropolis*, or "citadel," and is probably related to the Mycenaean form *ptolis*. Manville notes an ambiguity—*polis* could refer either to a "city" and not the surrounding countryside, for which there was also the Greek term *astu*, or to "a larger and more formal entity . . . implying a discrete but small political unit that comprised a central town and its adjacent territory," which is what is usually translated as "city-state" or "state."[264] As Starr notes, "Physically the *polis* was a definite geographical unit, in which public activities were concentrated at one point, the *asty* or *polis* proper."[265] However, the center was not sharply marked off from the rest either topographically—as Aristotle noted, there were not always walls—or politically.[266]

On the one hand, then, the *polis* is a site or a place, a definite location on a map. In Plato's *Laws*, for example, in the initial discussion between the Athenian and Kleinias concerning the founding of a *polis*, the first key issue is the place where it is to be situated. The Athenian asks whether it is on the sea or inland, about its harbor, about the surrounding land, and if there is a neighboring *polis*.[267] Following this discussion of the site, the old men turn to the inhabitants.[268] For Manville the political apparatus of the "state" and the notion of a "citizen body" are both at stake in the definition of the *polis*: citizenship (*politeia*), which has both legal passive and social active meanings, and which also means "constitution," and the *polis* were interdependent.[269] The *polis* was not separate from the *politeia*.

Manville's summary definition is therefore worth bearing in mind:
"The Greek polis was a politically autonomous community of people living in a defined territory comprising a civic centre with surrounding ar-
able countryside."[270] The *polis* therefore, while certainly encompassing a
political apparatus today thought of as akin to the state, did not see that
political apparatus controlling or policing a separate citizen body. Although Loraux is surely right that "conflict, barely domesticated into an
agōn, is already in the middle of the city,"[271] this is not a straightforward
divide between the rulers and the ruled. As Castoriadis notes, the *politeia*
meant both the political institution/constitution and the way people go
about common affairs.[272] Rather, the *polis* was a community with governmental features, within a demarcated area or place.

In many places this community aspect is stressed over the geographical. It is there in Aristotle, and, as Thucydides says, "Men make the *polis*,
not walls or a fleet of crewless ships," following the argument that soldiers
make a *polis* wherever they encamp.[273] Elsewhere, he reports the words of
Pericles: "What we should lament is not the loss of houses [*oikion*] or of
land [*ges*], but the loss of life. Men may acquire these, but they cannot acquire men."[274] A similar point is made by Aeschylus, in the *Persians*, when
he suggests that while men remain to a *polis*, its defenses are secure.[275]
As Stambaugh suggests, Thucydides's report of Pericles's funeral oration
shows that he did not mention monuments of the *polis*, but only private
houses, and rather stresses achievements and ideas, "a perception of Athens which is organic not material; political not topographic."[276] That said,
as the first part of this chapter outlined, the notion of autochthony—of being born from the very soil of the place that is inhabited—was extremely
important in both Athens and Thebes.

However their relative importance may be seen, it thus seems evident
that both notions of site and community need to be borne in mind when
thinking of the *polis*. There is a similar dual sense when thinking of the
term *demos*, as evidenced in Kleisthenes's reforms. However much support there is in classical and modern authors for this dual understanding
of the *polis*, in most modern translations the term is rendered as "city"
and/or "state." Occasionally the term *city-state* is utilized, and one of the
reasons behind this is the way that it can be used to describe later examples of political organization such as the Roman republic or Italy in the
fifteenth century. But this leads to confusion.

What the discussions of this chapter have shown is that the Greek *polis* was not always the same kind of political organization or association,

and that the weight given to its various aspects is different in different cases. In the myths from autochthony, for example, the people are directly linked to the soil of which they are born. Here the two key terms have an organic and unbreakable linkage. In *Antigone*, a play that is essentially about the rites and rights of burial, particularly within the borders of the *polis*, there is a playing out of tensions within the people too. From the ruler of the *polis* to another member of the royal household, through the mediation of the wider community and their representatives in the chorus, tensions between place and inhabitants highlight the very essence of the *polis*. Then in Kleisthenes's reforms, with the parallels of Pythagorean number somewhere in the background, there is perhaps the first concerted attempt to think of these attributes as divisible, controllable, demarcatable. In Plato's *Laws*, this calculable understanding of land and people is taken to a symbolic, yet still rather crude, level. And yet in Aristotle's *Politics*, this quantitative understanding is eclipsed by a more qualitative understanding of relation and balance.

As de Polignac argues, Athens may not be the best model of a *polis*, but rather the exception. Others had a looser structure and were not so centrally focused. Equally he suggests that to use the term *polis* to describe the Geometric period—a time of Greek art until the eighth century BCE—may be extremely misleading. De Polignac's analysis does not abandon the idea of a central point, but supplements it with a median point, and suggests that it is the mediation between the border and center, the frontier sanctuary and the central sanctuary, that is central to understanding the *polis*. He cautions against applying this indiscriminately, but suggests that "the political could not exist without the median point of the cultural agent that founded civil society," enshrined on the cusp.[277] He notes that extraurban sanctuaries were often placed on a position right on the edge, either of the territory as a whole or where the center joined the *khora*.[278] On Lesbos there was a sanctuary that was equidistant from and shared by four *poleis*.[279] De Polignac suggests that "it was above all the nonurban sanctuary monument that symbolised the transition from land to the status of a territory."[280] The situation and attitude to boundaries would be different in different types of *polis*. If they had large external possessions of land and no near neighbors, they would be less concerned by frontiers than a *polis* where the *khora* was more closely linked to the center, through the procession through it to the border sanctuary.[281]

However, none of these understandings can be simply equated with the modern understanding of the state with a population within a clearly demarcated territory. De Polignac notes that "'city' (*polis*) and territory" seem so familiar to us that no precaution appeared necessary, though they may be inappropriate to the time he is discussing.

> The concept of a "territory," in the classical sense of a bounded space within which an exclusive sovereignty was exercised, should be used with caution in connection with the Greece of the end of the Geometric period. At that time, the establishment of strict boundaries to "the citizens' space" and the political elaboration that this presupposes were not always as definite as in the late archaic and classical periods.[282]

Yet the translation of *khora* as "territory" even in these later periods is still assuming a simplistic link. Strategists may find modern notions of states and territory in Thucydides, but many of these key passages become much more cloudy on comparing to the Greek. Take, for example, the treaty between the king of Persia and Sparta: "All the *khoran* and all the *poleis* held now by the King or held in the past by the King's ancestors shall be the King's."[283] The *polis* and the *khora* are two terms that defy simplistic translation, and need to be understood in the different contexts of texts and time. Ancient Greek theories and practices cannot serve as a direct source of modern conceptions.

CHAPTER TWO

From *Urbis* to *Imperium*

From its humble beginnings as a small settlement in the Italian peninsula, the city of Rome came to dominate the western part of Europe,
the Middle East, and North Africa for a millennium. The opening line of
Tacitus's *Annales*—an account of Roman history—explains that "when
Rome was first a city [*urbem*], its rulers were kings [*principio reges*]."[1] Yet
it was a republic until the civil wars of the first century BCE, which is the
time when the account here will begin.[2] This will be with two key figures,
who were contemporaries and together embody the military might and
legal and political system of the late republic. Two brief remarks are worth
making first.

The first concerns the model of government. Roman politics is sometimes summarized in the phrase *senatus populusque Romanus*—the Roman Senate and people—which appeared as SPQR on coins and the standards of Roman legions, among other things. It showed that while the
Senate—an assembly of the elders (*senex* means "old man")—held considerable power, it was supplemented by the assembly of the people, which
reserved certain powers such as declarations of war. Nonetheless, the Senate massively outweighed the people in votes. The division between patri-
cian and plebeian stems from this division: a social and not racial divide.[3]
Chapter 1 noted how the Greek *demos*, in its sense of the people, could
mean both the citizenry or the many, the poor. The Latin *populus* equally
had this dual sense.[4] Over time the machinery of government became ever
more complicated and arcane in its procedures.[5] The Assembly was organized according to the ancient tribes of Rome, which were also the units
for tax, census, and conscription. These tribes were a mix of hereditary
principles and geographical ones, with initially three named after families

and then ones named after areas. By the third century BCE, there were thirty-five, four in the urban areas and the rest in rural areas. At the time of the late republic, they were increasingly blurred.[6] There were a number of reasons, but the principal one was the massive increase in population. The number of Rome's own inhabitants had, of course, increased, but there were also changes because of the Social War (91–88 BCE), which had extended citizenship to Italia more generally.[7]

This leads to the second point. Rome's government changed, but so too did the land over which it was exercised. Rome was initially within the *pomerium*, a sacred boundary, but this was extended over time, and it certainly exercised control over the surrounding areas. Beloch suggests that the sixteen old tribal areas at the end of the fifth century BCE constituted 822 square kilometers, expanding to 23,226 by the time of the Punic Wars.[8] Some of these wars could be construed as defensive, protecting Rome from enemies and attempting to secure its defenses; others were aggressive and more clearly designed for acquisition.[9] One of the key needs was agricultural land: Harris estimates that by the time of the Second Punic War, it would have extended to 9,000 square kilometers.[10] It is important to realize that the growing empire consisted of towns, with their additional agricultural lands—an expansion on the model of Rome itself. This provided a clear structure to the political command and taxation.[11] The city or town was known as an *urbs* or sometimes a *civitas*; the surrounding areas as *ager*, fields, or *pagus*, countryside.[12] These cities were the root of the politics, but citizenship was tied to descent and legal status, not place of birth or residence. As with Rome's own tribes, though, this became more complicated over time. The reward of soldiers at the end of their military service was central here. Money alone was not acceptable, and since land was often seen as the key source of wealth, this became a much more common source of remuneration.[13] In Plutarch's Greek account of the life of Romulus, for instance, Sabine lands (*khoran*) are distributed to Roman citizens.[14] In time, one of the key purposes behind the founding of further colonies was to reward legionaries, who were fighting the wars to gain that land in turn.[15] It is important to realize, though, that while the existing populations were denied their property in land and often expelled, the parcelization of land led to a fragmentation of landholdings. While there was a clear relation between land, wealth, and political power, this was in terms of influence back in Rome rather than the buildup of cohesive, contiguous land-based powers.[16]

CAESAR AND THE TERRAIN OF WAR

Gaius Julius Caesar is a remarkable figure by any measure. A politician, warrior, and writer of Latin that is often held to be the model of clarity and elegance, he was then murdered in one of the most infamous scenes of Western history.[17] He is of key interest here as a writer, of two works: the accounts of his time in Gaul between 58 and 51 BCE, *De bello gallico*,[18] and the civil wars that occurred on his return to Rome, *De bellum civili*.[19] The first were written as reports from the front line, with a debate about to what degree they were later embellished. As Adcock notes, however, they were almost certainly written before he embarked on the political machinations that put paid to the ideals of republican Rome.[20] The second text presents those struggles in detail.[21] In Cary's assessment, "Of the extant Roman historians, Caesar had an appreciation of geography which extended beyond the details of his campaigns."[22]

These geographical interests are signaled from the very opening lines of *De bello gallico*: "Gaul is a whole divided into three parts, one of which is inhabited by the Belgae, another by the Aquitani, and a third by a people called in their own language Celts, in ours Gauls."[23] It is clear from this and other accounts that Gaul is much larger than modern France, including parts of the Low Countries, Switzerland, and modern Germany. The southern part was already a Roman province. Many years later, after he had triumphed in the civil wars, there is a story that he commissioned four Greek geographers to travel to the ends of the earth to gather information, possibly for a world map.[24] The tale is recounted by Julius Honorius, with Nicodemus sent to the East, Didymus the West, Theudotus the North, and Polyclitus the South.[25] However, there is some doubt about the story, and the names, since the compass points we are now familiar with were not the directions known to ancients, which were based on the continents.[26] Either way, Caesar was dead before they returned, and on the *mappa mundi* found in Hereford cathedral (on which, see chapter 4 below), it is Augustus who is seen sending the geographers on their task.[27]

Yet Caesar did not work in the way that might be expected of a geographically minded historian. Rawson, for instance, suggests that it is remarkable that he spends so much time discussing the geography of Gaul, when it would have been so much clearer "if a simple map with the main rivers had been appended." Rawson notes that there is no indication that there ever was such a map, which had perhaps later been lost.[28] Many recent editions fill in this void.[29] But this is to expect that there was a shared understanding of things, and the evidence that the Romans used maps or required them for their geography is lacking.[30]

Caesar, for instance, never mentions using a map in any of his extant works, and if he had one, it is likely that it was "small and crude."[31] The way that Caesar describes landscapes is as he encountered them as a military leader, viewed from the ground; and the connection of places is that of an itinerary, with descriptions detailing what was seen of places as he and his troops marched through them.[32] In these descriptions, as generally, Caesar's language is precise and often unembellished. As Adcock notes, he had a tendency to decide upon the best word to use to describe things, and then stick to it, stressing a precision and repetition of diction.[33] This is certainly true of the use of geographical terms.[34] Despite what might be surmised from the translations, Caesar never uses the word *territorium*, but instead marks out the limits (*fines*) of lands and occasionally operates with a sense of *terra* as earth or ground that goes beyond economic concerns to operate with a military sense of terrain.[35] There is a very strong geographical sense to the descriptions of his battles, with space seen as a strategic medium rather than passive backdrop. To take one indicative example:

> The character of the natural terrain [*loci natura*] selected by our officers for the camp [*locum nostri*] was as follows. There was a hill, inclining with uniform slope from its top to the river Sambre above mentioned. From the river-side there rose another hill of like slope, over against and confronting the other, open for about two hundred paces at its base, wooded in its upper half, so that it could not easily be seen through from without. Within those woods the enemy kept themselves in hiding. On open ground along the river a few cavalry posts were to be seen. The depth of the river was about three feet.[36]

Yet when Caesar uses the word *spatium*, he means it as distance or extent rather than space in the sense of a container. Sometimes this is simply a span of time.[37] For example, he recounts a time "when the first legion had reached camp and the rest were a great distance [*magnum spatium*] away,"[38] and describes a method of wall building where each balk "is tightly held at a like space apart [*paribus intermissae spatiis*] by the interposition of single stones."[39] The use of *spatium* in this sense was the predominant one in classical Latin. It also meant a lap of a chariot race, usually used in the plural.[40] Only rarely does Caesar use *spatium* in a way that comes close to an area, describing the cutting down of forests as a defensive tactic: "With an incredible rapidity a great extent [*magno spatio*] was cleared in a few days."[41] One time he does talk about the extent of land and population of a place, in this instance Aquitania, but the phrase used is "*regionum latitudien et multitudine hominem*."[42]

Caesar regularly uses the word *fines* to mean borders, such as the crossing of the Helvetti through the land (*agrum*) of the Sequani and the Aedui, "into the borders [*fines*] of the Santones."[43] He is much exercised by the notion of these setting the limits of the land controlled by a particular group, be it the Gauls, the Germans, or the Belgae.[44] Occasionally he is specific that there is something contained by those borders, such as when he marched "into the land within the borders of the Ambiani [*ab eo loco in fines Ambianorum*]",[45] or uses the term in the sense of "the furthest parts of the lands [*ad extremos fines*]."[46] There is a clear opposition between the two senses when he stations legions in a range of places. Some are on the borders of peoples (*ad fines*) and others in their lands (*in finibus*).[47] He sometimes talks about moving outside the *fines* of an area, in either attack, retreat, or exile.[48] The unit to which these *fines* applied was frequently described by Caesar as a *civitas*, a grouping of people within a particular area, which can only be misleadingly translated as "tribe."[49]

The nesting of spatial terms can also be found: "These Druids, at a certain time of the year, meet within the borders [*finibus*] of the Carnutes, whose region [*regio*] is reckoned as the centre of all Gaul, and sit in conclave in a consecrated spot [*loco*]."[50] Yet at other times he is less precise, offering an analysis of the Helvetti and their lands, in which he suggests they are "closely confined by the natural terrain [*loci natura*]. On one side there is the river Rhine, exceeding broad and deep, which separates the Helvetian lands [*agrum*] from the Germans; on the other the Jura range, exceedingly high, lying between them and the Sequani; on the third, the Lake of Geneva and the river Rhone, which separates them from the Roman Province [*provinciam nostrum*]."[51] This region, sometimes also called Transalpine Gaul, is what is now called Provence of modern France.

In terms of his own conquests, this was a general theme. In his later praise of Caesar, Cicero invokes "the Rhine, the Ocean, the Nile" as the limits of his conquests, each of which he exceeded in his transgression into Germany, Britain, and the Alexandrian war.[52] As Maxfield notes, "The boundary of Caesar's conquests in Gaul extended from *Mare Nostrum* to *Oceanus*, from sea to sea, the Mediterranean to the Channel."

However, she continues, "in the north and east no such natural boundary was reached. It is clear from Caesar's commentaries that the *de facto*, if temporary, limit of land under direct Roman jurisdiction was regarded, by Roman and by barbarian, as the Rhine."[53]

This has significant consequences. As Mattern has noted, "It was in Caesar's commentaries on the Gallic war that the problematic concept of 'Germany' was invented."[54] While there are many problems with this terminology, and inaccuracies in the way Caesar applies it,[55] it has a purpose in his work that is at once descriptive and strategic. Caesar was able to describe all the people east of the Rhine as "German" for convenience, but it also lent these disparate groups an identity that provided a justification for his policy of stopping his expansion there.[56] Yet despite Caesar's use, and Tacitus's adoption of the term *Germania* in the first century CE, neither really thought there was cohesion to their cultural identity.[57] Indeed, the division of "Gaul" from "Germany" only really took on a sense with the limits of Roman expansion; yet it would be some time before this stabilized as either an ethnic or geographical entity.[58] Indeed, Caesar rejects the argument of the Sugambri that the "Rhine was the limit of the Roman people's *imperium* [*populi Romani imperium Rhenum finire*],"[59] crossing the river temporarily to further his objectives.[60]

Yet the Rhine was neither a particularly good strategic boundary nor a coherent cultural frontier.[61] Tacitus suggests that the division was not a barrier to Gauls themselves, surmising that they had crossed in the past, and to whom the river was merely a "minor obstacle."[62] Caesar may therefore have believed it was a cultural division, but he was unwilling to see it as a strategic one. Suetonius recognized him as "the first Roman to build a military bridge across the Rhine and cause the Germans on the farther bank heavy losses."[63] Sherwin-White thus contends that Caesar and Pompeius shared at least one thing in common: a refusal to see "a great river, Rhine or Euphrates, as the precise demarcation of her zone of influence. Caesar's reply to Ariovistus is the same in essence as Pompey's to Phraates: 'he would take as boundary what seemed fit to him.'"[64]

Perhaps the most significant boundary that Caesar is associated with is the river Rubicon. The river separated Cisalpine Gaul, where he had authority, from Italy, where he had none.[65] In 60 BCE the first triumvirate of Pompeius, Caesar, and Crassus had taken power, with geographically separated areas of command. The divide between Gaul and Italy was the point where the Senate had ordered that Caesar was to lay down his arms and don the toga. In crossing it to meet his army, he signaled his intention to challenge Pompeius, break the triumvirate, and set off the civil war.[66] In doing so, whatever his intentions, he destroyed the republic.[67] Unrest in the general population was one thing; but if armies could be turned against the *res publica* by their generals, then transformation was inevitable.[68] Crossing the Rubicon has become a cliché for a momentous decision, as have the words Caesar is supposed to have uttered: *iacta alea est*, "the die is cast" or "the game is begun."[69] Yet the accounts of later historians are much more dramatic than Caesar's own, where he barely mentions the event and the river not at all. Yet it was nearly not to be, since he did not have a map and got lost.[70] This crossing is significant, since it clearly demonstrates that the Romans did have linear boundaries, even if only to mark off different parts of the empire.

The external *limes* of the empire will be discussed later in this chapter. But it is worth making one remark here. In his account of *The Civil War*, Caesar recounts how he sent a letter to Metellus Scipio that sets out what is the only coherent account of his political aims: "tranquillity for Italy, peace for the provinces and security for the *imperium*."[71] As Gelzer suggests, Caesar's political activity exceeded Rome as city and "embraced the length and breadth of Italy and the provinces."[72] Yet it is clear that such aims could be achieved only by war elsewhere: a continual struggle to secure and expand the edges of the empire. His attempt at a peace overture in the civil wars is of similar construction. He claims that he is willing to do anything for the sake of the *res publica*: "My terms are these: Pompeius shall go to his provinces; we shall both disband our armies; in Italy all shall disarm; the regime of terror shall cease; there shall be free elections and the Senate and the Roman people shall be in full control of the *res publica*."[73]

On the defeat of Pompeius, Caesar similarly attempted to claim that his role was merely temporary, with an ultimate aim of restoring the *res publica*—the formulation being *dictator rei publicae constituendae*[74]—although he later became *dictator perpetuo*, less a dictator for life than one without the usual temporal limit specified. But this did not last long. In 44 BCE Caesar was murdered by a group of senators in the forum, offended by his personal supremacy and the challenge to their oligarchy.[75] Ultimately the wars that followed put paid to the system as a whole, but for a while it was thought that things could return. Gruen suggests that the wars started by Sulla a generation earlier had not destroyed the republic, and it was only the long sequence of wars that followed the civil wars in Rome that finally led to its demise.[76]

CICERO AND THE *RES PUBLICA*

Cicero was a key figure in late republican and early imperial Rome, living through tumultuous times and ultimately perishing as a result of his political convictions.[77] A politician, philosopher, and legal advocate, his extant writings cover many bases of Roman thought and practice. His speeches remain a high point of classical oratory, leading to the adjective *Ciceronian*.[78] In Grant's summation, "He was by far Rome's most enlightening political thinker, and perhaps its greatest."[79] Yet there was not great competition. Syme, characterizing and agreeing with Cicero himself, contends that "the Romans had an extreme distrust of abstract speculation, especially if it touched state and society. Political theory was foreign and Greek: idle and superfluous when not positively noxious."[80] This leads to his judgment that "apart from Cicero, the last epoch of the Roman Republic shows a dearth of political theory."[81]

Yet even here this was a by-product of his political action. He had been active as a consul, and had been instrumental in saving Rome from the Catiline conspiracy in 63 BCE. Cicero had refused to join the first triumvirate despite Caesar's overtures, because he believed Rome should remain a republic.[82] Cicero had characterized both Pompeius and Caesar as effectively the same: "Both seek domination [*dominatio*], not the well-being and fair fame of the *civitas*."[83] The republic was dead long before Caesar crossed the Rubicon.[84] His name was allegedly invoked by Brutus as Caesar lay dead, hailing the defense of the republic against the dictator; though as Cicero himself realized, while the monarch was dead, the monarchy was not.[85] As Syme has noted, Tacitus's phrase *"non mos, non ius"*—"no morals, no law"—characterizes the age.[86] Marcus Antonius, Caesar's ally and second-in-command, was allowed to survive and quickly maneuvered himself into a position of power. Cicero claimed that had he been "invited to the most glorious banquet on the Ides of March," there would have been no "leftovers":[87] had he been involved, Antonius would have been killed too. Cicero fiercely criticized Antonius, in works known as the *Philippic Orations*, named after Demosthenes's invectives against Philip of Macedonia.[88] He was eventually murdered by two officers in the service of Antonius, in December 43 BCE. This was about twenty months after Caesar's death. Antonius had Cicero's head and hands—which had criticized him in speech and writing—brought to him as trophies.

Although Wood suggests that "today Cicero is seldom taken very seriously except by classicists,"[89] he is of interest for two key reasons. The first is his political and philosophical writings, which include the works

De officiis (On Duties or Obligations);[90] *De oratore* (On the Orator);[91] and the twin works *De legibus* (On Laws) and *De re publica*, commonly translated as "The Republic."[92] The last two are perhaps the most interesting, but *De re publica* is only partly preserved, and *De legibus* is unfinished. Indeed, the former was lost entirely for centuries, until in 1819 parts were discovered to have been reused as a palimpsest for a copy of Saint Augustine's commentary on the psalms.[93] The second reason Cicero is of interest is his deployment of the word *territorium*, a word that is extremely rare in extant works of classical Latin.

As the titles of his two most political works might suggest, Cicero was heavily indebted to Plato, whose two great political works provided the model for these writings. However, Cicero had also taken much from the works of the Stoics, and the historian Polybius, especially his discussion of the constitution of the republic.[94] Indeed, an earlier branch of scholarship, *Quellenforschung*, made a virtue out of trying to hunt down all the sources of his claims. Douglas characterizes this work, not unfairly, as having been predicated on five assumptions that, he suggests, "may serve as hypotheses but must not be treated as axioms":[95] "(i) nobody ever said anything for the first time, particularly if he was a Roman and is extant, (ii) nobody had any general knowledge—he always used a 'source,' (iii) nobody ever compared sources if he could follow a single one, (iv) nobody read even a single reputable source if he could use a digest or handbook, (v) lost sources never make mistakes, while extant writers make egregious blunders."[96] Yet Cicero would have accepted as much: as Rawson suggests, he "made few conscious claims to originality."[97] He patently loved the Greek philosophers, and aside from the few fragments of direct translation that he made, much of his work is a rerendering of Greek ideas into a Roman context.[98] This forged a Latin vocabulary that set not just the terms but the very words of debate.[99] Jones notes that his Latin skill was also a limit to this project, suggesting that "he did not succeed in making Latin an adequate medium for abstract thought, and his failure was due precisely to his reverence for his own language, which he shrank from deforming and overwhelming with neologisms."[100] As Clarke notes, however, it is the combination of these ideas into a coherent whole that is Cicero's own achievement, noting that "the spirit of the dialogue is faithful to the setting in which it is placed, the circle of Scipio Aemilianus."[101] Scipio Aemilianus was a figure from the second century BCE, who had led the siege of Carthage and was a senator of the republic.

De re publica was the last work Cicero had completed before the outbreak of civil war between Caesar and Pompeius. Caesar's victory ended republican government, and so the book cannot be understood as a reflection on how things are. Rather, it is a diagnosis of why the republic had failed and prospects for the future.[102] In several places in the later *De officiis*, he notes that the *res publica* does not exist,[103] or bemoans it as "lost," "fallen," "overthrown," or the victim of "parricide."[104] Yet the point here is not simply that the "republic," as a particular form of government, has fallen, to be replaced by the *imperium*, but more profoundly that the common sense of things has been abandoned.[105] Yet as Stockton contends, this sense was long gone, that Cicero had been "born too late."[106] In *De re publica* Cicero therefore offers an idealized version of how Rome should be, not in theory but as restored to its past condition.[107] *De legibus* is the legal code for such a polity.[108] Thus, for Cicero, the two works should be seen together, and Schmidt even claims that they were originally part of a whole, *Politika*.[109]

The most important element of *De re publica*, from the perspective of this study, is the way Cicero defines the subject. *De re publica* literally means "on public things," "the public affair," or "the public property." The most important definition is the following:

> A *res publica* is the property of a people [*res populi*], but a *populus* is not any collection of humans brought together in any sort of way, but an assemblage of people in large numbers [*coetus multitudinis*] associated in an agreement with respect to justice and a community of interest [*iuris consensu et utilitatis communione sociatus*]. The first cause of such an association is not so much the weakness of the individual as a certain social spirit which nature has implanted in man. For man is not a solitary or unsocial creature, but born with such a nature that not even under conditions of great prosperity of every sort {is he willing to be isolated from his fellow men. In a short time a scattered and wandering multitude had become a body of citizens [*civitas*] by mutual agreement [*concordia*]}.[110]

The term *res publica*, occasionally used as a parallel term to *civitas*, relates to the Greek *polis*, *politeia*, and even *demokratia*, and English translations tend to be varied in their renderings, choosing "republic," "state," "commonwealth," "civic order," "constitution," and "government," among others.[111] Wood contends that Cicero is "the first important social and political thinker to give a succinct formal definition of the state," but this is surely premature, at least in terminology.[112] He adds that it is significant that its "major purpose" is conceived "largely in non-ethical terms, as the protection and security of private property."[113] Yet, the discussion Cicero offers in *De officiis* of the "natural principles of the community and human society," drawing on Aristotle's *Politics* in suggesting language as the thing that disassociates humans from animals, since it allows reason, "justice, fairness and goodness," is much closer to that kind of account.[114]

> For since it is by nature common to all animals that they have a drive to procreate, the first *societas* exists within marriage itself, and the next with one's children, where there is one house in which everything is shared. Indeed that is the principle of a city and the foundation of a *res publica* [*principium urbis et quasi seminarium reipublicae*]. Next there follow bonds between brothers, and then between first cousins and second cousins, who cannot be contained in one house and go out to other houses, as if to colonies. Finally there follow marriages and those connections of marriage from which even more relations arise. In such propagation and increase is the origin of *rerum publicarum*. Moreover, the bonding of blood holds men together by goodwill and love; for it is a great thing to have the same ancestral memorials, to practice the same religious rites, and to share common ancestral tombs.[115]

Yet even though Cicero appears to be invoking the role of the people, he had a very clear view of who was qualified to speak. Two passages from his letters to his confidant Atticus show that he retained an elite attitude. In one he describes the masses as "the wretched starveling rabble, the bloodsucker of the Treasury [*quod illa contionalis hirudo aerari, misera ac ieiuna plebecula*]";[116] in the other he characterizes Cato as speaking and voting "as though he were in the *Politeia* of Plato, not among the scum [*faece*] of Romulus."[117]

In extant materials, Cicero only uses the word *territorium* once.[118] It comes in his second *Philippic* against Marcus Antonius (44–43 BCE), in which he accuses him of having founded a colony at Casilinum where Caesar had already founded one. In "marking out the boundaries of the new colony with a plough [*ut aratrum circumduceres*]," Cicero claims he "almost grazed the gate of Capua and reduced the *territorium* of a prosperous colony."[119] *Territorium* here clearly means the lands belonging to Capua, the surrounding areas, of common formation to words such as *praetorium* or *dormitorium*.[120] Shortly before, and in a work addressed to Cicero (ca. 47–45 BCE), Varro derives *territorium* from *terra*, earth, describing it as follows: "The place which is left near a *colonis* as common property for the farmers is the *territorium*, because it is trodden [*teritur*] most."[121] As

Drummond and Nelson suggest, *territorium*, or more commonly *partum* (meadowland), also designates the lands used by Roman legions or other military units for sustenance.[122] *Territorium* became a word that was used of the agricultural lands surrounding any settlement.[123] This is the way it is later used by Seneca, in the one instance in his work: "You will see the mighty city itself, and its *territorium* spread wider than many a city's boundaries."[124] Seneca is thus clearly distinguishing between the city and the surrounding lands. There is one instance of the word in Pliny the Elder, which comes in the context of a discussion of a cloud of flies leaving the *territorium* in Olympus where a bull has been sacrificed, a sense that clearly means simply "area."[125] Of the classical Latin writers, these are the only instances of the use of the word. Of such innocuous beginnings great things result. As chapter 3 will show, the word becomes more common in early medieval Latin, yet still without the specific meaning we attach today. Only in late medieval Latin does the term start to be used in a more terminologically precise and politically significant way. As chapter 7 demonstrates, this is through an engagement with the law. Yet translations of Cicero use the English *territory* with rather more frequency. It is used not merely to render *terra*, land, belonging to a people or other political entity, but also to apply to the use of the term *ager*, field. For one instance, see his use of *agri* to refer to the lands of the Phliasians, translated as "territory."[126] It is clear, though, from this example, that *ager* did not mean merely agricultural lands for Cicero. As he notes, "The Lacaedaemonians asserted that all lands [*agros*] belonged to them which they could touch with a spear."[127]

The ownership of lands seized in war was a key political issue as Rome expanded, and a recurrent concern for Cicero. He was especially concerned with plans for the division of lands among the people, rather than retaining them as nominally public but largely farmed on behalf of absentee landlords.[128] Conquered lands, or at least some proportion of them, were initially *ager publicus populi Romani*, public lands of the Roman people. As Powell and Rudd note, there were two options: "It [lands] could be leased to tenants who would pay a rent to the treasury, or it could be divided up and allocated to private owners."[129] The issue appears several times in *De re publica*, dating back to the semilegendary king of Rome Numa Pompilius, who "divided up among the citizens the land [*agros*] which Romulus had won by conquest."[130] His grandson, Ancus Martius, is supposed to have done the same.[131] Cicero was a long-standing opponent of land being divided in his own time. In his reflection on old age, he praises

Quintus Fabius Maximus in the following way: "Nor was he less eminent in civilian life than in war. In his second consulship, though Spurio Carvilio was quiescent, he resisted as long as possible the proposal of the tribune Caius Flaminius to divide the lands [*agrum*] of the Picenians and Gauls individually in defiance of the Senate's authority."[132] Cicero clearly approved of the expansion of Rome's lands as a whole, noting the fundamental achievement of a people:

> Wisdom urges us to increase our resources, to enlarge our wealth, to extend our boundaries [*finis*]—what else is the reason for the praise carved on the tombs of the greatest generals that "he extended the boundaries of the empire [*finis imperii propagavit*]" if something had not been added from others [*nisi aliquid de alieno accessisset*] . . . our *imperium* now holds the known world [*orbis terrae*].[133]

Yet in one of his legal speeches, he compared the life of the general with that of the lawyer: "He spends his time enlarging boundaries. You spend yours defining them."[134] The first pertains to the political limits of Rome, the second to the private boundaries of clients' properties.[135]

In the Senate in 63 BCE, Cicero challenged proposals put forward by P. Servilius Rullus that would have put a great deal of power in the hands of a small committee of ten (the *decemvirs*) who were allowed to sell lands that had become state lands outside Italy, from the time of the consulship of Sulla and Pompeius—that is, in the previous twenty-five years.[136] These were the *legis agraria*, laws concerning *ager publicus*—that is, lands owned by the *res publica*, not private property.[137] The speech was persuasive, but also disingenuous, because Cicero purports to take a radical line in the interests of the people but actually is defending a conservative position.[138] Cicero is particularly scathing of a clause that notes that this can apply to "land which can be ploughed or cultivated," not land that has been already. In other words, as Cicero notes, it applies without limit.[139] Yet in the standard translation, *ager* is sometimes translated as "territory," although this largely appears to simply be to vary the words used: *fields, lands*, and *territory*.[140] It is worth noting that in *De legibus* he recalls that the fundamental laws of Rome have dictated that some land could never be privately owned: "The Twelve Tables have provided that ownership of a five-foot strip [along a boundary line] can never be acquired by possession."[141] As Keyes explains, "This strip was left free for the turning of the plough, and as a path. Ownership of it could never be acquired by a 'squatter.'"[142]

The geographical element is not pronounced in Cicero's work, aside
from this concern with land law. Yet there are a few instances that are
worth noting. In *De re publica*, for instance, he notes that

> the earth is inhabited in just a few confined areas. In between those
> inhabited places, which resemble blots, there are huge empty expanses
> [*solitudines*]. Those who live on earth are separated in such a way that
> nothing can readily pass between them from one populated region to
> another.[143]

While this is largely a comment on the separation of temperate zones in
different hemispheres, it also relates to other aspects of population spread.
In one of his defense speeches, he notes that Greek is understood across
the world, but "Latin is confined within narrow limits [*Latina suis finibus, exiguis sane, continentur*]."[144] He also invokes a range of geographical measures in the following passage from another such speech: "Gaius
Julius Caesar is now far away, in distant regions of the world's orbit, which
by his own achievements are the ends of the empire of the Roman people [*Sed quoniam C. Caesar abest longissme atque in eis est nunc locis,
quae regione orbem terrarum, rebus illius gestis imperium populi Romani
definiunt*]."[145] Like Caesar, he occasionally uses *agris* to specify the lands
owned by a group,[146] and in his praise of Caesar talks of the "widely separated lands [*disiunctissimas terras*]" and "unlimited places [*locis infini-
tas*]" he has conquered.[147] He is also greatly concerned with the situation
of Rome, praising Romulus's foresight in the site he chose.

> The location of a city is something that requires the greatest foresight
> in the establishment of a long-lasting commonwealth, and Romulus
> picked an amazingly advantageous site. . . . Consequently it seems to
> me that Romulus must at the very beginning have had a divine intima-
> tion that the city would one day be the seat and hearthstone of a mighty
> empire; for scarcely could a city placed upon any other site in Italy have
> more easily maintained our present widespread domination.[148]

If this seems scarcely credible, nor even particularly true, the description of the layout of the city in relation to the surrounding countryside
lends some purchase to the arguments.

> As to the natural defences of the city itself, who is so unobserving
> as not to have a clear outline of them imprinted upon his mind? The

line and course of its walls were wisely planned by Romulus and the
kings who succeeded him, being so placed on the everywhere steep
and precipitous hillsides that the single approach which lies between
the Esquiline and the Quirinal hills was girt about by a huge rampart
[*fossa*] facing the foe and by a mighty trench; and our citadel was so
well fortified by the sheer precipices which encompass it and the rock
which appears to be cut away on every side that it remains safe and
impregnable even at the terrible time of the advent of the Gauls. In
addition, the site which he chose abounds in springs and is healthful,
though in the midst of a pestilential region; for there are hills, which
not only enjoy the breezes but at the same time give shade to the valleys below.[149]

Yet there is another argument at stake here. *De re publica* and *De legibus* are attempts to describe the ideal constitution for a real city, whereas
the works of Plato that they are modeled on were ideal constitutions for
ideal cities. The exact way these might work out in practice is of course
complicated, but Cicero is clear on this point, suggesting that Plato's *Laws*
begins with the establishment of a *res publica* in "an unoccupied tract of
land [*aream sibi sumsit*]," and so is unrealistic for "men's actual lives and
habits."[150] If Plato was undoubtedly the greater philosopher, Cicero was the
more realistic politician.

THE HISTORIANS: SALLUST, LIVY, TACITUS

Most of the other key works of Rome that directly relate to politics that
have been preserved are historical studies. It is essential to underscore
that what texts are available are a tiny fraction of those initially produced.
As Mattern suggests, "After Caesar, none of the commentaries, memoirs,
or dispatches in which geographical information was usually reported
have survived. Important works that probably made use of these sources,
such as Pliny the Elder's history of the German wars and all seventeen
books of Arrian's *Parthica*, have also been lost."[151] Here the focus is on the
writers in Latin rather than the Greek histories that relate to Rome, such
as those of Plutarch, Polybius, Cassius Dio, Appian, and Josephus.[152] The
writings of Livy and Tacitus, and to a lesser extent Sallust, while themselves not complete, do offer some intriguing insights into these issues.
In some of these works, notably Tacitus's biography of his father-in-law,
Agricola, who was governor of Britain, there are geographical elements in
the narrative. Tacitus's account of *Germania*—for which the likely full

title was *De origine et situ Germanorum* (that is, the beginnings and situation of the Germans)—is one of the first ethnographic studies, and also includes some geographical information.[153] Yet these are not central to the analysis, and in his more properly historical accounts—the *Historiae* and the *Annales*—the geographical is distinctly underplayed.[154] Syme blames this on a widespread attitude: "Geography was held by Roman writers to be a difficult, abstruse, and rebellious subject."[155] It is therefore difficult to dispute Cary's assessment:

> Sallust and Livy give good descriptions of particular sites, but do not supply the general geographic background of their histories, Tacitus might be called the most ungeographical, as Mommsen has dubbed him the most unmilitary, of Roman historical authors.[156]

Nonetheless, these authors do yield some important insights concerning the ownership of land and conflict over terrain. It is worth immediately noting that, like Caesar, none of these writers—at least in extant works—uses the term *territorium*.[157]

Sallust has fared poorly in terms of the works preserved. His *Historia* is almost completely lost, and the most substantive texts are his accounts of the Catilline conspiracy and the Jugurthine war. Sallust is, however, of interest for a number of reasons.[158] One of these is that he was one of the first authors who could be properly described as "imperial," writing during the time of the first triumvirate.[159] In Sallust we find a recognition that an element of geographical description is essential to his work. As he notes: "My subject seems to call for a brief account of the geography of Africa [*situm Africae*] and some description of the peoples there with which Rome has had wars or alliances."[160] He also provides an insight into how the Romans understood the division of the world: "In their division of the earth's surface geographers commonly regard Africa as a third continent, a few recognising only Asia and Europe, including Africa in the latter."[161] In this he follows Strabo and Herodotus, rather than Varro, who understood it solely as twofold.[162]

Nonetheless, the geographical elements of the history itself are generally underplayed. As Handford complains, he "often leaves us ill-informed on geographical points such as the position of towns and battlefields and the direction of marches."[163] Two minor exceptions are, however, revealing. Sallust tells us that between the city of Cyrene and Carthage, there "lay a sandy featureless plain. There was neither river nor hill to mark the frontiers [*finis*], a circumstance which involved the two people in bitter and lasting strife."[164] Later in the same chapter, he tells the story of the Carthaginians, in which two brothers called Philaeni were "buried alive in the place which they claimed as the boundary [*finis*] of their country." At a later time altars were consecrated "on that spot [*in eo loco*]" to their memory.[165] The significance of this is that, like Caesar, Sallust is using *finis* as the word to mark out the limits of rule.

Livy too provides some interesting examples of terrain in battle,[166] and notes that almost all of Rome's land has been acquired by war.[167] He gives some instances of how the spoils of war were divided among the Roman people, with the story of sending people out from Rome to settle the lands of the Labici, with a grant of two *iugera* (about one and a half acres) each.[168] There is a similar account for the colony at Satricum, with two and a half *iugera* per person.[169] He also recounts disputes between the plebeians and senators over the allotment of land.[170] If he is treated here in much less detail because his remaining works are of an early period, a similar case could be made for his geographical language to that of Tacitus. His main word for lands controlled by a people is *agris*, which is usually distinguished from the cities to which these lands belonged.[171] He tells us, for instance, that the Ligurians felt enabled to act because "the Roman forces were nowhere near their towns or lands [*agro urbisque*]."[172] He uses a range of words to describe how peoples or lands were separated, talking, for example, of the river Larisus "which divides the lands of Dymae from those of Elean [*Eleum agrum ab Dymaeo dirimit*]";[173] of enemy *fines*,[174] the *fines* of a kingdom;[175] and the Taurus mountain range as "the limit of Roman rule [*finem imperii Romani*]."[176] Elsewhere Roman and Etruscan lands are simply described as adjacent (*adiacet*).[177] He recounts that "the Senate refused to admit that a dictator could legally be appointed outside Roman lands [*agrum*], that is outside the boundaries [*terminari*] of Italy";[178] and talks of the "furthest limits of the known world [*ab extremis orbis terrarum terminis*]."[179] Sometimes these terms come together in a phrase or analysis. Under the early king Ancus, he suggests that this was a "time of growth not only for the city [*urbs*], but also for her lands and boundaries [*ager finesque*]."[180] The Gauls are told to "keep themselves within the boundaries of their own lands [*agrorumque suorum terminis se continerent*]."[181] He provides details of a settlement for Macedonia, in which "the Macedonians were to be free, keeping their own cities and lands [*urbes easdem agrosque*]." Macedonia was to be divided into four districts (*regiones*), each with clearly set out geographical limits, especially rivers, and mountains.[182] The boundaries are variously described as *fines* or as ending (*terminaret*) in a specific place.

Two instances of founding will serve as interesting examples of understandings of legal rights over land. Livy notes that Romulus and Remus
conducted an *inauguratio*, a procedure in which they looked for auguries
to decide after whom the city they were founding should be named. The
ritual required a prayer, the naming of signs, and the delineation of the augur's field of vision. Romulus took the Palatine hill, and Remus the Aven-
tine. Remus received the first sign of six vultures; Romulus followed with
the sight of twelve. Each brother's supporters hailed this as the sign, "one
side basing its claim on priority, the other upon number." In the struggle,
Remus was killed. Livy reports another version of the story that has it that
Remus jumped over the unfinished walls of Romulus's settlement and was
killed by his brother.[183] Plutarch gives the story slightly differently: the
dispute is about whose site is preferred, and he notes that some contend
Romulus lied about the number of birds. Remus, on hearing about the de-
ceit, mocks Romulus's building plans and eventually jumps over them,
leading to his death at the hand of Romulus or one of his companions.[184]
Either way, the new settlement bore a name derived from that of Romulus.
The conquest of lands was partnered by a conquest of people, most notoriously the rape of the Sabine women in order to populate and perpetuate
the state.[185]

The second concerns Carthage:

> Then the debate began with the king's representatives on the subject
> of land [*agro*]. The Carthaginians rested their case on their boundary
> rights [*iure finium*], on the ground that the land [*agrum*] in question
> fell within the limits [*terminos*] set for Carthaginian jurisdiction by
> Publius Scipio. . . . The Numidians retorted that the Carthaginians
> were lying about the limits [*terminatione*] prescribed by Scipio; they
> asked what land in Africa was the rightful property of the Carthagin-
> ians, if one wanted to trace the genuinely original right of possession?
> The Carthaginians were immigrants who had been granted, as a fa-
> vour, the area [*loci*] they could encompass with a cut-up bull's hide,
> for the purpose of building a city; whatever extension they had gained
> beyond the confines of the Bursa [*Bursam*] was gained by violence and
> without right [*iniuria*].[186]

A few things are worth noting here. Again, there is the range of terms
used for land, boundaries, and limits. The bull's hide is cut up into strips
that could be joined as a kind of rope to demarcate an area, and the name
for the citadel, the Bursa, comes from the Greek word for ox hide. The no-

tion of law or right, *ius,* is applied to both boundary rights and the disregard for them.

Comparisons with Sallust and Livy are invidious now, given the paucity of the available texts, but Tacitus himself certainly felt exercised by the competition.[187] His works, like those of Caesar, are worth analyzing for the use of language.[188] The first thing to note is that Tacitus sometimes goes out of his way to avoid a geographical term such as *terra* or *ager.* He tells us, for instance, that "Caecina spent a few days among the Helvetians [*in Helvetiis*]";[189] and that there was a mutinous outbreak *in Chaucis.*[190] In both instances, and not without reason, translators have opted for "Helvetian territory" and "the territory of the Chauci." Tacitus, here, seems to opt for locating within a people rather than a place, and yet neither the Helvetians nor the Chauci seem important to the narrative. Related examples can be found in his description of "the extremity of the Bructeris [*ad ultimos Bructerorum*],"[191] when he surely means the edge of their land; or his use of the simple phrase *per hostis,* "through the enemy," when the meaning is of crossing through enemy lines.[192]

Interestingly, he tries to suggest that the subject matter may be part of the issue, explaining it in terms of the time he is treating. He suggests that "I am aware that much of what I have described, and shall describe, may seem unimportant and trivial."[193] In contrast to previous periods, in the recent past, "peace was scarcely broken—if at all. Rome was plunged in gloom, the ruler uninterested in expanding the empire [*proferendi imperi*]."[194] These, then, are not the most exciting of topics:

> What interests and stimulates readers is a geographical description [*situs gentium*], the changing fortune of a battle, the glorious death of a commander. My themes on the other hand concern savage mandates, continuous accusations, faithless friendships, innocent men ruined—a conspicuously monotonous glut of downfalls and causes.[195]

Yet, this is not to suggest that there are no geographical elements. Despite his avoidance at times, like earlier writers he very occasionally uses *ager* to refer to lands under the possession of a group of people. He talks, for instance, of "the lands of Cremona [*in Cremonensem agrum*]"[196] and in his account of action in Britain notes the incursion into "the whole district as far as the Trent and Severn . . . ravaging their lands [*agri*] and collecting extensive booty."[197]

More interesting are, in spite of his note above, his occasional geographical descriptions. There is, for example, an explanation of the geogra-

phy of the Rhine River,[198] and a depiction of Armenia in which he suggests
that the Armenians have "a national character and a geographical situation [*situ terrarum*] of equal ambiguity, since they have a wide extent of
frontier coterminous with our own provinces [*quoniam nostris provinciis
late praetenta penitus*] . . . the Armenians lie interposed between two vast
empires [*maximisque imperiis*]."[199] There are very occasional references to
the questions of terrain. He reports a speech from Germanicus, who suggests that "open country [*campos*] was not the only battle-field favourable
to a Roman soldier: woods and glades [*silvus and saltus*], are good too, if
he acts sensibly."[200] If no Caesar, his battle descriptions do occasionally
admit of a geographical element: "Caecina planted a strongly entrenched
camp between Hostilia (a village in the vicinity of Verona [*vicum Vero-
nensium*]) and the marshes of the River Tarturo, choosing a site protected
in the rear by a river and on the flanks by a barrier of marshland."[201]

Tacitus is interesting in terms of the ways he discusses the extent
of land. Like other writers of the time, he uses *spatium* in precisely this
sense. He declares, for instance, that "the distances were so great [*distan-
tibus terrarum spatiis*] that the advice arrived after the event"[202] or that
"this immense extent of land [*inmensum terrarum spatium*] is not merely
occupied by the Chaudi but filled by them."[203] In both these instances,
spatium is used as a qualifier, a modifier of *terra*. The same is true in his
description of "the huge stretch of land [*quantum ingenti terrarum*] between the Syria and the river Euphrates [*initio ab Suriae usque ad flumen
Euphraten*]."[204] Yet at other times key geographical terms are assumed
rather than stated. The emperor's troops, for instance, are said to have
"ravaged and burnt [land] for fifty miles around [*quinquaginta milium
spatium*]";[205] and "many Roman generals had recognised the sanctity, not
only of the temple, but [the land] for two miles round."[206] Then there is the
suggestion that during one of the many civil wars, the emperor lost almost
everything: "Nothing was left to Vitellius of a world-wide domain [*toto
terrarum orbe*] but that [land] between Tarracina and Narnia."[207] There is
the recognition of the quality of land rather than simply its extent. He
discusses the partition of Thrace by Tiberius between the brother and son
of the old king, Rhescuporis and Cotys. "The partition gave Cotys the ar-
able land, the towns, and the vicinity of the Greek cities [*arva et urbes et
vicina Graecis*], while Rhescuporis got what was wild, savage and adjacent
to hostile neighbours [*incultum, ferox, adnexum hostibus*] . . . but soon
Rhescuporis began to encroach and annex that allotted to Cotys, and to
attack when he resisted."[208]

Were these the only geographical elements in Tacitus, then this would be a meager yield. Yet he is of interest not for his descriptions of the spaces enclosed by boundaries but for his vocabulary on boundaries themselves. He notes that the Emperor Claudius was responsible not only for an enlargement of the Roman Empire, but for redrawing its internal boundaries. The province of Italia "herself has been extended [*promotam*] to the Alps, uniting not merely individuals but whole lands [*terrae*] and peoples under the name of Rome."[209] Tacitus notes that

> the Emperor also extended [*auxit*] the *pomerium* of the city. Here he followed an ancient custom whereby those who have expanded the empire [*protulere imperium*] are entitled to enlarge the city boundary also [*terminus urbis propagare datur*]. Yet no Roman commander except Lucius Sulla and the divine Augustus had ever exercised this right, however great their conquests.[210]

This raises the important issues of the *pomerium*, a strip of ground around the city that set its limits and formed its ritual boundary; formed, so Plutarch notes and Livy endorses, as a contraction of *post murum*, "behind the wall."[211] Aulus Gellius quotes the augurs, who defined it thus: "The *pomerium* is the area within the fields [*Pomerium est locus intra agrum*] along the whole circuit of the city outside the walls [*effatum per totius urbis circuitum pone muros*], forming a fixed determined region [*regionibus certeis determinatus*] and which forms the ends [*qui facit finem*] of the city auspices [*urbani auspicii*]."[212] Anything beyond this line was not really Rome, but what belonged to Rome. While the day-to-day meaning of the city and its people had long exceeded this line—it was no longer true that only those who lived within the *pomerium*, or later, within one mile, were Romans—it still exercised a powerful symbolism.[213] According to tradition, it had initially been set by Romulus, had been expanded by Sulla and Augustus, and probably by others, despite what Tacitus says, and now by Claudius.[214] Romulus's boundary, like the boundaries of new colonies, such as discussed by Cicero in his second *Philippic*, had been plowed by oxen.[215] Tacitus tells the route the stones took:

> The furrow to mark the town started from the *Forum Boarium* (a bronze statue of a bull is displayed there, because oxen are employed for ploughing), and ran outside the great altar of Hercules. Then there were stones at regular intervals [*certis spatiis interiecti lapides*] marked along the base of the Palatine Hill to the altar of Consus, the old *curiae*, the shrine of the Lares, and of the Forum. The Forum and Capitol

are believed to have been included in the city not by Romulus but by
Titus Tatius. Subsequently the *pomerium* grew as fortune expanded
[*pro fortuna pomerium auctum*]. The limits [*terminos*] established by
Claudius are easily traceable and are indicated in public records.[216]

The *pomerium* had a more than merely symbolic and historical purpose. It separated out different kinds of authority. Within the *pomerium*
was *domi*, "at home," or what we would now call domestic affairs, and
limits were set to the exercise of power. Military leaders exercised *imperium militiae*, leadership on campaign, but had to set this aside on passing
the *pomerium*, including changing from uniform to the toga and laying
down their arms.[217] Within the *pomerium* a more limited kind of *imperium*, *imperium domi*, was exercised by consuls and praetors.[218]

Tacitus's description of the route of the *pomerium* indicates one of the
words he used to suggest a boundary: *terminus*. He uses it, for instance, to
describe the Rhine[219] and suggests that the Clyde and the Forth can function as a *terminus* within Britain since they are separated by only "a narrow extent of land [*angusto terrarum spatio*]." Agricola defended this with
a line of forts, and pushed the enemy north of this line, into what he describes as "another island."[220] Many years after Tacitus, this would indeed
be the site of the thirty-seven-mile Antonine Wall, an attempt to move
the limits of Rome farther north than Hadrian's Wall. This is to get ahead
of the narrative. Two other instances are worth noting: another recognition of a river as a boundary, though without the use of a specific word—
"the river that flows between [*interfluit*] Raetia and Noricum"[221]—and the
use of the term *conterminae gentes* to mean bordering or "neighbouring
peoples."[222]

Like Caesar and Livy, though, Tacitus still uses the term *finis*. Like
Caesar, he can use this to mean the borders of a country, such as Britain,[223]
or a tribe, such as the Frisians.[224] It can also be used in a less country-
specific sense: "We also require you to put to death all the Romans in your
lands [*in finibus*]";[225] or "This was then the army with which Titus entered enemy lands [*finis hostium*]."[226] It can also be used in a sense of the
borderlands themselves more specifically: "The Treveri built a battlement
and rampart across their borders [*per finis*]";[227] or "The Mardi, experienced
brigands, with a mountain range to secure them against invasion harassed
him as he skirted their borders [*finis*]."[228] Other terms are used as well.
Under Augustus he suggests that "the *imperium* was fenced [*saeptum*] by
the ocean, or distant rivers."[229] He also talks of a sea that "is believed to
circle and girdle the earth [*cingi cludique terrarum orbem*] because the

last radiance of the setting sun lingers on here till dawn, with a brilliance that dims the stars."[230]

Yet Tacitus is important for his introduction of a new sense to the term *limes*. While *limes* had been used in Livy to designate a path or byway,[231] Tacitus uses it in a way that became very important: in the sense of the fortifications alongside this line. There are two key instances appearing in the *Annales*. The first is when he describes "the line of delimitation [*limitemque*] commenced by Tiberius," a word that Yardley, with some justification, translates as "fortified boundary."[232] The second is when he notes that "the whole region between Fort Aliso and the Rhine [*cuncta inter castellum Alisonem ac Rhenum*] was heavily fortified with new highways and earthworks [*novis limitibus aggeribusque permunita*]."[233] The first thing to note is that *region* is an extension of the Latin, in which *cuncta inter* simply means "the whole between" or "everything between." The second is the use of the term *limitibus*. The meaning of this term is disputed, here translated as "highway." A related use is found in the *Germania*: "After the *limes* was made and the guard posts moved forward, they [the *Agri Decumates*—the ten cantons] were considered to be a projection of the empire [*imperii*] and a part of the province [*provinciae*]."[234] As Whittaker argues, "The word *limes* here obviously meant not a border for frontier defense but a road, perhaps with its original sense of a road as an administrative limit."[235] Can, however, the same meaning be given to a passage from *Agricola*? "It was no longer the *limite imperii et ripa*, but the winter-quarters of the legions and the maintenance of possessions [*possessione*] that were in danger."[236] Can this be rendered as "the imperial roads and rivers"? Or does it mean something closer to "land and river boundaries"? This raises two themes for subsequent sections: the meaning of *imperium* and *limes*. A discussion of Roman cartography and science is found in chapter 4.

AUGUSTUS AND *IMPERIUM*

The year after Caesar's assassination, Antonius, Octavian, and Lepidus formed a *"tresviri rei publicae constituendae,"* a triumvirate for constituting the *res publica* in 44 BCE.[237] The plan was that the three men would have five-year consular authority, with different parts of the empire under their control.[238] Antonius took lands to the east, including Egypt; the west was divided between Lepidus and Octavian. Octavian held Italy, but this was seen by Antonius as "something of a poisoned chalice because of the disruption that was expected to be caused there."[239] Cassius Dio describes the areas each controlled, following the removal of Lepidus, using the word *khora* in its political-geographical sense.[240] The second triumvirate came to an end with Octavian battling and Marcus Antonius struggling for complete control, and finally the decisive sea battle of Actium in 31 BCE. With Octavian's victory over Antonius and Cleopatra, a return to the republic seemed unlikely. Yet he was sufficiently canny to realize that he had to construct his new regime in a different way. He resigned his offices, returning power to the Senate and people.[241] Fearful of his abandoning the *res publica*, the Senate implored him to continue its protection. He was tasked with military control of key areas for ten years, and given the honorary name of Augustus.[242] As Tacitus describes it, Octavian "organised the *res publica* neither as a kingdom or a dictatorship, but as the first citizen [*non regno tamen neque dictatura, sed principis nomine constitutam rem publicam*]."[243] By the name "*princeps*," he appeared to merely be preeminent among equals; yet this was dissembling. As Meier notes, "When Cicero declared that the republic was lost it was still in existence. When Augustus said it was restored it had come to an end."[244]

After Caesar's death, he took the name Gaius Julius Caesar Octavianus, to mark his adoption, usually dropping the last to hide his more lowly birth.[245] While there are questions about the legitimacy of his adoption, since Caesar did not make this clear during his life,[246] it carried him a long way. The subsequent deification of Caesar meant he could style himself not simply the son of Caesar, but the son of God, *divi filius*, worthy of veneration as holy himself.[247] He was given the title *Imperator* in 39 BCE. This was not merely a replication of the way that traditional generals were given this designation for their military victories, but as an honorary first name. Cassius Dio says that he was not *autokratos*, which is Dio's word for *imperium*, but possessing or being marked by power, *kratos diasemainousan*.[248] He thus became *Imperator Caesar divi filius*, adding *Augustus* in 27 BCE.[249] Indeed, it has even been suggested that 27 BCE is as good a break as any to mark the shift between the republic and the empire, and the characterization of rule, for this very reason.[250] *Augustus* is a word that has been linked to *augur*, "divination," and *augere*, "to increase."[251] It has been suggested that it deliberately linked him to the "august augury" at the time of the foundation of Rome, and that *Augustus* is also linked to the word *auctoritas*.[252] This had not previously been used as a personal name; it was the Latin equivalent of the Greek *Sebastos*.[253] As Gibbon notes, "*Augustus* was therefore a personal, *Cæsar* a family distinction."[254]

These names take on an important future significance. *Imperator* was a traditional acclamation given to a military leader following a victory. Scipio Africanus was hailed as this by his soldiers in the second Punic war, which is the earliest known acclamation.[255] Augustus notes in the *Res gestae divi Augusti*, the deeds of the divine Augustus, that he was "acclaimed *imperator* twenty-one times."[256] Augustus was important in terms of taking it as a permanent part of his name, although Vespasian (CE 69–79) was the first to become known by this in the sense of emperor.[257] Augustus preferred the use of *princeps*, the principal, the first citizen.[258] It is for this reason that his rule is known as the principate. Caesar remained a common part of imperial nomenclature. In addition, successors took the name *Augustus*, which became a title rather than a name.[259] Yet continuity was also important: *res publica* continued to be used of the polity; and leading citizens had long been referred to as *principes*.[260]

The same could be said of the notion of *imperium*. This was one of the most important terms in Roman political vocabulary.[261] It had originally been held by the king, and been used in the republic to mean "command" or "control," in distinction to the more ordinary *potestas*. Usually *imperium* was circumscribed and geographically limited, though not in an especially precise way. It would be restricted by a specific period, and there would be degrees of *imperium*; this was not absolute authority. At times the notion of *imperium maius*, an overarching *imperium*, would be granted.[262] By the third century CE, the jurist Ulpian could declare that "what the emperor decides has the force of a statute, because the people confers upon him all its *imperium* and power through the royal law, which is passed concerning his *imperium*."[263] Augustus was responsible for converting this notion from a limited understanding of command to the idea of empire with which it is now associated. Despite earlier precedents— not least those of Pompey and Caesar, who were both given the designation *imperium maius* at different times—the idea of the Roman Empire dates from the rule of Augustus.[264] In so doing, *imperium* shifts from being a measure of control to beginning to indicate the area over which that control was exercised. The *imperium Romanum* is both the command of Rome and, as we would understand it, the Roman Empire.[265] As Richardson suggests, "The idea of the Roman Empire as a territorial entity is an Augustan product."[266] The changing terms used can be traced from the coins issued during this period.[267] As Nicolet puts it, "The beginning of the Empire marks a series of mutations in knowledge, perception and mastery of the space over which power is exercised: both geographical space but also social and political space. In other words, lands and seas, and the people which populate them."[268] Yet in the *Res gestae*, the term used is not *imperium* but *auctoritas*, a term that is difficult to render into English. It is not simply "authority," since that implies more of an office-based position; rather, it covers terms such as *influence, prestige, wealth, birth,* and *connections.*[269] Fundamental to it was the ability to provide gifts of food and games—the "bread and circuses" of Juvenal's later stinging phrase—to the populace on an almost unprecedented scale.[270]

The *Res gestae divi Augusti* is a major source, even if unreliable. It begins: "The deeds of the Divine Augustus by which he placed the whole world [*orbem terrarum*] under the *imperium* of the Roman people, and of the amounts which he expended upon the Roman people and *res publica*, as engraved upon two bronze columns which have been set up in Rome."[271] These two columns were placed outside the Mausoleum of Augustus, as a lasting record of his achievements.[272] Clearly they are somewhat overblown, but the claims are important because of the terminology used.[273] He describes conquest as placing under the *imperium* of Rome, and that this was true of the known world, the *orbis terrarum*, a phrase Cicero also used, which the Greeks knew as the *oikoumene*.[274] Augustus describes this as an expansion in all directions:

> I extended the boundaries [*fines auxi*] of all the provinces of the Roman people bordered by tribes [*gentes*] not subject to our empire [*imperio*]. The provinces of the Gauls, the Spains, and Germany, bounded by the ocean from Gades to the mouth of the Elbe, I pacified. The Alps, from that region which lies nearest to the Adriatic Sea as far as the Tuscan Sea, I pacified without bringing an unjust war to any tribe. . . .
>
> The tribes of the Pannonians . . . I brought under the *imperium* of the Roman people, and I pushed forward the frontier [*fines*] of Illyricum as far as the bank of the river Danube. An army of Dacians which crossed to the south of that river was, under my auspices, defeated and crushed, and afterwards my own army was led across the Danube and compelled the tribes of the Dacians to submit to the *imperium* of the Roman people.[275]

These achievements need to be recognized on the terms he set out. He did indeed add more land to the empire than any other ruler, and some pride in that achievement—which was certainly seen as such by his contemporaries—is not unjustified.[276] In this, the contemporary poet Virgil played an important justificatory and expansionist role—we might almost say a propaganda role (noting that *propagare* is "to expand"). He has Jupiter promise Romulus: "I am imposing no bounds on his realm, no temporal limits [*his ego nec metas rerum nec tempora pono*] / I have given *impe-rium* without end [*fine*]."[277] Only a few lines later, the link to Augustus is made explicit: "There will be born of this splendid lineage a Caesar, a Trojan / He'll end [*terminet*] his *imperium* at the Ocean, his glory at the stars."[278] This fulfills the desire of Venus for those who "would hold all lands, all seas under their rule [*dicione*]."[279] His contemporary Horace suggested that the fame and majesty of the *imperium* would extend from the sunset to the dawn.[280] Writing later in Augustus's reign, Ovid declared that while "other peoples have fixed limits [*limite certo*] / the extent of Rome is both the city and the world [*Romanae spatium est Urbis et orbis idem*]."[281] The later Greek orator P. Aelius Aristides talks of the boundar-ies the Romans have established like a circle "beyond the outermost circle of the inhabited world, indeed like a second line of defence in the fortifica-tion of a *polis*."[282] These limits are "the Red Sea, the cataracts of the Nile, and Lake Maeotis, which former men spoke of as the ends of the earth," but for this *polis* they are "like 'the fence of a courtyard.'"[283] Aristides continues: "What a *polis* is to its boundaries and its lands [*horiois kai kho-rais*], so this *polis* is to the whole inhabited world."[284] Yet in many parts of the empire, especially in the South and Southeast, the efforts were more to gain power over neighboring tribes than to actively seek their land, what Lintott has called "a psychological ascendancy over peoples living at the fringe of the desert or wandering from the desert into cultivated areas." As he rightly suggests, the key area of expansion and campaign was in northern Europe—in Britain and Germany.[285] Realistically, no matter how ambitious Augustus was, there were areas even within the *orbis terrarum* that were never likely to be conquered.[286]

It is important to remember that just because the Romans believed that only the *orbis terrarum* was habitable, this did not mean that they believed that the world was flat. *Orbis* means "round" or "circle" rather than "orb," but several images from the time show globes. A coin from 44 BCE shows Julius Caesar wearing a crown, described as "perpetual dic-tator," with symbols of clasped hands demonstrating the relation between Caesar and the army, and a globe asserting claims to world domination.[287] Pompey's earlier victories had led to a globe being carried in a triumph, and another coin had a globe surrounded by wreaths.[288] Even earlier coins show Rome with a foot on the globe, "like a football referee before a game."[289] The symbolism of a foot on a vanquished opponent was a recurrent one, especially under Hadrian, with statues and coins showing him with his feet on a captured boy (possibly a Jew), a lion, and a crocodile.[290] Claudius was represented subduing Britain, depicted as a female figure with breast

Fig. 1A and 1B. Boscoreale Cups. From Héron de Villefosse, *Le trésor de Boscoreale* (Paris: *Monuments et Mémoires, Fondation Piot*, 5, 1899).

exposed.[291] On one of the Boscoreale Cups, Augustus is first shown holding out his hand in clemency to surrendering barbarians and then pictured between Venus and Mars holding a globe in his hand.

The people represented on the cups are from Africa, Gaul, and Spain and, as Kuttner explains, "are an emblematic catalogue of peoples now administered by Rome under Augustus."[292] Venus holds a statue symbolizing Victory.[293] Other examples can be found from the times of Pompey, Augustus, and Trajan.[294] Kuttner suggests that "the globe, as a symbol of the world, and so of world rule, was far more popular in Roman art than in Greek art, and I think that this is partly to be explained by the fact that it is a visual translation of the Latin *orbis terrarum*."[295] The *orbis terrarum*, or circle of the world, does indeed imply a clearer spatiality and sphericity than the Greek *oikoumene*, the inhabited world.[296]

Yet Augustus was actually more modest in his ambitions. Indeed, while his conquests were large, they also put an end to Rome's continual expansion. Most of the land gains were made during the republic; Rome had an empire before it became an empire.[297] Augustus did not initially set limits on the growth. In the east, his operations extended beyond the Rhine, but it is sometimes claimed that his aspirations may have even been toward the Elbe, and to the south the Danube. Germania therefore became a battleground, even if it is uncertain that he ever aspired to conquer it entirely.[298] As Wells notes, the idea that the Elbe and Danube were to be frontiers has no foundation: there is no evidence for this; they would be poor frontiers in any case; and the state of geographical knowledge at the time would have furnished little basis for thinking that they would have been.[299] Yet eastward expansion was continued, often at great cost. In September 9 CE, three Roman legions led by Varus were ambushed and destroyed in the Teutoburg forest. It was a defeat without compare in this time, and has been described as "one of the decisive battles of world history."[300] Augustus bemoaned their loss, letting his hair and beard grow as a sign of mourning, and crying "Varus, give me back my legions!"[301]

After this time periodic attempts were made to go beyond, led mainly by Tiberius, and then, when Tiberius himself was emperor, by Germanicus. Tacitus contends that the real reason was to avenge Varus rather than expand the *imperium*.[302] In the main, the reinforcements sent to the area began to fortify their camps and supply depots into a more permanent position. Tacitus reports that the *Res gestae* were read in the Senate by Tiberius, his adopted heir and successor as emperor. Augustus had added a note of advice suggesting that the "empire should remain within its boundaries [*coercendi intra terminus imperii*], either through fear or jealousy."[303] This was not a policy he had followed himself, which perhaps explains Tacitus's final clause. Questions have been raised about the validity of this. Gruen, for instance, suggests that the wording may have been Tiberius's own, giving legitimacy to a policy he intended to implement.[304] Nonetheless, it seems likely that episodes such as the loss of the three legions had led Rome to believe that there were, indeed, limits to its power.[305] There were always more tribes to the east, ready to take the place of those that the Romans defeated.[306] As Drummond and Nelson put it, "In this almost accidental way, the Roman frontier in the West was established, a fortified line that was to endure for almost 400 years."[307] Wells sums this up, suggesting that "the Augustan commanders did not have the Maginot Line mentality. They were not thinking about keeping the barbarians out, but of going out themselves to conquer the barbarians."[308] Thus, the initial intent was not defensive, and Augustus in particular was careful not to make it appear so, but events meant that it became this.[309] It remained a problem for successive emperors. In Gibbon's words, Augustus chose to "relinquish the ambitious design of subduing the whole earth," and "he bequeathed, as a valuable legacy to his successors, the advice of confining the empire within those limits which nature seemed to have placed as its permanent bulwarks and boundaries."[310]

THE *LIMES* OF THE *IMPERIUM*

The meaning of the term *limes* is widely disputed. One thing appears incontestable: that it was only after Augustus that it made sense to talk about frontiers as *limes* at all.[311] In the period of the republic, there was no clear sense of limits to Rome's expansion, although the coasts of Italy and the Alps to the north did provide some geographical constraints.[312] Some classical authors described the Alps as like a wall to protect rather than a line that they needed to respect themselves.[313] Cicero declares that once they were a "rampart," and had prevented Gauls from stopping the rise of Rome; now they could sink into the earth because there was nothing to fear until the ocean.[314] Livy calls them an "insuperable" barrier,[315] but Hannibal tells his troops that they are not impassable,[316] and later tells them that in scaling the Alps, they are entering not just Italy but the city of Rome itself.[317]

Under the republic, Rome certainly did not have clearly marked lines and defenses.[318] There were, of course, internal divisions, which were common. The reason is that they applied to taxable property.[319] The *pomerium* has already been mentioned, but there were also occasional lines drawn between administrative zones, the provinces. A *provincia* meant a role or task, but came to mean the area under the control of a magistrate. Islands such as Sicily, Corsica, or Sardinia were *provincae*, but they also existed within the peninsula and certainly in newly conquered lands.[320] Lintott claims that the first large-scale boundary was between two Spanish *provinciae*, Hispania Ulterior and Citerior. He suggests that the line was somewhat imprecise, and it was likely that boundary stones were only visible "on major roads or on high ground."[321] He suggests that over time the relation of magistrate control to a discrete area became more firmly fixed, and that some territorial sense of a *provincia* is not anachronistic, providing that it is not "interpreted too rigidly" and that no inference is drawn that they "were thinking in terms of absolutely precise boundaries."[322] It is also important to recognize that there were not homogeneous spaces of absolute Roman control. As well as the borders being imprecise, there would be overlapping jurisdictions, and enclaves of other rulers.[323] As Dilke notes:

> From the legal point of view, it was important to lay down exactly how far, in the topographic sense, jurisdiction extended. On the larger scale, every province had its boundaries with Italy, if adjacent, and with other provinces. The outer boundary of the Roman Empire was known by the same term, *limes*, as we find so commonly used in surveys. But there is a difference: in the case of a colony, for example, the *limites* were all within its territory, and its boundaries were called *fines*.[324]

The *agrimensores*, the land measurers or surveyors, were charged with this task. They did not simply measure land, but used various techniques to plan and lay it out, especially in the case of new colonies.[325] Setting out boundaries was their most important duty, followed by land allocation.[326] The grid over urban space was known as *insulae*; over the countryside the squares were known as *centuriae*.[327] There was a similar approach of meticulous planning employed by the army: when they set up a camp, it was exactly the same wherever it was, so that it was easy to navigate around. This required the leveling of the ground to establish the basis, and detailed work.[328] In the early second century CE, a military engineer recounted what he was required to do:

> After we first entered hostile territory, Celsus, our Caesar's earthworks began to demand of me the calculations of measurements. After a prearranged marching length had been determined, two parallel straight lines had to be produced (on the terrain) along which a large defensive structure of pallisaded earthworks would arise for the protection of communications. By your invention [Celsus], when part of the earthworks was cut back to the line of sight, the use of surveying instruments extended these lines. Regarding surveying bridges, we were able to state the width of the rivers from the bank close by, even if the enemy wanted to harass us. . . . After our supreme emperor [Trajan] most recently opened up Dacia for us by his victory . . . I returned to my studies, at leisure as it were.[329]

Of the surveyors themselves, a compilation of works exists as the *Corpus Agrimensorum Romanorum*,[330] which was compiled in the fifth century CE but later acquired more material. Land surveying was an ancient profession with a heritage back to Egypt and Thebes, where taxation was based on property size.[331] In Greece, as shown in chapter 1, there was some planning, but little surveying. Rome, while not developing new mathematics or techniques, put them into practice in a more rigorous way. This was a practical use of the mathematics.[332] In the words of the Old Latin playwright Plautus, "I shall now demarcate its regions, *limites*, and confines [*confinia*]; I have been appointed its surveyor [*finitor*]."[333] The use of *limites* here is important. As Dilke notes, "*Limes* properly means a boundary zone; in agriculture, a path or a balk. Where the technical writers speak of a straight boundary line having no width, they use instead the word *rigor*."[334]

Within this compilation are some important texts generally attributed to Frontinus, the military strategist and aqueduct builder from the first century CE, but possibly not by him and of later provenance.[335] The first of these, *De agrorum qualitate*, declares that "there are three types of land: first, land that has been divided and allocated; second, land that has been contained in a survey throughout its extent; third, land of uncertain boundary [*arcifini*], which is not confined in any survey."[336] As Campbell notes, "This may be the epitome of a longer version, and it is unclear what has been omitted or how far the surviving text has been altered."[337] In a related text, "Frontinus" suggests that there are fifteen kinds of land dispute, including the position of boundary markers, land ownership and possession, and what he calls *iure territorii*, "territorial jurisdiction."[338] These kinds of disputes, he suggests, are twofold. They can relate to areas within a town itself, and its surrounding agricultural land.[339] He adds that "a territory is something established for the purpose of terrifying the enemy [*territorium est quidquid hostis terrendi causa constitutum est*]."[340] This text, *De controversiis*, is the only text in which the word *territorium* appears in Frontinus,[341] which should give us cause to pause, especially since his writings on strategy might be supposed to deal with military control of land. Yet there he uses the more standard vocabulary of his time—he was born seventeen years after Pliny the Elder and sixteen before Tacitus—to describe such issues: *ager* and *fines*.[342] It is difficult to disagree with the judgment of Campbell that "Frontinus' role in land survey and his contribution to its study, though seemingly impressive, remain shadowy."[343]

There is also the famous definition of Siculus Flaccus, that "since the citizens are terrified and driven away, they call these *territoria* [*Territis*

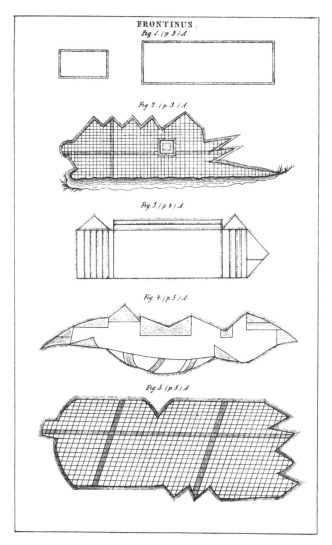

Fig. 2. Frontinus. From F. Blume, K. Lachmann, and A. Rudorff, *Die Schriften der Römischen Feldmesser*, 2 vols. (Berlin: Georg Reimer, 1848), 1:417.

fugatisque inde civibus, territoria dixerunt]."[344] (He notes that the interpretation of this word is disputed, and says he will return to it later, but this is unfulfilled in this text.)[345] Siculus Flaccus is interesting, because he does not simply argue the point etymologically but demonstrates it in practice too. He suggests that the question of land division is one that arises from conquest: "As the Romans became the masters of all nations,

they divided up the land captured from the enemy among the victors."[346]
He returns to this theme later, suggesting that "war created the motive
for dividing up land," because the spoils were given the soldiers who had
seized it.[347] This naturally provides both the reward and the incentive for
new conquests. There is an expulsion of existing people, a recoding of the
places conquered, and a new legal order: "Once the war is over, the victorious people expel the vanquished from their lands, and declare them *ager
publicus* and universal *territorium*; and within their limits [*fines*] exercise
the right of dictating the law."[348] Siculus Flaccus also provides a detailed
description of how these lands are divided, although it must be stressed
that these are within the wider Roman *imperium*.

> *Territoria* are demarcated between *civitates*, that is, between *muni-
> cipia*, and colonies, and *praefecturae*, sometimes by rivers, or by the
> tops of mountain ranges and watersheds, or even by the placing of con-
> spicuous stones, which differ in shape from the stones used to mark
> private boundaries [*terminorum*]; indeed sometimes between two
> colonies the boundary is marked by continuous *limitibus*. If a ques-
> tion arises about these, that is, about *territoriis*, the laws granted to the
> *civitates* are examined, that is, to colonies, and *municipia*, and *prae-
> fecturae*. I have often discovered in public records that *territoria* are
> distinctively described; for the description begins to go round the *ter-
> ritoria* with the names of several locations included.[349]

A couple of the early instances of the word in his text are simply
written as "p R toria," expanded by editors to "p R <terri>toria," and by
Campbell to "p(opuli) R(omani) <terri>toria," the "territories of the Roman people."[350] What is interesting is less the work of the translator than
that of the editors. Shorn of its prefix, the word appears as just the suf-
fix -*orium*, "surrounding." The lands were simply those surrounding the
Roman people. But as the text progresses, and the lands become farther
afield, the argument from expulsion and terrifying becomes more compelling. The date of this text is unknown: Campbell suggests that sometime
in the second century CE is possible;[351] Dilke suggests the third century.[352]
Campbell describes it as "amongst the most coherently argued and compe-
tently written of extant surveying material."[353]

The word *limes* was therefore a surveyor's term, meaning a track or
path. It was given a military sense as a supply route or road, a sense with
which it appears in surveyors' manuals.[354] Thus, *limes* were, at least at one
time, things that linked rather than divided, modes of connecting points

within the *imperium*. By the time of the *Historia Augusta, limite* is used to invoke the military areas Marcus Aurelius was fighting in.[355] Clearly something fundamental changed. There is a range of accounts. Piganiol suggests that the "*limes*, in the time of Domitian and Trajan, was a military road in the service of offensive policies. From the time of Hadrian, it was a frontier line whose significance was more juridical than military."[356] (It is worth a brief note on the succession of emperors and royal families. The dynasty begun by Augustus lasted for almost a hundred years—from Tiberius through Gaius, known as Caligula ("little boots"), to Claudius and Nero. Nero's suicide in 68 CE led to civil wars, with Galba, Otho, and Vitellius as emperors in quick succession.[357] Vespasian became emperor in 69 CE and ruled for ten years, establishing the Flavian dynasty of Titus and Domitian. This was succeeded by the Nervan-Antonine dynasty: Nerva, Trajan, Hadrian, Antonius Pius, Marcus Aurelius [co-emperor with Lucius Verus for the first eight years of his reign], and Commodus. This chronology takes us from the end of the first century BCE to the middle of the second CE.)

The construction of a frontier *system* is often dated to this time. Pelham contends that it was the Flavian and Antonine emperors who organized it, while Shotter suggests that the Flavians were concerned with the maintenance of the gains of earlier periods rather than their growth.[358] Domitian, in particular, strengthened the *limes Germanicus* and the *limes Raeticus* in the East and Southeast aside Germania.[359] In Frontinus, he is described as having advanced the *limitibus* by 120 miles.[360] Hadrian is often given credit for solidifying the system, with the wall bearing his name being only the most explicit evidence.[361] Hadrian styled himself as a new Augustus, spending much of his reign away from Italia and seeing almost the whole empire.[362] He almost certainly visited the region of the wall that bears his name.[363] He withdrew from the new provinces Trajan had added east of the Euphrates, and lands north of the Danube, lending weight to theories that he was pursuing defensible boundaries.[364] In justifying this, he was undoubtedly helped by his secretary Suetonius, who wrote his accounts of the emperors under Hadrian's reign, and in Birley's words credited "Augustus with a purely pacific policy, of a suspiciously Hadrianic character."[365] This is not to suggest that there were no attempts at expansion, but after Augustus, there were only two main conquests that led to long-term expansion: Britain in 43 CE and Dacia in 105–6 CE.[366] It can appear that Rome shied away from further colonization after the third century, but intermittent attempts were also made to expand into Germania and along other frontier zones.[367] These are sometimes explained as

an attempt to find the most defensible boundaries, with expansion to secure the lands of the empire, though how much this was planned and how
much the product of circumstances is disputed.

The most elaborate argument for a changing system is that of Luttwak,
who has suggested that until Nero there was an attempt to stabilize the
frontier; from Vespasian to the end of the second century, an attempt to
create "scientific" and static borders; and from then on, "defence-in-depth"
as the empire tried to prevent its collapse.[368] The second stage is in a sense
the most crucial: "The limits of empire were then demarcated very precisely, on the ground, so that all could tell exactly what was Roman and
what was not. The established client states had been absorbed, and with
several significant exceptions that illuminate the purpose of the rest, the
land borders of the empire were guarded by defended perimeters that com-
plemented the natural barriers of river and ocean."[369] His approach tries
to discern the strategy behind these periods through an examination of
"the actual dispositions of troops and frontier installations at successive
periods."[370] Luttwak's thesis has been criticized for being too conveniently
schematic and for having inaccuracies of detail.[371] Archaeology seems to
demonstrate that the lines were rarely seen as fixed and not absolutely
respected even when they were; and examination of texts and other documentary evidence suggests that the kind of "grand strategy" Luttwak proposes was simply not possessed by the Romans. The paucity of geographical information, including the absence of maps, meant that emperors and
generals would simply not have known when they had reached strategic
points or what lay on the other side.[372] As Isaac notes, "It is unlikely that
most Roman frontier lines were determined by choice and by a conscious
desire to halt indefinitely all further advance."[373] Luttwak's single biggest
flaw is that he reads Roman empire building and its policing in modern
territorial terms.[374] Nonetheless, as even one of his critics recognizes,
Luttwak's book "proved to be the catalyst that saved Roman frontiers from
the spades of the archaeologists."[375]

Some of these *limes* ran alongside rivers. The two most famous were
the Rhine and Danube, although others such as the Elbe and Neckar (a
major tributary of the Rhine in modern Baden-Württemberg) were used at
different times. In the latter case, at least, the *limes* was more of a strategic line of supply, using the river for transportation.[376] The rivers chosen
fit with these transportation issues: the Rhine flowed into the North Sea,
and the Danube could be used to make a link to the Black Sea and the
Mediterranean.[377] That even these rivers that did more clearly separate the

Romans from barbarians were used for transport shows that they required control of both sides. Whittaker notes the building of fortifications on the other bank, *"in solo barbarico* ('on barbarian soil')," and suggests that political control was common beyond the mere line.[378] This could be through direct military presence or tactical alliances.[379] As Maxfield suggests, rivers are, although common, not great boundaries, unless both sides agree. "They may be bureaucratically convenient . . . but they are lines which are difficult to enforce, they are militarily weak; they are highways which unite, not barriers which divide."[380] It is thus clear that the *limes* are not simple lines of zero width, but areas or zones either side of the river itself.[381] Only this understanding makes sense of claims such as that of the fourth-century-CE speech by Ausonius, where he declares that *"limes* of the Danube and of the Rhine are pacified";[382] or that of Fronto: "The *imperium* of the Roman people was extended by the Emperor Trajan beyond the hostile rivers [*trans flumina hostilia*]."[383]

Others were on land. The most famous is Hadrian's Wall, in northern England.[384] In the words of the author of the *Historia Augusta*, it was "to separate Romans and barbarians."[385] The author continues, suggesting something of an overall strategy on Hadrian's part: "In a great many places where the barbarians are separated off not by rivers [*fluminibus*] but by frontier-barriers [*limitibus*], he set them apart by great stakes driven deep into the ground and fastened together in the manner of a palisade."[386] Birley contends that the wall was made of stone, in part, because of the lack of suitable forests for a timber barrier, such as in Germania.[387] Yet the wall was never referred to as a *limes* in more contemporary accounts,[388] and indeed, it does not seem to have been intended to act as more than a temporary line, because archaeological evidence shows that forts were built north of it, and that it was left behind about two decades after it was built.[389] This was to move northward in 142 CE, under the Emperor Antonius Pius. The line that was aimed for was the one Tacitus mentions at the time of the governor Agricola, between the Clyde and the Firth of Forth, with the fortifications known as the Antonine Wall. This was the extent of Roman Britain for about twenty years before they fell back to Hadrian's Wall. Millar notes that there is evidence that the northern wall was destroyed on two separate occasions and probably finally abandoned in the mid-180s,[390] but more recently this has been revised to suggest abandonment as early as the 150s.[391] The archaeological and documentary evidence is much less for this second wall.[392] It was largely built of turf with a ditch and rampart,[393] although stone may have been intended, and was more of a series of linked forts than the more planned model of the wall farther south.[394] Hadrian's Wall seems to have been abandoned only shortly before the Romans left Britain, around the end of the fourth century.[395]

Both these walls were built by legionaries, and part of the purpose may have simply been to give them something to do.[396] More important reasons were to control, rather than prevent, movement. The walls were built with gates at each milecastle, which could of course be opened as well as closed, and allowed the supervision of movement and the collection of taxes.[397] While the walls alone were not enough to prevent invasion, and an army was needed to fight battles periodically, the walls did prevent raiding parties from crossing and, more important, easily returning with things seized.[398] The use of the army is important: soldiers would have advanced from the wall to fight on open ground, not used the wall as a fortification to defend.[399] According to the later Byzantine historian Procopius, the troops who manned these frontier posts were known as *limitanaious, limitanei*.[400] Procopius makes a distinction between *eschatia*, the frontier posts where the soldiers were stationed (i.e., the *limes*) and *horos* of the empire itself.[401] These forces were light troops intended to hold off invaders rather than a heavy army for future conquest.[402] Indeed, it has been suggested that the walls were "essentially a monument to failure," since the need to build them admitted that Rome's attempt to conquer the whole world had run dry.[403] This may have been a deliberate effort on Hadrian's part to show it was more important to preserve it for future generations than to continue to expand: a triumph of time over space.[404]

There were other lines in Britain. The Fosse Way, a road running through the Midlands, was used as a communication and supply line, but Collingwood argued that it also served a strategic purpose in the battles with the Iceni tribe: as a line to be held by some of the army while the rest was fighting the battles.[405] Even if this were true, it does not appear that it was of especial strategic significance, at least beyond this particular campaign, and should rather be seen as part of the network structure of Roman Britain.[406] In modern-day Wales there were various outposts and fortifications built, until the area as a whole was finally conquered under Vespasian.[407]

In other parts of the *imperium*, there is a similarly confusing picture. In North Africa and the region we currently call the Middle East, there were ditches, roads, and fortifications, but these appear to be at least as much concerned with transportation as defense.[408] The road that the Emperor Trajan constructed between Damascus and the Red Sea (known as the *via nova Traiana*) enabled the Mediterranean to be linked to Arab lands and, farther afield, Indian trade.[409] Earlier views that these were part of a systematic defensive system have tended to be rejected or at least modified in recent years.[410] In any case, with the possible exception of Hadrian's Wall, none of these structures were especially defensible. They may have withstood small incursions, but not larger attacks.[411] But they had a psychological impact: "The *via nova Traiana* and Hadrian's wall, for example, were vast, sophisticated structures in the middle of what was otherwise relatively primitive and undeveloped countryside. They must have seemed impressive, even terrifying."[412]

In the history of the later empire written by Ammianus Marcellinus, there is a sense of a siege mentality against barbarian invasions: "During this period throughout the whole Roman world [*universum orbem Romanum*] heard the trumpet-call of war, as the most savage peoples roused themselves and raided the nearest *limites*."[413] He tells of how "in Britain the wild tribes of the Scots and Picts broke their understanding to keep peace, laid waste to the country near the frontier [*loca limitibus*]."[414] Ammianus provides a few of the other instances of the word *territorium*. He suggests that barbarians were occupying the *territoria* of Strasburg, Brumath, Saverne, Seltz, Worms, and Mayencee, but avoided the *civitates* themselves "as if they were tombs surrounded by nets."[415] Once again there is the clear indication that the *territorium* is the lands surrounding a city, not the whole land occupied. Elsewhere, though, he begins to use *territorium* in an undifferentiated sense of lands in general. There are two instances: King Chonodomar escaping a site of battle and realizing he could only reach "his own lands [*sua territoria*]" by crossing the Rhine,[416] and Theodorus hiding in "a remote part of the country [*in devia territorii parte abscondito*]."[417] Other contemporary texts use the word in this sense.[418]

Another contemporary of Ammianus wrote the anonymous text known as *De rebus bellicis*.[419] One of the key concerns of this text is the problems of the frontiers, and in particular the situation of the people who live in these regions.[420] Part of the point is to suggest tax reform, so "with the abolition of abuses in the system of taxation, the frontier-dweller may do honour to the lonely stretches of the imperial boundaries in safety, after the erection of fortified defences."[421] The author suggests that "veterans enriched with imperial grants and farmers still powerful of limb" will till the fields, inhabit the *limites*, cultivating and defending the land, and will eventually become taxpayers.[422]

> First of all it must be recognised that frenzied native tribes, yelping everywhere around, hem the Roman empire in [*cirumlatrantium*], and

that treacherous barbarians, protected by natural defences, menace every stretch of our frontiers [*limitum*]. For these people to whom I refer are for the most part either hidden by forests or lifted beyond our reach by mountains or kept from us by the blazing sun. There are those who, defended by marshes and rivers, cannot even be located easily. . . .[423]

A proper concern for the frontiers which surround the empire on all sides is also to the advantage of the *res publica*: an unbroken chain of forts will best assure the protection of these frontiers, on the plan that they should be built at intervals of one mile, with a solid wall and very strong towers; the landowners would construct these defences on their individual responsibility, at no expense to the public, and of course watches and country patrols should be organized in them, so that the peace of the provinces protected by a kind of belt of vigilance, may rest unharmed in quiet.[424]

From this brief survey it should be apparent that the Romans did not have one system for how to control the frontiers of the *imperium*. Hadrian's Wall cannot be seen as the model for other areas; some were more akin to a road with constant patrols; others were between inhabited and desert regions, such as in North Africa.[425] It is therefore difficult to give too much credence to accounts that see the walls, frontier posts, and rivers as forming a coherent system.[426] The *limes* should be understood not as fixed lines, but as fluid zones, both for the rivers and fortifications on land, which meant that people lived in these areas.[427] Three points need to be underlined. First, the Romans secured the land on both sides of these notional lines: the far bank of rivers or lands beyond fortifications. It was therefore more a case of controlling, rather than preventing, passage. But in this way they were able to force passage through specific sites, and to extract taxes.[428] Second, they saw them as fortifications or temporary stopping points, not as static absolute limits to their expansion. What was on the other side was not the possession of another sovereign entity that was recognized as an equal, but merely separated what was Rome and what was not yet Rome.[429] The Romans regularly went beyond the lines, both in terms of seeking to expand and in their general practice. There is plenty of archaeological evidence for this.[430] Third, and following from these, it does not make sense to see the Roman Empire as having boundaries in anything like an unproblematic modern sense. This has been recognized by some, but by no means all, international relations scholars.[431] As Mattern summarizes, "Thus the Roman frontiers emerge as a complex phenomena, subject to interpretation only with difficulty. Rather than a coherent, mainly defensive system, we see variation, mutation, and ambiguity."[432]

⟨∽⟩

The Roman Empire succeeded in stabilizing control over large parts of the Mediterranean and adjacent lands for several centuries. Lintott rightly contends that, on its own terms, it was a success: it provided security and set up a system of government where delegation to provincial control allowed continuity even when there was instability in Rome, and weak emperors. This is assessing it as an empire: the republic was long gone. "It was not its function to be a *res publica*, to perform the socio-political functions of a civic community, a *polis* or a *civitas*."[433] This was, in a sense, a classic case of imperial overstretch: as Cunliffe puts it, "By the second century CE the core—the Roman empire—had grown so quickly that it had engulfed its periphery without fully integrating it."[434] The continual warfare meant that military power and political power became ever more intertwined.[435] Emperors could rule without spending much time at all in the center. "The locus of political power ceased to be the capital, to become the military camp of the frontier areas."[436] But Rome's political and military practices have less of a legacy in terms of the question of territory than its legal and surveying technologies.

It is essential to remember that the transportation systems of the empire were very slow. Troops could march at about 3 miles an hour when encumbered by equipment, averaging perhaps 20 miles a day on good roads; ships could average two to three knots with favorable winds, but of course could continue for much longer in a day.[437] These were of course the norm: exceptional speeds are reported in the literature. Tacitus claims that six Vitellian legions covered 30 miles in a day;[438] whereas Pliny the Elder recounts various running achievements in excess of 100 miles, also noting that Tiberius Nero apparently made it to his brother's deathbed by carriage covering 200 miles in twenty-four hours;[439] Julius Caesar averaged 100 miles a day for eight days, which Suetonius describes as "an incredible pace";[440] and a dispatch from Mainz made it to Rome in eight or eight and a half days, a speed of 160 miles a day.[441] There was thus a good reason that the empire expanded around the Mediterranean, even though this meant that lines of supply on land were rarely convenient. Transportation costs have been estimated that suggest a cargo could be transported the entire length of the sea for the cost of taking it 75 miles inland.[442] Jones suggests that this meant that transport of corn over 50 miles was uneconomical.[443] This is, of course, only part of the story. The risk of transporting things by sea sometimes outweighed mere cost or time. Nonetheless, there was a well-organized structure of roads, bridges, and other transportation infrastructure and technologies, mainly built by the military.[444] Seasons and weather also played a part; sailing was only viable in the summer, and in calm, but not too calm, weather. When winter came, "the passes filled with snow, the great flagstones of the Roman roads sank in the mud, the stores of fodder dwindled in the posting-stations, and the little boats rocked at anchor. The Mediterranean ceased to exist; and the distance between the Emperor and his subjects trebled."[445] The horizons of everyday life were truncated: Brown has estimated a radius of only 30 miles for the provision of food, shelter, and clothing, meaning that a winter of starvation awaited towns whose harvest failed.[446]

The technology therefore simply did not allow speedy communications or centralized decision making. Both military forces and civilian governors had to make immediate decisions without consultation.[447] The importance of presence—exchange of gifts, verbal communication—should not be underestimated.[448] Just as provincial governors could become detached from what was happening in Rome, so too could the capital become peripheral to events. This meant, unsurprisingly, that the logic of multiple sites became appealing. The fracturing of the West led to power rooted in diverse places rather than centralized through an *imperium* that was becoming too large to effectively govern. This division had long roots. After the pact of Brundisium between Antonius and Octavian in 40 BCE, the Roman Empire had been divided in two. Antonius took the East, including Asia, Egypt, and Greece.[449] While they were brought back together, Syme is right to suggest that the future break between East and West was anticipated here. He suggests that this was the frontier by "nature, by history, by civilisation and by language between the Latin West and the Greek East." He argues that "it is one of the miracles of Roman history that in subsequent ages the division between West and East was masked so well and delayed so long."[450] The linguistic aspect is worth stressing. In the eastern part of the empire, the commonly used language was Greek, even if local languages such as Coptic and Aramaic survived.[451] For a time the administration was bilingual between Greek and Latin, but as the split widened, the use of Latin became less important. In the western part, as chapter 3 will discuss in more detail, knowledge of Greek disappeared. This had significant consequences, for both politics and faith, in that Constantine's conversion led to the conversion of the empire, and the Christian religion—then a minority religion without wide support—was given a basis on which it could build.[452]

For a time, then, the *orbis christianus* and *orbis romanus* coincided.[453] This would be the aim of future attempts at restoration. Yet Rome was beset by problems. There was internal dissent, external conflict, and an exhaustion of political will. Its internal weaknesses were only one, and arguably not the major, factor in its collapse.[454] There was a dramatic change in terms of its military status. Gibbon notes that "the predecessors of Jovian had sometimes relinquished the dominion of distant and unprofitable provinces; but, since the foundation of the city, the genius of Rome, the god Terminus, who guarded the boundaries of the republic, had never retired before the sword of a victorious enemy."[455] In 406 CE the Rhine froze and barbarians crossed into Gaul;[456] Rome was sacked four years later by the Visigoths, the first time in nearly eight hundred years that the city had fallen to a foreign force.

PART II

CHAPTER THREE

The Fracturing of the West

The following four chapters attempt to treat the period that is often known as the "Middle Ages" in terms of the relationship between political power and the places over which it was exercised. A number of complexities arise in such a venture. Perhaps most important, there is the question of the term *Middle Ages* or the adjective *medieval*.[1] This was a term coined in the fifteenth century and applied retrospectively. Hay, for instance, suggests that in 1469 the preceding era was described as a "middle time" between the ancient and modern world,[2] and notes that around 1452 Flavio Biondo wrote a work known as *Decades of History from the Decline of Rome*, a history of the period 472 to 1440.[3] These broad dates—from the takeover of what was left of the Western Roman Empire by the Germanic general Ricimer to the foundation of the Prussian Confederation—were generally accepted, although the end date was sometimes revised to 1453 (the fall of Constantinople) or later (discoveries of the New World).[4] Yet many of the key aspects of this period owe much to antiquity, which continued to assert a major influence, and there is certainly no clear break. At the other end, Jacques le Goff, among others, has suggested that the Renaissance is crucial to the Middle Ages, as it forms a historical watershed.[5] Yet others have suggested that really we need to take the "whole period from 1150 to 1650 as a single era of essentially continuous development."[6] And, for Olson, "The exasperatingly labelled Middle Ages were not a 'middle' but a beginning for Europe."[7]

There thus needs to be a challenge to the periodization inherent in the idea of a Middle Ages. As Ségal notes, there are many dangers with naming periods and then investing them as preexistent historical beings.[8] Here I have tried to avoid a strict distinction between the before, the after, and the middle, while recognizing that it is both a useful shorthand and one

used by many historians of this period. Most especially, I have avoided
jumping from the classical period to the modern, or taking just a couple of
emblematic figures to stand for the millennium. In the history of political thought, this is an area that is traditionally neglected, with Saint Augustine and Saint Thomas Aquinas usually, at best, the two emblematic
figures. Yet, for Nederman, who bemoans the neglect of the Latin Middle
Ages, Augustine is "more properly a late classical figure" and Aquinas
only one current in the thousand years.[9] In the company of a host of tal-
ented scholars, these two figures—who are both treated in some detail—
are situated back within those multiple currents. In tracing the relation
between place and power—the prehistory of the concept of territory—the
approach is close to that which Woodward suggests is needed for the history of cartography. He suggests that, like the history of science, undertaking this work becomes less of a search for predecessors and more an
attempt to understand previous times in their own terms.[10]

Another crucial aspect, much remarked upon, is the connotation of
"medieval" with primitive or barbaric. The period, or at least its earliest
centuries, is often described as the "Dark Ages." As Dagenais and Greer
suggest, "The Middle Ages is Europe's Dark Continent of History, even as
Africa is its Dark Ages of Geography."[11] This description, needs, as Abu-
Loghod has noted, both historical and geographical specificity. Large parts
of Europe, especially those still in contact with the Byzantine, continued
to develop and flourish. Trade with Asia and cultural contact with a range
of civilizations endured.[12] It is easy to be so overwhelmed by certain images of the period—the Crusades, sieges, warfare, and the Inquisition—
that artistic, literary, and religious achievements, not to mention the more
prosaic aspects of everyday life, are forgotten.[13] The disputes can also be
found in the academic study of this period. Biddick, for instance, notes
how nineteenth-century "medieval studies" attempted to separate themselves from more popular studies, which they labeled "medievalism."
"Medievalism" as a label has been more recently reembraced, but Biddick
suggests that it always inhabits medieval studies as "an abiding historical
trauma."[14] *Neomedievalism* has been used to describe some of the aspects
of the contemporary "war on terror,"[15] globalization,[16] or in terms of political geography the changes of the European Union.[17] Or, in Alain Minc's
term, since the fall of the Berlin Wall, we have entered a "new Middle
Ages."[18]

A third issue is the geography of this time, a theme that will be returned to. Here the focus is generally on the "Latin West"—that is, largely
those lands that had previously been part of the Roman Empire in Western

Europe. Yet other large parts of the Roman Empire in the East continued to exist under imperial rule for another millennium. In addition the Muslim lands to the south and east exerted a considerable influence on the West in a range of ways, from politics and philosophy to mathematics and culture. And, at the end of this broad period, part a product of its demise and part accelerator of it, the "New World" to the west forced rethinking across a range of parts of human knowledge and practice.[19]

For Morrall, the ascent of religion over political power is central to the very definition of the period: "Instead of religion, as hitherto, forming the buttress for a communal political tradition, it was elevated essentially above the political sphere and from this position of transcendence it bestowed on political authority whatever limited justification the latter possessed."[20] Kimble sees it as a period of intellectual impoverishment for this very reason: "During the Middle Ages . . . the Greek tradition of disinterested research was stifled in Western Europe by a theological dictatorship which bade fair, for a time, to destroy all hope of a genuine intellectual revival."[21] Ullmann is more positive, noting that "the governmental and political ideas dominant in the Middle Ages have created the very world which is ours. Our modern concepts, our modern institutions, our political obligations and constitutional ideas are either direct descendents of those of the Middle Ages, or have grown up in direct opposition to them."[22]

The overall suggestion here is that we can only understand the legacy of the classical age, and in particular the rediscovery of Greek political thought and Roman law, in the light of the "medieval" period. The transformations in later centuries need to be situated in the context of the transformative relations between the church and secular political power. Neither the law nor the thought was brought back to life in an earlier form, but radically transformed in these new contexts. As Stewart suggests, it makes sense to think of this period as one of "silent preparation and steady self-teaching which must necessarily intervene between the death of an old world and the birth of a new."[23] In Western Europe several key elements are worth noting. Latin was the dominant language, the *lingua Romana* or the Roman language. This had largely erased the previous languages of those areas that had been part of the Roman Empire. Dialects developed on the basis of the distinction between spoken and written language, alongside barbarian influences. These became the Romance languages. Such linguistic hegemony was not achieved in Britain, Balkans, and North Africa, with the Latin influence in English coming from the Norman conquest rather than directly from Rome. Yet, as Hay notes, "In the older area of Roman domination the Rhine and the Alps remain to this day an enduring linguistic boundary."[24] Farther east, Greek remained the dominant language. That said, *Europe* itself is a misleading term, with few people having a sense of a scale beyond the village or parish. For those who did, it would have extended to "county or diocese, or kingdom"; and beyond that, people would have described themselves as "inhabitants of Christendom" (at least from the year 1000).[25]

The specifically geographical aspects have been receiving more attention in recent years, with a number of volumes devoted to such themes across a range of subjects. Tomasch and Gilles, for instance, have discussed the relation of texts and what they call "territories":

> It is the reciprocal interaction of two associated processes—the textualization of territories and the territorialisation of texts—that perhaps most clearly illustrates the pervasiveness and potency of geographical desire. Through these processes, land is re-presented as territory, and works are surveyed, explored, located and bounded; they become, as it were, texts.[26]

An analysis of the geographies of the "Middle Ages" is, these chapters suggest, useful in understanding the complicated relations of history, politics, and geography, and especially the relation between place and power.

AUGUSTINE'S TWO CITIES

Saint Augustine of Hippo was the most important Christian thinker of his time, referring back to classical sources, but fundamental for the development of later thought.[27] While he was familiar with a range of Latin sources, he knew almost no Greek. As he says in the *Confessions*, "Even now I cannot fully understand why the Greek language, which I learned as a child, was so distasteful to me."[28] As Brown notes, his "failure to learn Greek was a momentous casualty of the Late Roman educational system: he will become the only Latin philosopher in antiquity to be virtually ignorant of Greek."[29] What this meant was that his intellectual climate was formed from almost exclusively Latin sources, with Greek texts known to him only in what translations existed. This would be a trend for the next several centuries. As Kimble suggests, there were real problems of contact: "From the fifth to the twelfth century Greek scholarship might not have existed—so little influence did it exert on Western culture."[30] At Augustine's time, the Roman Empire stood as the "political embodiment" of the "intellectual and religious legacy of the ancient world,"[31] a world that Augustine was profoundly shaped by. His major political work, *De civitate Dei*, usually translated as *The City of God*,[32] has been described as, after the Bible, probably "the book most widely read in the west in the Middle Ages,"[33] and as having "a greater influence on subsequent medieval political thought than any other book written in the early middle ages."[34] Yet it is not a treatise primarily concerned with what we could now call politics.

Augustine was, famously, converted to Christianity at a fairly late stage of his life, writing the *Confessions* to outline the failures of his past and his conversion. The first sack of Rome, by the Visigoths under Alaric in 410 BCE, has been seen as the inspiration for *The City of God*,[35] though this is somewhat overstated. Augustine died with the Vandals at the gates of Hippo, and the collapse of the Roman Empire certainly formed a context within which his works operate. More important than the mere fact of the collapse, though, was that some pagan writers suggested that Rome had become weak because of its conversion to Christianity. To challenge that claim was one of Augustine's principal aims—a theme he shares with Paulus Orosius, who outlined the crises of the Roman Empire before Christianity, in order to contest the view that it was that belief that led to its downfall.[36] Orosius's is a more historical account than Augustine's theological one, and proposes the idea of a *translatio imperii*, a gradual movement of imperial domination from the east of the Persians to Greece to Rome. As the old Augustine writes in his *Retractationes*, "The worshippers of many false gods, whom we call by the customary name pagans, attempting to attribute its destruction to the Christian religion, began to blaspheme the true God more sharply and bitterly than usual."[37] As Saint Jerome said, quoting and then modifying Lucan: "What is enough, if Rome is too little? . . . What is safe, if Rome is lost? [*Quid satis est, si Roman parum est? . . . Quid saluum est, si Roma perit?*]."[38] Beyond this narrow purpose, then, and to make sense of it in broad scope, Augustine wanted to put forward an understanding of human relations to the divine, a "philosophy of history that could include and transcend the history of Rome."[39]

Augustine did not therefore write a book *about* the sack of Rome, but one that found its spur and audience in those events. Those events, of course, meant that both at the time and ever since it has been seen as a book with a political purpose and message, rather than simply a work of theology or exegesis. Indeed, Augustine's earlier commentary on Genesis had devoted considerable attention to the division between heaven and earth, suggesting that the idea of a book on this theme was already being considered.[40] This view seems much more plausible than Knowles's suggestion that it "had its origin almost as an occasional essay, a *pièce de circonstance*."[41]

The City of God is an unwieldy, rambling text that covers a great deal of ground. A lot of it is concerned with the superiority of Christianity over Neoplatonism and paganism.[42] There is therefore a danger of systematizing Augustine's work, to reconstruct something that was never really constructed.[43] Its problems begin with, but are certainly not exhausted by, its title. While the operative phrase certainly means the *civitate* of God, it is not at all clear that this should be straightforwardly rendered as the "city," much less "state."[44] As Brown notes, Augustine took the term *de civitate Dei* "as a technical term, taken from the Psalms, to express what we might call 'The Communion of the Saints.'"[45] Indeed, Augustine cites several psalms that use this term.[46] The more standard classical Latin term for a city, *urbis*, is absent from the title. "Citizenry of God" or—somewhat more of a stretch—"Commonwealth of God" have been suggested as alternative renderings.[47] Although "City of God" is well established as a translation, it is certainly crucial to remember that Augustine does not mean quite the same as a modern "city" and certainly has no mere urban environment in mind. Indeed, the "Community of God" may be closer to his intent. In the reading of Augustine that follows, *civitate* will be translated as "city," not least because of convention and the use of the term in the quotations from secondary authorities. But the plural senses of the term should always be remembered.

Augustine splits the human race into two parts, the two communities of men or, mystically, the two cities. He therefore distinguishes between the City of God and the earthly city (*civitas terrena*). These are both *societas*.[48] These two cities are, for Augustine, mixed in body for the time being,[49] but separated in heart, and running through history until the end. One is Jerusalem, which means "vision of peace" and whose love is God and whose joy is eternal peace; the other Babylon, whose name means "confusion," whose love is the world, and whose joy is temporal peace.[50] He stresses that the Roman Empire is "an earthly, not a heavenly, power,"[51] and thus challenges Rome's aspirations to be the eternal city, yet simultaneously opens the potential for a new Roman Empire that would be holy and represent less a place than the souls of Christendom.[52] Medieval papalist authors regularly took Augustine's notion of a city whose founder and ruler was Christ to mean a city ruled by the church, the institutional church, but Augustine intends to say something rather different.[53] The kind of justice Augustine has in mind would only be fully realized after this world had ended.[54] The *civitate Dei* was not church led, but the "society of grace: the entire community, past and present, of those who unfeignedly love God."[55] Augustine himself notes that the city comes to earth through faith and hope. The "House of God," "Temple of God," or the "City of God" mean the same thing;[56] and "it is in hope that the City of God lives, during its pilgrimage on earth, that City which is brought into being by faith in Christ's resurrection."[57] Thus, while Markus is right that "the archetypal society, where alone the human fulfilment can be found, is the society of the angels and saints in heaven: not a *polis*,"[58] it is crucial to remember that Augustine's role concerning this city is twofold: he seeks to defend the city "as it exists in this world of time . . . and as it stands in the security of its eternity."[59]

If the *civitate Dei* is thus not something that can be straightforwardly found in the contemporary world, or a program for a future political settlement, it is also essential to note that the "earthly city" similarly does not have a direct relation to any specific state or political situation.[60] The earthly city is rather a *mystical* entity, just as is the city of God.

> Though there are many great peoples throughout the world, living under different customs in religion and morality and distinguished by a complex variety of languages, arms and dress, it is still true that there have come into being only two main divisions, as we may call them, in human society: and we are justified in following the lead of our Scriptures and calling them two cities. There is, in fact, one city of men who choose to live by the standard of the flesh, another of those who choose to live by the standard of the spirit.[61]

Yet, inevitably, though this distinction is clear in a formal sense, it is continually overlapping when looking at particular human societies.[62] A somewhat more rigid distinction—between those who follow the one true god and those who worship the pagan gods of classical Rome—is a recurrent theme. Augustine declares that "we have learnt that there is a City of God: and we have longed to become citizens of that City, with the love inspired by its founder. But the citizens of the earthly city prefer their own gods to the founder of this Holy City."[63] This is more than merely a theological concern, but has important political consequences: "True justice is found only in that commonwealth [*re publica*] whose founder and ruler is Christ; if we agree to call it a commonwealth, seeing that we cannot deny it is the 'weal of the community' [*rem populi*]."[64] Although his main task is "defending the glorious City of God against those who prefer their own gods to the Founder of that City,"[65] he realizes he "cannot refrain from speaking about the city of this world [*terrene civitate*], a city which aims of dominion, which holds peoples [*populi*] in enslavement, but is itself dominated by that very lust of domination."[66] In doing so, he talks of how the "the Roman *res publica*" changed "from the height of excellence to the depths of depravity."[67]

The notion of the *populus* is worth looking at in a bit more detail. While it is a grouping of people, following Cicero, Augustine declares that it is not simply "any assemblage or mob, but 'an association united by a common sense of right [*ius*] and a community of interest.'"[68] Cicero himself suggests that the *res publica* is a *res populi*, the commonwealth is the property of a people, with a *populus* as "an assemblage of people in large numbers [*coetus multitudinis*] associated in an agreement with respect to justice [*ius*] and a community of interest."[69] The idea that a *populus* only truly exists when defined by right or justice is an intriguing one, given that Augustine thinks that *ius* is only truly present in the city of God. This has a number of consequences. First, it means that *populus* is a more specific term than *people*. Adams has described this latter term as "diffuse and broadly polysemic," which is less the heir of *populus* "than its descendent."[70] Second, that the *populus* is defined through membership in a political community, through laws, ritual, and consent, which could presumably be changed or taken apart as much as constituted.[71] This leads to the third consequence: that it is not necessarily ethnic or linguistic in affiliation.[72] It is a group of humans—Augustine follows Aristotle in seeing humans as distinguished by their rationality[73]—that comprises men, women, and children, even if the latter are excluded from political citizenship in other ways; diverse classes; and is only coincidently one that arises through birth or location.[74] The fourth is the most contentious conclusion, but it is one anticipated in Cicero himself.

> For Scipio gives a brief definition of the state, or commonwealth [*res publica*], as the "weal of the people" [*res populi*]. Now if this is a true definition there never was a Roman commonwealth, because the Roman state was never a "weal of the people," according to Scipio's definition. For he defined a "people" as a multitude "united in association by a common sense of justice and a community of interest" [*multitudinis iuris consensus et utilitatis communione sociatum*] . . . a *res publica* cannot be maintained without justice, and where there is no true justice [*iustitia uera*] there can be no right [*ius*]. . . . If there is no

people [*populus*] then there is no "weal of the people" [*res populi*], but some kind of a mob [*qualiscumque multitudinis*], not deserving the name of a people.[75]

That there was never a people in Rome and therefore never a *res publica*, since it had no true justice, appears absurd, especially since, if carried to its conclusion, it means that there will never be a *res publica* on earth.[76] Augustine effectively carries the idea that a *res publica* needs justice, as it does in Cicero, to the idea that justice can only be understood in a spiritual Christian sense.[77]

Augustine realizes this issue, and offers the modified definition: "A people is the association of a multitude of rational beings [*coetus multitudinis rationalis*] united by a common agreement on the objects of their love [*concordi communione sociatus*]."[78] He suggests that following this means that "the Roman people is a people [*populus*] and its estate [*res*] is indubitably a commonwealth [*res publica*]."[79] Despite their object of love, and their morality, "I shall not make that a reason for asserting that it is not really a people or that its estate [*rem*] is not a commonwealth, so long as there remains an association of some kind or other between a multitude of rational beings united by a common agreement on the objects of its love."[80] Nonetheless, he insists that "because God does not rule there the general characteristic of that city is that it is devoid of true justice."[81] This is continued in his analysis of secular societies:

> Remove justice, and what are kingdoms but gangs of criminals on a large scale? What are criminal gangs but petty kingdoms? A gang is a group of men under the command of a leader, bound by a compact of association, in which the plunder is divided according to an agreed convention. If this villainy wins so many recruits from the ranks of the demoralized that it acquires land [*loca*], establishes a base, captures cities and subdues people, it openly arrogates to itself the title of kingdom, which is conferred on it in the eyes of the world, not by the renouncing of aggression but by the attainment of impunity. For it was a witty and truthful rejoinder which was given by a captured pirate to Alexander the Great. The king asked the man, "What is your idea, in infesting the sea?" And the pirate answered, with uninhibited insolence, "The same as yours, in infesting the earth! But because I do it with a tiny craft, I'm called a pirate: because you have a mighty navy, you're called an emperor."[82]

It follows from this, crucially, that Rome should not be given especial credit for becoming dominant. Justice is clearly not a prerequisite for
success. Augustine notes that the Assyrian kingdom of King Ninus was
similarly powerful: "If this Assyrian kingdom reached such magnitude
and lasted for so long, without any assistance from the gods, why are the
gods given credit for the Roman Empire's wide extension in space [*locis
amplum*] and its long duration in time? Whatever is the cause of the one
is surely the cause also of the other."[83] This raises a somewhat different
question: is the *populus*, like the *polis*, something of determinate size? It
clearly needs to be a *multitude* above a certain size, although Augustine
does not specify this, and needs self-awareness of its unity. But, as Adams
notes, Augustine does not address two crucial questions that Aristotle provided clearer answers on: can a *populus* be too big, ceasing to be a *populus*;
and can a *populus* be made of other *populi*?[84] Nonetheless, it is clear that
Augustine thought himself a member of several different *populi*, which
would imply that membership is not exclusive: Christianus, the church,
and the local civil community. He also recognized that *populus Romanus*
and *populus Israel* existed.[85]

This has important implications for the understanding of the individual. Moral decisions are always social ones; the individual is always part of
the collective, the *populus*, and ultimately the city.[86] The *City of God*, then,
is a book about our mortal life in the world and the relation between our
spiritual soul and that world. One city "trusts in the things of the world,
the other in the hope of God," but both are concerned with the way we
deal with that choice in our shared mortality, which stems from Adam.[87]
As Brown puts it, "Far from being a book about flight from the world . . .
it is a book about being other-worldly in this world."[88] The key theme
then emerges: love. "The two cities were created by two kinds of love: the
earthly city was created by self-love reaching the point of contempt for
God, the Heavenly City by the love of God carried as far as contempt of
self. In fact, the earthly city glories in itself, the Heavenly City glories in
the Lord."[89] Gilson, in particular, has developed this theme, noting that if
the name *city* means "any group of men united by a common love for some
object, we say that there are as many cities as there are collective loves . . .
since there are two loves in man, there should also be two cities to which
all other groupings of men are reduced. The scores of men who lead the
life of the old man, the earthly man, and who are united by their common
love of temporal things, form the first city, the *earthly city*; the multitude
of men who are joined together by the bond of divine love form a second
city, the *City of God*."[90] Knowles pushes this point to its conclusion: "The

two cities are therefore two loves, and these are an inward and spiritual, not an outward and political distinction."[91]

It would therefore be easy to detach the idea of the two cities entirely from a spatial mooring. The communities are those of belief, the attachment one of love rather than location. Nonetheless, a number of times Augustine does stress the importance of place to associations of people.[92] He speaks of the lands of various peoples, such as "Chaldean lands [terra]."[93] One of the most significant of these is the land of Canaan: "For the promise spoke of the land [terra] of Canaan stretching from a certain river of Egypt to the great River Euphrates."[94] This was a promise "which concerned land [terrena] . . . the Hebrew people should continue in the same land in undisturbed stability, as far as temporal prosperity is concerned."[95] Joshua "led the people into the land of promise [terra promissionis] and settled them there by God's authorization, after he had crushed the nations who were then in possession of those places [loca]."[96] Biblical narratives thus concern lands of peoples, not some immediate source of the modern notion of territory. Attachment to place can take a more primitive form. Augustine recounts the story of Erichthonius[97] and suggests that it is not insignificant that in Genesis, God "fashioned man out of the dust of the earth."[98] Canaan is thus important for Jewish identity, as Attica was for Athenians, but not exclusively, and at times it appears more as a looser sense of a longing for a lost past.[99] This is particularly the case concerning the Jewish people, who were dispersed across the earth, but carried with them their holy books. Indeed, Markus notes that this is for him more of a symbol than the Roman Empire, used by Jerome and Ambrose to underline the "universal mission of the Apostles."[100] In distinction he suggests that Christ's rule is without limit, following the Psalm on Solomon: "He will rule from sea to sea, and from the river [Euphrates] to the ends of the earth [ad terminos orbis terrarum]" (Psalms 71 [72]:8).[101]

Augustine's attitude to the geography of the Roman Empire is ambiguous. On the one hand, he stresses the way in which the empire's grasp of land has been uncontested and must be due to special favor from the gods, or God:

> The race of Mars (that is the Roman people) will never give up to anyone a place [locum] that is in their possession; thanks to the god Terminus no one will ever move the Roman borders [terminos], . . . they have, in fact, been able to yield to Christ, without any loss of land [locorum] in the Empire.[102]
>
> Let us go on to examine for what moral qualities and for what rea-

son the true God deigned to help the Romans in the expansion of their empire; for in his control are all the kingdoms of the earth.[103]

He recognizes, of course, that at times Rome overstretched itself, and gives the example of Hadrian's redrawing of the *limes* in the East to the Euphrates in 117 CE, in order to end the war Trajan had begun with the Parthians in 114. Thus, Augustine notes, he altered the "eastern borders [*termini*] of the Roman Empire . . . [and] ceded to the Persian Empire three famous provinces, Armenia, Mesopotamia, and Assyria."[104] Augustine is interested in the idea of limits to authority, and provides a valuable account of boundary practices at the time, when he is discussing the properties of charcoal. He notes that it is remarkable in that it is brittle and can be broken easily, but equally durable in that moisture and age cannot destroy it. His example is that "it is customary to put charcoal under boundary marks [*limites*] when they are set up, to refute any litigant who might come forward at any time in the remote future and maintain that a stone fixed in earth was not a boundary stone [*limitem*]."[105] There is also the suggestion, quoting Justinius, and picked up by John of Paris many centuries later, that "the rule of any one man extends only to the limits of his own homeland [*suam cuique patriam regna finiebantur*]."[106] On the other hand, he asks, "What is a city but a group of men united by a specific bond of peace?"[107] "What is Rome but the Romans?" and suggests that "a city is its citizens, not its walls."[108] What we therefore find in Augustine is a continual wish to stress the interpersonal relations constituting political communities rather than their location or the abstract nature of these entities. He tends not to use words like *imperium* and *res publica*, but to talk about emperors and kings.[109]

Augustine's general comments on land in a political register are accompanied by some more general reflections on issues of place and space. He is concerned with the interrelation of the two Latin words *spatium* and *locus*, which we would today render as "space" and "place," but the former tends to retain its classical Latin sense of a distance, a stretch, or an extent rather than the modern idea of a container. Thus, when Augustine asks if God is "some kind of bodily substance extended in place [*per spatia locurum*], either permeating the world or diffused in infinity beyond it,"[110] the *spatia* defines the limits of the *locus*, rather than the reverse. Elsewhere in the *Confessions*, he elaborates on the theme:

If I tried to imagine something without dimensions [*spatiis*], it seemed to me that nothing remained, absolutely nothing, not even an empti-

ness [*inane*]. For if a body were removed from the place [*loco*] which
it occupied, and that place [*locus*] remained empty of any body whatsoever, whether of earth, water, air, or sky, there would be an empty
place, but there would still be an extent of nothing [*sed tamen sit locus
inanis tamquam spatiosum nihil*].[111]

One of the best-known passages of the *Confessions* concerns the difficulty of thinking the concept of time. Augustine declares that "this
world was not created *in* time but *with* time . . . the world was in fact
made *with* time, if at the time of its creation change and motion came into
existence."[112] Here too *spatium* and cognate terms are used to understand
temporality. "I see time, therefore, as an extension [*distentionem*] of some
sort";[113] and further that "it seems to me, then, that time is merely an extension [*distentionem*], though of what I do not know, and I begin to wonder if it is of the mind itself."[114] He discusses "infinite stretches of time
[*infinita spatia temporis*]" alongside "infinite stretches of place [*infinita
spatia locorum*]" and talks of "the boundless immensity of place [*inter-
minabilem inmensitatem locorum*] which stretches everywhere around
the world."[115] Here, as elsewhere, his idea of a place understood through
its extent in three dimensions is very close to the meaning later writers,
notably Descartes, would give to *spatium*. Indeed, he does occasionally
see *spatium* as having properties rather than simply as a property of something else. He talks, for instance, of the "qualities related to space, such as
density, sparseness, or bulk";[116] or of a fountain being within "a confined
space" or "narrow compass [*ampliora spatia*]."[117] Yet he is still some way
from this being its overall determination, tending to attribute a more Aristotelian sense to the properties of place:

> A body inclines by its own weight towards the place [*locum*] that is fit-
> ting for it. Weight does not always tend towards the lowest place, but
> the one which suits it best, for though a stone falls, flame rises. Each
> thing acts according to its weight, finding its right level. If oil is poured
> into water, it rises to the surface, but it water is poured onto oil, it sinks
> below the oil. This happens because each acts according to its weight,
> finding its right level. When things are displaced, they are always on
> the move until they come to rest where they are meant to be.[118]

Two themes come from Augustine's writings that will fundamentally shape the relation between place and politics in the centuries to follow. Even though he never explicitly clarified the relation between the

church and the worldly state, this was an issue that would be central to the church's intervention in politics in several registers. The crowning of the king of the Franks as emperor of the Romans is only its most obvious manifestation. Augustine's claim that even corrupt rulers receive their power from God links to his contention that subjects should obey their rulers.[119] To more formally cement that relation through practice was an inevitable step. It would undoubtedly be misleading to see Augustine's heavenly and earthly cities as a forerunner of the later distinction between spiritual and temporal realms, especially if this is an attempt to cast the work within that interpretative framework.[120] Yet in a sense the key here is not what Augustine meant, but how he was appropriated, and his ideas were undoubtedly reread in that later context.[121] The other key theme that Augustine stresses is the privileged status of Canaan, the Promised Land. While Augustine was not especially fixed on the geographical location, this would be more strongly emphasized in future centuries, especially in the Crusades. Both of these themes will be developed in chapter 4 and the remaining chapters of this part.

But before moving to those issues, it is worth looking at two other figures of this age of transition, who tried to hold back the advent of the intellectual dark ages with their attempt to preserve something of the classical tradition: Boethius and Isidore of Seville. This chapter will then move to a discussion of the political and cultural situation of this early part of the Middle Ages, looking at the "Germanic" tribes and use the *Beowulf* poem as a particular, limited, glimpse of the organization of land politics at the time.

BOETHIUS AND ISIDORE OF SEVILLE

Boethius is not an especially original thinker, and in his major work *The Consolation of Philosophy*,[122] he mixes Platonism, Aristotelianism, and Stoicism, with some of these filtered through later interpreters rather than the primary sources.[123] As Stewart notes, his "genius was imitative rather than initiative and nothing if not dependent."[124] Yet his importance is undoubted, both as a translator of Aristotle and as a mediator of classical thought. As Watts notes, he "stands at the crossroads of the Classical and Medieval worlds."[125]

The task he set himself was to translate "the whole work of Aristotle, so far as it is accessible to me . . . and all Plato's Dialogues. . . . When accomplished, I will venture to prove that the Aristotelian and Platonic conceptions in every way concord, and do not, as is widely believed, entirely contradict each other."[126] He achieved only a fraction of this work, but his accomplishments were by any standards extraordinary. He translated Aristotle's *Categories, Prior and Posterior Analytics, On Interpretation, Sophistical Refutations,* and the *Topics*—the collection of logical writings known as the *Organon.* Some of these works, especially those known as the "New Logic" (the *Analytics, Refutations,* and *Topics*), were lost until the middle of the twelfth century.[127] He knew other works of Aristotle, including the *Metaphysics, Physics, De Generatione et Corruptione, De Anima,* and *Poetics.* He also translated Porphyry's *Introduction to the Categories of Aristotle* (sometimes known as the *Isagoge*), which he used to preface the *Organon,*[128] and works by Ptolemy, Euclid (maybe including the first five books of the *Elements,* although this has not survived), Nicomachus's work on arithmetic, and Archimedes on mechanics.[129] These were only a fraction of the entire works he had intended, but until the early twelfth century, his translations were the key means of access for a Latin West that no longer had any Greek; and until the impact of the Arabic versions of Avicenna and Averroes, his were the key versions.[130] Notable absences include Aristotle's political writings in the *Politics, Ethics,* and *Rhetoric.* Given the history of these texts and the impact of their rediscovery in the thirteenth century, Kantorowicz raises the intriguing question of whether there would actually have been a Middle Ages, as we know it, had Boethius completed his work.[131]

It was his role as a mediator of Aristotle and as a preserver of classical thought that was his fundamental gift to later thought. He was, in Southern's terms, "the schoolmaster of medieval Europe."[132] His influence can also be found in less likely places. Alfred the Great, Chaucer, and Elizabeth I, for instance, all translated the *Consolation,* and Chaucer's work more generally, especially *Troilus and Criseyde,* owes an appreciable debt.[133] One of Boethius's pupils, Cassiodorus, took the division of knowledge still further, being credited as the inventor of the distinction between the *trivium* (grammar, rhetoric, didactic/logic) and the *quadrivium* (music, astronomy, geometry, and arithmetic).[134] Geography is notable by its absence, but so too is history. In the fifth century, Martianus Capella depicted these in *The Marriage of Mercury and Philology,* where Mercury's wedding gift is the seven servants of the arts.[135] The state of knowledge in these disciplines was based on summaries of leading historical figures: Donarus and Priscian for grammar, rhetoric based on Cicero, logic on Aristotle, music on Aristoxenus, astronomy on Ptolemy, geometry on Euclid, and Nicomachus for arithmetic. As Colish notes, Nicomachus's approach "emphasizes number theory, ratio, and proportion, rather than the technique of calculating."[136]

Isidore of Seville's principal work is the multivolume *Etymologiae*, of which the first seven volumes treat these same seven disciplines.[137] He was another thinker whose impact goes far beyond any originality of his own work.[138] Working in the early seventh century, he wrote studies of scripture, theology, and liturgy,[139] as well as a history of various barbarian tribes, which begins with a poetic description of his native Spain: "Of all the lands [*terrarum*] which stretch from the West to India, you are the most beautiful . . . queen of all the provinces [*provinciarum*] . . . the glory and ornament of the world."[140] Books 13 and 14 of *Etymologiae* were on geography. This work has been rightly described as being both "the last product of the Roman encyclopaedic tradition and the starting-point for most medieval compilations."[141] It was, in a sense, a forerunner of the *florilegium*, a "collection of extracts in which individual monks arranged the fruits of their reading for their own use and satisfaction . . . sometimes in an introductory passage we see the quiet, industrious, unambitious mind at work reducing years of reading to an orderly form. . . . The scholastic method was a development of the *florilegium*."[142] It is important to note that as well as these practices of citation and compilation, medieval writers often presented quotations in altered form as if they were faithful renditions.[143]

Isidore's labors brought together what was left of the classical Roman heritage of science, philosophy, and literature, to serve as a compendium and companion to thought. It is not always clear he understood the material that he was borrowing, and consistency across entries is not to be found.[144] Although he does frequently suggest etymologies for words, taking their meaning from their derivation, and often using Hebrew and Greek roots, he almost certainly did not have training in either language.[145] Hay has described it as "pitifully meagre, written in a Latin which betrayed at every turn the linguistic isolation of its author," yet recognizes that, along with *The City of God* and *The Consolation of Philosophy*, it gave the next six hundred years "the basis of its knowledge of the ancient world."[146]

Isidore was one of the first writers to provide a map of the world that served as the pattern for what became known as "T/O" maps.[147] These show the known world as a disk—the O—with the rivers Don and Nile running from left to right and the Mediterranean from the center to the bottom, forming a T shape within the O. Asia occupies the upper segment, Africa the bottom right, and Europa the bottom left. Each has the name of one of the sons of Noah: Shem (Asia), Cham (Africa), and Japheth (Europe).[148]

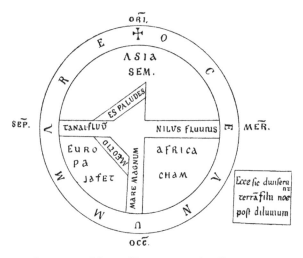

Fig. 3. Isidore's map of the world. From Konrad Miller, *Mappae Mundi: Die altesten Weltkarten*, 6 vols. (Stuttgart, 1895–98), 6:58.

Surrounding them is the *Mare oceanum*, the oceanic sea. Outside the sea are the compass points, with *Oriens*, the east, to the top, hence "to orient." The key to the bottom right reads "Behold, how the sons of Noah divided the world after the flood."

There is some dispute as to whether he believed the earth was flat or, more plausibly, that the known world was a disk floating in the sea on a sphere.[149] Some versions of the map have another continent below the disk, sometimes labeled *terra australis incognita*, unknown southern land. While T/O maps do look like a "flat earth," they are simply representations of the *oikonumene*, not the earth as a whole. Most modern projections render the earth in two dimensions, and contrary to the widespread belief, people in the Middle Ages did not think the earth was flat, with knowledge of its spherical shape since the third century BCE.[150] His *Etymologiae* was also influential as a source for a thirteenth-century Spanish text, *Semeiança del Mundo*, which was often wrongly attributed to Isidore himself.[151]

One of the most intriguing parts of the *Etymologiae* concerns his distinction between areas of different sizes. He makes a distinction between *Terra*, the earth; *terrae*, single parts of land such as Africa and Italy; places (*loca*), which are expanses of land (*terrae spatia*), which themselves contain provinces (*prouincias*), parts of which are regions (*regiones*), which are colloquially called cantons. Isidore notes that the "region was named for regulators; territories are parts of one [*a rectoribus autem region nun-*

cupata est, cuius partes territoria sunt]."[152] A similar model is found in his definition of measure, *mensorum*, more generally.

> Measure is whatever limit is set in respect to weight, capacity, length, height and mind [*animus*]. And so the ancients divided the planet [*orbem*] into parts, the parts into provinces, the provinces into regions, the regions into places, the places into territories [*loca in territoriis*], the territories into fields, the fields into hectares [*centuriis*], hectares into acres [*iugerebis*], the acres into *climata* [sixty feet square], the *climata* into hides [*actus*], perches, yards, grades [*gradus*], cubits, feet, spans, inches, and fingers. For so great was their ingenuity![153]

Henderson has described this as "nesting tables inside Chinese boxes within [a] Russian doll,"[154] and the hierarchies and container model is certainly interesting. The smaller measures are linear, which are squared to give measures of area as the size increases. But for our purposes here, it is the relegation of *territorium* to such a subordinate position that is remarkable: merely a part of a place within a region that is itself a part of province. It is also notable that places are described as *terrae spatia*, extents or expanses of land, where *spatium* has its meaning as a qualifier to the geographical term, not that term itself. Indeed, one of the few places where *spatium* appears as a substantive in Isidore is as the lap of a chariot race.[155] *Territorium* here clearly means an expanse of land, and not of especially great extent. It is a piece of land, somewhat larger than a single field, but predominantly agricultural. As he says, echoing Servius, "A *territorium* is so called as if it were a *tauritorium*, that is 'broken by a plow' [*tritum aratro*] and by a team of oxen (c.f. *taurus*, 'bull')—for the ancients used to designate the borders of their possessions and *territoria* by cutting a furrow."[156]

Henderson notes that Isidore sees that "the countryside fills in the space around cities,"[157] which hints at a sense that political power is concentrated in particular sites, weakening at a greater distance from these centers. Concerning this agricultural land there is once again the nested or parcelization model: country land is divided into fields (*agri*). Isidore contends that "borders [*fines*] are so called because fields are divided by cables [*agri funiculis sint diuisi*]. Measurement lines are stretched out in partitioning the fields so that dimensions can be equal. . . . *Termini* [ends or limits] are so called because they distinguish and mark land-measures [*terrae mensuras distinguunt atque declarant*]."[158] This notion of the importance of techniques for agricultural lands recalls his description of geometry, which, like Herodotus, he ascribes to the Egyptians measuring lands flooded by the Nile.[159] Yet despite the formal structure of the work that purports to treat all divisions of knowledge, there is little material on geometry in the *Etymologiae*. It has no examples, simply formal definitions of terms, and none of the geography that filled in the gaps in some of his immediate predecessors.[160]

Isidore says relatively little on what might be called political theory, although Wallace-Hadrill has suggested that he outlines the Visigothic view of kingship. The king should make laws, exercise power, *potestas*, over his people; he should control the church in his kingdom. This should be "real *potestas*, characterised by *virtus* and by *terror disciplinae*."[161] It is worth noting, as will be picked up in chapter 7, that though the Justinian *Codex* of Roman law was compiled a century before Isidore wrote, he was one of those who were unlikely ever to have heard of it. Instead, he would have been aware of some of the sources that Justinian's compilers used.[162]

THE BARBARIAN TRIBES AND NATIONAL HISTORIES

Boethius and Isidore are meager thinkers compared to what came before, but giants compared to what followed. Almost nothing that would fit contemporary notions of a text in political theory was written in the Latin West until the twelfth century. Analysis of political action needs to look elsewhere, at a range of rather different texts—constitutions, laws, histories, and religious documents.[163] As Markus has suggested, one of the changes after Augustine was that writers such as Isidore of Seville, Paulus Diaconus (Paul the Deacon), and Bede wrote their histories taking the Roman Empire simply as one polity among others rather than as a universal model. Its time had come and gone, and the barbarian tribes such as the Goths, the Lombards, and the English "had their own Christian destinies," and their history "could be written without their being brought into any essential relation with the Empire. Its existing remnant in the East, 'the Empire of the Greeks,' had no universal mission or claims in their eyes."[164] While attempts to reassert a universal model of Christendom happened later in the Middle Ages, there was a long period where fractured and fragmented lineages were the most coherent narrative. In a powerful analysis of four such historians—Jordanes, Gregory of Tours, Bede, and Paul the Deacon—Goffart suggests that they occupy the "trough of the curve" between the fall of Rome and the rising of Christian Europe from its ruins.[165]

Each of these historians touches upon issues of geographical detail in their narratives, rarely giving the politics of land explicit attention but often providing in passing some important indications.[166] Isidore, for instance, suggests that

> when, after seizing despotic power, he [Athanagild] had long tried to deprive Agila of his kingship, he had asked the Emperor Justinian to help him with the soldiers, whom he later was unable to remove from the borders of the kingdom [*finibus regni*] despite his efforts.[167]
>
> Not content with the kingdom of Aquitania, he [Theudered] rejected the Roman peace treaty, occupied most of the Roman free towns nearby, and attacked Arles, the most famous town in Gaul, besieging it with many troops.[168]

These indicate some sense of the geographical extent of the lands occupied by the various kings, even if somewhat vaguely defined.[169] Indeed, it is tempting to provide a more concrete specificity to the descriptions, as Donini and Ford do in their translation, rendering *nearby* in the second quotation as "near his territories." A similar thing can be found in Hydatius, a historian of Spain. Hydatius is examining how the barbarian tribes, having invaded Spain, eventually made peace and "alloted to themselves areas of the provinces for settlement [*sorte ad inhabitandum sibi provinciarum dividunt regiones*]."[170] Goffart translates this as "divided the territories of the provinces among themselves by an apportionment for [permanent] residence," and takes various other translators to task for their readings of parts of this passage.[171] But "territories of the provinces" implies greater geographical specificity than is actually found in the passage, the continuation of which shows that it is provinces that are allocated, not more tightly circumscribed regions within them.

Similar indications are given in the work of Isidore's contemporary Gregory of Tours, who wrote a book of *Histories*, often described more specifically as "a history of the Franks."[172] The Franks were originally a barbarian people, who moved progressively westward, although unlike other tribes such as the Ostrogoths they retained their roots in their homeland, expanding contiguously.[173] In 406 Clovis, king of the Franks, converted to Christianity, and the Merovingian dynasty united the Frankish lands. For three hundred years his dynasty endured, with Pepin the Short the final figure. Pepin's sons were Charles, better known as Charlemagne, and Carloman.[174] With these figures of the Carolingian dynasty, a new chapter opens, but the Merovingian period bears a little analysis. What we know about Clovis is largely filtered through Gregory of Tours,[175] which is why his work is so valuable.

Gregory, like Tacitus before him, had struggled to understand Frankish or German practices with the Latin concept of a *rex*, the king.[176] The particular issue is less important than its consequence: that there are at least two lineages of the notion of a king in Western Europe. One derives from the Latin, the king as he who rules (*regere*); the other is the Germanic the *Cyning* or *kuning* as "the man of, or from, or representing the kin."[177] The king was understood through a notion of *majestas*, being major, superior, to other people of his kin.[178] Kings were kings of a people, not of a kingdom. As Ejerfeldt suggests, "The king was *rex Francorum*; the king of a certain country or geographic territory is a later conception."[179] This is not to say that exchange of land, conquest of it, and its cultivation were not key themes of the period. Wallace-Hadrill puts it thus concerning the Merovingian dynasty of the Franks:

> Territory, lands old and new, endowments, run like a refrain through Frankish documents of the time. New land is being won for cultivation . . . new families are settled on property; new churches and monasteries look for endowments; and royal initiative lies behind much of it. The Merovingians endow, confirm, witness, sanction.[180]

The use of the word *territory* by Wallace-Hadrill is not entirely anachronistic. Gregory uses the word *territorium*, in a range of spellings, but this is clearly in a looser sense of "land," rather than the more specific sense it would later acquire.[181] As Wallace-Hadrill himself notes, Gregory does not find it necessary to explain his use of the term *territorium*. While for Wallace-Hadrill the point is that Latin is still widespread, this could be taken another way: in that the term still has the unstressed meaning of surrounding lands or areas.[182]

There are various instances in Gregory's history where the term means simply the neighborhood of a town: "in the Dijon neighbourhood [*in Divionensi territurio*]";[183] "somewhere near Avignon [*in Avennico terreturio*]";[184] "the region round Auxerre [*Audisiodorensim territurium*]";[185] "in the Senlis area [*in Silvanectinse vero terreturio*]";[186] "in the neighbourhood of Tours [*infra Toronicum territurium*]";[187] "coming to his monastery in the lands of Limoges [*eius Limovicino in termino*] . . . then I came to the lands of Trèves [*territurium Trevericae*]."[188] To translate *territorium* in these as "in the territory of Dijon" and so on would be to distort the mean-

ing seriously. At times there is an indication of the political-economic relations at stake, especially concerning ownership, such as when he speaks
of the "crown lands in the Soissons area [*quas ei rex a fisco in territurio
Sessionico*]."[189] One interesting issue is where he sees the *territorium* as
distinct from the town or settlement it belongs to. For instance, he talks of
"the town and the neighbouring district [*urbis territurio*]";[190] "an island in
the Loire, in the lands of the city of Tours [*terreturium urbis Toronicae*]";[191]
and events happening "in the city and the vicinity [*civitatis territurio*]";[192]
or "in the neighbourhood of the town of Chartres [*ab urbis Carnotinae
territurio*]."[193] There are a few places where Gregory pairs *territorium* with
termini, to invoke a bounded area. For instance, he says that someone
"lived within the borders of the Trier lands [*intra Treverici termini territurio*]"[194] or that action took place "within the borders of the lands of Angers
[*infra Andegavensis territorii terminum*]."[195] In these instances, the meaning is certainly closer to the modern sense of territory. Something similar can be said of various strategic concerns, such as when King Guntram
"ordered all the roads through his realm to be closed, in order to prevent
the passage of all persons from Childebert's kingdom through the lands of
his realm [*per eius regni territurium*]."[196] And, finally, it is worth noting
that something quite close to the modern sense can be seen even when the
word itself is not used, in the terms for kingdom and province: "At this
time the two brothers Gundobad and Godigisel ruled the kingdom [*regnum*] around the Rhône and the Saône and the province of Marseilles."[197]

The geographical analysis can be made of other barbarian histories.
Jordanes's history of the Goths, though crude, recognizes the importance
of place to the account, especially in its introduction, which discusses the
location of the different tribes.[198] So too, especially in the preface, does
Saxo Grammaticus's account of the Danes.[199] In the chronicle of the Langobards, or Lombards, written by Paul the Deacon, for example, there is
an intriguing moment in the narrative that recalls Cicero's claim about
the Lacaedaemonians. This is when King Authari reached the foot of Italy
and touched a column in the sea with the point of his spear and said, "The
bounds of the Langobards will be up to here [*usque hic erunt Langobardorum fines*]."[200]

It is not uncommon to see the period between the collapse of the Western Roman Empire and the establishment of Charlemagne as emperor of
the Romans as a complete break from the imperial period, but it would
be more appropriate to view it as the product of a complicated set of overlapping practices, ideas, and cultures, owing much to Roman influences

alongside those from the different tribes. This suggests that there was no straightforward "fall" in 410 or end in 476, but a fusion and dilution over time.[201] As Goffart argues, it only makes sense to talk of a Germanic age with the Carolingians, and of course that time owed much to the idea of a reestablishment of Rome.[202] Long before Rome fell, indeed, much land had been granted to settlers from the barbarian tribes.[203] One key element of this enduring legacy was the continuation of Roman administrative centers as urban points, with their agricultural and rural surrounding areas, as a network across their kingdom.[204] Between these political units there would be no clear division of land, but a gradual weakening of effective control. The frontier rather than the border predominated.[205] These city-based polities, for which the term *civitas* was often used, would become the focal point of later political struggles. The personal, rather than geographical, element of rule also informs the barbarian tribes' understandings of inheritance, which was not solely patrilineal, and in which lands or possessions were sometimes divided between all surviving heirs.[206] In a number of studies, P. D. King has discussed how the Visigoths understood their law territorially, but suggesting that until the middle of the seventh century it was much more dependent on a kinship or what we might call a national basis.[207] If Gregory of Tours can be seen as the father of French history, Paul the Deacon that of post-Roman Italy, and Jordanes's *Getica* fundamental for Germans, Goffart is clear that "their works pale alongside the array of virtues that all agree are found in Bede's *Ecclesiastical History of the English People*."[208] Bede's attempt is to provide a narrative of England's cultural, political, social, and religious history in a broader context concerning Rome, world history, and Christianity.[209] This work is one that has a strong geographical element, mainly descriptive, of the places and peoples of Britain, taking into account topography, wildlife, and natural resources.[210] Bede is an innovator for a range of reasons, including two concerning the measurement of time. It is to Bede that we owe attempts to set the date of Easter accurately concerning the full moon and the vernal equinox, and using a dating system deriving from the birth of Christ.[211]

Three elements of his work are worth special attention. The first is that we are told that Bede was someone "*qui natus in territorio eiusdem monasterii*," "who was born in lands belonging to the monastery."[212] In Alfred the Great's Anglo-Saxon translation of Bede, he was born "*on sundorlonde* of the monastery," outlying lands, lands sundered from the estate itself but under its possession, and thus it has been claimed that this is the basis for the name of the town in northeast England, although it is not clear that it was this *sundorlonde*.[213] But this gives a clear indication that *territorium* in the early eighth century retained its sense as an extent of land under possession of some entity—in this instance a monastery—rather than as an object of political rule, which is a sense it would take on at a later time. It is telling that all of the other instances of the word *territorium* in this work are as *territoria*, the plural, in the phrase *"territoria ac possessiones"* or *"possessiones et territoria,"* as lands and possessions being given by the king to religious orders.[214]

Second, the main measure of land that he uses is that of a "hide," defined as a *terra unius familiae*, the amount a land that would support one family for a year. This extent clearly varied depending on the quality of the soil, rather than simply quantity, not to say the size of the family, and was thus a fairly vague term.[215] Nonetheless, Bede suggests, for example, that Thanet, east of Kent, "by English reckoning is six hundred hides in extent."[216] He also talks about hides in terms of grants of land, noting that King Æthelwealh gave Bishop Wilfrid "eighty-seven hides of land [*terram LXXXVII familiarum*]."[217] And he does occasionally use the notion of land, *terra*, to speak of the hides under the possession of a people. Instances include his description of "the kingdom [*prouinciam*] of the South Saxons, which stretches south and west from Kent as far as the land [*terram*] of the West Saxons and contains 7,000 hides."[218] He also speaks of "Pictish lands [*terra*]" and "English lands [*terras*]."[219]

As Fradenburg has suggested, in medieval England there was often a blurring of specific terminology:

> *Terra* is land, country, region, sometimes in collocation with *patria*. *Terra, patria,* and *natio* are often synonymous or apposite, and frequently appear in collocations with other. Each emphasizes a particular aspect of the notion of territory (land as such, my father's land, my birthplace), but these are usually nuances, not striking differences. But the very intimacy of these terms indicates the richness of the concept of territory.[220]

Third, he offers early indications of how when kings converted to Christianity, their provinces did too—for example, in 653 CE the lands of the Middle Angles under King Peada.[221] Given later conversations between different Christian confessions and the implications of this for the universal aspirations of the papacy and emperor, the notion that the conversion of a king would have an effect on his lands and the people within them is an important theme.

LAND POLITICS IN *BEOWULF*

The clash between emergent Christianity and pagan beliefs is a subtheme in the Old English poem *Beowulf*.[222] This is one of the earliest surviving works of English literature, or indeed in any northern European vernacular language. *Beowulf* is a complicated text, and can be read both as a work of creative literature and, with caution, as a glimpse into the context in which it was forged. Its geographies have been less well explored. In his final book, the distinguished medievalist Nicholas Howe made the claim that *Beowulf* "is profoundly a work about place."[223] Although Howe makes a number of suggestions about how that might be the case, his analysis is of some very particular passages rather than the poem as a whole. In addition, while his claim concerns place in general terms, the explicit analysis is of the notion of *epel*, homeland. Yet the geographies of this poem cannot simply be reduced to places as sites but can be seen to structure the drama, and have important political aspects, especially in the question of land.[224]

Many modern editions of *Beowulf* have a map of its geography at the beginning, showing the Geats in what is today southern Sweden, the Swedes and the Heathoreams to the north, and the Danes in the Zealand area of modern Denmark. The Old English *Gēat* is often taken to be the equivalent of the modern Swedish *Göt*, that is, the region of *Götaland*, and the historical tribe of the Götar, who lived in that region.[225] Various other tribes are loosely distributed across northern Germany, Poland, and the Low Countries, and the modern village of Lejre, near Roskilde, is often given as the location of Heorot.[226] While some of this work is valuable, an attempt to find a historical basis for the sites of the events is not the purpose of this section, as *Beowulf* does not say anything particularly concrete about those places that are mentioned in the text.[227] Nor does it seek to generalize from what insights there are in the text that can be tied to archaeological evidence.[228] Rather, the interest is in the exchange geographies implied and presented in the poem itself, especially concerning the question of land.

The story is structured around three battles with monsters, although there are many references to battles between tribes. A great mead hall has been built named Heorot, where Hrothgar rules the Danes. The hall is attacked by the monster Grendel, a descendant of the race of Cain. Beowulf of the Geats travels to Heorot and kills the monster with his bare hands. The next night the monster's mother seeks vengeance; Beowulf pursues her to her lair, a *mere*, or pool, and slays her too. Beowulf returns home, and in time becomes king of the Geats. Some fifty years later, Beowulf's people are attacked by a dragon. Beowulf meets the dragon in combat, and though he kills it, dies in the battle, which takes place at a burial mound filled with treasure. Without the protection of their king, the Geats are overrun by the Swedes from the north and tribes from the south.

Each of the battles—with Grendel, his mother, and the dragon—takes place in a particular site: the hall, the *mere*, and the burial mound.[229] These three sites are given detailed descriptions, mixing elemental, symbolic, and material geographies. Yet, crucial though they are to the drama of the poem, the question of land exceeds these particular sites and can be seen to be a significant theme throughout. In the Anglo-Saxon *Maxims*, the aphorism "The king desires royal power: keeping land [*londes*] he is hated, giving much he is much loved" is sometimes seen as indicative of land politics of the period.[230] In *Beowulf* the Christian poet suggests that God:

> distributes wisdom, land [*eard*] and nobility [*eorlscipe*] among man-kind. . . . He will grant him earth's bliss in his native land [*ēþle*], the sway of the stronghold of his people, and will give him to rule regions [*dǣlas*] of the world, broad realms [*rīce*]: he cannot imagine, in his folly, that an end will come.[231]

However, in *Beowulf* there are two rather distinct economies at work. The first is the distributive politics of Hrothgar. We are told at the beginning of the poem that Hrothgar was going to distribute "the gifts God had given him . . . apart from common land [*folcscaru*] and lives."[232] Although the second exception is supposed to show that he is no tyrant, the former—while certainly open to that interpretation—indicates something more. That is supported by his actions later in the narrative.

After the death of Grendel, Hrothgar showers Beowulf and his retainers with gifts, and even names him as his son, but the queen intervenes to ensure that the realm itself passes to *her* sons. The speech is notable, since though this is a patriarchal society, she still has an important role as the reproducer of a line:

> Heorot is cleansed, the ring-hall gleams again: therefore bestow while you may these blessings generously, and leave to your kinsmen the realm and its people [*folc und rīce*] when your passing is decreed.[233]

Hrothgar follows his queen's advice, and does not skimp on the treasures, while reserving any gifts of land. Even after the end of Grendel's mother, and Beowulf's departure from the Danes, Hrothgar presents him with twelve new treasures, and Beowulf departs "proudly gold-adorned [*goldwlanc*]."[234] But he does leave, retaining no ties to that land other than friendship and loyalty. When Beowulf returns to the Geats, he in turn presents the treasures to the king, Hygelac, his maternal uncle.[235]

Hygelac receives these gifts with pleasure and then reciprocates. Alongside Hygelac's father Hrethel's sword, which is given to Beowulf, the gift of land then comes from *this* king:

> He bestowed on him seven thousand hides of land, a princely throne and a hall. Inherited land, a domain by birthright, had come down to them both in the Geat nation; ancestral domain, the greater realm [*rice*] to the higher born of them.[236]

Very quickly, a matter of lines later in the poem, Beowulf inherits the "broad realm [*rice*]" of Hygelac's lands when the latter is slain.[237] Somewhat later in the poem, the full story is told. Hygelac's widow offers Beowulf "hoard and realm [*rice*], rings and a princely throne," because she does not trust that her son is strong enough to repel foreign invasion. Beowulf refuses this honor, staying merely as an adviser and ally, until the son himself is killed, "which allowed Beowulf to hold the princely throne and rule the Geats."[238] In these passages two key things are stressed: land through inheritance on death, and land through gift. While Beowulf owned land by birthright, he is given the seven thousand hides before Hygelac dies, only inheriting the balance later. Elsewhere we are told Hygelac made a gift of "land and linked rings worth a hundred thousand" to the retainers Eofor and Wulf for their deeds in battle against the Swedes.[239]

Much later, when Beowulf has been king for many years, the death of Hrethel—Hygelac's father and Beowulf's grandfather—is recounted as a prelude to the war between the Swedes and the Geats, but also as a prelude to Beowulf's own imminent demise. It is also mentioned in terms of inheritance. We are told Hrethel "left to his sons his land and towns at his life's faring forth, as the fortunate man does."[240] Beowulf adds that he was able to repay Hygelac in battle "for the treasures he had given me. He had given me land, dwelling and delight in homeland [*eard ēðelwyn*] to leave to my heirs."[241] In addition, when Beowulf does confront the dragon, and all his companions flee, the one who returns is Wiglaf. Of all the favors he remembers from Beowulf, to whom he owes allegiance, he explicitly recalls "the wealthy dwelling-place of the Waymundings, confirming him in the common landrights his father had held."[242] Here, then, it is not so much the gift of land but the support for the property rights of the commons that is important.

On Beowulf's death, lacking an heir, things are more complicated. Wiglaf tells them that because too few came to Beowulf's aid,

> now there shall cease for your race the receiving of treasure, the giving of swords, all satisfaction of ownership, all comfort of home [*eðelwyn*]. Each of your kin [*cynne*] shall become wanderers without land-rights as soon as athelings over the world shall hear the report of how you fled, a deed of ill fame.[243]

This comes to pass, but not quite in the way anticipated. It is less because of a loss of prestige than because with the death of Beowulf his overseas enemies become emboldened.[244]

As well as these senses of distribution, the word *land*—an English word directly linked to the Old English *land, lond*—has multiple meanings. Many of these are indicated in the poem. Land can be used in a straightforward, unstressed sense, in opposition to sea, such as when sailors sight or make land at the end of a voyage,[245] or when Beowulf surfaces from the *mere* after the fight with Grendel's mother.[246] It can be limited or marked, with a boundary, although here that is only used of the coastline (*landgemyrcu*).[247] It can be used in a way that means little more than place, such as the "mysterious land" Grendel and his mother inhabit,[248] or with a sense of region, speaking of the strongest in a land,[249] or the people of a land.[250] It can be plural, with Hrothgar's hall having a "radiance that shone over many lands."[251] It can also be an advantage, in that someone "knowing the land well" might "escape with his life."[252]

Yet even concerning distribution, *land* is used not simply to designate the property of a person or a king, such as Scyld Scefing being hailed as the "beloved leader of the land [*lēof landfruma*],"[253] but also of a people more generally. The land of the Danes (*land Dena*) is mentioned when Beowulf arrives on his quest against Grendel,[254] and again when he and his retainers depart.[255] When Beowulf is recounting his swimming contest with Breca, for instance, he claims that the sea currents carried him "to the land of the Lapps [*on Finna land*]."[256] There are other examples in the poem—the land of the Brondings and the land of the Frisians.[257] It is this sense of land and its relation to a people that is the topic here, for it leads to the question of conflict over this resource, both as the object and terrain of struggle.

Howe notes that "Anglo-Saxons conceived of the land itself and all that grew on it as more enduring than anything human beings could build on it and thus as more useful for legal purposes."[258] While he does not pursue all the resonances of the term within the poem, he does offer an important illustration of his earlier-cited claim that it is "profoundly a work about place."[259] Howe shows how the "breakneck chronology" of lines 2200–2214, discussed above, is "a political genealogy . . . [a] dynastic progression," showing the passage from Hygelac to Heardred to Beowulf, but that this is not simply a line of kings, but establishes the land they rule over: "These lines from *Beowulf* clearly demarcate a kingdom by offering its line of dynastic succession."[260] He suggests that the poem is in part "a political poem that asks what it means to be an *epelweard*," a guardian of the homeland, "then it must also be a poem about place, about the meaning of *epel*."[261]

The poet uses *epel*, homeland, and *rice*, kingdom or realm, throughout the poem, often juxtaposed with no clear distinction. A prince's son, for instance "should prosper, succeed to his father's rank, guard the people, treasure-hoard and stronghold, the realm [*rice*] of heroes, homeland [*eðel*] of the Scyldings."[262] Alongside the use of *epelweard*, Beowulf describes Hrothgar as "*rice weard*," guardian or ward of the realm,[263] and there are a couple of instances where Hrothgar or Beowulf is described as a *rices hyrde*, which effectively means the same thing.[264] What is intriguing, Howe suggests, is that there are only two instances of the verb *ricsian*, derived from *rice*, which means "to rule." But these are not used of humans. They are used to describe the dragon ruling over the hoard and, earlier, of Grendel effectively ruling over the hall until Beowulf arrives.[265] In both instances a monstrous rule is opposed to a heroic leadership of the homeland—in the case of the dragon, the lines in the poem come immediately after the invocation of Beowulf's role as *epelweard*.[266]

When dying, Beowulf recalls that he has ruled his people for fifty winters, and that in that time "not a single king of all the neighbouring peoples [*ymbesittendra*] about has dared to affront me with his friends in war or threaten terrors."[267] Yet the whole of the second part of the poem—that is, Beowulf's return to the Geats, the death of Hygelac, and the passing of the realm to Beowulf—concerns three key things: the fight with the dragon, the war between the Geats and the Swedes, and unrest on the southern borders.[268] These stories are continually intertwined, so that on Beowulf's death in combat with the dragon, it is not surprising that the full force of the other conflicts is unleashed: the poet has continually prefigured it.

Beowulf's predecessor as king, Hygelac, had actually died in an ill-fated raid on the southern tribes of the Frisians and the Franks. We are told that "fate carried him off when, out of pride, he went looking for

trouble, a feud with the Frisians."[269] Beowulf escapes from this battle by
swimming away.[270] Jack and Swanton note that there is historical evidence
for the battle and death of Hygelac, at least, in Gregory of Tours's histories, and the *Liber historiae Francorum* (sometimes known as the *Gesta
Francorum*).[271] In the poem it is clear that Beowulf has prevented the destruction of the Geats, and the recounting of the Frisian raid anticipates a
"baleful future."[272] Recounting the history of his line, Beowulf recalls that
it was after the death of Hrethel, Hygelac's father, that "there was hostility
and strife between Swedes and Geats, a mutual grievance across the broad
water."[273] Yet Hygelac makes this worse by inflaming tensions to the south
too, with the Frisians and the Franks, thus presenting the Geats with a
war on two fronts. On Beowulf's death, Wiglaf recognizes that "the people
can expect a period of conflict, once the fall of the king becomes openly
known abroad among Franks and Frisians."[274] A few lines later, having offered a detailed account of the stages up to this point, Wiglaf notes that
"this is the feud and the enmity, deadly hatred of men, for which I expect
the people of the Swedes to come looking for us, once they hear that our
lord has lost his life."[275] The loss of the king, *cyning*, produces a vulnera-
bility for his kin, *cynn*; the absence of the *eþelweard* or *riceweard* removes
the protection from the homeland and realm. In this the geopolitical conflict parallels Beowulf's previous three battles with monsters: holding the
forces of monstrosity, wilderness, and disorder at bay from a human world
of pockets of isolated order. And yet the poem continually insists on the
interrelation, and the crossing between such arbitrary borders.[276] What
happens with Beowulf's death is that a much wider world intrudes: distance and proximity are reordered.

Land, as an indicator of a set of relations that mix economic and political concerns, is the operative geographical question of this text, with
the interrelation of the people with the land they inhabit a key theme.[277]
"Territory" is a much later category that does not really make sense in
the period and place of *Beowulf*.[278] Sometimes, though barely hinted at in
the poem, land becomes a political-strategic rather than simply political-
economic question, which implies what we might think through the question of "terrain." *Terrain* is, of course, itself a complicated term with a distinct etymology and lineage to that of land, and it is a term that is, strictly
speaking, foreign to this text. But the interrelation of the site and stake of
struggle is important: land is not simply where battles take place, but often the focus of the struggle itself. This text thus gives a partial glimpse of
the political-strategic issue alongside its very particular political economy
of land.

CHAPTER FOUR

The Reassertion of Empire

THE DONATION OF CONSTANTINE

In the Vatican, one of the rooms is named after the first Christian Roman emperor, Constantine. The works in this room are attributed to the school of Raphael, his pupils, who continued work after their master's death in 1520. On the window side of the room is one of the most powerful, *The Donation of Constantine.*

In it Pope Sylvester sits on a throne on a raised dais, and Emperor Constantine kneels on the steps at his feet. Sylvester has his right hand raised in a blessing, while with his left he receives a gift from Constantine, who holds his left hand to his breast. The gift is supposed to be the text of an epistle signed by Constantine, the first part of which is the confession of his life before conversion; the second is more properly the "donation." The scene is set within Saint Peter's Basilica as it existed at the time of painting, and Sylvester takes on the visible attributes of the then current pope, Clement VII.[1] But it is not simply the setting and the pope that bear little resemblance to places and people at the time of Constantine. The story of the donation of Constantine is itself a fabrication, and the text a discredited forgery.[2]

Around the middle of the eighth century, a text purporting to be that of the "Donation of Constantine," or the *Constitutum Constantini* as it was known, was forged. Debate continues as to by whom, and for what exact purpose, but the story opens up some important issues, for this donation was a gift of land, a text that claimed to show that the first Christian emperor of Rome had transferred a great part of the empire to the church, thus giving the basis for the temporal powers of the popes. This

Fig. 4. Detail from *The Donation of Constantine*, school
of Raphael, 1520s. © Photo SCALA, Florence.

text was incorporated into the Pseudo-Isidorian Decretals (a handbook of
church documents, many of which were forged) in the middle of the ninth
century, and parts were included in medieval canon law compendia.[3] For
a long time it was accepted as a genuine document, and its use and interpretation are important in terms of the early history of the Holy Roman
Empire. Although its influence was not immediate and only minor in the
ninth century,[4] Dante, for instance, believed it to be genuine, though re-
grettable, criticizing the way that it had "transformed the pope into a rich
man."[5] Even though it was forged, McIlwain is correct when he suggests
that this does not diminish its importance, since it was genuinely believed
to be a historic document for many centuries.[6]

The process of donation does bear particular relation to a genuine donation, the Donation of Pepin of 756, where the king of the Franks promised lands to Pope Stephen II, as the representative of Saint Peter, lands
that the Franks had reclaimed from the Lombards.[7] This donation did indeed give the pope lands over which he exercised not simply religious, but
secular, power. The Vatican City is the last remnant of this, lands that
once belonged to the Roman Empire that were never again conquered.[8] Although it is not known for certain, it is possible that the document of the
"Donation of Constantine" was used in the negotiations between Pepin

and Stephen II. In that case, as the church was eager to later point out, Pepin's donation was simply a restoration of a previous situation.

The "Donation of Constantine" contains some remarkable passages, which—though not of fourth-century origin—are nonetheless revealing of eighth-century motives.

> The Emperor Constantine yielded his crown, and all his royal prerogatives [*omnem regiam dignitatem*] in the city of Rome, and in Italy, and in western parts to the Apostolic [See]. . . .[9]
>
> In order that the supreme pontificate may not deteriorate, but may rather be adorned with glory and power even greater than is the dignity of earthly rule [*terreni imperii*]; behold, we give over and relinquish to the aforesaid our most blessed Pontiff, Sylvester, the universal Pope, as well our palace, as has been said, and also the city of Rome, and all the provinces, places and cities of Italy and the western regions, and we decree by this our godlike and pragmatic sanction that they are to be controlled by him and by his successors, and we grant that they shall remain under the law of the holy Roman Church.[10]

This is the basis upon which the account is created, with Constantine yielding not simply his crown but his lands and "all his royal prerogatives," that is, all the attributes of his rule. The pontificate is opposed to "earthly rule," and the legal jurisdiction transferred to the law of the church. "Constantine" continues:

> Our *imperium* and the power of our kingdom [*regni potestatem*] should be transferred in the regions of the East, and that in the province of Byzantia, in the most fitting place, a city should be built in our name, and that our *imperium* should there be established, for where the supremacy of priests and the head of the Christian religion has been established by the heavenly Emperor, it is not right that there an earthly emperor [*imperator terrenus*] should have power [*potestatem*].[11]

Constantine is thus not simply transferring rule over Rome, Italy, and the western provinces, he is saying that he will retire to Byzantium—the Greek city on the banks of the Bosporus—so that there is not a conflict between secular and religious authority. The city he talks about founding, "built in our name," is, of course, Constantinople.

In the early twelfth century, Honorius of Augsburg wrote an analysis that drew on the donation. He suggested that while Sylvester received the

crown from Constantine, this was based on two key things: Sylvester took
the political authority but knew he only held it because of the material
support of the emperor; and future emperors received what political authority they had back from the *sacerdotium*. There were two swords: the
gladio verbi Dei and the *gladio materiali coerceri*, the sword of the word
of God and the material sword of coercion. Sylvester needed both, and the
latter could only be provided by Constantine.[12] The *sacerdotium* was the
ecclesiastical hierarchy, the priesthood, and was the spiritual equivalent
of the *imperium*.[13] Yet for papal theorists, both were parts of the same
body, the church, and the latter owed its power to the former.[14] But there
is crucially more to Honorius's account than simply this recognition of
the political realities of Sylvester's position and the subsequent reversal
in power relations. Honorius sees Constantine as simply confirming what
he regards as the appropriate order of things: the deference of the secular
power to the church. In doing so, he anticipates later arguments that temporal and spiritual authority together are held by the church and pope. For
the Carlyle brothers, this is the first explicit affirmation of this belief.[15]
Innocent IV had claimed that the "Donation" was to be read more as a recognition of the church's power and that previously Constantine had "exercised a usurped and unlawful power."[16] Yet Honorius is also crucial in the
doctrine of the "two swords," explored in much more detail in chapter 5,
below.

The text of the "Donation" was used by Gregory IX in the mid-
thirteenth century in an attempt to provide a justification for papal power
over Italy, and the right to crown emperors.[17] There were thus two parts
to the transfer: the terrestrial empire and the imperial insignia.[18] The latter meant that the pope "acquired the temporal plenitude of power, hitherto held by the emperors alone."[19] To simply ground it on the exchange
of the empire was a weak argument, since it opened the possibility of a
subsequent emperor simply reversing Constantine's proclamation, which
is what Frederick II tried to do.[20] The claim to have the plenitude of power,
and the right to appoint the emperor, was stronger. At that time, the text
was largely thought genuine, though legally suspect. Although it was long
rejected, on political, scriptural, and even religious grounds, it was not until the Renaissance that it was properly discredited, most importantly by
Lorenzo Valla in his *De falso credita et ementita Constantini donatione
declamatio*, written in 1440. Nicholas of Cusa, in his important *De concordantia catholica*, written some seven years earlier for the Council of
Basle, had made many similar arguments to Valla.[21]

Valla's arguments against the validity of the text are multiple. He thinks it is psychologically implausible, historically suspect, and textually corrupt. The first part of his exposure shows that it is unlikely that Constantine would have made the donation; the second that the text is philologically corrupt, and therefore from a later date, making it an obvious forgery. He begins with incredulity, asking if Constantine would have really given away that much of his empire, considering that rulers sought to increase their lands. Would Constantine have deprived "himself of one of the two eyes of his empire [*Imperii*]? That any one in possession of his senses would do this, I cannot be brought to believe."[22]

Was it, then, because of Constantine's conversion? Valla is having nothing of this: "I suppose it was a crime, an outrage, a felony, to reign after that, and that a kingdom was incompatible with the Christian religion! . . . If this be your idea, Constantine, you must restore your cities to liberty, not change their master."[23] Politically it is also questionable, he suspects, suggesting to Constantine that

> it is most unseemly for you now as a Christian emperor to have less power as a prince [*principatu*] than you had as an infidel. For the rule of princes [*principatus*] is an especial gift of God, to which even the gentile princes [*principes*] are supposed to be chosen by God.[24]

While it is tempting to render *principatu* as "sovereignty" and *principes* as "sovereign," this risks introducing a later term into the discussion—a term that would have had some resonance at the time Valla was writing, but that only really was realized in a recognizably modern form with Jean Bodin. Yet the circumscription of rule is important, since Constantine was transferring something that he only owned through the appropriation of these lands from the Senate and people of Rome. As such, his donation was already illicit.[25]

Valla also thought the donation implausible from the pope's side, claiming that Sylvester would have refused it as not befitting a holy ruler. For Valla, Jesus made it clear that "he had nothing to do with secular rule [*regnum saeculare*],"[26] and Valla quotes the Gospel of John to prove his point: "My kingdom is not of this world; if my kingdom were of this world, then would my servants fight."[27] Then, quoting the famous words from Matthew's gospel on the division of religious and secular rule, he has the pope say the following:

> "Render unto Caesar the things which are Caesar's; and unto God, the things that are God's." Accordingly, therefore, your Majesty, you

must not surrender the things that are yours, and I must not accept the
things that are Caesar's; nor will I ever accept them, though you offer
them a thousand times.[28]

In addition, Valla questions why no more durable record of this momentous event endures. He suggests that many less important texts are
preserved in far better state, in inscriptions or gold letters, yet we are supposed to believe that "Constantine signed a donation of the world on paper
alone and with ink."[29] Given the fragility of such actual texts, and the importance of the process of copying for medieval writings, Valla is unable to
authenticate any single document. Thus, his arguments have to be against
the content of the text.

In this, Valla is pioneering, in that he uses philology and grammatical arguments as tools of analysis, informed by a "perception that historical development was reflected by changes in linguistic style and word
meaning."[30] There were linguistic uses in the text that can be demonstrated as postdating its putative composition.[31] Among its historical infe-
licities are that it mentions Constantinople[32]—implausible given the dates
of Constantine's conversion, that of the supposed donation, and the fact
that Constantinople had not been founded, not becoming a chief seat until
two centuries later. As Valla exclaims, "It was not yet a patriarchate, nor
a see, nor a Christian city, nor named Constantinople, nor founded, nor
planned!"[33] And indeed, as noted above, and as Valla also points out, the
"Donation" itself later refers to Constantine's intention to found a new
city in Byzantium, not something that already existed.[34] Even without the
historical evidence, the text is contradictory. There are also uses of vocabulary that are flawed, such as the use of the term *satraps* for a lower-level
ruler, since this was a term used only from the eighth century.[35] Valla's arguments are so powerful, and the contemporary status of the "Donation"
so questionable, that some think that the criticism was partly in order to
illustrate a method rather than prove a point.[36]

Nonetheless, Valla did not meet many of the standards that would later
be expected of this form of analysis. As Delph suggests, his "treatise revealed the striking limitations of earlier Renaissance textual criticism."[37]
Delph shows how Agostino Steuco (1496–1549) was critical of the philological, grammatical, and historical arguments Valla used, in particular
Valla's failure to find the very best version of the text that he could to analyze, a *vera lectio*, instead of relying on a "truncated, corrupted recension
of the text."[38] Given the problems of textual copying, this is important,
since such processes could create problems of spelling, grammar, and eli-

sion, as well as the potential for other texts to be incorporated into the body of earlier works. Indeed, though today it is the textual basis of analysis that is seen as the most remarkable aspect of Valla's work, it is the political aspects that are in a sense more interesting. In combination, they are extremely powerful, because the "Donation" is not just a forgery in fact, but intrinsically flawed in theory, in that it operates with a model of land exchange that was entirely foreign at the time. By the time of its writing in the eighth century, land was regularly exchanged as a commodity, whereas in the Roman period, land was accumulated by the empire, and not exchanged or donated in that way.

While the argument that it was intended to demonstrate his powers of textual analysis cannot be entirely discounted, it is important to note that Valla was not without another motive. Valla's patron, Alphonse of Aragon, had had his accession to the throne of Naples and Sicily challenged by Pope Eugene IV. To challenge papal power in temporal matters more generally through a discrediting of this text would be an effective check in this specific instance.[39] Valla therefore is powerful not simply in showing it to be a forgery, but in demonstrating one of the ways in which there was a transition in the relation between place and politics. This was a shift from an early church that would not have accepted the role of Caesar to one that would rely on forged documents in order to do so. Indeed, while he goes on to suggest that a dual role was alien to Sylvester, Valla does recognize the temporal power, symbolized by the sword, of the pontiff, but also his ecclesiastical, spiritual power (*ecclesiastico*).[40]

THE ACCESSION OF CHARLEMAGNE

Constantine is a significant emperor for a number of reasons. His conversion to Christianity is one, but so too is the split he effected between the Eastern and Western Empires. This was first made in 293, but did not become permanent until 364. The dividing line was to the east of the foot of Italy.[41] When Romulus Augustus, the son of Attila the Hun's deputy Orestes, was made emperor of the West in 475, only to be deposed by one of his father's officers, Flavius Odovacar, the following year, this is seen to be the end of the Western Roman Empire.[42] Various attempts by the Eastern Empire to recapture the lands, especially under the Emperor Justinian, did not last long. While it did not exist for three centuries, the idea of the empire did not disappear: it continued to exist in the imaginary realm as a powerful political force.[43]

The barbarian kingdoms, discussed in the previous chapter, often became Christian. Thus, the Western church continued to have a powerful political presence even after the Western Empire had ceased to be. Pepin of the Franks, mentioned above, was one of the rulers who built the strongest links to the papacy. The transfer of lands from the secular ruler to, or back to, the pope cemented a relation that had begun when Stephen II had crowned Pepin king of the Franks. It continued after his death. Stephen had appointed Pepin's two sons, Charles, later to be known as Charlemagne, and Carloman, as his heirs; and Pepin's lands were divided between them on his death.[44] Carloman himself died in 771, and the lands came back together under a single ruler. Pepin, Charlemagne, and Carloman had been given the title *patricius*—"protectors of Italy and the see of St. Peter—the highest secular and honorary post the Pope could bestow."[45]

That was, until the year 800 on Christmas Day, when Charlemagne was in Rome for the celebrations. This was the day when he was crowned as emperor in Saint Peter's Basilica. As the chronicler Einhard suggests:

> Charlemagne really came to Rome to restore the Church, which was in a very bad state indeed, but in the end he spent the whole winter there. It was on this occasion that he received the title of Emperor and Augustus [*imperatoris et augusti*]. At first he was far from wanting this. He made it clear that he would not have entered the cathedral that day at all, although it was the greatest of the festivals of the Church, if he had known in advance what the Pope was planning to do.[46]

The plan of Leo III was that in crowning Charlemagne as emperor, he could accomplish a number of things. First, he was able to recognize what was already the situation—that Charlemagne was the dominant ruler in Western Europe, controlling an empire that included modern-day France, the Low Countries, Germany, and northern Italy—but in such a way that it appeared it was in his gift. What was clever about Leo III's actions was that he made it appear that it was through the crowning of Charlemagne that the emperor was created, in a sense beginning the idea that it was the gift of the pope.[47] Imperial and papal interpretations of the significance of this event would focus on this point. Did the pope play a merely ceremonial role at the crowning, or did he really have the authority to transfer title, according a temporal power?[48] Subsequent political theory concerning papal-imperial relations would take this as one of its key arguments to be debated.[49] As Folz nicely puts it, through the acclamation, "Charlemagne was thereby officially recognised as the newly created emperor, or, better still, it was a public proclamation—as in an Epiphany, in the full sense of the word—of the imperial status of the King of the Franks."[50]

Second, Leo was able to bind the secular power of this empire to the church, and thus effectively reestablish the ancient Roman Empire, but with a much stronger link to Christianity. This was the idea of Christendom given a material, political, and geographical form. As Charlemagne wrote to Leo III, he saw that his task was to defend the church from the pagans by armed strength, while the pope was to attend to their spiritual needs and pray for them.[51] In a sense, then, this was a means to dignify the relationship that had existed between the Franks and the pope for a century.[52] Third, in doing this, Leo was able to wrong-foot the rulers of the Eastern Roman Empire. Effectively what happened with the crowning of Charlemagne was that the imperial crown passed from Byzantium to Rome. Constantine had transferred the capital of the empire to Constantinople; Pope Leo III transferred the empire from the Greeks to the Franks, and then it was later transferred to the Germans. Einhard notes the hostility of the Roman emperors in the East to this transfer, but suggests that Charlemagne himself attempted to be conciliatory to them.[53] Other states, such as Spain, England, and later France, did not recognize the authority of this, but the pope and the empire did not think these kingdoms were independent in anything other than a de facto sense. *De iure* they owed their allegiance to Rome, to the pope, and the emperor who had been appointed by him. Yet he also wrong-footed Charlemagne. Charlemagne might have expected to be emperor, but not emperor of the Romans: that was the one in Constantinople. Charlemagne saw himself as a king over other kings, an emperor, coequal to the one in the East, not his replacement. The pope had other ideas.[54]

The new designation was of the *Respublica Romanorum*, the *res publica* of the Romans, not the Roman *res publica*. As Ullmann notes, the "latter term was always synonymous with the empire. The former was no doubt modelled on the old term and yet conveyed an entirely different meaning."[55] The difference is indeed significant. The new term meant "republic" or "commonwealth of the Romans," that is, those of the Roman faith, Latin Christians. Thus, the geographical link to Rome, the city, was not essential. Indeed, the pope continued to maintain political rule over Rome itself. Charlemagne was, for the pope, *Imperator Romanorum*, emperor of the Romans. The designation "of the *Romans*" is crucial because it means a designation of peoples, not of land, and the Frankish empire was thus, in an important sense, distinct from the crown that Leo III had put on his head.

Charlemagne's own designation was somewhat different. He did not use the title *Imperator Romanorum* but rather stressed the notion of *Augustus*, a direct link back to the emperors of antiquity.[56] The full title was *Carolus serenissimus augustus a Deo coronatus magnus et pacifus imperator Romanum gubernans imperium, qui et per misericordiam Dei rex Francorum et Langordorum.*[57] In English, this would read as "Charles most serene Augustus, crowned by God, great, pacific emperor governing the Roman empire, and who [is] also by God's mercy king of the Franks and the Lombards."[58] Several subsequent emperors used similar wordings, with the designation *imperator augustus* being common. The crucial phrase was *imperator Augustus Romanum gubernans imperium,* Emperor Augustus governing the Roman Empire.[59] In other words, Charlemagne included the Romans in his Frankish empire but did not center it on them.[60] His official seal bore the title *Renovatio Romani imperii*, restoration of the Roman Empire, which also appeared in his charters and on the coins from the period 806–14.[61] It was thus a sequence of formulations that balanced imperial, papal, and Byzantine wishes and worries.[62]

It was not a resurrection of the empire as a political concept, as the Roman Empire in that sense was in the East; rather, it was of a religious *imperium*, of the Romans as Christians.[63] Yet, as Wallace-Hadrill stresses, the point was broader than this. The Romans made God a magnified form of their own royal power, as part of an attempt to provide a divine legitimation for rule, and to use that legitimation to construct a very powerful scope to that rule.[64] While a direct challenge by Leo to Byzantium was intended, Rome did not consider it a usurpation, but rather that Byzantium had lost the right to rule. In 797 the Byzantine emperor had died, and his widow had sought to continue to rule without him. This was against Byzantine law, and Colish suggests that this meant that the throne was legally empty.[65] Thus, the idea of the *translatio*, a resurrection in the West without putting the East to death.[66] The continuity was crucial, since it allowed a claim that there had been no real break between Romulus Augustus and Charlemagne, merely that the power had been exercised elsewhere. This would have profound implications for the later centuries, since it effectively implied that this transfer was in the gift of the pope. The Western emperors, the Byzantine ones, and now Charlemagne, owed their position to the gift or acquiescence of the pope. Later claims for supreme papal temporal—that is, secular—power as well as spiritual power were thus based as much on the *translatio imperii* as the donation of Constantine itself.[67]

By these actions, Rome again became the center of the Roman Empire, and according to papal theorists, Byzantium had forfeited the right, meaning that now the emperor there was a "mere Greek king or emperor."[68] This had the result that the "Greeks" became seen as more foreign, thus increasing the distance between their historical lineages. The Eastern Empire was not part of Western Christendom, which effectively meant that it was not part of Europe. The Western Empire was, in Ullmann's phrase, a "wholly Latin-Christian body."[69] This would have an immediate impact, in terms of strategic alliances, with Charlemagne allying himself with Muslim caliph Haroun-al-Raschid, and the Byzantines attacking papal states.[70] But it would have a much longer term impact with regard to the relation between the Crusades and the Eastern Church. The relation between Charlemagne and the Muslims is a major topic, and has given rise to a significant body of literature. The debate began with Henri Pirenne's thesis that it was Islamic conquest, rather than the Germanic invasions of the Western Roman Empire, that ended antiquity and should be seen as the beginning of the Middle Ages. His argument was both economic, which has since largely been discredited, but more importantly cultural, stressing the continuity and adoption of Roman cultural norms by the Germanic tribes. He does not see the Germanic invasion as the fundamental break with the Roman West, and suggests that because the Byzantine Empire was weak, the Islamic conquests really challenged the Roman dominance.[71] Yet, on the other hand, Pirenne questions the idea of a Carolingian renaissance, in both economic and literary terms.[72]

Pirenne argues that it was the emergence of Islam that fundamentally altered Europe because the south, east, and west of the Mediterranean were conquered, and the European empire, in distinction to the Roman Empire of antiquity, became a much more land ordered society.[73] The link to Charlemagne is important, because Pirenne claims that without Islam the empire of the Franks would never have come into existence, and "without Mohammed Charlemagne would have been inconceivable."[74] This was in part because Europe was a power vacuum, following the Islamic challenges, and the Franco-papal power axis was a way of securing things. It was also an attempt by Pope Leo III to protect what he still had. In this analysis, the inauguration of Charlemagne was the completion of the break between East and West, because it showed the rejection of the old empire, which continued to exist in Constantinople.[75] Geostrategically this is a powerful analysis: he at least posed the right questions.[76]

As Pirenne notes, "Before the Mohammedian epoch the Empire had had practically no dealings with the Arabian peninsula."[77] In time, this would change, because Christianity, now embodied in a new, or resur-

rected, empire was under threat from East and West. This was the sense
in which Pope Gregory VII in the eleventh century was forced to confront
the discord between the spiritually limitless *ecclesia,* and the geographical limits to actually existing Christendom,[78] which he referred to as the
fines Christianitatis.[79] To the west, there was the Muslim presence in
Spain, which prevented even Western Europe being identified with Christianity.[80] To the east, there was Byzantium, and Islamic lands, now including Jerusalem. In le Goff's terms, this paved the way for "a geographical
confrontation," one "between Christianity which was supposed to spread
over the entire world and Islam which had snatched a vast tract of land
from it."[81]

After his death, Charlemagne became a figure of heroic legend, increas-
ingly detached from history.[82] Perhaps the most famous example is *The
Song of Roland,* a poem about the Spanish campaigns against the Moors.[83]
As Einhard recounts, the inscription on Charlemagne's tomb is telling:
"Beneath this stone lies the body of Charles the Great, the Christian emperor [*orthodoxi imperatoris*], who greatly expanded the kingdom of the
Franks [*regnum francorum nobiliter ampliavit*] and reigned successfully
for forty-seven years."[84] There are many parallels here to the emperors of
old, not least in the praise of his conquests of land. Einhard also discusses
the geographical settlement between the Franks and the Saxons: "Our borders [*termini*] and theirs were contiguous and nearly everywhere in flat,
open country, except, indeed, for a few places where great forests or mountain ranges interposed to separate the lands [*agros*] of the two peoples by a
clear demarcation line [*certo limite disterminant*]."[85]

Yet the land settlement of Charlemagne did not last long. His kingdoms were divided between his sons in a proclamation of 806, pending his
death. The text was entitled "Divisio Regnorum," and was on geographical
lines, but the phrasing is telling. In clause 4, for instance, it is decreed that
"Charles shall receive Ivrea, Vercelli, Pavia, and thence along the river Po,
following its course [*termino currente*] to the border [*fines*] of Reggio, and
its kingdom, Cittanuova and Modena up the boundary [*terminos*] of Saint
Peter." Despite how the English translation renders *fines* and *terminus* as
"territory"—"the territory of Reggio. . . . The territory of Saint Peter"—
the Latin text only uses *territorium* in the next line, when it says, "These
cities with their suburbs and territories [*civitates cum suburbanis et territoriis*] and the counties which belong to them" should be allocated to
Charles.[86] At the beginning of the ninth century, it is clear that *territorium* is still a possession of a city—that is, lands outside it, surrounding it,
belonging to it—and not a larger area that includes the city itself.

In the end, such specification was unnecessary, because only Louis survived him. Tellingly, Charlemagne crowned Louis himself, rather than the pope. Louis then continued to hold the lands and title until his own death, when his sons inherited it. It was divided into three, with Lothar inheriting the title and much of the land, but Charles and Louis receiving large areas. The relation of the empire to the Franks—which became the lands of the French kings—became ever more distant. This was cemented when the Carolingian dynasty lost control of France and German lands in the tenth century. As Folz suggests, terms such as *empire of the Franks* and *kingdom of the Franks* fell out of use at the end of the eleventh century. Yet, he suggests this was a crucial shift: "As the Frankish tradition faded, the memory of the Roman Empire gained by contrast in intensity, particularly from the time of Otto II and the Salian dynasty."[87] Nicholas makes a similar point, suggesting that from the time of the coronation of the German king Otto I in 962, there ceased to be any direct link between this empire and that of Charlemagne. From the eleventh century, the empire referred rather to the lands in Germany, Italy, and Burgundy that were ruled by the dominant king of the Germans.[88] This is important to stress: there were periods without an emperor. There was a gap in the early years of the ninth century, preceding Otto I, and again in the period known as the Great Interregnum, from 1250 to 1273 (between Frederick II and Rudolf of Habsburg). There were periods when there were two claimants— Otto IV and Philip of Swabia in 1198, followed by Otto IV and Frederick II between 1212 and 1218.[89] And there was often a significant gap between their election in Germany and the coronation in Rome, sometimes of several years.[90]

One of these kings, Frederick Barbarossa, denied that the emperor had received the imperial crown as a gift of the pope.[91] Rather, he laid claim to universal authority over other kingdoms, and self-designated himself as "Emperor and always Augustus," who held through time and by divine providence, "*Urbis et Orbis gubernacula*" over a "sacred empire and divine *res publica* [*sacro imperio et divae rei publicae*]."[92] This phrasing was reminiscent of ancient Roman claims, of Rome as *urbis* and *orbis*, city and world. Frederick had his status as king of Germany, Italy, and Burgundy. For Swanson these were "kingdoms whose territorial bounds set the limits of his actual power, regardless of any universalist connotations of 'empire.'"[93] If the phrase "territorial bounds" is still a little premature, it is a telling recognition of the geographical extent of the power claimed. Frederick's chronicler, Otto of Freising (1114–58), not only wrote in praise of the emperor (who, by marriage, was his nephew) but also tried to make a

case for the continued relevance of Augustine.[94] Otto contrasts Jerusalem
and Babel as the two kingdoms, both existing on earth but distinct in the
allegiance they had to either God or man.[95] The implication was that Jerusalem was being made real as an earthly kingdom, and indeed the term *Sa-
crum (Romanum) Imperium*, the Holy Roman Empire, was a designation
that dates from the mid-twelfth century.[96]

The term *Holy* was thus added at the point when its decline had become inevitable, yet the empire has been increasingly described as an *Imperium christianum* in liturgical manuscripts.[97] The idea of an empire of
the Romans, rather than a Roman Empire, had, in the pope's understanding, always implied this. Increasingly, as subsequent chapters will show,
the divergence between the pope's vision and that of the emperors would
grow. As Morrall notes, since Charlemagne, there had not been a united
Western Europe, "but the ideal of a revived Western Roman Empire had
lived on under the aegis of the German monarchy."[98] The Holy Roman
Empire endured until 1806—and with the fall of Byzantium in 1453 became the only claim to a Roman legacy—but it had long since ceased to be
more than a notional entity. Samuel von Pufendorf described it as "like a
monster [*monstro simile*]" defying easy categorization;[99] and Voltaire famously claimed that it was "neither Holy, nor Roman, nor an Empire."[100]
Yet this was not always the case.[101]

It is crucial to underline that there was an important difference between the election of an emperor and that of a pope. While both were appointed by men, the church believed that in the case of the pope, they
acted only as his ministers, and that the conferring of power actually
comes from God himself. For an emperor, the electors possess the power
themselves, and transfer it to one among them.[102] Yet some emperors, notably including Frederick II, had other, more concrete, sources for their
power and used these to effectively challenge the pope. As Keen puts it,
in Frederick II, "the medieval papacy faced the most formidable political
adversary it ever encountered. . . . In terms of sheer territorial authority
he was the most powerful emperor the west had seen since the days of
antiquity."[103]

Nonetheless, the idea of *an* emperor, a single one in the West, was also
challenged. The idea that the Holy Roman Empire was of the German people became dominant in the sixteenth century, and this exclusivity took
away its claim to universality. But there is another significant shift that
occurs at this time: the shift from the kingdom of the Franks to the Frankish realm, a move from people to land; from race to place. McIlwain calls
this the most important aspect of the Frankish period: "the change from

the personality to the territoriality of law."[104] Yet this took many centuries to work through in detail.

CARTOGRAPHY FROM ROME TO JERUSALEM

The second half of this chapter treats two themes: representations of land and possession of it. The first of these backtracks to Rome, to offer a broad account of cartography until the late medieval period. It leads into a discussion of the way the place of Jerusalem in maps relates to the politics of land and the Crusades. The second looks at feudalism and the Domesday Book. With both of these themes there is a huge amount of very good material that already discusses similar questions, so the treatment here can afford to be relatively limited. But they are crucial issues in thinking about the relation between place and power.

Roman cartography was of limited scope, with a lack of technical innovation.[105] What advances the Greeks had made in mathematics and geometry were copied, without necessarily being understood, and certainly not developed.[106] Even one of the most detailed accounts of Roman achievements accepts that "there seems to be little doubt that with few exceptions the standard of cartography declined under the late Roman Empire."[107] Nonetheless, there are some crucial issues that need a little attention. The key discussion revolves around the simple question: did the Romans have maps that were similar to our own, which they used in the same way, but which were subject to their technological limitations; or did they think about the relation between the mode of representation and the object represented in a quite different way?[108] Crudely put, Oswald Dilke and Claude Nicolet take the former line; Pietro Janni, Tønnes Bekker-Nielsen, Richard Talbert, and Kai Brodersen the latter.[109] Taking the former path means the study of ancient cartography will allow us insights into views on geography; the latter requires a rather different approach.

In a range of important studies, including his contribution to the first volume of the *History of Cartography, Greek and Roman Maps,* and *The Roman Land Surveyors,* Dilke has cataloged as much evidence as exists for Roman cartography.[110] Dilke believes that the technical skills of the *agrimensores* allowed the survey of large extents of land, and thereby the conceptualization of large-scale spaces.[111] A related argument can be found in the work of Nicolet, who offers a "historical reflection on space."[112] Nicolet contends that the Romans must have had perceptions of space in order to have conquered as much land as they did, to have had a sense of its boundaries, and to have had a means of continuing to control it.[113] He therefore undertakes his study through "two corresponding approaches: a historical retrospect of the geography of the Romans and a history of the administrative use of geographical space."[114] It is a remarkable study, but has been accused of taking the case too far: "Nicolet has created an exaggerated picture of the complexity and sophistication of both ancient geography and thinking about space."[115]

In distinction, writers like Janni, Bekker-Nielsen, and Talbert suggest that we need to avoid transferring modern assumptions about space and geography to the ancient world.[116] While our views of space are linked to perceptions made possible by modern technology, the Romans did not have such tools at their disposal. Instead, we need to realize that they would have experienced, and therefore understood, the places they visited and the land they fought over rather differently. This view is also held by more mainstream strategic scholars, who note that the Romans lacked many modern techniques and understood terrain through military campaigns.[117]

This does not mean that they had *no* sense of spatial relations, just that they would not have necessarily understood them in the same way.[118] In a very helpful summary of the evidence, Kai Brodersen has argued that there are good indications that large-scale maps of land and cities existed, and some proof of small-scale depictions of larger areas; there is nothing to indicate that there was anything in between. This means that practical maps of regions, provinces, or roads probably did not exist.[119] Janni takes a similar position: the Romans were unable to produce a map of the Mediterranean world; conceptually it was simply beyond their capacities.[120] Nonetheless, they were able to conquer and control large extents of land, which raises the question of how this could have been possible. The key, for Janni, was the road, and the representation of it, the itinerary.[121]

The road was the connection between known points, through a surrounding landscape that was often barely understood. Roman roads were commonly straight lines, which requires some degree of technical capacity. But it also made possible ways of thinking about the geography of their conquests that transferred easily into representations. All travelers needed to know was practical information such as the distance between points and the things they might encounter along the way. For Janni, this means that the Romans often conceived of what we now call space with one fewer dimension. In distinction to cartographical space, *spazio cartografico*, he describes their mentality as *spazio odologico*, hodological space. Taken from the Greek word *hodos*, path or way, it refers to a more linear space of connections rather than a plane of more varied possibilities.[122] He therefore opposes tessellation—the filling or tiling of the plane—to networks; we might say space to line. In the Roman way of thinking, those things outside the trajectory effectively did not exist.[123] Coastlines rather than the interior, rivers not banks, roads not landscapes, streets not the town characterize their thinking.

It should, of course, be noted that there is very little evidence to decide. So little has survived, it is difficult to make firm judgments. This has not stopped Dilke:

> The Romans used maps for various purposes. There were world maps; maps illustrating geographical treatises or works of literature; road maps and itineraries to help travellers find their way about; official and military maps; detailed town plans; and surveyors' maps. Other technical maps and plans are likely to have existed but have not survived.[124]

As Brodersen rightly contends, those artifacts "which have been adduced by scholars so far as material evidence for a 'Roman tradition of scale maps,' fail to prove the existence of this tradition, and resorting to the 'must have been' variety of logic does not solve the problem of how geographical knowledge was presented."[125] As some of the critics of Dilke's work have suggested, his extensive survey of ancient cartography actually produces the opposite outcome to what he intended. Its very comprehensiveness shows how little evidence there is, and reinforces a sense of the "almost total lack of map consciousness" for most Romans.[126] Talbert, in particular, is critical of his tendency to interpret every scrap of evidence to suit his claims that maps were very important and counters with the compelling argument that as well as lacking the techniques, they did not have the material means to produce and keep maps in difficult conditions, papyrus being too fragile and silk being used by the Chinese but not by the Romans.[127]

There are certainly some unwarranted assumptions, such as Thomson's dubious suggestion that the "Romans must have seen the maps of Greek geographies, and used travel and campaigning maps, as later, though clear mentions happen to be lacking for this time."[128] Yet it is also dangerous to assume that because they no longer exist, they never did. It is not impossible to try to think how the Romans could have proceeded without them. Brodersen's *Terra Cognita* argues that the Romans had no real maps, and that they did not need them in their organization and administration of empire. He notes that historians tend to assume that the Romans must have had them, and that they died out in the Middle Ages, but questions the basis for such an assumption.[129] He accepts that they could survey small areas, but suggests that it is not clear if they could do this for larger regions. On a Roman tradition of scale maps, he is deeply skeptical, contending that there is "no evidence at all: not even the surveyors' maps are drawn to scale."[130] Most scholars of Roman strategy therefore contend that the Romans did not have a sufficiently developed topographical sense in order for their cartography to have been much use in warfare.[131] The succession of similar landscape features and the inability of human recall on the one hand, and the lack of any detailed representation of this on the other, make claims to the contrary difficult to sustain.[132] This runs against Nicolet's assumption that "to govern it, it must be known, measured, and above all drawn,"[133] making maps an essential administrative tool.[134] "It is clear that geographical space, within the boundaries of the empire, is also and perhaps above all an administrative space."[135]

Two important examples from the Roman period should be discussed. The first is the *Mappa Agrippa*, a world map that was displayed in Rome on the wall of a portico. Unfinished when Agrippa himself died, it was supposedly completed by Augustus. It is unknown if it was a painting or engraving.[136] It supposedly portrayed the whole of the *orbis terrarum*, not simply the area controlled by the Romans.[137] Nicolet makes a case for Augustus's *Res Gestae* to be seen "at least in its second half as a genuine geographic survey,"[138] with the text appearing "almost as a commentary to a map and to require the guidance of a drawing," which he suggests is, in fact, the Mappa Agrippa.[139] His argument is that the text "claims the conquest of the world for Rome's advantage under Augustus's influence," but that makes this claim "without the slightest reference to symbolic concepts, and without metaphors. On the contrary it uses geographical, historical, and political concepts that were precise and, I would say, indisputable."[140] The data that Pliny and Strabo use in their gazetteers have been traced back to this map,[141] although Strabo never mentions this as a source. Nonetheless, it fits with the attitude of the time: as Nicolet notes, "It is not a coincidence that the most complete geographical work handed down from antiquity, that of Strabo, is from the Augustan period."[142] Pliny acknowledges it as a careful source, makes use of its measurements, and simply describes it as "a survey of the world [*orbem terrarum orbi spectandum*]" that was displayed.[143] It is generally acknowledged that accompanying the map was a *commentarii*, a written text from which it was apparently derived, leading to attempts to reconstruct the map either from fragments of the text or sources such as Pliny or Strabo.[144]

But this gives rise to a key question: was there actually a map at all?

Even Dilke acknowledges that Pliny might have derived his references from the commentary alone;[145] but Brodersen has taken this further to suggest that there was only the inscription of details of the known world and its constituent regions, a monumental description rather than depiction. It is not at all clear that *spectandum* in Pliny means a "map"; it is a word he uses elsewhere to refer to a text.[146] Brodersen also notes the ancient sources that discuss the building where this was displayed, but make no mention of a map.[147] This skepticism has been attacked, by Benet Salway in particular, who suggests that it perpetuates a "stereotype of theoretical Greeks versus practical Romans";[148] and while he concedes that in this particular case the evidence is equivocal, he counters that "the public display by the Romans of cartographical images—both world maps and detailed local plans—cannot seriously be denied."[149] His main example is of the surviving fragments of an "enormous marble plan of Rome once displayed near the forum," but this dates from the third century CE.[150] Another example comes from Livy, who tells us that a tablet (*tabula*) was set up in a temple, that it was dedicated to Jupiter, and that it had the form of Sardinia. Yet he tells us nothing that would indicate that the form had any other details on it, only that "representations [*simulacra*] of battle were painted on it."[151] Brodersen has replied in detail to Salway, and seems more convincing.[152]

The second example also hinges on the mode of representation. These are the itineraries, for which there is more extensive evidence. The most important is the *Itinerarium Antonini*, which gives lists of places along a route and the distances between them.[153] This has been taken to be the model for such documents and, depending on the author's view, confirms either the skills of Roman cartographers or the belief in "hodological space." As Bekker-Nielsen contends, these itineraries help to make sense of what appear to be somewhat peculiar routes of transportation, which did not always take the most direct course: "The only possible explanation is that they somehow saw their world as something different from what we see on our map."[154] The point is more general. If we look at modern maps of Roman expansion, they seem to illustrate a vulnerable empire with long frontiers either side of the Mediterranean. Yet this is only the case because of the information we now have and our way of representing it. As Moynihan plausibly contends, "The Roman leadership in the Augustan Age saw their situation far differently. . . . The Empire was not a vulnerable band of land circling the Mediterranean at one tip of the Eurasian land-mass. Rather, it was the larger portion of the habitable earth, poised on the verge of world rule."[155]

A graphic representation of an itinerary does exist, but this is an indisputably medieval document from around 1200 CE, the Peutinger Table or Map.[156] The debate is to what extent it is a copy of a Roman original and thus the "sole surviving witness to the character of ancient Roman, and in particular Latin, mapping of the entire known world."[157] Talbert believes it dates from around 300 CE but acknowledges other views that it was of Carolingian origin.[158] The Peutinger Table is a long strip of parchment, almost seven meters long but only thirty-four centimeters high.[159] This has given rise to the idea that the original was a papyrus roll, which may have been designed to be portable, and whose constraints dictated the way that the map was presented. Other possibilities include that it was a copy of the Mappa Agrippa, or derived from it; or that it was designed to represent only what was known and downplay the unknown.[160] It is therefore extremely distorted in its representation of distances, especially on the y-axis, with lands frequently compressed or not depicted. Seas and lands without Roman roads fare particularly badly. It is a graphic representation of a journey; distances are added, which in turn takes away any need for a scale on the x-axis.[161]

Yet it is more than simply a diagram. The place names that are provided are often in a seemingly strange form, which, as Salway has noted, is because they are derived from itineraries where the cases would have formed parts of phrases, such as "from X to Y, Z miles."[162] Yet while it is not merely illustrative in the sense of being decorative, it is difficult to be sure to what extent the artifact that exists is the product of some Roman original and what from a later copyist. In other words, the copy that exists may be an embellishment of a more functional original.[163] The same is true when we look at editions of Ptolemy's *Geography*, which while deriving from a second-century original text, have only survived in manuscripts with later maps.[164] The scholarly consensus is that the maps in Byzantine and Renaissance manuscripts were generated from the figures and measurements in Ptolemy's texts rather than copied from earlier versions.[165] As such, it is impossible to tell what the original maps would actually have looked like.

Maps from the medieval period tended to follow the T/O model pioneered by Isidore of Seville. This put Jerusalem right in the center of the map, often with a cross superimposed, as it was in the Hereford Map, a *Mappa mundi*, a "cloth of the world."[166]

This beautifully preserved map, however, can be misleading. Medieval Europe was a society that had little regard for maps, which were used only by a few and not widely known.[167] They were, however, more widespread than they had been in the time of the Roman Empire or the barbarian ages

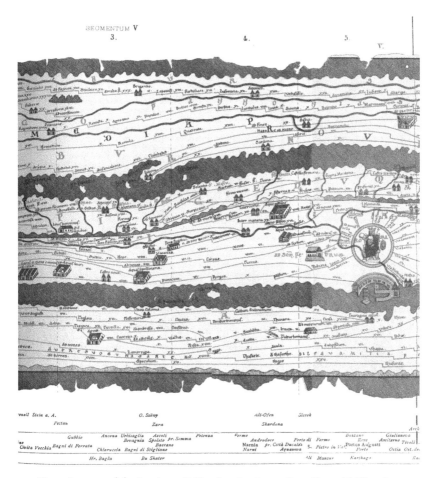

Fig. 5. Segment of the Peutinger Table, showing Rome. From Konrad Miller, *Die Peutingersche Tafel*, expanded ed. (Stuttgart, Germany: Strecker und Schröder, 1916).

that followed in the West.[168] While Harley and Woodward's general suggestion that "maps are graphic representations that facilitate a spatial understanding of things, concepts, conditions, processes, or events in the human world"[169] remains true, as Edson has noted, medieval maps were often as concerned with questions of time as they were with representations of space.[170] This is because they were often used to show journeys or itineraries. While some astronomical and other techniques were known, they had little impact on the cartography of the time.[171] Rather, cartography was much more explicitly linked to theology,[172] as can be seen from the situation of Jerusalem to stress its spiritual—rather than actual geographical—

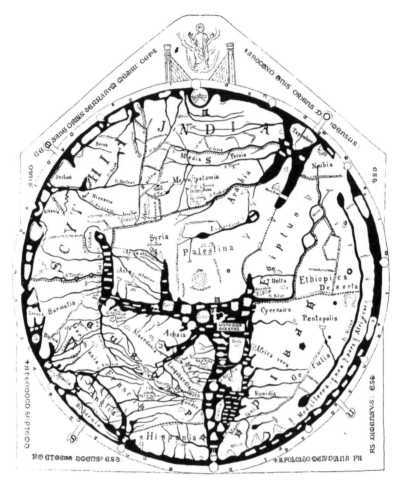

Fig. 6. Hereford Mappa Mundi. From Konrad Miller, *Mappae Mundi: Die altesten Weltkarten*, 6 vols. (Stuttgart, Germany, 1895–98), 4:2.

centrality, and Eden appearing in the East at the top of the map.[173] As Colish puts it, "Medieval cartographers were not interested in displaying the period's geographical knowledge with scientific precision or to scale. Maps for them, had a didactic and spiritual value."[174]

These considerations are related to the Crusades, the attempt to claim, or reclaim, the Holy Land for Christianity. They built on earlier pilgrimages to sacred sites, within a religion for which, in Hay's words, "the New Testament was not only a devotional work but also a geography lesson, almost a guide-book."[175] Hay suggests that the Crusades were only a short step from such voyages, but there were other factors at stake. Not least was the conflict between the Western, Roman and Eastern, Byzantine churches, with the former keen to establish its supremacy. For Pope Gregory VII, who called for a crusade in 1074, this was the key factor, even though he dressed it up as going to aid the Eastern Church.[176] Gregory's call was unsuccessful, due in part to the Investiture Controversy, but the call was made again by Pope Urban II some twenty years later, which led to what became known as the First Crusade.[177] The key point to be underlined here is that the Holy Land was understood itself as a relic—clearly some way ahead in its importance over any of the simple objects invested with religious and symbolic significance. Christ and the disciples had walked on these lands; they had been baptized in the river Jordan; their blood had soaked into the soil; their bodies had been interred in the earth.[178]

It is sometimes suggested that the Bible contains a sense of territory, especially in the idea of the Promised Land.[179] Yet the Vulgate—the key way the text was known for the Middle Ages—never uses the word *territorium* to render any of the geographical terms.[180] It is debatable whether the concept of territory is appropriate to understand the questions at stake. There are discussions of land, borders, and division certainly, as there are throughout the Greek and Latin texts examined in previous chapters, but these terms need to be carefully differentiated. The Vulgate uses *fines*, *terra*, and *terminus* much as the Latin historians had done. Interestingly, the medieval political texts discussed in this and subsequent chapters grappling with questions of political power, authority, and place do not make reference to the Bible to substantiate the point of what political power is exercised over: the Western, modern notion of territory that this book examines seems to emerge from a different set of debates and arguments. Indeed, in his *Security, Territory, Population* lectures, Foucault actually suggests that the Judeo-Christian notion of a flock deriving from the biblical Israel model gives rise to an entirely different history of government that is not territorially based.[181] In the late nineteenth and twentieth centuries, the biblical account gets reread in the light of contemporary debates about (and concepts of) territory to provide the basis for modern Israeli claims, but this is a rather different question.[182]

Like Gregory, Urban II positioned his call as an attempt to aid the Eastern Church, which feared the loss of Constantinople to Muslim forces.[183] This was in many ways the most successful of the crusades, leading to the establishment of the Kingdom of Jerusalem. The possession of Jerusalem endured until its capture by Saladin in 1187: Pope Urban III is rumored to have died upon hearing the news.[184] The Third Crusade was an unsuccessful attempt to regain it, led by Frederick Barbarossa, Philip II Augustus of France, and Richard I of England, on the call of Pope Gregory VIII. The kingdom itself lasted until 1291, for much of the time having Acre as its capital. Even when Jerusalem itself was not held, this provided an important strategic position for the Western church and the secular rulers who marched under its banner. And though Richard was unable to recapture Jerusalem, he was able to negotiate safe passage for Christian pilgrims to the holy sites. Yet the impact of the Crusades went beyond the religious and strategic considerations. In large part they reintegrated Western Europe into a larger "world system" that it had not been part of since the collapse of the Western Roman Empire several centuries before.[185] The opening up of trade routes and cultural contacts was important in terms of the subsequent development of the West.[186]

The justifications offered for the Crusades were many, but few repay much attention in terms of their importance for the story of the development of a sense of territory. Certainly there is a strong sense of a sacred space profaned by the presence of infidels, and a strategic importance cannot be underestimated. Both themes are found in one of the important works that takes the Holy Land as its focus, Pierre Dubois's *De recuperatione Terre Sanctum* (The Recovery of the Holy Land), written around 1305–7. He had attended lectures by Thomas Aquinas and was part of Philip the Fair's court.[187] Dubois stresses the importance of the "the Holy Land so sanctified by the precious blood, the acts, and the bodily presence of our Lord Jesus Christ,"[188] which gave this particular place a crucial significance. Innocent IV asked the question of whether Christians are allowed to "invade *terram* that is possessed by infidels." His answer is a qualified yes, citing Psalm 24 in defense. Psalm 24 declares that the earth, everything in it, and all who live on it are the Lord's, and only those who have "clean hands and a pure heart" can "ascend the hill of the Lord" and "stand in his holy place." But Innocent notes that "lordship, possessions, and jurisdiction [*domino . . . dominium & iurisdictionem*]" can exist among infidels.[189] Yet the cost they had was considerable: primarily in terms of what happened to the "infidels," but also at home. In Gibbon's terms, "Conspicuous place must be allowed to the crusades" in undermining the "Gothic edifice" of the entire structure of medieval society: "The estates of the barons were dissipated, and their race was often extinguished, in these costly and perilous expeditions."[190]

One other instance of cultural contact is worth noting here. As the Mongols moved west, even reaching the gates of Vienna in December 1241, they pushed many other groups ahead of them.[191] As Keen has noted, this had a profound influence on European geography, since it made Europeans realize that beyond the Islamic lands, there were "a countless multitude of pagans and unbelievers." This had two key results. The first was that the Crusades lost some of their religious significance, given the revaluation of priorities.[192] The second was that Europe realized its geographical position as, in Braudel's memorable phrase, "an Asian peninsula."[193] Keen has even suggested that "the revelation to Europe, in consequence of the Mongol invasions, of the vast Asian hinterland was of comparable impact, in terms of geography, to the later discovery of the Americas."[194]

THE LIMITS OF FEUDALISM

After Frederick II, various subsequent emperors—or kings claiming to be emperors—would attempt to reunite the German peoples, but the link between spiritual and secular power would remain fragmented and disputed.[195] As early as 1231, regional powers in the empire were being described as *domini terrae*, "lords of the land," for which the German term was *Landsherr*, and the overall principle one of *Landsherrschaft*, landlordship.[196] Lordship was attached to a person and inherited; it was not a right over land.[197] In other words, the land was a secondary, rather than determinate, aspect of the lordship. It is essential to note that the Latin *terra* accords to the German *Land*. While the notion of *Landeshoheit*—land supremacy—was the principle eventually conceded at Westphalia as *iure territorii et superioritatis*, territorial right and supremacy, at this time the land was still a possession of the ruler.[198] *Landesherrschaft or Landeshoheit* was understood as an expansion of the feudal relation of manorial lordship.[199]

A brief discussion of the notion of feudalism is thus in order. The literature on this topic is vast.[200] It has become a major component of historical materialist accounts of sociopolitical transformation.[201] In recent years there has been something of a backlash against the usefulness of the term.[202] It is essential to remember that *feudalism* is a retrospective term applied by historians rather than a description from the time.[203] It is, as Strayer and Coulborn point out, "an abstraction derived from some of the facts of early European history, but it is not itself one of those facts."[204] At its heart is an attempt to make sense of the political economy of land. As Carlyle and Carlyle have suggested, feudalism is perhaps the most difficult subject in medieval history, with its origins "still obscure and controverted."[205] The word *feudalism* comes from the Latin *feodum*. In the eighteenth century, Brussel suggested that the Latin *feodum* was the equivalent of the French *mouvance*, and McIlwain suggests *tenure* is closest to that meaning.[206] *Feodum* is customarily translated as "fief": an area of land, but also the infrastructure of agriculture and dwellings, and the people working on it. These areas could be inherited, but were in the grant of the king through conquest or other acquisition.

The debate is to what extent the political forms develop directly from the economic system, or whether feudalism is better understood as a system of power relations between lord and serf, which develop in economic, legal, military, and other directions.[207] The most sensible approach is to seek a middle position between seeing it as encompassing everything and seeing it in purely legal terms.[208] What appears relatively uncontroversial is that land was a major source of power.[209] The importance of land did not mean that other economic factors were unimportant. People were obviously needed to cultivate the land, and to get it to produce wealth. Land was plentiful but laborers comparatively scarce. Often the greatest problem was ensuring that land was cultivated.[210] What this means is that it created a strong relation between people and place. As population increased and political control developed, large areas of previously open or common land were enclosed, with farmers or lords claiming exclusive rights over them.[211] But there was nothing inherently fixed, since lands were often redivided on the death of a ruler, with marriage, or following changes in political power. This model of political power was based on complex hierarchies of royalty, nobility, knights, and serfs, and the power of the church, but in practice worked through decentralization. Hay has argued that this means that before the thirteenth century, history in Europe is local history.[212] The limited means of communication and the difficulties of long-distance travel meant that power may have been flowing from a central source, but that its actual workings were only ever diffuse.

The Norman conquest of England in 1066 was unusual because William considered all of England to be his property by conquest and distributed it out as fiefs. Elsewhere in Europe kings did not own all of their kingdom's land,[213] and so William's introduction of feudal relations into England was enabled in a rather purer form than elsewhere.[214] What this meant, in part, was that the overall structure changed, but at a local level there was much continuity. A new hierarchy of Norman lords ruled over an already existing structure of Saxon land relations.[215] Twenty years after the invasion, the grand land survey that became known as the Domesday Book was undertaken. While this gives an extremely detailed view of England at that time, as Hooke has suggested, it actually "reveals a landscape that is a palimpsest of many periods, a society in evolution."[216] The name "Domesday Book" derives from it being the "book of judgment," the final reckoning of land ownership, which allowed no appeal. It was a very quantitative document, looking at ownership of land and structured by "how many" and "how much" questions.[217] William was cataloging the realm he had won, through modernization and technical innovation.[218]

In the discussion of Bede in chapter 3, the notion of a hide was mentioned, which is a measure that springs from a subsistence economy.[219] The *mansion* in Gaul was a similar extent,[220] as was the German *Hufe* and the *manse*, which Charlemagne tried to introduce as a military and tax measure.[221] The vast proliferation of measures, many of which were regionally specific and not understood elsewhere, was an important element in political and economic power. For extents of land, the English also used the yardland and the oxgang;[222] and the medieval boundary between the shires of Huntingdon and Cambridge ran through the *meres*, or wetlands, "as far as a man might reach with his barge-pole to the shore."[223] Le Goff has suggested that "measures of time and space were an exceptionally important instrument of social domination. Whoever was master of them enjoyed peculiar power over society."[224] Some measures had an element of utility, but others were chosen for other economic reasons. The hide, for example, later shifted from the unit of sustenance to a more abstract unit of taxation.[225] There was often resistance to change—Edward III tried to standardize units of measure in England in 1340, but was largely unsuccessful.[226] The importance of land led to some minor developments in terms of the establishment of boundaries and techniques for the calculation of wealth.[227] There are examples of border stones being used in this period, and of the pursuit of deer not being allowed between woodlands of differing jurisdictions.[228]

Feudalism generally is looked at as emerging in the tenth century, and beginning to decline in the thirteenth and fourteenth, although there are many debates about its dating.[229] There is, of course, a danger in generalizing from the model of land power in England, the empire, or France.[230] During this time there was a plurality of legal codes, which increasingly applied to people in the place that they were, rather than where they were from or who they were. But this geographical focus of the law was scattered, vague, and often overlapping in terms of jurisdiction.[231] Uniting areas of land, and the legal codes that operated within them, was beyond most rulers of this period. The Magna Carta in thirteenth-century England, for instance, was a compromise between the king and the landowning barons of his kingdom, and protected their power rather than allowed its central-ization.[232] Feudalism, then, however it may be described, simply does not get past the conundrum that the Middle Ages, which put so much empha-sis on property in land, did not have a territorial system, and lacked an articulated concept of territory. For this we should look elsewhere.

CHAPTER FIVE

The Pope's Two Swords

JOHN OF SALISBURY AND THE BODY OF THE REPUBLIC

In the twelfth century, a bishop of English birth changed the terms under which discussions of politics operated. He did so in a rather understated way, but his major political work, *Policraticus*, provides many of the themes that will shape debates from that point on.[1] John read only Latin, and so was forced to rely on translations of classic Greek works. This limited his access to just a handful of texts, notably Plato's *Timaeus* and Aristotle's *Organon*, although he had some knowledge of Aristotle's *Physics*. Aristotle's moral and political writings, and indeed many of his most important philosophical ones, would not be available for another century.[2] The main source for John's arguments is, therefore, Holy Scripture, but he also drew on other classical sources, including the writings of Saint Augustine and Saint Jerome. While he has been described as one of the best-read men of his time, knowing a wide range of classical sources, it is clear that some of his sources are actually *florilegia*—compilations of quotations—rather than the texts themselves.[3] His other major work besides the *Policraticus* was the *Metalogicon*, a work of pedagogy and speculative philosophy.[4] It was a defense of Aristotle's views of logic, which he based on his knowledge of the *Organon*. Although this probably makes him the most important twelfth-century scholar of ancient thought, a significant distance still separates him from that thought itself.[5]

Born in the cathedral city of Salisbury sometime between 1115 and 1120, John studied in France under Peter Abelard, who had himself had a substantial impact on the shaping of scholastic thought. John served as the secretary to both Theobold and Thomas Becket, successive archbishops of Canterbury. In part through his relation to Becket, John was involved in political turmoil in England, and left for periods of exile.[6] One of these exiles lasted from 1163 to 1170, with his returning around the time Becket was murdered by King Henry II's knights in Canterbury Cathedral. Although his *Policraticus*, which was completed in 1159 and dedicated to Becket,[7] predates the murder, its themes clearly relate to the conflict between monarchy and church, and offer thoughts about the limitations of royal power.[8] John was made bishop of Chartres in 1176 and remained there until his death in 1180.[9]

Policraticus is a significant work, essentially the first substantial medieval handbook of political thought.[10] Although disjointed and digressive, it outlines what an ideal ruler should be and do, and how this differs from a tyrant.[11] He is much less interested in the private aspects of the prince, and more in the prince's public role.[12] As the subtitle indicates, he is scornful of the frivolities of courtiers and thinks that rulers should follow more in the footsteps of philosophers.[13] One of the reasons *Policraticus* is significant is because it was written before Aristotle's *Politics* was rediscovered, and thus represents a purely medieval tradition, which it explicitly labels itself as "political [*politica*]."[14] According to John, the prince's authority comes from the church, for which the symbol of authority is the sword, a term that will become extremely significant. John's most important, and widely noted, innovation is to compare the polity to a body, the notion of the body politic. Earlier models of this symbolism appear in Greek thought, and John was not the first to use it in the Latin Middle Ages. However, he did develop the model in important and significant ways.[15]

John begins this famous discussion with a reference to Plutarch:

> For a republic is, just as Plutarch declares, a sort of body which is animated by the grant of divine reward and which is driven by the command of the highest equity and ruled by a sort of rational management.[16]

However, the source for this, claimed to be Plutarch's *Instruction of Trajan*, is dubious. Aside from John's reference, there is no other evidence that such a work ever existed, written by either Plutarch or someone pretending to be him. Most scholars now claim that it never existed, and was probably invented by John in order to lend his ideas a classical credibility, or—perhaps more important—to deflect accusations of innovation, which was not appreciated at that time.[17] This is the now common opinion, but the previous assumption was that the work he referred to was either a

Latin translation of a compendium of Plutarch's work or a Latin original pretending to be Plutarch.[18] John continues:

> The position of the head in the republic [re publica] is occupied, however, by a prince subject only to God and to those who act in His place [locum] on earth, inasmuch as in the human body the head is stimulated and ruled by the soul. The place of the heart is occupied by the senate, from which proceeds the beginning of good and bad works. The duties of the ears, eyes and mouth are claimed by the judges and governors of provinces [provinciarum]. The hands coincide with officials and soldiers. Those who always assist the prince are comparable to the flanks. Treasurers and record keepers . . . resemble the shape of the stomach and intestines; these, if they accumulate with great avidity and tenaciously preserve their accumulation, engender innumerable and incurable diseases so that their infection threatens to ruin the entire body. Furthermore, the feet coincide with peasants perpetually bound to the soil [solo], for whom it is all the more necessary that the head take precautions, in that they more often meet with accidents while they walk the earth [terra] in bodily subservience; and those who erect, sustain and move forward the mass of the whole body are justly owed shelter and support. Remove from the fittest body the aid of the feet; it does not proceed under its own power, but either crawls shamefully, uselessly and offensively on its hands or else is moved with the assistance of brute animals.[19]

What we have, then, is not so much an idealized state but rather a living organism, the parts forming a greater whole. As an 1159 letter makes clear, each part is integral and strictly irreplaceable. All parts of the body "are united to secure the body's health; they differ in their effects, but, if you consider the health of the body, they are all working for the same end."[20] The body is not therefore simply an ordering principle, but the question of connectivity of the parts is certainly key.[21] The republic as an organic body bears relation to the idea of a church as a body of the individual members, and clearly there is an attempt here to find a way of relating the appropriate conduct of both.[22] For Lieberschütz, John's "attitude is that of pastoral care for the men who have taken the responsibility of command."[23] As Ullmann puts it, "It is the orbis Latinus, the congregation of the faithful in its corporate nature."[24] John indeed talks of the orbem latinum—the Latin world—as having heard of the death of Becket, for instance.[25]

It is important to underscore that while the model of the republic as a living creature is powerful, he is not arguing that it is wholly independent. The prince is head, but is ruled by the soul, in which they receive direction from God. John describes the "imperial vicar"—that is, the pope—as "the rector of the body politic [*corporum*]."[26] This, of course, is precisely the problem: what happens when the prince as head is no longer operating in terms of agreement with the soul, when you have the bicephalous pope and the emperor, and the period as a whole was much more shaped by their disagreement than their agreement.[27]

A source for John's idea of the body may well have been Calcidius's commentary on Plato's *Timaeus*, although this does raise the question of why this was unacknowledged.[28] Calcidius was a translator of the *Timaeus* into Latin, and his was the main version circulated in the Middle Ages. Most versions of the translation also included his commentary on the text.[29] This was a commentary that went far beyond mere exegesis of Plato, but incorporated a whole range of views on related topics from a range of authors.[30] In it he provides a reading of the tripartite model of society, with the leader, the soldiers, and the masses and traders conforming to the head, the heart, and the lower regions of the body. In turn they coincide with reason, vigor, and more earthy desires. As Dutton notes, his glosses on *Timaeus* 17c and 44d are particularly important.[31] Because John refers to Calcidius's commentary in *Metalogicon*, we know that he had access to this version.[32] Nonetheless, the parallels are rather vague, and bear only passing resemblance to work on the *Timaeus*, before he elaborates in much greater detail. Indeed, Dutton notes that he might equally have got this model from a much more general understanding of the shape of ancient cities: with the senate situated up high, the noble class below, and the farmers and hunters on the outskirts.[33] John's model for an ideal does indeed appear to be classical Rome:

> If all the histories of all people are reviewed, nothing shines more brightly than the magnificence and virtue of the Romans. This is proclaimed by the fullest splendour of their empire, so that the human memory can record no empire which was lesser at its beginnings nor which proceeded to more greatness through continual enlargement and extension.[34]

What is interesting about this is the model of enlargement and extension, one that underlies visions of the development of Christendom. Nonetheless, John does understand the limitations, speaking of "the populace of a foreign region [*extraneae regionis*],"[35] and noting the expansionist tendencies of other tribes. The Britons of Snowdon, for example, "attack and extend their borders [*terminos*] and, leaving their retreats and forests, occupy the flatland [*plana*], and assault, take by storm, and demolish or retain for themselves the fortifications of our chief nobles."[36]

John is, in some senses, a detour on the path. But in many more senses, he is a significant figure, because elaborated here is the basis for an account of politics that is, in part, independent of both antiquity and Christianity. Its debt to the past is apparent, but filtered through many layers of interpretation. Its Christian aspects are important, but the beginning of the basis of a secular approach to political thought can begin to be discerned. In Kantorowicz's terms, there was a shift from Christ-centered kingship to law-centered kingship.[37] One of the aspects of this comes in his theory of tyrannicide, a topic that has been widely discussed in the literature.[38] Although there is not an especially coherent theory in his work, it is clear that this is not simply a defense of, or argument for, regicide. Rather, any element within the body politic that fails to meet key criteria needs to be removed. This is irrespective of the position within the organic whole, and temporal and ecclesiastical power can both be abused.[39] The measure is the accord between their rule and notions of right and justice. If different parts are not operating in pursuit of a common goal, then the whole will fail.[40] As Forhan puts it, "All tyrants will suffer . . . virtue is essential from head to toe."[41] In addition, it is crucial that each part does its specific duty, and does not seek to take over the role of other parts.[42] What is novel about this is its reference to criteria independent of position: justice and virtue. Rule not in accord with those criteria is an abuse of authority, and tyranny. The difference between a prince and a tyrant is their accord with these ideals. It is this, rather than a straightforward argument for political violence, that is distinctive. It of course raises the question of how right, justice, and virtue are to be defined, but it allows us to make sense of Dickinson's claim that "the modern world is the direct heir of mediæval institutions and ideas, while it is the heir of classical antiquity only indirectly. The *Policraticus* has more light to shed on the issues of 1688 and 1789 than either the *Republic* of Plato or the *Politics* of Aristotle."[43] Yet more than this, the clue is in the title of the book. *Policraticus* means both the power (*kratos*) of the *polis*, but also implies plural powers, from *polus*, meaning "much" or "many," and the text recognizes that without a division of power, and appropriate conduct in each of the parts, tyranny will result. Yet in a different sense, Dickinson is profoundly mistaken, since it is the rediscovery of those classic texts, especially Aristotle's *Politics*, that shapes the next century of political thought.[44]

TWO SWORDS: SPIRITUAL AND TEMPORAL POWER

The relationship between the church and political power is, at best, vague in the Bible, and so selected passages were often pulled out of context, interpreted, and reinterpreted by all parties. Although Watt is correct to suggest that one of the "axioms of medieval politics" was that Christ himself had "separated the functions of king and priest," what this meant in practice was singularly disputed.[45] In Luke's Gospel, Jesus is asked if taxes to the secular powers should be paid. He asks whose head appears on the denarius coin. When told that it is the emperor's, he instructs those listening that they should "render unto Caesar the things which are Caesar's, and unto God the things which be God's" (Luke 20:25; see Matthew 22:21).[46] If that seems to set up a clear distinction and a clear division of loyalty, it was, for the papacy, in part countermanded by a passage two chapters later. This is when Jesus is readying the disciples for the conflicts and persecution to come, and he tells them to sell their goods to buy swords (Luke 22:36). They tell him, "Lord, behold, here *are* two swords," to which Jesus replies, "It is enough" (Luke 22:38). Later, when Jesus is betrayed by Judas, the disciples ask, "Should we strike with the swords?" (Luke 22:49), and one of them—according to John, it was Peter—cut off the ear of the high priest's servant. He is told, "Put up thy sword into the sheath" (John 18:10), or, more fully, "Put up again thy sword into his place: for all they that take the sword shall perish with the sword" (Matthew 26:52; see Mark 14:47).[47]

The straightforward interpretation is that two swords are sufficient to defend Jesus and themselves, and that Jesus initially explicitly suggested that they needed one sword each. Nonetheless, popes and others pointed out that Jesus notes that "it is enough." He does not say "It is too many" (or "too few"), and thus the church is said to have a right to two swords. What are these two swords, if read in a nonliteral way? Le Goff has claimed that the two swords as spiritual and secular power can be traced back to the early church fathers,[48] but at the time of Charlemagne, Alcuin of York argued that one sword was for internal use (against heresy) and the other for external use (against pagan enemies). According to Canning, it was only from the middle of the eleventh century that the spiritual and secular power use became current.[49] Whichever interpretation is correct, it is unquestionable that the swords as symbolizing spiritual and secular power became dominant, and regularly cited. The subsidiary interpretation is that while both swords belong to the church, the sword of secular power is to command, not to use. This is why Jesus tells Peter to sheath the sword: it is his by right, but not to use directly. The *locus classicus* is Bernard of Clairvaux's instructional manual to Pope Eugene III, where he suggests that "this sword also is yours and is to be drawn from its sheath at your command [*nutu*], although not by your hand. . . . Both swords, that is, the spiritual and material, belong to the Church; however, the latter is to be drawn for the Church and the former by the Church. The spiritual sword should be drawn by the hand of the priest; the material sword by the hand of the knight, but clearly at the command [*ad nutum*] of the priest and at the order [*jussum*] of the emperor."[50]

The distinction is thus initially between spiritual and material, and in particular spiritual admonishment and material punishment, which only later becomes one between spiritual and temporal power.[51] In this it combined this division with one that Pope Gelasius I had made in the late fifth century between temporal *potestas* and spiritual *auctoritas*—temporal power and spiritual authority—suggesting that the latter was the source of the pope's *superioritas*, and that the emperor had merely *regia potestis*.[52] Ullmann notes that these terms came from Roman constitutional law, where "the 'authority' of the Ruler was over and above mere 'power.'"[53] As Scanlan suggests, "The history of medieval political thought makes it abundantly clear medieval culture could not understand *auctoritas* apart from the *potestas* to which it was typically opposed."[54] Although the terms of power and authority are not strictly policed in distinctive meanings from this point on, the distinction is important in hierocratic understandings of the relation between papal and imperial roles. A medieval constitution from 1159 made this even more explicit:

> When in the passion Christ says he is content with two swords, this meant the Roman church and the Roman Empire for it is by these two heads and leaders [*capita et principia*] that the whole world is ordered in divine and human things. For there is one God, one pope, and one emperor ought to be enough, for one Church of God.[55]

In John's Gospel, Jesus makes the apparently clear declaration that "my kingdom is not of this world" (John 18:36), yet in Matthew's, he tells Peter, the rock (*petrus*) on which the church will be built, that "I will give unto thee the keys of the kingdom of heaven; whatsoever thou shalt bind on earth shall be bound in heaven" (Matthew 16:18–19). In the first he appears to be dividing the human authority of Caesar—that is, the Roman emperor—from the divine authority of God. Temporal authority therefore has a specific legitimacy, even though it is distinct from divine. As Monaghan notes, although various Christians might assert the superiority of spiritual rule, they did not deny secular rule entirely.[56] Yet the second passage, invoking the *keys* to heaven, again raises the question of the linkage, especially since it suggests that binding on earth is also on heaven. In addition it suggests the other terms by which spiritual and secular power would be known: eternal and temporal, or earthly. Spiritual power is over men's eternal souls, in preparation for heaven; while secular is over their temporal span on the earth. In what follows, the standard distinction will be between spiritual and temporal power, since it is the most often used, but the earthly nature of the latter will be returned to. The link between the two powers was also important: as Jesus said to Pilate in the Gospel of Saint John: "You would have no *power* at all against me, if it had not been *given* to you from *above*" (John 19:11).[57]

The spiritual and temporal could, of course, exist in a range of combinations; it was not simply one or the other in isolation. Francis de Meyronnes (1285–1328) outlined four possible societies: the pagan, entirely secular society, with no spiritual element; a civil community, temporal in essence but with a spiritual element, *temporalis per essentiam et spiritualis per participationem*; an ecclesiastical society, spiritual in essence but with a link to temporal concerns, *spiritualis quidem per essentiam, sed temporalis per participationem*; and, finally, the celestial kingdom of heaven.[58]

This blurring of spiritual and temporal concerns was particularly pronounced when it came to control over land. While the church was temporal ruler of some lands directly, especially in the Italian peninsula, the question that exercised popes was the extent to which their temporal power exceeded those areas. In 1076, in *Dictatus Papae*, Pope Gregory VII had attempted to reverse the idea that clergy should be subordinate to rulers, by suggesting that rulers were denied sacrality.[59] He used a passage from Matthew (16:19) as one of his foundational texts, claiming on textual authority that the pope alone "can use imperial insignia."[60] The *Dictatus Papae* became notorious, since it set out in clear and largely unequivocal form the powers of the papacy, although many of the things he claimed were lifted from the early Pseudo-Isidorian Decretals, as chapter 4 noted. The decretals were a compendium of texts, many of them forged, although it appears clear that neither Gregory nor his contemporaries knew this to be the case, and took them as accurate records.[61] The implications of the

denial of sacrality were important, since it provided a clear separation between religious and political rule. In a sense, that was more important
than the attempt at subordination: the latter could and would be long contested, but the idea that they inhabited different spheres would soon become uncontrollable. The division between the temporal, as secular rule,
and the spiritual, as religious rule, would be significant.[62]

Nonetheless, the church still felt a right to command the temporal
powers, seeing them as wholly subordinate to the church. The full articulation of this came somewhat later, but writers such as Bernard of Clairvaux urged the pope that "I believe it is time for both to be drawn in defence of the Eastern Church."[63] While there is no doubt that Bernard said
this, the debate is whether this was written in 1146 to justify the Second
Crusade or in 1150 to launch a new expedition.[64] Aquinas followed this up:
"The Church has either of the two swords. . . . It has the spiritual for the
sake of what needs to be done by its own hand. But it also has the temporal
sword at its order [*jussum*], which must be drawn at its command [*nutu*],
as Bernard said."[65] Nonetheless, Aquinas saw Jesus's instruction that the
sword should be returned to its sheath as the reason why priests themselves should not fight.[66] In *De consideratione*, Bernard suggested to the
pope that he had "no ordinary princely power [*Non mediocris iste prin-
cipatus*]: you must expel evil beasts from your boundaries [*terminus*] so
your flocks may be led to pasture in safety. Vanquish the wolves, but do
not dominate the sheep."[67] Bernard, then, is suggesting the powers due to
the church, but also their limits. He suggests a particular set of relations
rather than an absolute hierarchy.[68] As Carlyle and Carlyle note, though
these arguments were later used in more rigid and extreme form, this does
not mean that they were thought of in that way at the time.[69] As they suggest, at this time the idea that temporal authority was derived from spiritual authority was the view of a few writers in their private thoughts, not
church policy: "It must not be represented as having any official authority
in the Church, and as being generally or widely held. It received no sanction from any Council or from any Pope."[70]

At the beginning of the thirteenth century, though, Pope Innocent III
did begin to make just such claims.[71] Innocent, like Boniface VIII after
him, made use of the passage concerning the keys to heaven (Matthew
16:19),[72] and laid claim to temporal authority as the vicar of Christ.[73] The
pope had previously been described as the vicar of Saint Peter, but this
was taking things a stage further, setting him far above other men, and
crucially seeing him as a successor not merely to Peter, but to Christ himself.[74] On this understanding, the vicar of Christ was a direct mediator of

the rule of Christ, a view found in writers such as Aegidius Romanus.[75] As Kantorowicz notes, "Image of Christ" and "Vicar of Christ" were often used to designate medieval rulers. But while the "Image of Christ" was concerned with the being of the ruler, "Vicar of Christ" was concerned with the ruler's doing, his actions.[76] Yet for the story here, Innocent is most important for a ruling that he gave concerning a request to have bastard children legitimized.

This had previously been granted to the king, but now the request concerned a count's offspring, within the kingdom of France. In the bull *Per venerabilem*, Innocent dismissed the claim. The facts of the case are much less important than the general principle. Innocent importantly claims a degree of temporal power for himself as pope:

> Deducing from both the Old and the New Testaments that, not only in the patrimony of the Church where we wield full power in temporal affairs, but also in other regions [*regionibus*], after having examined certain cases [*certis causis inspectis*], we may exercise temporal jurisdiction occasionally [*causaliter*]. It is not that we want to prejudice the rights of anyone else or to usurp any power that is not ours.[77]

In some lands, therefore—the patrimony of the church—the pope is the temporal and spiritual ruler. In other regions, the pope has a degree of temporal rule, alongside spiritual authority, but this is neither absolute nor uncomplicated. Innocent notes the importance of rendering to God and to Caesar their distinct dues. Crucially he sees the king of France as having the same kinds of powers: in his kingdom, "the king therefore recognises no superior in temporal matters [*rex ipse superiorem in temporalibus minime recognoscat*]."[78] The implication here was that the emperor's jurisdiction did not exceed the bounds of the empire, even if the pope's spiritual authority—embodied in this ruling—did. There is a hierarchy of authority: the king must submit to the church, but the king was not below the emperor, so "had no secular superior."[79] What this meant is that in the case of the king's offspring, the pope was the right person to judge, because the king knew no temporal superior. In the case of the count's offspring, there was a temporal superior—namely, the king—so the pope dismissed the claim. The empire's limits therefore limit the power of the emperor. Yet, as Scott has noted, this assertion of the independent rule of the particular kingdoms, intended in part as a challenge to imperial universalism, also paved the way for a later challenge to the universality of the church itself.[80]

THE REDISCOVERY OF ARISTOTLE

After the death of Boethius, the lack of knowledge of Greek in the Roman West had meant that, for many centuries, key classical works were unavailable. Some had been translated into Latin in antiquity, and of these some remained in limited circulation. But many more were completely out of reach, including most of Aristotle's and Plato's major writings. Two key things enabled the recirculation of these ideas. The first was the pioneering work of Arab scholars, who translated Greek texts into their own language, which had a secondary effect in that this was better known in the West and therefore translations could be made through it. In Sicily, for instance, Greek, Latin, and Arabic were often used, and knowledge of at least two was essential for members of the royal court.[81] The second was the reemergence of Greek scholarship itself, and a greater circulation of ideas.[82] While some of this was benign, the sack of Constantinople in the Fourth Crusade in 1204 had led to the transfer of manuscripts to the West,[83] as had the fall of Toledo in 1085.[84]

The Arabic contribution to the gradual renewal of classical scholarship and learning more generally is a vast topic, and only a very few indications can be given here.[85] The work of Arabic scholars was particularly felt in mathematics, with the development of astronomical measurement and parallel developments in cartography.[86] As Kimble notes, the need to correctly orientate mosques was one reason for their work on this topic.[87] Pines has even argued that a semi-Newtonian distinction between absolute and relative space can be found in Abu Bakr al-Razi.[88] Euclid's *Elements* may also have been translated into Arabic in the ninth century. This text endured as the most significant work on mathematics throughout antiquity and the Middle Ages.[89] Yet, before the twelfth century, it was only known in Boethius's translation, of which only fragments survived past the eighth to ninth centuries.[90] But, in a rebirth of interest, it was translated three times from Arabic and once from Greek in the twelfth century alone; and the *Data* was translated once each from Arabic and Greek in the same period.[91] This practically founded the modern study of geometry,[92] with advances and developments following in its wake. As Nicholas notes, before this, geometry was mainly practical, and craftsmen in the late Middle Ages had "developed modes of measuring surface that were based on figures and used very little arithmetic."[93]

In philosophy too, the work of thinkers such as Avicenna (Ibn Sina) in the eleventh century and Averroes (Ibn Rushd) in the twelfth was significant. Averroes developed a reading of Plato's *Republic*, since Aristotle's

Politics was relatively unavailable in the Islamic world, including Spain, then under the control of the Moors.[94] Yet he knew the other works of Aristotle, including the *Rhetoric* and the *Poetics*, then considered to be part of the collection of logical works known as the *Organon*.[95] From these texts Averroes developed a stronger, more materialist reading of Aristotle, where the material world triumphed over the world of spirit.[96] His commentaries on a range of Greek texts, especially those of Aristotle, were particularly significant, and the mediating role played by Islamic scholars as a mode of access to pagan texts cannot be underestimated. The intellectual life of Islam far outstretched Latin Christianity at this time.[97]

The reading of Plato was significant in this period, though he had never disappeared from discussion. The Platonists of late antiquity had meant that many of his ideas were long in circulation, albeit in diluted and mutated form. Fundamental among them was Plotinus (ca. 205–70), who has been described as "a soul so close to his ancient master that in him Plato seemed to live again,"[98] yet his debt was probably as much to Aristotle.[99] His major work was the six-book *Enneads*.[100] Plotinus's disciple Porphyry (232–ca. 305) was more formally trained. His *Introduction* was long the first text for philosophy students. More generally, he developed ideas Plotinus took from Plato and turned them into a system.[101] The work of Plotinus and Porphyry was not one that made much claim to originality; rather, they saw themselves more simply as disciples of Plato. Historically, they have become known as Neoplatonists, but as Brown notes, they saw themselves as "'Platonists,' *Platonici*, pure and simple—that is, the direct heirs of Plato."[102] Yet, as Klibansky argues, to see it either as Platonism (a direct continuation) or as Neoplatonism (a development and departure) is to underestimate the complexity of the medieval reading of Plato.[103]

Yet one text above all was important: the *Timaeus*. It was this, rather than *Republic*, that held sway, as John of Salisbury's debt to it for his organic metaphor of the body politic illustrates. But its appeal was much wider than this. One of the most significant elements was, as Klibansky outlines, the "attempted synthesis of the religious teleological justification of the world and the rational exposition of creation," which provided the basis and starting point for attempts to outline "a scientific cosmology."[104] Other than *Timaeus*, the *Meno* and *Phaedo* became available in Henricus Aristippus's twelfth-century translations, and part of the *Parmenides* was included in Proclus's commentary, translated by William of Moerbeke in the thirteenth. Yet despite the changing availability of texts, Klibansky does not accept a clear break between classical, medieval, and Renaissance interpretations of Plato, and similarly cautions against thinking that every time Plato is mentioned there is a genuine interpretation taking place.[105]

In the thirteenth century, the ascent of Aristotle supplanted the twin figures of Plato and Augustine. This was both in terms of metaphysics and cosmology, on the one hand, and in politics, on the other. A graphic representation is Raphael's famous painting of the school of Athens, where Plato is holding a copy of the *Timaeus*, and looks to heaven, while Aristotle is holding a copy of the *Ethics*, and looks to earth. The materialist Aristotle proposed by Averroes is juxtaposed to the idealist Plato. Yet Kantorowicz has suggested that this shows that the either/or choice actually became a combination: Plato *and* Aristotle.[106]

The translations of this period were erratic and still somewhat piecemeal, and Aristotle's impact was similarly episodic and gradual, at least more so than the idea of an immediate impact.[107] Nonetheless, there was a remarkable concentration of effort. Three figures were key: Albert the Great (Albertus Magnus, ca. 1200–1280),[108] Herman the German (Hermannus Alemannus, ca. 1202–ca. 1272),[109] and William of Moerbeke (ca. 1215–ca. 1286).[110] Alongside them, and as an original thinker far surpassing them all, was Thomas Aquinas (ca. 1224–74). Aquinas, William, and Albert were all Dominicans. The context for their work was Gregory IX's bull *Parens scientiarum* (1231), which forbade the study of Aristotelian works in Paris.[111] This prohibition was against public or private instruction, but notably not the private reading, of his works on "natural philosophy." These included the *Metaphysics*, the *Physics*, and *De Anima*, but excluded the *Organon* and the *Ethics*.[112] Gregory's bull was not the first attempt to limit their impact. It followed the 1210 provincial synod of Sens, which had proclaimed that "neither the writings of Aristotle on natural philosophy nor their commentaries are to be read at Paris in public or private, and this we forbid under penalty of excommunication."[113] This had been renewed in 1215 by Robert de Curzon, cardinal legate in statutes for the university, for the school of the arts in Paris. Yet theologians and others outside could read him.[114] Gregory wanted the texts first to be "examined and purified from all suspicion of error"[115] before their contents were more widely circulated. It is important to recognize that this was not just because of the ideas being at odds with Christian thought, but also that there were issues with the translations and at times the Greek text was muddled with later commentaries from pagans or Arabs.[116] A three-man commission, headed by the Paris theologian William of Auxerre, and including Simon of Authie and Stephen of Provins, was set up to examine the texts, but William died that same year and the commission never reported. The works being taught outside was enough to mean the prohibition was largely ineffective.[117] While the church was thus reluctant to embrace Aristotle, it did acknowledge the useful material in his work. From the other side, as Ullmann has noted, there were distinct phases in the reception of Aristotle in this period: hostility, use of his work within a Christian system, and then reassertion of his distinctiveness as a thinker alone.[118] It is the second that is the crucial moment to be examined here, in the work of the greatest philosopher of this period, Thomas Aquinas.

It is worth noting the respective dates of the key political works. The *Nicomachean Ethics* was translated entire, though from the Arabic, around by 1240 by Herman.[119] He also translated the middle commentary of Averroes on this work alongside it. The first full translation direct from the Greek was made by Robert Grosseteste only a few years later. In the 1250s this was revised by William of Moerbeke.[120] The *Rhetoric* replicated the process of the *Ethics*—translated first from the Arabic by Herman in the 1250s and then by William in the 1260s.[121] Herman was also the translator of two of Aristotle's works on animals (the *Movement* and *Progress*), of the *Poetics*, and new or revised translations of works previously in circulation.[122] Many of these translations were made for the court of Emperor Frederick II.[123]

The *Politics* was thus one of the last of Aristotle's works to be known, and the William of Moerbecke translation of 1260 was direct from the Greek.[124] For the development of political theory, it is difficult to overestimate the impact that this text had. Coleman is skeptical of the idea of a Europe "taken by storm" by Aristotle's translation generally,[125] but there is some truth in Ullmann's suggestion that the *Politics* did cause a "conceptual revolution."[126] This was not a revolution simply in what people thought of politics, but in *how* they thought politics. The rediscovery changed both the language and the substance of political thought. This was not simply, as Ullmann argued, to articulate a more populist view against a hierarchical system, nor, as is now sometimes proposed, the reverse. Rather, it was because the clarity of Aristotelian thought made the articulation of different and more definite positions possible.[127] Yet, in addition, conceptualizing the *polis* in the very different circumstances of the time forced a rethinking of its very scope. Aristotle's terms allowed a new way to think the collective organizations of the time, but they also required his terms to be adapted to different situations. The kingdom, the realm, and even the nascent state could all be thought in Aristotelian categories.[128] It is worth underlining that these medieval commentators were more creative in their appropriation than mere repetition. If Aristotle's political or ethical texts were not able to provide guidance to contemporary concerns, ideas from his other texts were used to generate answers.[129]

The early translations were often poor or misleading, and were bound to replicate the errors, interpolations, and paraphrases of the Arabic sources.[130] The nonetheless remarkable efforts of the likes of Herman were swiftly superseded. The work of Albert and William was crucial in this—the intention being to produce workable, accurate renderings directly translated from the Greek texts. Aquinas used both these translations and earlier ones, although the belief that this was a coordinated division of labor where they provided the translations, recognizing his interpretative faculties, seems unlikely.[131] Yet even these translations were often excessively literal, sometimes merely transliteration.[132] This was an issue with particular relevance to the notion of politics, where the translations were constrained by the Latin terms available at the time. Greek terms were sometimes simply transliterated (*politeia* and *politicus*) and sometimes rendered simply as variants of *civis* (*civitas, civis,* and *civilis* for *polis, polites,* and *politicos*).[133] The crucial point is twofold: that *civis* and related terms now appeared to refer back to Greek antecedents;[134] and that a new vocabulary for thinking public affairs, citizenship, and communal life now existed. For the first time, these activities were seen as having a specifically political context in a way that endures today; and the very notion of the human as a "political animal" was similarly transformative of how people thought.[135] Yet the imprecisions concerning the specifically political geographical vocabulary continue.

THOMAS AQUINAS AND THE *CIVITAS*

The supreme achievement of Aquinas was the attempt to reconcile Aristotle with Christian thought, using the philosopher to provide a rational basis to divine revelation, recognizing that without the latter, Aristotle had taken unaided thought as far as it could go.[136] Whether his reading was true to Aristotle is much disputed, yet following his labors, by the fourteenth century the papacy came to accept much of what Aristotle could bring to Christian thought.[137] Hemming has described it as "the astonishing synthetic achievement and spectacular baptism of many of the texts of Aristotle."[138] Yet Aquinas was not merely an Aristotelian, and his thought cannot simply be described as "Christian Aristotelianism."[139] Primarily he was a theologian, working within the framework of Scholasticism. In its specific sense, Scholasticism is a means of articulating philosophical truths through questions, arguments, and conclusions—*quaestio, disputatio,* and *sententia.*[140] Other elements work their way through his thought, including Stoicism and Neoplatonism.[141] Nonetheless, the affinity between Aquinas and Aristotle is remarkable. Initially looking to interpret the texts that his Dominican brothers were translating, he was able to build on them; correct inadequacies of transmission or translation; supplement them with other material, theological or philosophical; and clarify their import in a very different context to that in which they were written.[142] Scholastic political theory can be found in three distinct types of work: "(1) commentaries on the *Nicomachean Ethics* and on the *Politics* . . . [;] (2) didactic literature, addressed to a specific ruler; (3) the discussions which occur in theological works."[143] Aquinas wrote all three types of work.

Nonetheless, attempts to find the political theory in Aquinas have been difficult. While he did indeed write commentaries on Aristotle's *Ethics* and *Politics,* they are incomplete and doubts are raised about the authenticity of the parts that remain.[144] With the *Ethics,* he worked only up to book 3, chapter 8; with the *Politics,* to book 3, chapter 6—and both were completed by Peter of Auvergne (d. 1304). Catto contends that while the *Politics* provides some key formulations for Aquinas, he made "little use of its substance" and that the *Ethics* remained the more fundamental text.[145] The treatise *De Regno* or *De Regimine Principum,* dedicated to the king of Cyprus, also has a disputed authorship. Most scholars now suggest that the first book, and a small part of the second, can be attributed to Aquinas, the remainder to Ptolemy of Lucca (d. 1326). This text is of fundamental importance, however, regardless of its actual authors, since for years it was taken to be a Thomist exposition of politics. It is in his theological works that some of the most important discussions are found, notably the discussions of law and justice in the *Summa Theologiae.* Indeed, it was the monumental efforts that went into the *Summa Theologiae* that led to the abandonment of these other projects.

The commentaries do allow some fundamental insights. Following Aristotle, Aquinas sees politics as a branch of applied ethics, suggesting that "it is evident that political science [*politicam scientiam*], which is concerned with the direction of human beings, is included in the sciences of action which are moral sciences, and not in the sciences of making, which are mechanical skills."[146] He endorses Aristotle's contention that "the philosophy that deals with human affairs is completed in politics"[147] because of the threefold structure of ethics:

Moral philosophy should be divided into three parts. The first studies men as individuals as ordered to a certain end: this is called individual (monastic) [*monastica*]. The second is concerned with the domestic community [*multitudinis domesticae*] and is called domestic
[*oeconomica*]; and the third studies the action of the civil community
[*multitudinis civilis*] and is called political [*politica*].[148]

What this means is that, for Aquinas, the political and the civil are
closely bound together. In part this is a product of translation, where Greek
terms such as *polis* and *politeia* are sometimes translated as *civis* and *civitas*, and sometimes simply transliterated. What this means is twofold:
the broadening of Latin vocabulary to include new, but closely cognate
terms; and the implied Aristotelian echo whenever terms such as *civis* are
used. The relation can be seen in passages such as this: "It is necessary,
then, for a complete philosophy to institute a discipline which will study
the city [*civitate*]: and such a discipline is called politics or civil science
[*politica . . . idest civilis scientia*]."[149]

The famous Aristotelian descriptions of the human as the *zoon logon
echon*, the *zoon politikon*, and *oikonomikon zoon*[150] are picked up by
Aquinas in a range of ways. William of Moerbecke uses the phrase *animal
civile*, which is used repeatedly by Aquinas in his commentary on this
translation. In discussing the key passage of the *Politics* on speech and political community, he declares that because human communication "produces the household and the political community [*domum et civitatem*],
so human beings are naturally domestic and political animals [*animal do-
mesticum et civile*]."[151] As he outlines Aristotle's argument:

> Then he shows that the association of the civil community is natural,
> regarding which he does three things. First he shows that the *civitas* is
> natural. Secondly that human beings are by nature political animals
> [*animal civile*]. Thirdly, he shows what is prior by nature, whether it is
> the individual human, the household, or the *civitas*.[152]

Aristotle's *zoon politikon* corresponds therefore to the *animal civile*,
but also to the *animale sociale* and *animale politicum*, used in the *Summa
Theologiae* and other works.[153] The use of plural adjectives to translate Aristotle's one is important, and the human animal is thus domestic, social,
civil, and political. Yet there is also a stress on the notion of *humanitas*,
the essence of the human, which, while not original to Aquinas, is important.[154] Indeed, Remigio Dei Girolami, pupil of Aquinas and probable

teacher of Dante, claimed that "if you are not a citizen [*civis*] you are not a man, because 'man is naturally a civil animal [*animal civile*].'"[155] Taken together they show that, for Aquinas, the political action of humans is part of their nature, but also that politics is rooted in social life rather than imposed by individual will.[156] This understanding of the political community as rooted in humanity's essential nature was a radical break from previous theology, which had followed the church fathers in seeing political authority as a necessary evil because of humanity's inherently corrupt nature, a means of dealing with sin.[157] This meant that political institutions, or what would become thought of as the state, were natural rather than needing external justification.[158]

Social, civil, and political, certainly, but Aquinas is at pains to stress that "a human being is not only a citizen of the earthly city [*civis terrenae civitatis*] but also a member in the heavenly city [*civitatis caelestis*] of Jerusalem, which is governed by the Lord and whose citizens comprise the angels and all the saints, whether they reign in glory and are at peace in their homeland, or are still pilgrims on earth." In this, like Augustine, he is trading upon Paul's letter to the *Ephesians*, where he argues that "you are citizens with the saints and members of God's household" (2:19). The hierarchy is clear: while being a "citizen of an earthly city" is possible through being human alone, to be a citizen of the other city must be through God's divine power, or divine grace.[159] Thus, while Aquinas does not discuss the Holy Roman Empire and largely leaves the direct political power of the pope or the church unmentioned,[160] this does not mean that the spiritual dimension was neglected. Rather, Aquinas's merit lies in the fusion of classical terminology with contemporary religious ends.

What, then, was the political community for Aquinas? At times he renders the Greek *polis* as a *politiae*, a "polity,"[161] but also as a *civitas*.[162] He suggests that "*civitas* is nothing but the congregation of men [*congregatio hominum*],"[163] but its more common meaning was a city. Following Aristotle, it has to be larger than simply a family or a small group, in terms of being self-sufficient. There is thus a qualitative element in the construction of a perfect community, the end to which other natural associations tend.[164] As he says in his commentary on the *Politics*, "Among such associations there are different grades and orders, the highest being the political community [*communitas civitatis*], which is so arranged as to satisfy all the needs of human life; and which is, in consequence, the most perfect."[165] He makes a similar point in the *Summa Theologiae*: "Just as a human is a member of a family [or household, *domus*], so a household forms part of a *civitas*: but a *civitas* is a perfect community, as is shown in the first book of the *Politics*."[166] His use of *communitas multitudine* is to render Aristotle's *koinonia*; but he also used a notion of the *Universitas* and the *societas*, which was inconsistently distinguished from *communitas*.[167] On one level, then, the *polis* is equated with the *civitas*, which might imply the city, but it also extends in size, to the *provincia* or even the *regnum*, the realm or kingdom.[168] Aquinas does not quite make the final step, of seeing a universal *Imperium* as the outcome, but it is an understandable consequence. For Aristotle, such political entities would have been beyond the viable size, but the experience of the Roman Empire and the more recent growth of kingdoms had demonstrated something rather different to those following him by so many centuries. Aquinas was, of course, thinking through Aristotle's terms in the light of the current realities in Italy, France, and the empire.[169]

The constitution of a polity through its people is crucial to Aquinas's determination. It is possible for a city to be divided, with some people in one location (*locus*) and others in another.[170] Aquinas thus underplays geography, in that this is not determinant of the polity, although one key issue, just as it was for the Greeks, was what happened to the city if the people moved or left. Further, Aquinas recognizes that Aristotle "says that a *respublica* is simply the organization of a *civitatis* with respect to all the rules that are found in it but principally with respect to the highest rule [*principatum*], which governs all the others."[171] Thus, the city is its land, its people, and its laws.

The importance of the law cannot be underestimated. Aquinas suggests that "positive law may be divided into the law of nations [*ius gentium*] and civil law [*ius civile*]." He argues that the former are the immediate conclusions of natural law, giving "buying and selling and other similar activities which are necessary to social intercourse," based on the fundamental definition of humans as social animals. "Those norms which derive from the natural law as particular applications, on the other hand, make up the civil law which any *civitas* determines according to its particular requirements."[172] What is significant here is that political power is implicitly understood as the authority of law, and its foundation is to be found in social organization or community.[173] The divine nature of this natural law has profound consequences. It effectively substitutes for any mention of a universal empire. As d'Entrèves suggests, the common shared sense of this law is what provides the basis for any particular legal systems that may arise from it. Aquinas thus links the *unus populus Christianus*, a unified

Christian community of faith, uniting and transcending all individual kingdoms and political units, with the whole church as the *unus corpus mysticum*, the one mystical body.[174] As Ullmann puts it, the community is a *"corpus politicum et morale*, a body politic with moral ends which took into account the social habits and customs of its citizens. The Church on the other hand was a mystical body (*corpus mysticum*)."[175] Every human was, either in actuality or in potentiality, a member of this multitude.[176] The universality thus comes, not from political organization, but from a shared humanity and divinity. Nonetheless, just as each individual church has one head, the bishop, so too does the church as whole have one head, one vicar, the pope.[177] Because Christ conferred this position on Peter, and through him to his successors, the successor of Peter the pope is thus the vicar of Christ and the "first and greatest of all bishops," with a primacy over the universal church.[178] Therefore, as Bigongiari suggests, "the *ecclesia* includes the *res publica*."[179] In his commentary on Matthew's Gospel, Aquinas offers a gloss on the famous passage of the keys in Matthew (16:19), where Peter is given universal power over the heavens (*in caelis*), and the later passage (18:18) where he is given power in heaven (*in caelo*). In the former, Aquinas suggests, the apostles are given power over churches, but in the latter Peter is given power over the universal church.[180]

It is the distinction and relation between the two that is crucial, both for Aquinas, but also for the political struggles and thought that came in his wake. Aquinas himself is clear on the relation between the two kinds of power:

> Both the spiritual and the secular powers are derived from the divine power, so the secular power is only subject to spiritual power to the extent ordered by God, that is, in matters that relate to the salvation of the soul. In such matters the spiritual power is to be obeyed before the secular. In those matters that pertain to the civil good the secular power is to be obeyed rather than the spiritual, according to Matthew 22:21: "Render to Caesar the things that are Caesar's." Unless perhaps the spiritual and secular powers are conjoined, as in the Pope, who holds the summit of both powers: that is, the spiritual and secular, by the mediation of he who is both priest and king in eternity.[181]
>
> The secular power is subject to the spiritual power as the body is subject to the soul. Therefore the judgment is not usurped if a spiritual authority enters into temporal matters on points where the secular power is subordinate to it or which it has relinquished.[182]

Aquinas thus considers that Peter was given universal power by Christ, and as Shogimen suggests, "The 'universality' of Peter's power has, in Aquinas, a territorial meaning."[183]

Nonetheless, the notion of a *territorium* is used very rarely by Aquinas. In his discussion of tithes, he notes that "common law requires that the parish church receive titles of lands in its *territorio*" but that individual clerics and churches sometimes have lands "within the *territorio* of another church."[184] In terms of the tithe of a flock of sheep, he considers that it is due to the church in whose *territorio* is the pasture, not the sheepfold.[185] There is also a discussion of the preservation of fruit trees for the use of the people for whom "the city and its *territorium* were to be assigned" as the spoils of war.[186] In each of these instances, it is clear that the meaning can only be the lands of, or belonging to, a city or church, a possession of these communities, rather than "territory" in the sense of an object of rule. Nor does Aquinas use other terms or hint at practices close to how we would today understand territory. It is the same with the notion of *spatium*. Relatively rarely used by Aquinas, it generally has its earlier meaning of an extent, stretch, or distance rather than itself being a container.[187] Nonetheless, in his discussion of Aristotle's *Physics*, he provides a hint of how these ideas will be developed:

> The ancients thought that place is the extent [*spatium*] which is between the termini of a container which has the dimensions of length, breadth and depth [*quod est inter terminus rei continentis, quod quidem habet dimensiones longitudinis, latitudinis et profunditatis*]. Nevertheless this *spatium* does not seem to be identical with any sensible body. For when different sensible bodies come and go, this *spatium* remains the same. According to this, therefore, it follows that place is separated dimensions.[188]

In the *Summa Theologiæ*, again drawing on Aristotle's *Physics*, he defines a void (*vacui*) as "a *spatium* capable of, yet not, holding a body. Our contention is that before the world existed there was neither place nor extent [*locum aut spatium*]."[189]

In *De Regno* or *De Regimine Principum*, the text of disputed authorship, the emphasis is much more on the narrowly political.[190] The most plausible account is that Aquinas did indeed have a hand in its early parts, which he abandoned in 1267, when its intended recipient, Hugh II of Cyprus, died. Ptolemy completed the study early in the fourteenth century.

It is certainly the work of two authors, with profound differences between the parts, although there are questions as to how much interference is present in the first book and the part of second that is commonly attributed to Aquinas. Here it is treated as a work that is of interest regardless of its author rather than as a definitive text of Aquinas. Despite being dedicated to a king, this text criticizes kingship as tyranny, and sees the Roman Republic, rather than the empire, as the ideal form of government.[191] While Aristotle's *Politics* provides much of the conceptual terminology and structure, Cicero and Augustine are also very important. Many of the claims bear close relation to ones that can be found in other texts by Aquinas, such as the fundamental determination that "human beings are by nature political and social animals [*animal sociabile et politicum*], living in groups [*multitudine*], as natural necessity requires."[192] It is the same with the hierarchy between spiritual and secular power:

> So that spiritual and earthly things may be kept distinct, the ministry of this kingdom was committed not to earthly kings but to priests, and especially to the Highest Priest, the successor of Peter, the Vicar of Christ, the Roman Pontiff, to whom all kings of Christian peoples should be subject, as to Lord Jesus Christ himself. For those who are responsible for antecedent ends should be subject to and directed by the one who is responsible for the ultimate end.[193]

Even more fundamentally, the text declares that under the new law, the law of Christ, "kings should be subject to priests."[194] Nonetheless, it recognizes that kings do have some degree of autonomy, and it is geographical determination. It quotes the words of Solomon from *Ecclesiastes* (5:8): "The king commands [*imperat*] over all the lands [*universæ terræ*] which are subject to him."[195] The geography of the ideal kingdom is the topic for discussion as Ptolemy takes over the work. The writers are discussing the situation of the city or kingdom at the moment of its founding, touching on issues such as locating it in a temperate region;[196] where there is healthy air;[197] and with a plentiful food supply, other amenities, natural resources, and artificial ones such as gold and silver.[198] Ptolemy goes further than Aquinas in thinking the materiality of the political community, drawing upon organic models:

> Further, there is an analogy between any kingdom, city, camp, or association [*regnum sive civitas sive castrum sive quodcumque collegium*] and the human body, as Aristotle and the *Policraticus* tells us. In the

latter, the common store is compared to the king's stomach, so that just as food is received by the stomach and distributed to the members, so also the king's store is filled with a treasury of money, which is shared and distributed to pay for the things that are necessary for the subjects and kingdom.[199]

Aquinas's writings seem to confirm the view that his political work is merely part of a larger project of morality. Ptolemy, though, was a more specifically political writer, and is the likely author of another important text from the late thirteenth century, probably written around 1281, *Determinatio Compendiosa de Jurisdictione Imperii* (A short determination of the jurisdiction of the Roman Empire). Here he makes the crucial claims that "while in temporal matters the Emperor has no superior in the person of any temporal lord, yet the validity of his jurisdiction comes from the Pope's jurisdiction as from a fountain of authority [*dominii*], from which as in the place of God all jurisdiction arises."[200] While this work was concerned with the power of individual cities in relation to the empire, it clearly clarifies the relation of the pope to the emperor. For its author, spiritual power ultimately exceeds temporal. This issue of the relation between kings, the emperor, and the pope will be the theme of the next chapter.

CHAPTER SIX

Challenges to the Papacy

UNAM SANCTUM: BONIFACE VIII AND PHILIP THE FAIR

At the end of the thirteenth century, the relation between the church and a king became the dominant issue in European politics. It pitted the powerful Pope Boniface VIII against Philip the Fair of France, and their conflict was both a fundamental moment in the church's relations with secular rule and the source of a wide range of important political theory—theory that would long outlast the particular context of its framing.[1] Boniface had come to the papacy in unusual fashion, following the abdication of Pope Celestine V in 1294, only six months after he had taken office. The resignation of a pope—even one as utterly unworldly as Celestine, formerly known as Peter the Hermit—was unprecedented, and questions were raised as to whether he was even allowed to do this.[2] Boniface thus became pope with doubts over his legitimacy, and regarding his role in Celestine's resignation, but nonetheless he has been described as the "last of the great medieval popes,"[3] and he was certainly one of the most important. While Boniface was a forthright defender of the power of the papacy, ultimately his intransigence created the cause, and his arguments paradoxically opened the way, for a fundamental break in the relation between spiritual and temporal power.

Rumors began to circulate that Boniface had claimed *"Ego sum Caesar, ego sum imperator"* while seated on a throne, wearing Constantine's diadem, and holding the keys to heaven and the sword of temporal power.[4] Hostiensis in the mid-thirteenth century had proclaimed that the church had one head, two swords, and noted that Jesus "did not say 'key' but 'keys' so that there were two, one which opens and closes, binds and looses in spiritual affairs, the other which is used in temporal affairs."[5] Thus,

Boniface, the one head, laid claim to ultimate power in both spiritual and temporal affairs. The long-standing issues of whether secular rulers could tax the clergy without permission from Rome and who had jurisdiction over crimes committed by clergy were the breaking point. Boniface was clear that the answer to the first was no, and to the second that it was the church. Philip fundamentally disagreed, with one of his supporters claiming in the *Quaestio de Potestate Papae* that France had no superior in temporal authority, "neither Emperor nor Pope."[6] This text also made the explicit equation of "the king in his kingdom and the emperor in the empire [*rex in regno et imperator in imperio*]" as effectively equivalent temporal lords.[7] And in retort to Boniface's claim that "we have universal power," Philip's emissary Pierre de Flotte is said to have retorted: "Certainly lord, but yours is verbal [*verbalis*] whilst ours is real [*realis*]."[8] Boniface sought to rule on the argument by issuing the patronizing bull *Ausculta fili* (Listen, my son) in late 1301,[9] which Philip promptly burned. Philip also issued a false bull, *Deum time* (Fear God), which claimed that spiritual and temporal authority rested with the pope, and those thinking otherwise were heretics.[10] Boniface was accused of a range of sins, including an array of sexual misdeeds; he was also alleged to have said that he "would rather be a dog than a Frenchman." This was not merely offensive but heretical, claimed the French, because Boniface was implying the French had no souls.[11] This was the context into which Boniface's notorious bull *Unam sanctum* (One holy) was issued.

Unam sanctum was not particularly new, compiled from a range of other sources, but its mode of expression made its claims unambiguous. The key argument was that temporal authority must be subject to spiritual power, because for salvation all must be subject to the Roman pontiff.[12] There must be one church, which is one body, with one head. Boniface here echoes the Nicene Creed: "one, holy, catholic and apostolic Church."[13] He notes that when the apostles told Christ there were two swords, he said that was sufficient, enough, not too many. The temporal sword is Peter's too, even though Christ tells him to replace it in his scabbard. Boniface thus claims that "both, therefore, are in the power of the Church, that is to say, the spiritual and the material sword, but the former is to be administered by the Church but the latter for the Church; the former in the hands of the priest; the latter by the hands of kings and soldiers, but at the will and sufferance of the priest." Then comes the crucial passage:

> However, one sword ought to be subordinated to the other and temporal authority, subjected to spiritual power. For since the Apostle said:

"There is no power except from God and the things that are, are ordained of God" [Romans 13:1–2], but they would not be ordained if one
sword were not subordinated to the other and, being inferior, were not
led to the highest things by the other.

Boniface contends that this is obvious. "We must recognize the more
clearly that spiritual power surpasses in dignity and in nobility any temporal power whatever, as spiritual things surpass the temporal." This leads
to the clear hierarchical nature, in terms of who is judged by whom: "If the
terrestrial power [terrena potestas] err, it will be judged by the spiritual
power; but if a minor spiritual power err, it will be judged by a superior
spiritual power; but if the highest power of all err, it can be judged only by
God, and not by man." Thus, terrestrial powers are subordinate to spiritual powers, who themselves are in a hierarchy, but the only sanction of
the supreme spiritual power—that is, the pope—is God. Boniface is saying
that nobody but God can judge him, whereas he can judge Philip.

Boniface then turns to the power of the keys, given to Peter, which bind
both on earth and in heaven. He closes: "Therefore whoever resists this
power thus ordained by God, resists the ordinance of God [Romans 13:2],
unless like the Manicheans he imagines two beginnings, which is false
and judged by us heretical, since according to the testimony of Moses, it
is not in the beginnings but in the beginning that God created heaven and
earth [Genesis 1:1]. Furthermore, we declare, we proclaim, we define that
it is absolutely necessary for salvation that every human creature be subject to the Roman Pontiff." Boniface thus accomplishes what Philip had
accused him of in the false bull: those who contest this model are heretics.
Ullmann has therefore claimed that Unam sanctum is the "magnificent
swan song" of the papal-hierocratic doctrine.[14]

The details of what followed are relatively well known. Philip and his
minions attempted to get Boniface to face a general church council to rule
on the issues at stake, and various salacious rumors spread about his conduct, but the plan backfired. Boniface was captured and mistreated by the
king's soldiers in 1303, and though he was released and returned to Rome,
he died shortly afterward. Boniface VIII's death can be seen as the end in
practical terms of papal claims for "universal temporal authority."[15]

This dispute is not of interest just for the political events but for the
fundamentally important works of political theory that were generated at
the time. The Quaestio de Potestate Papae offered a fundamental challenge
to the standard interpretation of the Donation of Constantine. It claimed
that it was "not sufficient to prove that the pope is lord of all Christians in

temporal affairs [*temporalibus*]." This is because the Eastern Empire was not donated, and there were indubitably Christians in the East. Similarly, those Western kingdoms that are not part of the empire are equally not under the pope in temporal matters: "The kingdom of France is not subject to the Roman empire. On the contrary, there are clearly defined frontiers [*limites*] by which the kingdom is separated [*dividuntur*] from the empire, and these have existed for as long as anyone can remember."[16]

The hint to look to the limits of the extent of power is the fundamental question of this period. It is worth noting here an intriguing text entitled *Disputatio inter clericum et militem* (A dispute between a priest and a knight), probably written around 1297 at the time of the dispute and promoting the position of Philip.[17] Much of the text is an unremarkable rehearsal of the different positions, but it is notable for including an important gloss on the notion of the *rex in regno* idea:

> Just as everything within the boundaries of the Empire [*infra terminos imperii*] is known to be subject to the Empire so everything within the boundaries of the realm is subject to the king [*infra terminos regni regno*].[18]

As the editor notes, the text was not discovered by other writers until about seventy years after it was written.[19] Thus, even though prescient in setting a spatial determination to secular rule, it was all but eclipsed by later texts.

The two fundamental thinkers of this conflict were John of Paris (ca. 1240–1306) and Aegidius Romanus (ca. 1247–1316). The former was an advocate of the king's position, specifically tasked with contesting the universal aspirations of *Unam sanctum*,[20] while the latter moved from a position of adviser to the king to papal apologist. Their mature arguments stand for the two key positions in the debate.[21] John of Paris, or Jean Quidort, articulated his main arguments in *De potestate Regia et Papali*, (On royal and papal power).[22] He began with Aristotle's definition of man as a political or civil animal, adding the notion of being a social animal,[23] and then followed it with Cicero's suggestion that humans come together into a community ruled by one.[24] He suggested that the *regnum* alone was the appropriate size—that the household or village was insufficient.[25] *Regnum*—usually translated by Monahan as "kingship" and occasionally as "kingdom"—is understood as "rule over a community [*regimen multitudinis*] perfectly ordered to the common good by one person."[26]

The limits to power extend upwards too. John thought that it only made sense to talk of universal power in a spiritual sense. It is possible to have a universal church with supreme authority, but impossible to conceive of universal or world government. Temporal power requires the use of the sword, and since this cannot be done over large distances, there are limits to temporal power.

> One man is not enough to rule the entire world in temporal matters, although one man is adequate to rule in spiritual matters. For spiritual power can easily exercise its censure, which is verbal, on all persons near and far; but the secular power cannot so easily apply its sword, which is manual, to persons who are distant. It is easier to extend a word than a hand.[27]

As Monaghan puts it, this implies that "efficient exercise of temporal authority entails a feature of geographical limit."[28] John is therefore endorsing an idea of a plurality of temporal rulers, each in a sense equal and distinct, an idea that can be seen in germ in both Aquinas,[29] and, as John notes, in Augustine's *City of God*, where there is the suggestion that there would have been a plurality of kingdoms if they had not provoked war with Rome.[30] John's writings are therefore full of references to the lands of the pope, to the *terram Sancti Petri, terra propria, terra non subiecta sibi*—that is, the lands of Saint Peter, his own lands, and lands not subject to him.[31]

John's role in these disputes forced him to confront many of the key themes of this period. He suggests, for instance, that the idea of the two swords "is only a kind of allegorical interpretation from which no valid argument can be made."[32] He quotes Paul's letter to the Ephesians: "Take the breastplate [*loricum*] of justice . . . and the sword of the spirit, which is the word of God" (6:14–17). Thus, the two swords are either the Old and New Testaments or "the sword of the word and the sword of present persecution." Similarly, he notes Jesus's words in the Gospel of Matthew: "I did not come to bring peace to the world, but a sword" (10:34–35). John pointedly underlines that he said "sword," not "swords."[33] Yet it should be clear from his work that he did not straightforwardly take the royal line. Rather, he was interested in mediating between the different views. He recognized two powers but thought they were separate, working in different registers. He therefore does not deny Christ's power, but argues that because it is not of the temporal world, his kingship is spiritual.[34]

John of Paris anticipates an argument that will be found more fully worked through in Marsilius of Padua: that the contract between ruler and ruled was one that could be canceled if the ruler failed or was unjust. John also engages in the issue of property, which will assume center stage in William of Ockham's work. John claims that "to have proprietary right and ownership over property is not the same as having jurisdiction over it. . . . Princes have the power of judging even though they do not have ownership of the property in the question."[35] He equally felt that the church should neither be excluded from all ownership of property—as the Franciscans contended—nor lay claims to absolute ownership. It undoubtedly owned some as the result of individual gifts, but it, as with temporal powers, did not own everything by virtue of its universal jurisdiction.[36] The question of property and claims to power was, of course, fundamental to the validity of the Donation of Constantine. John pointedly notes that "it also seems surprising that the Emperor Constantine is said to have given political authority over Italy and the entire temporal jurisdiction to the Church, and that the Church received this is as gift, if it also held this *de iure*."[37] John further notes that Constantine gave the pope not the Western Empire but particular places (including Italy, but not France) and transferred the empire to the east.[38] Thus, neither the empire itself, nor even its western half, belongs to, and is therefore subject to, the papacy.

Similar arguments were made in an anonymous tract, also written around 1302. This suggests that man is divided into two parts—body and soul. The body concerns temporal things, and the king has "power over bodies and over the bodily things which have to do with bodily life"; the soul concerns spiritual things, and "the pontiffs have spiritual jurisdiction in relation to those things which pertain to the government and health of souls."[39] This separation of powers is related to the idea of the two swords. The writer agrees, but suggests that the "princes use the material sword . . . [and] the apostles used the spiritual sword. . . . These two powers are distinct, therefore, and must not trouble one another."[40] It is a clear articulation of separate spheres of operation: "Pontifical authority and royal dignity are two distinct powers, divinely instituted . . . clearly divided and separate, so that the one may not usurp the jurisdiction of the other."[41]

Almost directly contrary arguments, yet often using the same sources and terminology, can be found in Aegidius Romanus. Aegidius, or Giles of Rome, had been a pupil of Thomas Aquinas and Ptolemy of Lucca, and in his earlier work *De regimine principium* had made similar arguments to them.[42] *De regimine principium* was in some senses the exemplar of the medieval *speculum principum*—the mirror for princes, a guidebook for rulers.[43] Yet what is interesting and unusual about this text is that it has no discussion of the role of the church in the kingdom.[44] Dedicated to

Philip the Fair, just before he became king,[45] it was translated into Middle English by John Trevisa in the late fourteenth century, and was used both by kings and those opposing them in disputes over royal power.[46] Yet in *De ecclesiastica potentia* he offers the doctrinal form of the argument made in more polemical form in *Unam sanctum*,[47] and has been described as "perhaps the most extreme papalist of all."[48]

This was not Aegidius's first venture into ecclesiastical politics. In 1297 he had written a short text justifying the resignation of Celestine V and the accession of Boniface VIII, entitled *De renuntiatione papae*.[49] Such a move undoubtedly helped in his position within papal circles and his probable role in drafting *Unam sanctum*.[50] Whether the bull preceded or was consequent to *De ecclesiastica potentia* is uncertain, but the arguments of these two texts are closely related, and the claims of the shorter text are certainly developed in *De ecclesiastica potentia*. Aegidius provides a discussion of the way in which different powers are judged, suggesting that inferior temporal lords are judged by higher temporal ones and that ultimately by the spiritual: "If the earthly power goes astray, therefore, it will be judged by the spiritual power as by its superior. But if the spiritual power, and especially the power of the Supreme Pontiff, goes astray, it will be able to be judged by God alone."[51]

De ecclesiastica potentia suggests that just as the soul controls the body,[52] so too must the church control the faithful. In this, Aegidius agrees with Henry of Cremona, who suggested that because the pope has power over souls, and the soul over the body, ultimately the body is under the power of the pope, which means that the pope does have ultimate temporal power.[53] Secular rulers thus hold power only in trust, and it can be recalled. Hostiensis made a related point, commenting on the description in Gregory IX's *Decretales* of the pope as "*dominus temporalis,*" suggesting that this should be understood in the specific sense of having the power to remove corrupt or unjust rulers for the welfare of Christendom as a whole.[54]

Aegidius makes a related point: "There is no lordship with justice [*dominium cum iustica*], either of temporal things or . . . of lay persons except under the Church and instituted through the Church."[55] This gives rise to a particularly hierarchical form of power. Aegidius continually stresses the superior lordship of the church over all temporal things. In the days of Adam, mutual covenant was all that was needed, not law, for things such as "the partition and division of the earth."[56] Now, though, Aegidius argues that the church was established over peoples or nations (*gentes*) and kingdoms. For this he draws upon Jeremiah 1:10: "I have today placed you above nations and kingdoms, to uproot and destroy and disperse and scatter, to build and plant." He sees an instance of this in the transfer of the empire from East to West, suggesting that the temporal was subordinate to the spiritual.[57] Aegidius therefore proclaims that "both the Church and the faithful have lordship of this [temporal] kind: but the Church has universal and superior lordship, whereas the faithful have particular and inferior lordship."[58]

Temporal power, for Aegidius, is entirely subordinate but has full integrity.[59] Even though he stresses the importance of predominance of papal authority as fully superior, and suggests that temporal power derives from spiritual power,[60] he nonetheless clearly articulates the scope of temporal power. "earthly and temporal power [*terrana itaque et temporalis potestas*], then, as earthly (that is, inasmuch as it receives the fruits of the earth) and as temporal (that is, inasmuch as it possesses temporal goods)."[61] This is important to note, even as he provides a justification for papal claims of a plenitude of power,[62] and suggests that "the priestly [*sacerdotalis*] power precedes royal and earthly power in dignity and nobility."[63] In making these claims, he continually refers to the argument of the two swords. He provides many justifications of the importance of this idea and its structuring relations.[64]

> What, therefore, does this mean—that while there were two swords, the one was drawn and other remained in its sheath—if not that the Church has two swords: the spiritual to use, which is represented by the drawn sword, and the material not to use, but to command, which is represented by the undrawn sword?[65]

Aegidius draws upon Romans (13:3–4)—"For he who bears the sword is a minister of God, an avenger to execute wrath upon him who does evil. For princes are not a terror to good works, but to evil"—in order to show how temporal power can be used as an arm of the church.[66] It is worth noting that in the coronation of the emperor, the pope gave him an unsheathed sword from the altar, "signifying by the sword the care of the whole empire."[67] The relation between these two swords, these two powers, is also crucial. The relation is hierarchical: "The temporal sword, therefore, as inferior, must be led by the spiritual as by a superior, and the one must be ordained under the other as inferior under superior."[68] He is dismissive of arguments about the separation of temporal from spiritual power: "The temporal sword must be under the spiritual, kingdoms must be under the Vicar of Christ, and, *de iure*, even though some may act contrary to this *de facto*, the Vicar of Christ must have *dominium* over temporals themselves."[69]

Similar arguments were made in the work of James of Viterbo. Like Aegidius, James sees the Donation of Constantine as something of a liability, since the idea that the church had received its temporal jurisdiction as a gift was the opposite of that which they wanted to argue.[70] James therefore tries to argue that Constantine was merely confirming the previous situation in divine law. James is interesting for two key points. The first is the attempt to provide an extensive basis for papal involvement in temporal matters. Popes had long claimed ultimate jurisdiction over temporal matters through the notion of *ratione peccati*, reason of sin. Because the pope had jurisdiction over sins, then anything that involved sin immediately became a concern of the popes. In his decree *Novit Ille* from 1204, Innocent III had justified intervention on three grounds: sin; to keep the peace; and oath breaking, because oaths too were the preserve of the pope. Clearly, all three, especially sin, can be very broadly defined, and by the pope himself, thus providing a widespread basis for intervention.[71] James tries to articulate the scope of spiritual power in a hierarchical and extensive way. He attempts to define holy, *sanctus*, through its Greek term *hagios*, which he claims "means 'not earthly [*sine terra*],'" although this etymology is almost certainly erroneous.[72] And, like Aegidius, he suggests that "one temporal power is governed by another temporal power, and the temporal by the spiritual, and one spiritual power by another, and every spiritual power by the one primary one, the Supreme Pontiff."[73]

The second is the way in which James articulates the spatial extent of the church. For James, Christ is "the king and head and the Church, His kingdom and body."[74] He goes on to argue that the church is a kingdom, but also "a kind of community [*communitas*], because she is a congregation or association [*adunatio*] or convocation of many believers [*convocation multorum, fidelium*]." Following Augustine, he notes that such a community can be of three kinds: household, city, and kingdom (*domum, civitatem et regnum*). He is at pains to underline that each of these is defined by the people who make it up, rather than the dwelling place, the stones, or the land of the kingdom. "Though an extent of land [*terrestre spacium*] and collection of many towns [*multarum urbium collectio*] may sometimes be called a kingdom, a kingdom properly so called is nonetheless an association of peoples and races [*adunatio plurium populorum et gentium*]: so named from the one who governs the multitude, that is from the king." What is significant here is that, even though James suggests that a kingdom is really the people, he recognizes that a geographical determination is sometimes used. He recognizes the scale of comparison of kingdoms: "One also exceeds another in area [*locorum amplitudine*], for many people need a greater extent of land [*maiori spacio terre*] than do a few."[75] Within those regions, a king governs the people, a point he underlines by citing Psalm 147:14: "He maketh the peace in thy borders [*fines*]."[76]

DANTE: *COMMEDIA* AND *MONARCHIA*

At the same time that the papacy was attempting to assert its continued dominance, a number of writers were advocating the basis for secular rule. Three will be treated at length here: Dante Aligheri, Marsilius of Padua, and William of Ockham. If it is the case that in the early fourteenth century the arguments of the papalists were much better than those of their opponents, very quickly this was no longer the case.

Dante is of course best known for his three-part epic *Commedia*, the *Comedy*, which has long been known as *The Divine Comedy* following the judgment of Giovanni Boccaccio.[77] D'Entrèves has argued that the *Commedia* "is as much a political as it is a religious poem," and that "the reason lies perhaps first and foremost in the particular quality of Dante's religion."[78] Dante believed that religion could change the world. Yet while that work is allegorical, Dante was also the writer of an important, and much more conventional, political work entitled *Monarchia*, or Monarchy, written at a disputed date between 1308 and 1318, possibly 1310–13.[79] But we should note that while his political writings are of importance, and should be read alongside the poetry, it is likely that had he not been a great poet, then the *Monarchia* might have been forgotten.[80] Indeed, so dominant is the *Commedia* that Ferente at one point describes the *Monarchia* as "his other major political work."[81]

Dante is concerned with trying to justify the importance of empire, to contest the power of the papacy, particularly in the wake of *Unam sanctum*, but also to improve the politics of the city. Dante's angry denunciations of Florence are undoubtedly in part shaped by his exile, but cannot be reduced to that status.[82] He argues that political units come in three broad sizes: city, kingdom, and empire. Cities, as the *polis* was for Aristotle, are the smallest units that can meet basic human needs, kingdoms are extensions of these, but, as Ferente puts it, "only a world-monarchy, an empire, can control the greed and aggressions of individual cities and princes."[83] These three units are mirrored in the attitudes and in the structure of the *Commedia*, but there both cloaked in poetic language and put more forcefully because of the power of his allusion.[84]

The political reading of the *Commedia* in these terms is that hell, or Inferno, is the corrupt city; Paradise is the "ideal city-empire"; and Purgatory is "neither city nor empire, but a loosely knit kingdom without a centre, like Italy."[85] Italy is, he suggests, like a "ship without pilot in the rage of the tempest," a horse with "saddle empty."[86] Paradise then avoids both the corruption of the city and the disunited chaos of the kingdom, and the empire is the best model for worldly government. All smaller types of government are brought together "within a single unified whole."[87] The Inferno, on this reading, is Florence; Purgatory is contemporary Italy; and Paradise the Roman Empire, at least as it should be.[88] As Ferente has noted, there are also parallels with Augustine's two cities,[89] with the damned and heavenly cities as poles, and overall his work combines the Aristotelian view that the political community is a natural or rational unit, designed to satisfy particular collective needs, with a more religious motivation, including acting as a "remedy against the infirmity of sin."[90] What Dante calls *civitas et regnum*—city and kingdom—are his understandings of the Greek *polis*, the second being an extension of the model: "The end of the kingdom is the same as that of a city, but with greater confidence that peace can be maintained."[91] It is worth underlining here that Aristotle's text comes back into circulation in the West, in a political landscape that was considerably transformed from the time of its writing. Instead of the "city-states" of ancient Greece, or the empires of his time or those that came later, there were kingdoms. No wonder that Dante, as others, tried to think through Aristotle's arguments in that new context.

For Dante the key problem is that of Italy. While it existed as a well-defined geographical, and to some extent linguistic and cultural, unit, it was a long way from being a unified political body, and certainly not one that measured up to Dante's standards of the *civitas et regna*.[92] Rather, the model is to be found in the past, in the classical Rome that was, in Ovid's terms, *Romanae spatium est urbis et orbis idem*—Rome's extent is both the city and the world.[93] As Ferente puts it, "To be at once the smallest and the largest political unit is a paradox which should exist only in heaven and yet it was achieved by Rome on earth."[94] Dante's political theory is thus in part an exercise in the writing, and interpreting, of history.[95]

Yet while the kingdom is an element in the movement from city to empire, and this helps makes considerable sense both of the poem and of Dante's politics, in the tripartite structure of *Monarchia* it is city-world, empire, and church that figure as the main factors of his political theory. Three questions are raised; each part is devoted to one of them. At the end of the text, Dante himself summarizes it in this way:

Whether the office of Monarch is essential for the well being of the world; whether the Roman people obtained the Empire by right; and last, whether the Monarch's authority derives from God directly or through some intermediary. Yet the truth upon this last issue is not to be narrowly interpreted as excluding the Roman Prince [the Emperor] from all subordination to the Roman Pontiff, since in a certain fashion our temporal happiness is subordinate to our eternal happiness. Caesar, therefore, is obliged to observe the reverence towards Peter which a first-born son owes to his father; so that when he is enlightened by the light of paternal grace he may the more powerfully enlighten the world, at head of which he has been placed by the One who alone is ruler of all things spiritual and temporal.[96]

Essentially Dante wants to interrogate if a supreme authority is necessary, how the Romans achieved this, and where authority comes from. He is concerned with these questions within the temporal sphere, suggesting that "the temporal monarchy [*Monarchia*] that is called the Empire [*Imperium*] is a single Command exercised over all persons in time, or at least in those matters which are subject to time."[97] Dante here is making a distinction both between a philosophical concept and a political reality, and between temporal and eternal demands. He thus recognizes that in some respects there is a necessary hierarchy between the eternal and the temporal. But in pushing these issues within the temporal sphere, he is immediately challenging some of the power of the papacy. Indeed, as Canning puts it, "Although Christ was lord and governor of both spiritual and temporal matters, it did not follow that his vicar was as well."[98] While there may be a role for religion, and Dante did not challenge the spiritual authority of the pope, the unification at a political level is crucial: "A Monarch or Emperor is necessary for the well-being of the world."[99] It is important to note that here *imperium* is intended to mean absolute authority rather than a spatial extent. While in *Purgatorio* he does use the term to describe Italy as the "garden of the empire [*'l giardin de lo' imperio*]," this is not the standard use of the term.[100]

Dante's model for the Holy Roman Empire is the classical Roman Empire, although he may have had in mind more recent powerful emperors like Frederick II.[101] For Dante, Rome's success in domination was not simply due to the weakness of others or its brute force: "I maintain then, that it was by right, not by usurpation, that the Roman peoples acquired that monarchical office over other men which is called the Empire."[102] He had earlier held a different position, that it was by force alone.[103] But by

the time of writing *Monarchia*, Dante was convinced of a divine purpose
to Roman expansion, that the nobility of the Romans was "intended by
nature to rule the world."[104] Essentially book 2 is designed to refute the
idea of the *translatio imperii*—the translation or movement of the empire,
where Constantine gave the empire to the pope, and the pope was therefore
able to give it back to Charlemagne, the Franks, or the Germans.[105] Rather,
the Romans ruled by direct God-given right. He makes these arguments
throughout the book, moving toward a more religious basis for this position in book 3. Nonetheless, the church for him must take a subordinate
role. The strength of the empire must be, in temporal matters, absolute.
Neither challenges from below (cities or kingdoms) nor those from above
(the pope) can be allowed. The church must be pure, poor, and without
temporal power.[106] Dante thus contests the idea of two swords, especially
in book 3 of the *Monarchia*. The imagery of two swords is appropriate,
he suggests, but the interpretation of it is problematic, especially the hierocratic nature. Dante argues that the straightforward, and nonallegori-
cal, meaning of the passage should be taken. Christ initially suggests all
the disciples need a sword, but accepts that two are sufficient.[107] While
in the *Commedia* Dante offers the prospect of two suns—one political,
one religious—it is crucial that both belong to Rome, and that they light
separate paths.[108] Indeed, when he discusses this in the *Monarchia*, Dante
explicitly rejects the idea that the sun giving the light to the moon means
that spiritual authority empowers the secular.[109] From this is it clear that
Dante's *Monarchia* is designed not as an elegy to the empire but rather as a
very concrete proposal for its future.[110]

Dante argues forcefully that the temporal ruler is not bound to the
pope, and does not derive his authority from him. His most powerful statement is that

> the temporal realm [*regimen* or *regnum*] does not owe its existence to
> the spiritual, nor its power (which constitutes its authority), nor even
> its operation as such—though it certainly receives from the spiritual
> the energy to operate more effectively, by the light of grace which God
> infuses into it in heaven and which is dispensed to it on earth by the
> Supreme Pontiff.[111]

Many of his arguments contest papal claims, grounded on quotations
from scripture. Based on his reading of Aristotle, Dante contends that
while many of the premises are true, there are flaws in the syllogistic logic
where predicates are not the same, extension of sense is inappropriate, or

there is a willful misreading of the claim at stake.[112] He denies the validity of the Donation of Constantine, not because he is able to refute it philologically, but because he considers it improbable, indeed impossible, that the imperial power was divided and destroyed in such a way. Scott suggests that he had "little choice but to accept the forgery as genuine,"[113] so he does not dispute the history so much as its application.[114] He offered strong arguments against it: Constantine was not empowered to give the empire away, because "to divide the empire is contrary to the office delegated to the emperor"; and the church was not right to accept it.[115] In reference to the impossible division of Jesus's clothes after his crucifixion (John 19:23), Dante implies that the empire is a "seamless garment."[116] There are several passages in the *Commedia* where Dante is critical of the Donation. In *Paradiso* he notes that Constantine "turned the eagle's wing against the course of heaven," suggesting that in transferring the empire from West to East, he was reversing the direction of the sun—that is, against God and nature.[117] In *Purgatorio* a voice from heaven is heard to say, "O my ship, what an evil cargo you have taken on board";[118] and the pope is mournfully suggested to have become the "first rich Father."[119] Indeed, Dante, unlike Ockham, does not discuss poverty, at least in the *Inferno*, but rather he highlights avarice.[120]

Dante, then, while politically conservative in terms of the defense of the empire, is also radical in terms of his challenge to the papacy. *Monarchia* was heavily criticized at the time of its writing, and early manuscripts are no longer extant. The earliest that exist, from the second half of the fourteenth century, were either hidden among other writings or anonymized, on the one hand, or have marginal comments, suggesting both that it was a text that was dangerous to own and that it was one that provoked reactions.[121] Shortly after its appearance, Guido Vernani wrote a powerful polemic against it. He agreed with Dante that there should be a single monarch, but said that this should be the pope. The empire should exist, Vernani conceded, but it was not divinely ordained and the emperor received his authority only through the pope, not directly from God. For Vernani, obeying the pope and living in accordance with the Gospels would produce the true perfect monarchy.[122] The church condemned *Monarchia* around 1328–29, and restrictions lasted until 1881.[123]

MARSILIUS OF PADUA AND THE RIGHTS OF THE CITY

Marsilius (or Marsiglio or Marsilio) of Padua is often described as one of the most important political thinkers of the Middle Ages. He was certainly one of the most radical,[124] although many of the claims for his importance are because of what he putatively anticipates rather than his contemporary role: others later accused of heresy claimed that they had got their ideas from the "accursed Marsilius."[125] Marsilius's key target is the pope, and he offers some powerful arguments against papal plenitude of power, the *plenitudo potestatis*, in terms of both its control over temporal power and indeed its absolute spiritual authority.[126] Nonetheless, he has relatively little to say concerning the focus of temporal power, even as he provides justifications for its existence.

Marsilius was born sometime around 1275 and died in 1342. His key works are the *Defensor Pacis*,[127] the *Defensor Minor*, and *De translatione imperii*.[128] The first, the *Defender of the Peace*, is by far the most substantial; it was completed in 1324 and—unusually for a medieval text—exists in a manuscript corrected in the author's own hand.[129] It was dedicated to Ludwig of Bavaria. The editor of its Latin text has described it as "one of the many tedious polemics in the last weary contest of the medieval Empire and the Papacy," and "intolerably long, diffuse and cumbrous, full of repetitions."[130] It is comprised of three discourses, of widely varying length. The first discusses the powers of what Marsilius calls *regnum*, the realm; and the second, by far the longest, those of the church. The very short third discourse is effectively a summary of the preceding ones, although it also provides some concrete proposals for those combating dangers to the peace.[131] Although nominally separate—perhaps more so in the literature on the text than in the text itself—these topics are of course inherently interrelated.[132] But it is significant that discourse 1, taken alone, offers a distinct argument concerning temporal government. As Nederman has noted, this was unusual for the time, and while Marsilius distinguishes between secular government and ecclesiastical powers, most of his contemporaries "integrated their secular political theory into writings which were primarily concerned with the relation between spiritual and earthly realms."[133] The key criterion for a secular, independent political unit is peace.[134]

Defensor Pacis is often claimed to be an inherently modern text, and various commentators have tried to find forerunners of a whole range of ideas that would later become central to political discourse.[135] Previté-Orton suggests that Marsilius is often struggling for words to describe things, because these terms do not exist. He suggests that this is a mark of Marsilius's originality, and that one of these terms is "absolute sovereignty."[136] He goes on to suggest that "in the fourteenth century the very idea of legal sovereignty was unfamiliar, save perhaps veiled in technical glosses among the civilians or in the *plenitudo potestatis* of the Pope."[137] Marsilius certainly anticipates the problems that notions of sovereignty would later emerge to address. As Maiolo notes, in making a distinction between intrinsic and extrinsic danger, he begins to address what we would understand as "the core of the modern conception of internal and external sovereignty."[138] Yet we should be extremely careful in suggesting that what Marsilius really meant was something other than what he actually said.[139]

The other two texts are rather different. Much shorter, they are also closer to other writings from the period. *Defensor Minor*, or the *Minor Defence*, is focused on the relation between temporal and spiritual power, and is therefore part of a very particular genre of which several earlier examples have already been discussed. It provides a discussion and updating of key claims in *Defensor Pacis*, written probably at a distance of fifteen to twenty years. The relation between the *Defensor Pacis* and *Defensor Minor* is disputed. The first clearly advocates the rights of small political units against the church, but the second appears to suggest that instead of seeing what we might now call a republic as the best political system, there is an advocacy of empire. Others suggest that Marsilius was essentially an apologist for empire all along. Nederman contests the foundation of both readings: that the *Defensor Minor* promotes empire. Rather, he suggests that both texts are ambiguous and suggests that while universal empire can be legitimate, this does not always mean it is the best outcome. Nederman plausibly argues that compared to other advocates of empire, such as Dante or Lupold of Bebenberg in *De iuribus regni et imperii*, the *Defensor Minor* is "not primarily a work of imperial political theory at all."[140] Rather, he suggests, Marsilius sees empire as justified if it accords with certain key precepts.[141]

A work that is more clearly proempire is the third key political text, *De translatione imperii* (*On the Transfer of the Empire*), which looks at the legitimacy of transfers of imperial power. The two most significant are, of course, the putative Donation of Constantine and Leo III's crowning of Charlemagne—transfers from the Romans to the Greeks and the Greeks to the Franks. Concerning the first, Marsilius lacks the necessary knowledge to prove it a forgery, and so does not so much attack its validity as a document as criticize the way it had been appropriated.[142] Indeed, Garnett has suggested that "it provided too valuable a weapon for Marsilius to be ambivalent about its authenticity."[143] This was a theme that also figured in more muted form in *Defensor Pacis*.[144]

The imperial power [*imperium*] remained at Rome without being moved through the reigns of thirty-three emperors and for 354 years and five months, right up until the time of Constantine the Great. Constantine, in the seventh year of his reign, changed the imperial seat, transferring it to the East, to the city of Byzantium which is now called Constantinople and which in accordance with the Empire's laws enjoys the prerogative of ancient Rome.[145]

Marsilius effectively reverses questions of legitimacy and religion. It is the community that provides the justification for temporal rule; religion is utilized by that political unit.[146] The empire is, he suggests, legitimate, but not because of the pope; nor does it need the pope to crown emperors, even if this has become customary. The contemporary German emperors are legitimate through a series of transfers of power. As he says of the crowning of Charlemagne, "The extent of the strength, rectitude [*iuris*] and firmness which this transfer of the Empire had is revealed in the final chapter of my *Defensor Pacis* and can be clearly perceived by all."[147] Crucially, Marsilius recognizes that there is some ambiguity about terms.

It is necessary to note that, in one meaning, the term "Roman Empire" [*Romanum Imperium*] sometimes signifies the monarchy or royal rule only of the city of Rome or the Roman civic body [*civitatis*], such as was the case at its origins. . . . In another sense, "Roman Empire" signifies a universal or general monarchy over the whole world, or at any rate over the majority of the provinces, such as was the government and city of Rome as these emerged.[148]

Like other writers, Marsilius deploys what resources he can to make his arguments, including Franciscan writings. In *De translatione imperii*, Marsilius's tactic is to take a propapal work—Landolph of Colonna's *Tractatus de translatione Imperii*—and to reexamine and rework its sources such that it reaches entirely different conclusions.[149] Aristotle is widely used in *Defensor Pacis*, and is seen as almost infallible: a citation from him is the strongest of arguments.[150] In the *Defensor Minor* the sources are more traditional: mainly biblical, some of the church fathers, rarely medieval sources, and neither Aristotle nor the Franciscans because his audience was different.[151]

There are, Marsilius notes, two modes of living, and living well: "one temporal or worldly, but also another, which is customarily called eternal or heavenly."[152] He provides a detailed discussion of what is meant by each. Temporal applies in two main registers. It first applies to "all corpo-real things (both natural and artificial) apart from man, which—being in man's power in some sense—are ordained to supply his uses, needs and pleasures in and for the status of this worldly life."[153] Crucially, this im-plies property and has an important relation to questions of poverty. Sec-ond, "temporal" applies to things done by humans—that is, "disposition, action or passion"—on themselves or others for an end in this world. Mar-silius narrows this a little to those things that affect someone other than the doer, for these "are what legislators of human laws are mostly con-cerned with."[154] In this specific sense he is undoubtedly right, and we get the clear sense that "temporal" concerns are those that affect this earthly life, the duration of time in which we live physical lives. This becomes clearer if the definition of spiritual is looked at. Marsilius notes that it is said of "incorporeal substances and their actions"; of things humans do that remain within themselves (cognitive actions or appetites); of di-vine law and "all the gifts of the holy spirit that order us toward eternal life." From this derives the sense of anything humans do as actions or have as passions "for the sake of meriting the blessed life of the world to come." There are various other meanings that Marsilius considers inap-propriate and improper.[155] The point is clear: temporal things concern our life on earth; spiritual, those directed toward our souls, orientated toward eternity.

In doing so, and in excluding the spiritual from the temporal, Mar-silius defends notions of poverty but breaks their relation to charity.[156] Rather, for him it is a question of a broader "temporal disendowment," universalizing the specific Franciscan question to apply to the church as whole.[157] Denying power over temporal goods is an effective denial of tem-poral power. Marsilius makes a distinction between some temporal goods like food, drink, and medicines that are "consumable in a single act or use" and others such as clothes, tools, servants, fields, or houses "which last and which are of a nature to serve several uses."[158] Nonetheless he is clear that "the simple use or licit having of a thing is distinct . . . from any dominion of it as stated previously, i.e. the power of claiming it or an aspect of it for oneself or barring it from another."[159] He goes on to sug-gest that "it is an insane heresy" to claim the contrary, thus implying that Pope John XXII is himself a heretic.[160]

Marsilius also contests notions of *plenitudo potestatis*. This view was that the pope had unlimited power over the flock. Aegidius and, earlier, Ber-nard of Clairvaux (1090–1153) had been important in developing this idea, the latter suggesting to the pope that "others are called to share part of the care for souls; you are called to plenitude of power [*plenitudinem potesta-tis*]. The power of others is bound by definite limits: yours extends even over those who have received power over others."[161] As Tierney notes, this notion of *plena potestas* or *plena auctoritas*—full power or authority—was crucial to the "canonists' doctrine of representation," and was an effective forerunner of notions of sovereignty.[162] Yet its very nature as unlimited means that it cannot simply be equated with more modern notions of sovereignty, since it knows no limit to its extent, and is better understood as "papal supremacy," since it does not deny that there is temporal power, but rather asserts that its power operates above that.[163] Marsilius takes the most extreme instances of this view as his target: writers like Aegidius and Augustinus Triumphus.[164] As Canning notes, only in texts such as theirs was there a theory of the papal plenitude of power that provided "the kind of dragon which Marsilius sought to slay."[165] Marsilius is careful to note that the issue is not what power Christ had, nor what power and authority he could have granted to Peter and his successors, but what power he *did* grant to them.[166] His suggestion is that it is clear that Christ excluded them from all "principate [*principatu*] or worldly government."[167] He therefore glosses Christ's retort to Pilate that "my kingdom is not of this world" (John 18:36) as "I did not come into this world to reign with temporal government or dominion [*regimine seu dominio temporali*], in the way that the kings of the world reign."[168] It is this that informs his discussion of the notion of "two swords."[169] His response is that when Christ says, "It is enough," his utterance is "metaphorical . . . mystical." Marsilius cites a gloss (which he attributes to Saint Ambrose) that states: "Two swords are allowed, one of the New, the other of the Old Testament, with which we are armed against the attacks of the devil. And he says 'it is enough' because nothing is lacking to one who is armed with the teaching of both testaments."[170] Even if taken literally, Marsilius suggests, "The material sword is not principate nor judgment of secular acts."[171]

Like many writers in the late medieval period, using the *Politics*, Marsilius finds it necessary to expand the notion of the *polis* beyond the small community envisaged by Aristotle in that text. As Dante showed, it was possible to extend it further to the *monarchia* or *imperium*.[172] For Marsilius the political community is most commonly the *civitas* but can also be understood as the *regnum* or *provincia*.[173] Marsilius knew the *Politics* in Latin translation, where *polis* was rendered as *civitas*, and he would undoubtedly have been aware of the notion of a *res publica*.[174] As Gewirth notes, "Marsilius' Latin is itself in many respects a translation, so that in translating him we are also translating translations." Thus, any rendering into English of *civitas* is not just a translation of that Latin term, but also of Aristotle's *polis*. That Gewirth himself uses the misleading "state" to translate these words does not detract from the substantive point.[175] Nonetheless, it is in the notion of the *regnum* that Marsilius makes perhaps his most interesting contribution. He notes that *regnum* is generally understood to be a kingdom, but he wants to broaden its meaning to encompass more regimes than simply this.[176] A *regnum*, he says, can mean "a plurality of cities [*civitatum*] or provinces [*provinciarum*] contained under one regime [*regimine*]. In this sense a *regnum* does not differ from a city [*civitate*] in terms of the form of polity [*politiae*], but rather in terms of size."[177] He goes on to suggest that it can also imply a particular kind of regime, which Aristotle calls "temperate monarchy [*monarchiam temperatam*]."[178] He thus relates it to the *civitas*, but it cannot be rendered as a simple equivalent. Brett translates the term as "realm" to recognize that it is neither as modern as a state nor as limited as a kingdom.[179] Marsilius sometimes discusses the *regnum Italicum*, but this is neither the Italian state nor kingdom, but rather an as yet unfulfilled promise for the realm or region of the Italian peninsula. He bemoans the fact that it is "divided and lacerated" and therefore "can be more easily oppressed."[180]

Gierke therefore notes that "the differences between *civitas*, *regnum* and *imperium* became mere differences in size instead of being joints in the organic articulation of a single body."[181] Yet it is not simply a scalar differentiation: crucially, they are formally equivalent and share the same logic of power, which for Marsilius means that they derive their legitimacy from the community and bow to no external superior power.[182] His work therefore provides a strong justification for the independence of even the smallest political units from papal interference, at the same time as the overall argument appears directed toward a defense of the empire and in particular Ludwig of Bavaria. It was for that that he was condemned as a heretic by John XXII in 1327,[183] but his significance goes beyond this. In defending the rights of the small units, he is providing a justification that could be used for their independence from imperial edict, a question that would reemerge some time later. This importance is often overdrawn in the literature, with d'Entrèves suggesting that because Marsilius articulates the subordination of religion to the state, he anticipates the Reformation and that his "vindication of the right of several petty governments to exercise the supreme authority and control of religious matters within the boundaries of their several states, and of the duty of the subject to conform to the religion of the prince" articulates "the principle of territorialism."[184]

It is the last that is of course crucial here. But d'Entrèves is guilty of assuming that this argument must be in Marsilius because it seems so evident. Nowhere does Marsilius articulate a spatial limit or extent of power. And, as Garnett notes, Gewirth's suggestion that for Marsilius the state's sovereignty was "within a definite territory," or, in Garnett's gloss, "territorially restricted," finds absolutely no relation to the text.[185] Marsilius is thus not advocating control within "boundaries," much less in "territories."[186] Previté-Orton glosses Marsilius's use of *"civitate vel regno,"* "in city or realm," as "territory," but this is surely a step too far.[187] Rather, Marsilius is advocating this control over a much more loosely conceived *civitas*, a community that seems simply to comprise the people. Nor does he yet argue for the discretion of the prince in setting the religion, rather for political direction of the priesthood. To argue for a check to papal power does not, yet, anticipate the ruling at the Diet of Augsburg.[188] Nonetheless, he does see unlimited papal power as the key problem facing the individual political community, and as destructive to human happiness.[189] In response he provides a justification for an almost unprecedented secular jurisdiction, which Gewirth suggests exceeds both papal plenitude of power and the *majestas* that Jean Bodin would later articulate.[190] Yet in his discussion of temporal matters, Marsilius notes that in its broadest sense, "everything that begins and ends in time is customarily called 'temporal.'" His reference here is Aristotle's *Physics*, book 4.[191] In this book Aristotle is concerned with time as the number of motion, and specifically movement through place.[192] Barely hinted at here is what will become increasingly significant and is, arguably, already implied in the temporal as worldly or earthly relation: temporal matters are those that take place *somewhere.*

WILLIAM OF OCKHAM AND THE POLITICS OF POVERTY

William of Ockham is undoubtedly one of the most important philosophers and theologians of the late Middle Ages. Yet his status as a political thinker is more complicated. McGrade, for instance, has described him as "the most formidably technical of the great scholastics," who became an activist, and then "a political thinker of major rank."[193] There is a clear break: before 1324 almost no political writings; after 1328 almost nothing but political writings.[194] The last theoretical work, largely composed in Oxford, but the final version probably completed 1324–25 in Avignon, was *Quodlibetal Questions*. This is a work that has no reference to the pope, empire, poverty, power, or property, and only a brief mention of authority: all key themes in the later political writings.[195] These writings were largely completed during the years he spent in Munich, but were sparked by a time in Avignon.[196] This gives a good starting point: Ockham *became* a political thinker, and this largely independently of his contributions to other fields of philosophy. The magisterial intellectual biography of Marilyn McCord Adams, for instance, covers five distinct areas: ontology, logic, theory of knowledge, natural philosophy, and theology, but not the politics.[197] For McGrade, the philosophical and theological writings are not obviously political; nor are they utilized in the more political, or polemical, writings.[198] Coleman, by contrast, argues that the kinds of views Ockham previously developed concerning epistemology were crucial to his later political works.[199] Nonetheless, there is general agreement that Ockham's first political writings were direct polemical engagements in the issues of his time, even though he eventually came to write more reflective contributions.[200] What was the spark for this dramatic change of vocation?

Ockham was a member of the Franciscan order, whose theory of the poverty of Christ had been ruled heretical immediately before his arrival in Avignon, in the papal bulls *Quia nonnunquam* (1322) and *Cum inter nonnullos* (1323).[201] This was a direct retraction of an earlier bull, from Nicholas III, *Exiit qui seminat* (1279), which had tolerated their views.[202] While the Franciscans believed that Christ and the disciples had not owned anything, but had merely used things, Pope John XXII rejected the permanent separation of ownership and usufruct, in his *Ad conditorem canonum* (1322), noting the absurdity of it applying to "an egg, or a cheese, or a crust of bread, or other things consumable by use."[203] While the Rule of Saint Francis clearly states, "Let the brothers appropriate nothing to themselves, neither house nor land [*locum*] nor any thing,"[204] the Franciscan view of poverty is nonetheless complicated, made especially so by the fact that in his own work Saint Francis himself does not even mention the notions of *dominium* or *proprietas*.[205] These were key terms, along with *possessio* and *usus*, and their interrelation is complicated. *Dominium* was understood to be ownership, often absolute and usually in a passive sense, and was often distinguished from the active right to use, *usus*, although they were frequently conflated, as were *dominium* and *possessio*—that is, ownership and possession.[206] Yet as Coleman notes, "The question of *dominium* was to become one of the exercise of political authority."[207]

While Francis was unequivocally against money, and for poverty, this does mean that he advocated a complete denial of property. Nonetheless, subsequent Franciscans, notably Bonaventura in his *Apologia Pauperum* of 1269, made a fundamental distinction between *dominium* and *proprie-*

tas, on the one hand, and *usus*, on the other. As he said, they "relinquish earthly possessions in respect of *dominium* and *proprietatem*, and do not reject *usum* entirely, but restrain it."[208] An individual could use, and benefit from, something, without having a claim to its ownership. Bonaventura did not distinguish between *dominium*, *proprietas*, and *possessio*; what was important was all being set apart from *usus*, sometimes glossed as *simplex usus*.[209] *Quo elongati* (1230), a papal bull of Gregory IX, had made a similar point: it rejected the claim that the apostles had held common or individual *proprietas*, but they had had the *usus* "of utensils, books and movable goods."[210] Yet while a distinction between *dominium* and *possessio* seems plausible, to distinguish *possessio* from *usus*, especially in the case of consumables such as food, is more difficult to swallow.[211] *Exiit qui seminat* also made use of Bonaventura's ideas, but did two crucial things. First, it not only codified the distinction between Bonaventura's terms, but it also introduced a separation between *ius utendi* and *simplex usus facti*—the right to use and the simple fact of use. *Ius* is thus important to four terms—*dominium*, *proprietas*, *possessio*, and the *ius utendi*—which are henceforth seen as rights, *ius*, whereas *de facto* use is "juridically indifferent."[212] Second, it dealt with the doctrinal basis of Franciscan ideas, not merely their way of life.[213] As Tierney notes, it effectively functioned as "a sort of Magna Carta for the Franciscans."[214]

Over a period of twelve years, Pope John XXII incrementally destroyed the doctrine of poverty.[215] He had initially tried to reconcile positions, claiming in 1317 that "poverty is great, but integrity is greater; and obedience is the greatest good."[216] The maintenance of the church as whole, through obedience to Rome, was not heeded. As Coleman has noted, John drew on Aquinas's views on property, and canonized him.[217] The key bulls date from the years 1322 and 1323. *Quia nonnunquam* was issued on March 26, 1322, and suggested that it was acceptable to revoke previous bulls if they were "harmful rather than profitable." The strong Franciscan response to this led to *Ad conditorem*, first issued on December 8, 1322. Petitions against that led to an even stronger version, issued on January 14, 1323, but given the same date as the first version, not therefore merely updating it but replacing it. On November 12, 1323, *Cum inter nonnullos* was issued (a work that interestingly omits mention of the terms *dominium* and *proprietas*), and this claimed that believing that Christ and the apostles had not owned anything was heretical.[218] In part this attitude of the papacy was sparked by the way that some Franciscans, notably Ubertino de Casale, had tried to criticize the church's policy and the papacy more generally from a Franciscan viewpoint. No longer simply an alternative way within the broader church, the proclamation that this should apply to the church as a whole sparked a wider conflict.[219] John's final word came in 1329 in *Quia vir reprobus*, where he ruled that perfection and possessory rights were commensurate because without rights there could be no justice.[220] It is essential to realize that this is not simply a debate about Christ's poverty, or even whether as a consequence the church should be poor. Rather, it concerns the question of what papal authority should actually be, and whether it should have an earthly component.[221]

Ockham had been summoned to Avignon because of papal concerns about his own, earlier, nonpolitical work, especially his *Commentary on the Sentences*.[222] John Lutterell, chancellor of Oxford, had gone to the pope alleging fifty-six heretical theses in Ockham's writings, and in 1326 fifty-one of these were ruled open to censure, although he was never formally condemned.[223] One of the people who examined him on this, Cardinal Jacques Fournier, later became Pope Benedict XII.[224] While Ockham was in Avignon, his Franciscan brothers, including notably Michael of Cesena, ordered him to look at some papal writings.[225] In so doing, he became convinced that John XXII was himself a heretic, and therefore could not be a legitimate pope.[226] Reflecting on this a few years later, Ockham notes that in the pope's writings he found "a great many things that were heretical, erroneous, silly, ridiculous, fantastic, insane, and defamatory, contrary and likewise plainly adverse to orthodox faith, good morals, natural reason, certain experience, and fraternal charity."[227] He was forced to remain in Avignon under house arrest, but eventually left in 1328 under the protection of Ludwig of Bavaria, recently elected emperor, though not recognized by the pope. For the last twenty years of his life, Ockham devoted his attention to books and tracts arguing against these papal decisions and for the removal of John and his successors.[228] He apparently said of this arrangement, "Protect me with your sword O Emperor, and I shall protect you with my pen."[229] Whether apocryphal or not, it is an accurate reflection of the relation. While some Franciscans attempted to suggest that the 1279 bull could not be retracted, some deferred to the pope, vainly trying to reconcile the 1279 and 1321 pronouncements. Ockham was unusual in that he tried to analyze the doctrines in the bulls themselves theologically.[230]

Ockham's first significant political text was *Opus Nonagintan Dierum* (The work of ninety days), so called because it was apparently completed in three months.[231] This interminable text—two hundred pages of which are devoted to eating bread without it being property—has been described as a "work of overwhelming erudition," which together with its scope "virtually precluded reply."[232] His other major texts include the multipart *Dialo-*

gus, which discusses heresy in part 1, and theological disputes in part 3;[233] writings against John XXII and his successor, Benedict XII;[234] *Octo quaestiones de potestate papae* (*Eight Questions on Papal Power*);[235] and the *Breviloquium de principatu-tyrannico* (*A Short Discourse on Tyrannical Government*).[236] His final text is known as *De Imperatorum et pontificum potestate* (*On the Power of Emperors and Popes*), which concerns spiritual and temporal power generally.[237]

The question of poverty, as noted above, also inherently implied the question of the temporal, earthy power of the church. Ockham, following Bede, notes that Christ's kingdom was eternal and in heaven and "not earthly and temporal [*terrenum et temporale*]"; suggests that "no one who teaches us by word and example to despise temporal kingdoms should be thought to be a king in temporal matters"; and that "blessed Peter did not receive from Christ any temporal kingdom."[238] Ockham is clear that if the pope lays claim to property, it is not as a successor of Peter. Peter was able to fulfill Christ's injunction to "feed my sheep" (John 21:15) without gold and silver (see Acts 3:6). As Coleman notes, *dominium* in its narrow sense was *dominium in rebus*, and thus easily led to *ius in re*.[239] She suggests that it is clear that rights to something that can be defended, alienated, or transferred depended on a notion of a profit economy.[240] Thus, following Bernard's *De consideratione*, Ockham notes that in "abundance of riches" the pope is the successor of Constantine, not Peter.[241]

Ockham's arguments are multiple. He makes use of what resources he can, drawing on Aristotle's *Politics* in places, notably on the types of polity.[242] He also notably uses Aquinas at times to attack the papal doctrines.[243] While he was not trained in canon law, he was strategic enough to recognize that John XXII, who was a professional canonist, was highlighting the key texts that he, Ockham, needed to address.[244] The other way of putting this is, of course, that Ockham is forced to operate continually on the pope's terms. But it is important to note that he is not—at least at first—opposed to the papacy as such, rather to heretical doctrines and those who hold them. Unlike Marsilius, then, he is not inherently advocating a limited papacy, rather advocating that it hold to the correct doctrine.[245] Yet as Shogimen points out, it would also be reductive to see him simply as a Franciscan ideologue, merely defending their doctrines, because his theological arguments and examinations led him to challenge papal heresy far more generally. He became convinced, for example, that John XXII's heresy was rooted in papal ignorance of theology and an overtly legalistic interpretation of scripture.[246] As Lambert puts it, John was "a jurist of the highest calibre but not a theologian."[247]

Aside from poverty and heresy, among the issues repeatedly raised is the question of papal fallibility. The retraction of the 1279 bull clearly implied that either the previous pope or the current one was mistaken. Various means were used to try to reconcile this state of affairs. Guido Terrini suggested that while a pope as a person could be mistaken, *as* pope he was infallible. This was the distinction between *ut magister*, the academic theologian, and *ut papa*. While the former made a probable argument that could, theoretically, be in error, the latter's pronouncements were binding and could not be questioned. Ockham's view was different. All such pronouncements must meet the same test: they are binding because they are true and papal, not simply because they are papal alone.[248] As he suggests:

> The sentence of the pastor who is true (not merely according to human opinion) is to be feared, namely, that sentence which by its own right and deed is not void so as to be unqualified to pass judgment; the sentence of the false pastor, however, no matter how good a pastor he is according to human opinion—even the majority of Christians—is not to be feared.[249]

Ockham's view is that unless John XXII accords with scripture; or that it can be *proved* that the pope cannot err; or even if he can, that he must be obeyed anyway, then he will continue to be challenged.[250] Ockham often underlines that he himself is open to being challenged and that if anyone can prove his mistake "by means of an argument or an authority which I am bound to accept; and if I have no reasonable answer to make, I shall confess that I have been in error."[251] But he expects the same openness of others: Ockham argues forcefully that the pope *can* become a heretic, and this deprives him of *power*.[252] One of his most furious responses is to the idea that the pope can rule on doubts or questions concerning faith, even when he is one of the disputing parties, since this would effectively mean that faith rests on the pope, rather than on Christ.[253]

Thus, around 1337, Ockham shifts from the narrower question of heresy to look at papal authority more generally.[254] One crucial argument was his challenge to the idea of *plenituda potestatis*, the plenitude of power.[255] Ockham considers that papal power is indeed limited, and there are things that the pope can legitimately do, but equally those he cannot.[256] In making these engagements, Ockham necessarily touches on the question of the relationship between the empire and the church. He first does this in *Tractatus Contra Benedictum*, book 4, which moves from polemic against the papacy to begin to think through these broader issues. This was not

simply due to his patron, but more importantly because of theological and
political theoretical concerns. He saw the empire of his time, which he
simply calls the "Roman Empire," as the continuation of the Roman Empire of old, and notes that because it is obvious that this empire existed
before Christ, "the Empire is not from the pope."[257] By *from* here, he does
not mean historical origin, but rather jurisdiction, a feudal structure of
power.[258] He claims that "it is not from divine authority, but only from
human authority, that the pope has regulating power over the temporal
things [*temporalibus*]."[259] Like Marsilius, Ockham contests the reading
of the "two swords" doctrine. He allows that "he does not carry a sword
for nothing" (Romans 13:4), but suggests that while the sword can signify
power, it can also mean the word of God (Ephesians 6:17). He therefore
suggests that the two swords, following the same gloss Marsilius quoted,
and again attributing it to Ambrose, are the New and Old Testaments. Importantly he suggests that neither interpretation is proved, but that this
simply means seeing the two swords as spiritual and temporal power need
not be accepted. Similarly, even if the two swords are two powers, there
is no reason that they are spiritual and temporal: they could the powers
of preaching and miracles—that is, two spiritual powers; or two temporal
powers, either pure or mixed. Ockham also notes that is it neither said nor
implied that the same disciple had both swords.[260]

There is also an implied question of where these powers extend. In the
Short Discourse he notes that popes have greater temporal power in some
regions than in others. His reference is a papal decree from Innocent III,
dating from 1199, which claimed that "in the lands [*terris*] subject to our
temporal jurisdiction we ordain that the goods of heretics are to be con-
fiscated; and in other [lands] we order the same to be done by secular rulers and powers [*potestates et principes sæcuares*]. If they neglect to do so
we wish and command that they be compelled to it by ecclesiastical censure."[261] In doing so, the pope is suggesting a limit to his direct temporal
powers, which only operate in some places, but an unlimited extent to
the ability to command. The first is the ability to ordain a law; the second to order others to do so, or, failing that, to use ecclesiastical pressure.
Significantly, Ockham explicitly rejects the idea that the pope has more
temporal power over the empire than other realms.[262] This was, of course,
one of the issues at stake following Boniface's dispute with Philip: if the
pope was unable to command such things in France (where the king has
"recognised no superior in temporal matters"), then what implications did
this have for things within the empire?[263] Consequently, the issue then

becomes what *is* the relation between the papacy and the empire? For Ockham there is a clear distinction: the empire is not from the pope.[264] In addition, he does not see the empire as inherently superior to other temporal powers, paving the way for recognition of independent and nominally equal jurisdictions: "It is not beneficial for the totality of mortals to be ruled by one monarch of the whole world [*totius orbis*], but it is, as a rule, beneficial for it to be ruled by several where none of whom is superior to another."[265]

Yet Ockham's view is not simply concerned with demarcating the scope of papal and other power. He was also an important early advocate of rights and liberties of the individual, seeing that those provided by God should be safeguarded from external interference. This too was a limit to papal *plenitudo potestatis*.[266] Indeed, for Ockham, right or law, *ius*, was a form of *potestas*.[267] Ockham thus, in Shogimen's words, "not only severs the spiritual order from the temporal but also separates the moral domain from the political."[268] In doing so, he is forced to consider further the relation between the spiritual and temporal domains of power. As Bochus puts it, for Ockham, spiritual power is autonomous, comes "directly from God and [is] absolutely independent in its own realm." Secular or temporal power is "from God through the people, who decide on or elect their ruler, and it also is dependent on its own realm and governed only by natural law."[269] It follows from this that while rulers may be elected or hereditary, fundamentally they owe their power to God. While this means that rulers can usurp power, on the other hand, it means that people cannot remove their ruler arbitrarily.[270] His concern is thus with the scope and limit of power, particularly that of the pope but more generally of all rulers. Like Ockham, Marsilius sees the fullness of power as the challenge.[271] Yet while Marsilius felt that spiritual power needed to be subservient to temporal, in order to maintain civil order, Ockham found the balance through a focus on the individual, an individual subject to both spiritual and temporal laws.[272] As Coleman tellingly puts it, "Like Marsilius, Ockham does not glorify 'states': he legitimates them."[273]

Yet there are significant weaknesses in this argument. McGrade has highlighted the "non-metaphysical character of Ockham's political thought. . . . Ockham was without a grand scheme of history, mistrusted allegorical interpretations of Scripture, and showed little fondness for the biological, psychological, astronomical, and other analogies in terms of which his papalist opponents developed their theories of government and society."[274] Judged by later standards, he might be thought a limited political theorist, since he neglects many of the issues that might be considered essential to a worked-through schema. Perhaps most crucially, in making the distinction between spiritual and temporal power, he focuses much more on the limits to the papal rather than the extent of the secular. Ockham, drawing on book 7 of Aristotle's *Politics*, suggests that "no less wisdom or virtue is needed in one who rules the whole congregation of the faithful in spiritual matters than in a king who presides over his subjects in temporal matters."[275] The interest is not in the question of wisdom, but in the object of rule: the "whole congregation of the faithful" as opposed to the subjects.[276] The latter remains resolutely unspecified—its temporal scope its only limit, its extent otherwise undefined. Yet he was a formidable polemicist, engaged in the fundamental question of the relation between state and church, or the papacy / Roman Empire specifically.[277] As Tierney has noted, however unlikely it might seem, "the murky theological problems" of these debates "gave rise to intricate debates on topics that would later become central issues of Western political theory."[278] While for Tierney this is especially the case for notions of natural rights, we can make the case for notions of power and authority too.

In Gramsci's words, Dante is a transitional figure, bringing "a phase of the Middle Ages" to an end. By the time of Machiavelli (1469–1527), he suggests the questions raised are those of the modern world.[279] As he argues elsewhere, Dante "opposed the anarchy of the communes and of feudalism, but he looked for a semimedieval solution. In any case he posed the question of the church as an international problem, and he pointed out the need to limit its power and its activity. . . . Dante is really a transition: there is the assertion of secularism but still couched in medieval language."[280] Yet it is his linguistic developments that are perhaps most important. His decision to write the *Commedia* in his own Tuscan dialect of Italian, rather than Latin, was crucial in shaping the formation of the modern Italian language. (*Monarchia* had been written in Latin.) Dante was the author of an important text—*De vulgar eloquentia*—justifying the worth, and eloquence, of the vernacular languages, a text Gramsci calls "an act of national-cultural politics."[281] Yet this created a paradox. As Reade put it, Dante "destroyed for ever in the realm of language the very creed which he championed so gallantly in the realm of politics. Linguistically the *Divinia Commedia* creates a single and united Italy, which expels from itself all invaders from beyond the Alps, and absorbs into itself imperial Rome."[282]

Dante's critique of the power of the church is thus allied to Marsilius, although his hope is rather different from Marsilius's defense of the small, autonomous political unit. Dante and Ockham both see a role for the church, even if corrupt or heretical people currently fill its offices. While Marsilius's criticisms are certainly provoked by the particular individuals holding the office, his claims go further and address the basis of its power itself.[283] Marsilius may have been primarily concerned with the denial of papal power over "temporal polities" like Padua, but his work had the subsidiary effect of narrowing the scope of that influence more broadly.[284] Taken together, Dante, Marsilius, and Ockham show that the power of the papacy was under considerable threat at this time. It was the defense of secular political rule—of whatever size—from the interference of the pope, whose power was increasingly seen as confined to the spiritual alone. Texts from both Marsilius and Dante were used, for instance, in the dispute concerning the accession of Ludwig of Bavaria in the 1320s.[285] This was, of course, the situation in which Ockham became closely involved.

But it is some time before Machiavelli—almost two hundred years before *The Prince* was written, in 1513. The next generation of thinkers now operate in a breach created by the fracturing of the church/empire relation and the renewed splintering of secular rule. What is revealing here is twofold. First, that it is the papacy that propounds the view of spiritual and temporal power, claiming it owns both, but thereby clearly articulating a view, purpose, and scope of power that could be later used against it. And, second, that those who challenged the linkage, essentially on behalf of kings and the emperor, laid open the future potential for challenges to their own rule on effectively the same grounds. Interference from a pope in secular affairs shared the same logic as imperial involvement in the politics of a smaller unit. In these three thinkers, a crucial problematic is opened up. Temporal power is articulated as a distinct type of rule, which owes its legitimacy to something other than papal edict. In 1356, for example, the Golden Bull provided for imperial elections without papal involvement, and actually divested much power down to the levels of the individual components of the *Reichstag*, including the free cities.[286] What was crucial was that the possession of the rights of the electors would be explicitly tied to their possession of specific lands. Never again would there be the kind of issue that was at stake with Ludwig's nonrecognition, though this is not to suggest there were no disputes. But the distinct object of government of temporal power remains to be adequately theorized. Brett claims that many thinkers of this time "showed very little interest either in the concept of territory, or in place more generally."[287] But subsequent thinkers would grapple with these issues. Though the figures are different, Knowles agrees: "With the death of Aquinas a whole age began to draw to a close, and when William of Ockham died it had ended."[288]

PART III

CHAPTER SEVEN

The Rediscovery of Roman Law

THE LABORS OF JUSTINIAN AND THE GLOSSATORS

While the Roman Empire was overrun and its structures collapsed in the fifth century in Western Europe, in the eastern provinces of the empire, the Byzantium system endured for another millennium. Yet the longer-lasting legacy of the empire was its law, arguably Rome's most original contribution.[1] The foundation of Roman law comes from the Twelve Tables, written in the fifth century BCE. The tables themselves are lost, but they survive in part through quotations in other texts. For Livy they were "the source of all public and private law, running clear under the immense and complicated accretion of modern legislation," and Cicero notes that children were made to memorize them.[2] Over the next thousand years, Roman law developed and adapted, and an enormous volume of literature was produced commenting on it. After the fifth century, Roman law continued in unbroken form only in the East, but it endured in the West too. The barbarian tribes brought their own laws, but in places where the majority of the people were still Roman, the previous legal system continued, since the law of the victorious was too limited and crude to supplant it. What this meant was that law became tied to the citizenship or identity of the individual, not to the political system he or she lived under. This is a system known as the personality of law, where Roman and barbarian law existed side by side.[3] This was a return to an earlier system that had existed before the *constitutio Antoniniana* (212 CE). In order to deal with the Romans who were now their subjects, some barbarian kings issued summary legal codes. The most important ones are *Lex Romana Visigothorum*, *Lex Romana Burgundionum*, and *Edictum Theoderici*.[4] Then there was the codification in the *Codex Theodosianus*.[5]

Surpassing them all, however, were the labors initiated by the eastern Roman emperor Justinian in the sixth century.[6] Justinian had initially been coemperor with his uncle, but shortly after he became sole emperor in 527, he began a process of compiling the documents that are known as the *Corpus Iuris Civilis*, the Body of Civil Law. The project was initiated on February 13, 528, and comprises four key elements. The first is the *Codex*, sometimes misleadingly known as the *Code*, which comprises the statements or edicts on the law between the time of Hadrian in the second century and the time of compilation.[7] The first edition of this was completed in 529, but a second edition appeared a mere five years later. Only the latter survives.[8] The second part is by far the longest, and is known as the *Digest*, which was an attempt to preserve the best of the classical Roman jurists.[9] Thirty-nine of these writers were analyzed and excerpts taken from their work. The third was the *Institutes*, which looks most like a statute book but was designed as a student textbook, and was heavily based on the work of Gaius.[10] Last was the *Novellae*, which compiled the new imperial pronouncements of Justinian's own time.[11] The legal codes should be seen as constituting a whole.[12]

Tribonian was Justinian's minister responsible for this work, and given the speed with which he produced these texts, it is clear that he had a large team working with him. Indeed, it seems likely that there were three committees working on the *Digest*, with their separate labors then put together.[13] This accounts for some of the inconsistency and the peculiarities of its arrangement. There are no additional comments from the compilers, though there are some excisions and other adjustments.[14] Most of the *Novellae* were written in Greek, which was the administrative language of the East, but the remainder and all the other works were in Latin, since they were compilations from the Roman texts. Thus, the compilation of the *Corpus Iuris Civilis* shows the transition from the language of the old Rome to that of the new. Justinian was, incidentally, the last eastern emperor to be a native Latin speaker.[15] But this also meant that the texts were not widely read at the time. They were unreadable for many in the East, and largely unknown in the West. One of their signal achievements was the preservation of the Roman laws, which might otherwise have been lost. Most of those laws and legal texts not included are indeed no longer extant. However, the *Corpus Iuris Civilis* may have actually precipitated this, since it was designed to replace older collections and individual works, which as a consequence were less copied and preserved.[16] The material in the *Digest* is about 5 percent of the original material worked with, the idea being to preserve the best and most useful parts, with repetition and redundancy removed. Justinian also saw these works as definitive, with commentaries on the *Corpus Iuris Civilis* initially prohibited.

In time, however, this changed dramatically. Although it is of almost no use in understanding the Roman law of the old Western Roman Empire,[17] the *Digest* produced a volume of commentary that is rivaled only by the Bible.[18] This was both because it was the basis of the Byzantine legal system for another nine hundred years and because, when it was rediscovered in the West, it provided the basis for the civil legal systems of much of the continent.[19] Although it had a short life in Italy in the sixth century following Justinian's reconquest of the peninsula, it was quickly forgotten and largely unknown in the West for five hundred years. It is worth noting, in passing, that the comparison with the Bible has another interesting resonance. As Ullmann contends, Roman legal terms found their way into Jerome's Latin translation of the Bible, the Vulgate, where he used them to render legalistic phrases. This is often overlooked, but it meant that people who read the Bible took on something of Roman jurisprudence at the same time as theology, perhaps providing a fertile ground for the later rediscovery.[20] It is the rediscovery and reinterpretation in the West half a millennium later that is the basis for this chapter. Read anew in an entirely different context to its original writing and Justinian's compilation, it produced a fundamental shift in the way the relation between power, people, and place was understood.

In 1077 a complete manuscript of the *Digest*, probably written around the end of the sixth century, was discovered in Pisa.[21] Being in Latin, it provided a ready-made and internally consistent legal system that could potentially replace the fragmented and outdated customs and laws that existed at the time. As Mousourakis puts it, "Compared with the prevailing customary law, the works of Justinian comprised a developed and highly sophisticated legal system whose rational character and conceptually powerful structure made it adaptable to almost any situation or problem, irrespective of time or place."[22] Of course, it was not simply the discovery of the manuscript. This was a time of profound social and economic change, with the slow resurgence of scholarly endeavors, and the law was a tool that secular monarchs could use—if not entirely in their favor—at least as a basis for making pronouncements and as a system to resolve conflicts.[23] It was, of course, a system of law that owed its origins to a time before the Roman Empire had become Christian, and was based on secular texts, even if Jerome's translation had provided an anticipation of some claims.

The law was, however, neither immediately accessible nor usable. It was over five hundred years since Justinian and Tribonian had compiled the *Corpus*, and that was of course on the basis of laws that were sometimes already four hundred years old. Enormous academic endeavors were needed to make sense of them. In part, this work led to the University of Bologna's founding in 1088, and Irnerius produced a version of the *Digest* to serve as a student text. Evans has even suggested that the European Renaissance begins at this moment.[24] Irnerius (ca. 1050–ca. 1130) is a largely unknown figure, but he founded the school known as the Glossators.

A painting of Irnerius, by Luigi Serra from the nineteenth century, on the ceiling of one of the rooms of the Palazzo d'Accursio in Bologna, demonstrates the interrelation of legal, political, religious, and geographical themes in the Glossators' work. The background shows a city, landscape, and armies, with a priest blessing the troops as they prepare to go to battle. This is the context of the legal work Irnerius is doing in the foreground. Yet it is the work of Irnerius and those who followed him that provides the basis for the politics of land and conflict that is taking place at the same time. The lawyers were working on behalf of kings and cities, and the influence of their work continues into later legal theorists such as Francisco de Vitoria on the Spanish conquest of the New World and Hugo Grotius in his work on the rights of war and peace.

Irnerius pioneered the addition of explanatory notes and elucidations— *glossae*—to the text, initially between the lines and then expanding into the margins.[25] The Glossators were concerned with the restoration of the Roman legal texts, and undertook detailed analytic work on problems in the texts, which was necessary given the state they were in and the time since their composition and compilation.[26] The glosses could range from simple explanations of meaning and clarifications of syntax to much more lengthy commentary and analysis.[27] The work they undertook relates closely to the Scholastic method and to the progression of the liberal arts in the *trivium*—from grammar to logic or dialectic to rhetoric. Thus, there was a linear sequence from clarification to the resolution of inconsistencies to application.[28] While the Glossators' task was, in part, necessary because of the time lapse, their work was largely ahistorical, not fully appreciating the different context of their reinterpretation.[29] By the time of the mid-thirteenth century, their work had reached near-definitive form. At this time, Accursius produced the *Glossa ordinaria* (not to be confused with the gloss on the Vulgate that bears the same name), which compiled

Fig. 7. Luigi Serra, *Irnerio che glossa le antiche leggi*, 1886, tempera su intonaco staccato (tempera on detached plaster), 484 × 334 cm, MAMbo, Museo d'Arte Moderna di Bologna © photo Sergio Buono.

the work of his predecessors, and supplemented this with his own *Glossa magna*. Rabelais was critical of this work, seeing the law books as a "fine cloth-of-gold robe, marvellously grand and costly but trimmed with shit," declaring that the gloss was "so foul, stinking and infamous that it is no better than filth and villainy."[30]

BARTOLUS OF SASSOFERRATO AND THE *TERRITORIUM*

Following these labors came the Post-Glossators, sometimes known as the Commentators. They took a much more interpretative and philosophical approach to the laws, though their work was only possible because of the labors of the Glossators.[31] Though their work had many characteristics, the fundamental difference between them and the Glossators was their challenge to one major principle: that the law was immutable and facts changed. Instead, the Post-Glossators suggested, if there was a disjuncture, then the law should be adapted to the new facts. This was the fundamental methodological foundation for the shift between these two ways of approaching legal study.[32] As the key figure of the Post-Glossators, Bartolus of—or perhaps more accurately "from"—Sassoferrato declared, "It should not be a matter of surprise if I fail to follow the words of the Gloss when they seem to me to be contrary to the truth, or contrary either to reason or to the law."[33] Such work had been prepared for by earlier thinkers. Aquinas in particular had offered his own text on law in the *Summa*.[34]

While the Post-Glossators were not political thinkers, but jurists, they necessarily touched on issues that were political.[35] This is especially the case in the work of Bartolus and his student Baldus de Ubaldis. Bartolus was born around 1314 and died in 1357.[36] While he only had a short life, he produced an enormous volume of work, with his collected works running to around ten volumes in the different editions.[37] These volumes comprise his commentaries not only on the Roman law of Justinian but also on canon law, the statutes of Italian cities, and post-Justinian additions to the *Corpus Iuris*. But in his final years, he also wrote a number of shorter works that are closer to theoretical treatises, on topics such as insignia, witnesses, rivers, tyranny, the government of cities, and the rival factions of the Guelphs and Ghibellines.[38] These treatises are distinctive because they transfer arguments from the political advice literature, often written by theologians, into guidebooks for law students.[39] Bartolus's method was to outline all the evidence before him, and then to offer his own opinion. This meant that he quoted a lot, but his arguments usually come through clearly in time as a resolution to the problem.[40] Skinner has described him as "perhaps the most original jurist of the Middle Ages," suggesting that the key accomplishment of his work was to lay the basis for a range of independent polities, separate from the empire and each other.[41]

Bartolus had been taught by Oldradus de Ponte, who had been a canon lawyer in Avignon, and one of the first *consilia* theorists.[42] *Consilia* were legal opinions on specific matters, the likes of which are found in the *Di-*

gest, but this was a form revived in the twelfth century. Oldradus was one of the first to collect his *consilia* together, and this shaped the way those who followed him, including Bartolus, operated.[43] Oldradus had also been responsible for the advice that led to an important ruling of the church concerning the jurisdiction of the empire. This was the bull *Pastoralis cura* (1313), which ruled on the status of the kingdom of Sicily.[44] This bull was issued by Clement V (1305–14), a pope who had consolidated papal power in the light of the new realities following the effective defeat of Boniface VIII by Philip the Fair.[45] After the short-lived papacy of Benedict XI, Clement had been effectively approved by the French Capetian dynasty, and moved the seat of the papacy to Avignon.[46] In the bull, Clement declared that Sicily was outside the jurisdiction of the emperor because it was outside of the lands of the empire (*extra districtum imperii*). Kings had authority and were not subjects of other kings or rulers, and therefore could not be cited before a tribunal. It thus set the papacy over the empire, and further clarified the limited role of the empire, spatially and jurisdictionally, with the latter dependent on the former. By extension it further strengthened the independence of France from papal involvement, either directly or mediated through the emperor.[47] Canning has suggested that this ruling was the logical continuation of *Per venerabilem*, which accepted the temporal superiority of kings, because the papacy here was again stressing the claim of another monarch over imperial superiority.[48] Ullmann has claimed that it stresses territory and sovereignty, but in doing so, he slips between lands and territory, on the one hand, and between supremacy/superiority and sovereignty, on the other.[49] While the former terms within each pair certainly anticipate the latter, the latter terms were not used in the bull.

Yet Bartolus's commentary on it makes the move from *terris ecclesiae* to *alieno territorio*,[50] and it was a truly significant ruling that would have a profound impact on political realities as well as on the legal work of the Glossators. One of the problems that they had concerned political units, since the independent kings that existed at the time they were working had no basis in the *Corpus Iuris*. The question was whether these kingdoms, or nascent states, were *imperium*, or whether they were closer to the Aristotelian *civitas* or *regnum*, the terms used in the medieval translations. France could clearly not be understood simply as a *provincia*, nor could a free city simply be thought of as a *municipium*.[51] The hierarchy of political units had been a cause for concern and debate for some time. Engelbert of Admont in the early fourteenth century had suggested that Aristotle had the sequence of *communitas* as *domus*, *vicus*, *civitas*, *gens*, *regnum*; whereas Augustine has *domus*, *urbs*, *orbis*. Engelbert chose to adopt a variant of his reading of Aristotle: *domus, vicus, civitas, regnum,* and then finally *imperium.*[52] Bartolus similarly adopted a scalar model, discussing *provincia* and *vincia* (neighborhoods) in terms of tyranny, which can exist from empire, to province, to city, a house, and an individual, though not in a neighborhood.[53] More important, he advanced a solution to these kinds of issues. The independent polity had two origins: an Aristotelian *polis* and an empire on a reduced scale.[54] This fusion of Greek political thought and Roman law is crucial, since it brought together two strands of ancient thought in a modern context. Its importance is fundamental to the developments that followed from this time. While Aristotle is mentioned only rarely by Bartolus, his influence is readily apparent.[55]

Two key themes in Bartolus concern the relation between place and power. The first is the extent of law, of jurisdiction; the second is the relation between different political units. The first came up in several cases Bartolus wrote *consilia* on. One of these concerned a case where a murderer had not been prosecuted. Bartolus was asked to rule whether the community, a collective entity, could be found responsible for this failure.[56] Was the city as a whole to blame? This raised the question of jurisdiction and its spatial extent. While Bartolus was writing on behalf of the city, and thought that the case for negligence was unproved, the question of jurisdictional extent is important and was a recurrent theme in his work. Bartolus explores the relation between jurisdiction and *territorium,* and concludes:

> *Dominium* is something that inheres in the person of the owner [*domini*], but it applies to the thing owned. Similarly jurisdiction inheres in an office [*officio*] and in the person who holds the office, but it applies to a *territorium,* and [jurisdiction] is thus not a quality of the *territorium,* but rather of the person.[57]

Bartolus's use of *territorium* here is significant. He is taking the notion of land, or land belonging to an entity, as the thing to which jurisdiction applies, thus providing the extent of rule. The *territorium,* then, is not simply a property of a ruler; nor is jurisdiction simply a quality of the *territorium.* Rather, the *territorium* is the very thing over which political power is exercised; it becomes the object of rule itself. It thus becomes something sufficiently close to the modern sense of territory that we can begin to translate the term in that way. Another *consilium* concerns the compensation due when a foreigner assaulted a Florentine citizen. At stake for Bartolus was the question of status, rather than the facts of the assault.[58]

In a commentary on one part of the *Codex*, Bartolus quotes the gloss that says that "if proceedings are brought at Modena against a man from Bologna, judgment should not follow the local legislation of Modena, to which he is not subject." This is another indicator of the difficulties of deciding which law applies: is it the place where an action occurs that determines the jurisdiction, the person taking the action, or the person sued? Bartolus suggests there are two key questions: "whether local legislation [*statutum*] extends beyond the territory [*territorium*] to non subjects; second, whether the effect of such legislation extends beyond the territory [*territorium*] of the legislator."[59] It becomes clear that the territory is the essential object of rule, and that things that happen within, and people located within, are subject to the jurisdiction. A similar inquiry concerns jurisdiction of an invading army: "Suppose the army of one city is occupying the *territorio* of another and one foreigner kills another there; may he be punished by the authorities of this city?"[60]

This particular example is of interest because of the two definitions of *territorium* that Bartolus gives. The first is that it is a *res immobilis*, an "immovable thing."[61] Roman law made a distinction between *res mobilis* and *res immobilis*: immovable things, like land and buildings; compared to movable things, which covered other types of property.[62] A *territorium* is thus different from other kinds of property, but certain key aspects are shared with the wider laws on this issue. It is easier to get clarity if we look at Roman law on land. The Romans made a distinction between land and other kinds of property, but this was not especially significant, and as Miller notes, "Rights in land did not dominate private law as they did in subsequent European feudal systems." The law on property broadly applied to both kinds of things; the differences are more matters of detail.[63] As Robinson suggests, "For all of us in the western (feudal) tradition it is hard to grasp that the Romans had no land law. Land was (in classical law) just one of the *res mancipi*; there were no special forms of conveyance for land, no special kinds of security over it, no special rules about succession to it."[64] This shows that there was actually another distinction at stake: *res mancipi* and *res nec mancipi*. Land that was in Italy (south of the river Po) and certain other things—slaves; horses, mules, and donkeys; and houses on Roman land—were classed as *res mancipi*, which meant that certain legal procedures were needed to obtain or transfer *dominium*, full title, over them.[65] It therefore predominantly applied to agricultural issues. The distinction was abolished in Justinian's time, and Bartolus is therefore operating simply with the *res mobilis* and *res immobilis* separation.

The second definition of *territorium* is given in passing when gives his answer as to the jurisdiction. "*Territorium* is so called from terrifying [*ter-rendo*] [see *Digest*, L.16.239, §8]. So long as the army is there, terrifying and dictating to that place [*terret et coercet illum locum*], an offence there committed will properly be able to be punished by the authorities of the city as if it had been committed in their own territory [*in eius territorio*]."[66] This relation of territory and terrifying has been discussed elsewhere,[67] but this is a fascinating example of how the exercise of authority becomes *de facto*: an occupying force, because it is terrifying, can treat the land as its territory. But the reference Bartolus gives to the *Digest* is worthy of more attention. While other extracts in the *Digest* use the word—such as Ulpian's suggestion that "it is customary to sentence certain persons to be barred from remaining within the *territorium* of their native land, or within its walls"[68]—this is a definition. Pomponius declares that:

> the *territorium* is the sum of the lands within the boundaries of a *civi-tas* [*Territorium est universitas agrorum intra fines cuiusque civita-tis*]; which some say is so named because the magistrate of a place has, within its boundaries, the right of terrifying, that is expelling [*quod ab eo dictum quidam aiunt, quod magistratis eius loci intra eos fine ter-rendi, il est summouendi ius habent*].[69]

In his commentary on this passage, Bartolus makes an explicit link between the *territorium* as that over which the *civitas* exercises military force, and describes it as the power to punish or fix the limits of the laws over the terrified place.[70] The source for this is the sole book of Pompo-nius's *Enchiridium*, the *Manual*.[71] Pomponius died in 138 CE, and was working in the reign of Hadrian: he was the earliest of the major sources in Justinian's *Digest*, and the fourth most cited authority.[72] A piece of his *Enchiridium* in the *Digest* provides important indications on the earliest Roman law.[73] Several things are worth noting of this passage. The *territo-rium* is defined by the aggregation of lands, or fields (*universitas agrorum*). It is seen to have discrete boundaries (*fines*). The power of the magistrate, which is *terrendi*, or *summouendi ius*, is within those boundaries. The first fits with Isidore's later scalar model of the place of *territorium* within units of measure. The second provides a clear sense of the spatial extent of a polity at this time. The third takes that spatial extent as the limits of the magistrate's legal power. Yet this understanding was not the dominant one in Hadrian's Rome. At that time this would have been understood only in terms of constituent parts of the empire, rather than as discrete political units. Yet, for Bartolus in the fourteenth century, it allowed something very different.

Bartolus can thus be understood as trying to reconcile the universalist rule of the empire with the particularist rule of the individual rulers within and outside the empire. In his argument, the key to understanding the relation was not imperial permission—that is, the granting or delegation of powers—but rather the question of jurisdiction. This point was also important when there was conflict between local laws and the *Corpus Iuris Civilis*. Bartolus crucially did not always believe the latter prevailed; rather, it was a question of the scope of the former. This was the beginning of the discipline known as the conflict of laws—the question of competing jurisdictions, within what is now called private international law.[74] Maiolo therefore contends that "Bartolus paved the way for the modern conception of territorial sovereignty."[75] This requires some nuance, and care that modern notions are not read back into the text.[76] Maiolo is contending that the territorial state is there in his work, but not necessarily in a form we would recognize. Rather, he suggests there are the key elements of such an understanding. *Iurisdictio* and *persona iuridica* are two of those elements.[77]

This raises a significant point. Bartolus is drawing a distinction between rightful and effective jurisdiction, or *de iure* and *de facto*. On the former side, there is the empire, the *imperium romanum*, a political organization that can, in theory, extend across the world. On the other hand, there are the independent political units that actually exist within and beyond that *imperium*.[78] There may be a range of reasons for this, including the lack or limitations of imperial presence.[79] Bartolus is explicit about operating with the situation he finds: his challenge is not to the pretensions of the emperor, but to try to adjudicate on the actual basis of political power. He therefore splits jurisdiction (*iurisdictio*) into two elements: *imperium* and *iurisdictio* in a more restrictive sense. He equates power and jurisdiction, following the gloss in deriving the meaning of the latter from *ius*, law, and *ditio*, power (*potestas*).[80] This is how he can declare that "whoever would say that the emperor is not lord and monarch of the entire world [*dominum & monarcham totius orbis*] would be a heretic"[81] while still ruling on everyday legal matters.

> I say that the Emperor is the lord of the entire world [*dominus totius mundi*] in a true sense. Nor does it conflict with this that others are lords in a particular sense, for the world is a sort of *universitis*. Hence

someone can possess the said *universitas* without owning the particular things within it.[82]

Thus, ultimate obedience to the law of Rome was not the test of absolute authority. There were those who obeyed parts of the law, and those who held their independence to be dependent on imperial privilege or mutual agreement. This did not prevent them from being under the absolute authority of Rome. And there were those who had none of these links, but as long as they did not themselves claim to be lord of the entire world, they were not conflicting with imperial claims.[83] Thus, when it came to the relation between the pope and the emperor, Bartolus was clear. Their power should not simply be understood as different jurisdictions, but as operating in distinct territories, which a summons cannot pass between.[84] When it came to temporal power, their territories are similarly distinct— the lands of the church (*terrae ecclesiae*) and those of the empire (*terrae imperii*).[85] Again, there is a recognition of the limitless theoretical extent of the latter, since what is not in the first must be in the second, with the exception of independent cities.[86] The pope therefore has spiritual jurisdiction generally, and temporal jurisdiction in the lands of the church; the emperor has temporal jurisdiction only in the lands of the empire.[87] Bartolus therefore recognized the *specific* temporal jurisdiction of the papacy rather than its universal aspirations.[88] Woolf calls this the universal and territorial conceptions of powers—the papacy and the empire—but this is not quite right, since the papacy is similarly territorial in its temporal scope.[89] This distinction extends to the remit of the law: civil law applies in the territory of the empire; canon law applies in the territory of the church, but also extends to the empire for spiritual issues.[90] The pope has universal spiritual lordship, but the emperor does not have *de facto* universal temporal lordship. Canon law therefore applies in the lands of the church in temporal matters and universally in spiritual matters; but civil law only applies in the lands of the empire.[91] Civil law, then, crucially, is entirely limited in its spatial extent; whereas canon law is only limited when it pertains to temporal matters. Rather than universal and territorial conceptions of power—the former belonging to the pope and the latter to the emperor—we have a distinction between *laws*, with canon law claiming universality and civil law territorially bound:[92] a universal empire in theory, and territorially bound polities in practice.[93]

Bartolus made a distinction between cities that did acknowledge a higher power and those "cities which recognise no superior."[94] Those that

did owed their allegiance to the Roman Empire, and most Italian cities
tended to be dependent in this way on an imperial overlord. Yet some had
independent constitutional rights—that is, a legal basis.[95] In his analysis
of tyranny, Bartolus discusses a "city [*civitas*] or fortified place [*castrum*]"
that might have the right to choose its own ruler, but might not.[96] Those
cities that do not recognize the emperor as their superior were independent, a situation Bartolus described in the following way: "In such a case
civitas sibi Princeps, the city is a prince [or even emperor] unto itself."[97]
The designation of the *civitas* as having no superior, of not recognizing
a lord (*dominum*), is deliberate: a parallel is being drawn with the king of
France. The *civitas sibi princeps* should be understood to mean "the city
is a prince itself," not, as it was later reinterpreted, "the people are the
prince."[98] In his critique of tyranny, he similarly stresses this: "Those who
hold the *res publica* by tyranny detain it by force from the *res publica* itself [that is, the *civitas*] or its superior lord, not from any private person."[99]
Although Bartolus does ultimately think that the citizens themselves
have the authority, which is clear in his commentaries on the *Digest*,[100]
he should not be predominantly seen as a nascent theorist of democracy.
Skinner has suggested that both Marsilius and Bartolus see sovereignty
as tied to the people, who might delegate power but do not alienate it.[101]
While this is important, arguably his most significant contribution lies
in seeing the cities as independent of the empire on a legal basis, and as
exercising jurisdiction within their *territorium*. In his stress on the *res
publica*, he is ultimately closer to Cicero than to Locke.

Kantorowicz has described the situation resulting from this as hierarchical, with power working downward from the empire to the kingdoms and *civitates*. He therefore stresses the relation between *rex imperator in regno suo* and *civitas sibi princeps*.[102] Yet it is crucial to note that
Bartolus does not use the *rex in regno* phrasing, even though he would
surely have been aware of it. Woolf claims that the *civitas sibi princeps*
phrase is functionally the same, and that therefore the logic of the relation
holds. His suggestion is that it was just that Bartolus applied it to the issues he was concerned with, which were those of the *civitas* and not the
regnum.[103] This legal power is, crucially, just as it was for the king, territorially restricted. As Canning has noted, the attribution to the city of
the same powers as the emperor within its territory is Bartolus's "juristic
masterstroke."[104] This is the key: the law is territorially determined. Canning has suggested that this effectively means sovereignty can be understood as "the powers of the *princeps* within its own territory (*civitas sibi*

princeps)."[105] Though the term *sovereignty* is a little premature, there is certainly a recognition of a territorial scope and target of power, which would in time become sovereignty:

> If the *Princeps* concedes to you a *territorium* as a whole [*universaliter*], it seems that he has conceded to you jurisdiction as a whole too; because if someone gives you a thing he has given you *dominium* over it, and so it seems that whoever concedes the *territorium* as a whole has conceded jurisdiction over it as well, which [the relation between jurisdiction and *territorium*] is the same as *dominium* and some particular thing.[106]

Yet the most important development is that the *territorium* becomes not simply a possession of power, nor incidentally the extent of that power, but the very object of political rule in itself, and, as a consequence, that rule is over the things that take place within it. The *civitas sibi princeps* is a question of object as much as extent and hierarchy. Bartolus defined the urban as signifying "a quality more than a place," at least, more than simply a place bounded by walls.[107] Bartolus even talks of the extent (*spatium*) of a *civitas*'s control of the sea off its coast, what he calls the *territorio mari*, maritime territory. He suggests that this should extend one hundred miles, and no farther than two days' travel.[108] As Miceli notes, this is an echo of the land claimed by some cities: two days' travel from the city gates.[109]

Bartolus is similarly important because of a distinction he draws between different *territoria*:

> Some *territoria* are distinct, but are nevertheless all under the same lord [*domino*], just as the Roman Empire is divided into provinces [*presidatus*]. . . . Some *territoria* are distinct and separate and not under the same lord [*domino*], as is the case with the *territorium* of the empire and the pope, and then no summons runs [*non potest fieri citatio*] between one *territorio* and the other.[110]

In many respects, this was simply a recognition of the continuation of a preexisting situation: the basis of *Pastoralis cura*. While cities are *de jure* subject to the empire, they are *de facto* independent for day-to-day tasks. Internally they recognize no superior, but externally relations are still subject to the empire.[111] But it is the description of the object of their rule as a *territorium*, which gives both the object and scope of jurisdiction, that is key. Bartolus offered a view on this too, supporting the view of *Pastoralis cura*, in opposition to the emperor's *Ad Reprimendum*. The pope laid claim to certain lands as their direct temporal ruler, not simply by nature of his role as spiritual ruler, and therefore within those areas people were bound to him temporally, not to the emperor. The transfer of the Roman Empire to the church meant that "the vicar of Christ possesses both the spiritual and the temporal swords."[112] Bartolus argued, effectively, that the emperor had no spiritual jurisdiction, since the pope had this universally. Both had temporal jurisdiction only within geographically circumscribed areas.[113] For Bartolus, even though the pope had transferred authority to Charlemagne, the key elements of sovereignty had been retained. In some areas under direct papal control, there was temporal and spiritual control; in other areas, the emperor had temporal power that was bestowed on him by the pope.[114] So this denied a rejection of any clear separation between the two powers. Bartolus therefore grudgingly accepted the validity of the Donation of Constantine, because he was a Christian in the land of the church.[115] That said, his defense is equivocal, largely using quotations from others and presenting both sides of the case.[116]

Bartolus was not an elegant writer. The humanist Lorenzo Valla—who was discussed in chapter 4 in relation to the Donation of Constantine—was particularly scathing. The Latin of Bartolus, Accursius, and Baldus was, he declared, "not the language of the Romans at all, but barbarian," ignorant, ineptly written, and it was more like honking geese (*vocem cantum habere cygnorum*).[117] Bartolus was not even close to Cicero: instead, he spoke like an ass.[118] More substantially, Valla rejected the approach of reinterpreting laws in new contexts.[119] Emerton suggests that the first of these criticisms misses the point. Bartolus, he concedes, has "no style at all." But this is unimportant, and probably would not have concerned him: "Latin is for him only a sort of code, required to make him intelligible to his colleagues, and his only decoration is found in his continual references to passages of the civil and the canon law. [Coluccio] Salutati's work was written to be read; Bartolus's was written to be used."[120]

Before we leave Bartolus, two other aspects of his work are worthy of attention. First, and merely in passing, there is his work on insignia, *De insignia et armis*.[121] While this was a work that was written to rule on who could use coats of arms, with questions of appropriate design and the laws of their inheritance, it had a long afterlife as a work on rules of heraldry more generally. As Groebner suggests, it was "no mere juristic formalization of heraldic categories but rather a key text on late-medieval semiological practice."[122] The second is much more significant. This was a text that Bartolus claims to have written while on holiday in 1355. He was walking along the banks of the river Tiber and became interested in some of the property issues in land that would arise if a river changed direction. How should alluvial deposits be divided between the landowners of the banks of a river? How should an island that emerged in a river be apportioned? And who owned the rights to a dried-up riverbed?[123] This text is the *Tractatus de fluminibus seu Tyberiadis*—the treatise on rivers or the Tiber. It comprises three parts: *De alluvione*, *De insula*, and *De alveo*. *De alluvione*, on how to divide up alluvial deposits; *De insula*, on how to divide up an island; and *De alveo*, on the dried-up riverbed.[124]

In the treatise, Bartolus claims that he was visited by a figure in a dream who said to him, "Look, I brought you a reed pen for writing, a compass for measuring and drawing circular figures, and a ruler for drawing straight lines and making the figures."[125] This is the remarkable feature of this work. Although it is a discussion along the lines of many of his other works, it also makes use of a rudimentary geometry to demonstrate how the principles he puts forward could be put into practice. Several figures accompany the text, to demonstrate the use of parallel and nonparallel lines, types of angles, and ways of bisecting angles and dividing areas, such as alluvial deposits. In the original they were drawn by Bartolus himself, though of course those that exist are largely copies, some more embellished than others. The first part of the text has twenty-two diagrams; the second seventeen; whereas the final part simply makes use of the methods already outlined.[126]

As Cavallar puts it, "Although deeply rooted in the medieval exegetical tradition of the commentary, this tract displays an unprecedented feature: a mixing of disciplines—that is, law and Euclidean geometry."[127] Whether the dream figure is genuine or, more likely, a literary device is in a sense unimportant. What is significant is that Bartolus feels the need to justify the use of diagrams in the work, since using geometry in the interests of a legal process was unusual. As Franklin notes, "It remains one of the few legal treatises to contain geometrical diagrams";[128] indeed, Cavallar has even described it as "the first tract devoted to legal geography."[129] It acts as a foundation for the three texts on political issues: on rival factions, the city, and tyranny.[130] What is significant here is that for the understanding of property rights over land—an economic question—Bartolus shows the importance of the law and technique. These would be key to the development of territory as an object of political rule. Indeed, the word Bartolus usually uses for land in this work is *territorium*.[131]

ceam, vt in puncto d. fiat circulus ; qui tangat punctualiter prædium
Titij tantum ; occupabit partem de prædio Caij, vt oftendit circulus
azureus A F G. Refpondeo , hoc, quod alicui accreuerat iure propinquitatis, habet locum , quando illud ; de quo acquirendo agere-
tur, neutrius fundo coheret, vt Infula, quando vni coheret, & alio
non , femper illi cedit, cui coheret , nec attenditur proximitas alte-
rius, vt. ff. de acquir. rerum domin. l. Infula, in princip.

PROPOSITIO VII.   FIG. XII.

I G V R A ifta facta eft, vt oftendatur quòd prædia
quædam poffunt effe ; quibus de alluuione non de-
betur vfq; ad flumen, fed ante intermoritur ; Vt fi
alluuio contineatur inter duas lineas , fcilicet a b ,
& b c; quæ faciunt angulum in puncto b. qui angu-
lus cadit quafi in medio prædij Titij . Primò ergo
fiat linea rubea diuidens per mediü, vt linea b F. vt probatum eft fupra in octaua figura . Ex quo patet, quòd quicquid eft ab ea linea
fupra verfus caput fluminis, cedit illi ripæ a b. & quicquid eft ab

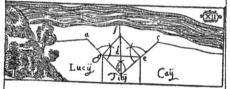

illa linea infra verfus finem fluminis, cedit illi ripæ b c. & hoc ratione
propinquitatis, vt dictum eft ; Nec prædium Titij habet diuidere
fuperiori vicino, & inferiori . At fi bene refpicias , punctus diuidens
prædia cadit in lineis rectis fecundum fe , ergo debet diuidi dicendo
lineam rectam fuper eam , vt oftenfum eft fupra in feptima figura, fi-
cut oftendit linea e f. & linea g f. Et fic in puncto f. portio contin-
gens agrum Titij finitur, & intermoritur : Si enim tranfiret vltra, vel

si linea

tur alius punctus ; qui magis fit proximus extremitati fecundæ infulæ ; quæ plus proxima eft infulæ primæ, & fit b. Deinde pone pedem
circini in puncto a, & extende ita quòd pertranfeat aliquam partem
infulæ fecundæ, & volue femicirculariter . Deinde pone pedem circini in puncto b. & per tantundem fpatium extende , & nota vbi
dicti duo circuli fe profcindunt, fcilicet punctos c, d, & ab illis duo-

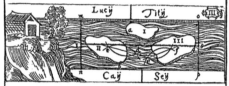

bus punctis duc lineam rectam rubeam C D. ex quo concludo, quòd
quicquid eft infulæ fecundæ fupra dictam lineam c d, totum cedit infulæ primò relictæ : Et eodem modo comperies in infula tertiò refi-
cta, vt lineæ ductæ oftendunt , quod fic probatur . Si ab æqualibus
æqualia adimas ; quæ remanent , funt æqualia : fed conftat, quòd tan
tum diftat circulus vnus à fuo centro a. quantum diftat alius circulus
à fuo centro b. & conftat , quòd linea c d, adimit ab vtraque parte
æqualiter: Ergo patet, quòd linea illa æqualiter diftat ab vtroque: &
fic quod eft fupra illam lineam, cedit puncto a. iure proximitatis,
quod verò eft infra , cedit puncto b, iure proximitatis .

FIGVRATA PROPOSITIO IIII.

V P R A D I C T V M eft de figuris ; quæ habent lineas
rectas iuxta ripam : Hic videndum eft de illis : quæ
habent ripas tortas, & curuas ; veruntamen hoc con
tingit dupliciter . Nam quandoq; funt retortæ ab
vtraq; parte fluminis, & ita vna ripa inflectitur verfus aliam, q̃ vbiq; ipfæ æqualiter diftant, vt in hac figura oftenditur, licèt hóc rarò contingat: Quandoq; ripa eft recta ab

vna

Fig. 8*A* and 8*B*. Bartolus, *Tractatus de fluminibus*, ed. Hercule
Buttrigario (Bononiae: Ioannem Roscium, 1576), 39, 79.

BALDUS DE UBALDIS AND THE *CIVITAS-POPULUS*

Baldo degli Ubaldi (ca. 1327–ca. 1400), better known as Baldus de Ubaldis,
was Bartolus's most famous pupil.[132] Franklin has described him as "the
most philosophical of the medieval legal writers," particularly stressing
his extensive knowledge of Aristotle.[133] Like his master, he wrote commentaries on the *Corpus Iuris Civilis*, but unlike him he did not write
tracts on specific political topics. Thus, while his works include a lot of
political discussions, these are solely in his juristic works.[134] He is most
important as a writer of several thousand *consilia*, of which sixteen hundred are still extant in the manuscripts for the period 1380–1400.[135] This
makes it very difficult to discern the overall structure of his argument.
As Pennington notes, "Reading Baldus can be exasperating. He could be
opaque, even purposefully obscure."[136] Baldus often proceeds in a manner

that bears comparison to Scholasticism: he states the question; provides an answer, which he critiques; and then provides the true response.

Baldus held that the pope had supreme spiritual jurisdiction, and the emperor supreme temporal jurisdiction. While he considered the pope to be the emperor's superior, this was an otherworldly superiority, with the emperor dominant in this world.[137] Like many of his predecessors, he recognized that the *terrae imperii* and the *terrae ecclesiae* were separate, and in the latter the pope held both spiritual and temporal jurisdiction.[138] While he held that spiritual power was above temporal power, he declared that "temporal jurisdiction is in every way separate from the church's jurisdiction, except in the lands of the Church [*terris ecclesiae*]."[139] While the popes were trying to stress their own temporal power, and the exclusion of the emperor from the papal lands, in so doing, this stressed the spatial limitation of temporal power. This gave independence both to the emperor and to rulers outside of the empire. Yet it is crucial to note that the popes created this situation, precisely because of the stress they put on their own temporal power in the *terris ecclesiae*. In doing so, they excluded the emperor from control in those areas, thus limiting the geographical extent of the empire,[140] a situation underlined in the bull *Pastoralis cura*. Baldus declares that "whatever the king can do in his kingdom [*rex in suo regno*], the pope can do in the ecclesiastical monarchy."[141] This papal power—and its explicitly geographical location and limitation—was one of the key foci of Baldus's work.[142] This is interesting in that it reasons from the secular case back to the power of the papacy. Indeed, as Canning has suggested, the key principle of his political thought was "the acceptance that universally sovereign authorities, in the form of the emperor and the pope, coexist with territorially sovereign entities, that is independent city-republics and kingdoms."[143] In this he is building on and developing the claims of Bartolus. Baldus stresses the functional equivalence of the different polities, and his claims to one do extend to all.[144] These attributes are determined by territorial extent.

Under the Roman legal system, the alternative for the upper class to death or forced labor was banishment. There were different degrees of this: people could be expelled simply from Rome or from Italia as a whole. There was a more extreme form of deportation, which was perpetual exile, with the loss of citizenship and property.[145] Similar dispossession could be caused by capture. If someone was captured by brigands, this does not apply; but it does apply if captured by an enemy of Rome—that is, someone on whom Rome has declared war, or who has declared war on Rome.[146] Discussing the family relations that come from the capture of a father, the *Institutes* notes that the status of the children is held in suspense, because in time the father may come back: the right of rehabilitation (*ius postlimini*).

> The word *postliminium* comes from *limine* [threshold] and post. When a prisoner of war crosses back over our borders [*fines*] we can say he has crossed the threshold again [*postliminio reversum*]. The ancients chose to see the boundary of the *imperium* [*imperii finem*] as a threshold, as though the limit of a home [*domibus finem*]. The word *limes* [a limit], indicating an edge or end [*finis . . . terminus*], has the same origin. *Postliminium* comes from the prisoner's later re-crossing the same threshold over which he was lost. If he is recovered by our going out and defeating the enemy he is still said to have re-crossed the threshold.[147]

In his discussion of banishment and extradition, Baldus claims that "jurisdiction inheres in a *territorio* . . . but a *territorium* has its own boundaries [*fines limitatos*]."[148] Territory and jurisdiction go together, he suggests, "as mist to a swamp [*sicut nebula sup palude*]."[149] Baldus did not give *plenitudo potestatis* to cities, and effectively gave them sovereignty below that of monarchs.[150] Kings have the right of *rex in regno* because they are separate from the empire, but cities are part of the empire.[151] There is thus a hierarchical model in Baldus that Bartolus was trying to collapse. In distinction to Bartolus, Baldus does not say that the city is a *princeps* itself, but rather that it holds that position through delegation. A judge is elected "by a people [*populo*] in the emperor's place [*vice imperatoris*], because it is the *princeps* in its *territorio*."[152] Baldus gives cities rights because they "fill the place [*vice*] of the emperor in their *territorio*."[153] Thus, the outcome reached is in some respects the same as Bartolus, but the route taken to get there is distinct. For Baldus the territorial aspect is the key definition of the power of cities. As Canning puts it, "The sovereign city replaces the emperor within its territory. Territory defines as much as it limits a city's sovereignty."[154]

Alongside the territorial dimension, Baldus stresses the importance of the people, and the endurance of the arrangements beyond the lives of those ruling it:

> A realm [*regnu*] contains not only the material territory [*territorium materiale*], but also the peoples of the realm [*gentes regni*] because the collective *populus* [*populi collective*] is the realm. . . . And the uni-

versitas or *respublica* of the realm does not die, because a *respublica* continues to exist even after the kings have been driven away. For the *respublica* cannot die.[155]

In the opening book of the *Digest*, Gaius makes the distinction between the laws particular to a *civitas*—that is, *ius civile*, civil law—and those laws, deriving from natural reason, that are the *ius gentium*, the law of nations or peoples.[156] The point with the first is that it is peculiar to a specific polity or peoples; whereas the second applies to all people, or more specifically the type of law Rome developed to mediate between Romans and others.[157] Commenting on this passage, Baldus defines the *populus* as those people within a territory, because jurisdiction "adheres to a *territorio*, and those inside possess the *territorium*."[158] While Baldus's phrase *iurisdictio coheret territorio*[159] is a development made possible by Bartolus,[160] the crucial step that Baldus takes is that the *populus* itself is a territorial entity, defined in its corporate state by its bounds.[161] Territory is thus not just the limit of the jurisdiction but its very definition.[162]

This is crucial, because the *civitas* is thus political, territorial, and comprised of a population. Taking a phrase from Aquinas, but rendering it political, Baldus suggests that the people transcends the individuals that make it: "Separate men do not make the *populus*, and the *populus* is properly not men, but a collection of men as *unum corpus mysticum*, taken as abstract, the significance of which has been revealed by the intellect."[163] The dividing and combining faculty of the intellect is important in the development of the idea of the polity as a corporation: a fictive legal entity that comprises but exceeds its members.[164] Bartolus and Baldus thus argue that the *populus* was a *universitas* within the city; that could be a *persona*, which could be a *princips*.[165] Canning has suggested that the *civitas-populus* can therefore be understood as a corporation that works both as an abstract entity "distinct from its members and government" and as a "body composed of natural, political men." He claims that this means it is therefore a state, under Skinner's definition of that as "an abstract entity distinct from its members and government." Canning therefore dates this two centuries earlier than Skinner, who saw it as emerging only in the late seventeenth century. Canning suggests that Baldus grafts "an abstract dimension onto an Aristotelian idea of the state" through his use of legal ideas.[166]

What is important in this is that it is not simply the Aristotelian point that the *polis* is its members, a congregation.[167] Marsilius had seen the political community as distinct, but here the political unit has a territorial extent and an existence beyond, and separate from, its members. While in practice Baldus really only applies this to cities, it was clear that this principle could be extended to kingdoms as states too,[168] which he admits almost in passing as a prelude to his argument about cities. Baldus makes this distinction on the basis of recognizing the pope's supremacy in the other world, and as a mediator of God; and the emperor's superiority in this world. Canning sees this as a *de iure* and *de facto* distinction, which does not simply apply to papal-imperial relations, but to all levels of political authority. Canning argues that this can be understood as sovereignty, but only if universal sovereignty and territorially bound sovereignty are understood hierarchically and not exclusively.[169]

REX IMPERATOR IN REGNO SUO

At the end of the fourteenth century, the papacy was in a profoundly different position. If at the beginning of the century, with Boniface, it had truly aspired to a universal supremacy in both spiritual and temporal power, it had quickly moved to Avignon. The initial aims of the Avignon popes were to pacify Europe and mend the damage caused from the unrest with the French, and, more grandly, to recover the Holy Land and restore the Papal States.[170] Wilks contends that for some the liberation from Rome was a blessing, since Rome continued to be seen as an Italian rather than a universal city, with a tendency to localize the papacy and its aspirations, and especially to tie it to the resurgent empire. The move to Avignon could thus be seen, from a certain perspective, as creating "a truly universal monarchy. The 'Babylonish captivity,' often regarded as being in fact the nadir of the medieval papacy, was in theory its crowning triumph."[171] Yet even if this were the case, the exile to Avignon paved the way for the Great Schism (1378–1417), when there were competing popes, each with their own power base and location. This was a profound crisis, from which the papacy has never fully recovered. If, in the first half of the century, the controversy was the church versus secular rule, in the second half the church turned more in on itself, with 1378 as the culmination of this split.[172] The popes, henceforth, were concerned as much with their own standing as with any claims to universal temporal rule.

In contrast, the secular rulers of Europe were keen to adopt that other legacy of Rome, its law. This was uneven, of course, but in time most kingdoms in continental Europe would utilize it. German emperors saw themselves as successors of Roman Caesars, so they were particularly eager to use the law. The Holy Roman emperor Lothar had adopted parts of the law in the twelfth century, leading to the designation of the Holy Roman Empire of the German Nation.[173] Later emperors, notably Frederick I, saw it as important for the distinctive legal and political system they were establishing.[174] Outside of the empire, the imperial origins of the law were, in distinction, a barrier to its adoption. However, as Nicholas notes, the supremacy of the emperor became a benefit to kings and princes when they realized they could adapt the idea that they were emperors within their own territories. Suddenly all the texts propounding the absolute power of the emperor—that is, both indivisible and inalienable—could be used to their benefit.[175] The *Digest* declares, for instance, that "a decision by the *princeps* has the force of law. This is because the *populus* commit all their *imperium* and power to him and on him, doing this with the *lex regia*, which gives him the *imperium*."[176] The idea that the ruler was not bound by the law, as he makes the law, was of great appeal to absolutist rulers.[177] The reason for this is that the *constitutions*, the enactments or edicts of the emperors, found their way into the *Codex* (and the *Novellae*).

Looking at Bartolus and Baldus alongside the work analyzed in chapter 6 demonstrates that the fourteenth century is one of considerable importance in the history of political thought. This acts as an effective rebuttal to the claim by the Carlyles that the thinkers of this century did not add much to those of the previous two. They claim that these thinkers had no "appreciable influence" except that they confirmed that all authority in the state came from the community.[178] It is actually much more complicated. Not only did Aegidius argue for the continuing importance of the papacy, but Ockham and Dante put the emphasis on the empire. John of Paris and others made the case for the king. Only in Marsilius and Bartolus, and Baldus to a lesser extent, can anything like the power of the community be seen. And in the jurists this is arguably not the most significant aspect of their thought. Rather, it is the stress on the object of political rule, which both reconciles the conflict over jurisdiction and gives the very extent of power, that is the key development. If Skinner's definition of the state as an abstract entity separate from both the governor and the governed is accepted, then Canning is on solid ground in seeing this in the late thirteenth and fourteenth centuries, in the works of these thinkers.[179] Yet this notion is not in the *Corpus Iuris Civilis* itself: it required the work of the Commentators, building upon the labors of the Glossators.[180]

While Bartolus and Baldus are infrequently discussed today, they were extremely influential on succeeding generations of thinkers. The central aspect of their influence is that the previous distinction between spiritual and temporal power gets recoded as a question of jurisdiction, not merely in canon law, but in civil law.[181] Canon law, the embodiment of papal claims to supremacy, was geographically limitless and unrestricted, but this no longer provides the basis for either the exercise or legitimation of secular power. Civil law, understood as the exercise of political jurisdiction, was the privileged form of power within the empire, the kingdom, or the independent *civitas*, but its very basis was that it was restricted to the territory in which it was exercised.[182] In 1346 Pope Clement VI recognized this, describing Emperor Charles IV as head of a *status sacri imperii*, which was *"spatiose dilatatus,"* a sacred imperial state of great, but not unlimited, spatial extent.[183] A distinction that was initially on the basis of time—the temporal span of human life on the earth or the immortality of human souls—is recoded spatially. Temporal power becomes territorial, spatially limited but supreme within its extent. Eternal or spiritual power can be understood as operating extraterritorially, with its extent unlimited but its scope severely restricted.

The source of the phrase *rex imperator in regno suo* is much debated. Woolf suggests that it can be found in Oldradus in the early fourteenth century, when he declared *"quilibet rex de facto teneat locu imperatoris in regno suo."*[184] Yet even Woolf recognizes that earlier figures might be given: Cino de Pistoia, Gulielmus Durandus, Andrea d'Isernia, Marinus da Caramanico, or Jean de Blanot. As Armin Wolf notes, this formula was sometimes written in the form of *rex imperator in terra sua*—"the king is the Emperor in his lands," rather than kingdom.[185] John of Salisbury, for instance, in an 1168 letter notes how King Henry thinks that he has the pope and the cardinals "in his pocket" and has achieved the position of his grandfather, "who in his own land was king, papal legate, patriarch, emperor, and everything he wished [*qui in terra sua erat rex, legatus apostolicus, patriarcha, imperator et omnia quae volebat*]."[186] As early as the mid-thirteenth century, Henri de Bracton had declared the supreme power of the king of England. For Bracton, "The king has no equal within his kingdom [*parem autem non habet rex in regno*]. . . . The king must not be under man but under God and under the law [*sub deo et sub lege*]."[187] Morrall suggests that the link to the kingdom comes from at least a century earlier than Oldradus, with the dispute being whether it was French or Neapolitan lawyers who formulated it.[188] Ullmann believes that it was both, independently: the French formulating it in terms of the absolute authority of the king's courts with no further right of appeal; the Neapolitans in terms of Sicily. As was recognized in *Pastoralis cura*, the king of Sicily was not a subject of the emperor. It has also been suggested that the idea could work at a more local level as well: a man's house is his castle.[189]

Yet Woolf is undoubtedly right when he stresses that the authorship is not the key question: "Whoever it was, Oldradus or some other lawyer, who took the step, he did a work of the highest importance in the history of political thought and one deserving the highest praise."[190] By the time of Bartolus, it had become a commonplace: the term is neither his invention nor his formulation. Woolf nonetheless gives Bartolus credit for applying it consistently.[191]

The king is equally he who recognizes no superior in temporal affairs—a notion that dates from *Per venerabilem*. This is a formula that stresses the personal standing of the ruler: a supremacy over others without the possibility of legitimate judgment over him.[192] The granting of the right of supremacy within the kingdom means that the king has what we would now call exclusive internal sovereignty. But this implies nothing about what became understood as equal external sovereignty. It did not imply that the emperor would treat the king as a formal equal in foreign or "international" issues. Thus, the kings have supremacy within the kingdom, but remain subservient without. The distinction cannot be simply collapsed into *de iure* and *de facto* sovereignty. As Wilks underscores, this "explains the perplexing tendency of so many writers to describe the emperor as a world ruler in one sentence and the king as his own emperor in the next."[193] Wilks helpfully stresses the other crucial point. The invocation of *imperium*—from this point on, if not from much earlier—has two connotations. There is a universal empire in theory, but an actual empire in practice. The empire that actually exists and that exercises power is much smaller than the hypothetical empire that has no limit. Wilks suggests that this means that "*imperium* has both an authoritative and a territorial connotation: there is an empire within an empire."[194]

Thus, two formulas of quite distinct beginnings are fused, and added to another of judicial source. The king is he who recognizes no superior in temporal affairs, and his standing in his kingdom is functionally equivalent to that of the emperor in the empire. The first stresses the personality of the ruler; the second gives both a spatial definition and limit to that rule. Put together, they form the composite "*rex imperator in regno suo, superiorem in temporalibus non recognoscit.*"[195] Emperor Frederick II put this into practice in 1231, when he published a collection of Sicilian constitutions. He was king of Sicily as well as emperor, and stressed the similar standing each position gave him in relation to the pope. According to Kantorowicz, he was "the only monarch of the 13th century who literally acted in accordance with the new maxim *Rex est imperator in regno suo.*"[196] It becomes interesting when used to justify independence from within the empire.[197] Yet it is the juristic understanding of *iurisdictio coheret territorio*, when added to these previous ideas, that completes the relation between supremacy, spatial extent, jurisdiction, and territory.[198]

The basis for the absolutist monarchies is already here.[199] Each king saw himself as emperor in his kingdom, recognizing no superior within those bounds. The territory became the object of political rule, and people within it, and actions that took place therein, were under the king's jurisdiction. No higher authority could be appealed to.[200] This shift had been developing for some time. There were economic reasons, with the development of the feudal system being key to the establishment of a more developed system of land.[201] There were strategic reasons with the sedimentation of previously nomadic groups and the relative stability of a number of states: England under the Normans, France under the Franks, and some of the constituent parts of the Holy Roman Empire in central Europe. There were developments in the law, necessitated by the kinds of conflicts that were arising but enabled by the adaptation of the centuries-old *Corpus Iuris Civilis*. Indeed, while Arnold sees "territorial lordship" as long existing within Germany, he concedes that it was Roman law that articulated and codified this.[202] Yet if the theoretical basis was there, and the political will evident, the practicalities were still largely undeveloped. While the kings were basing their rule on an equivalence to the emperor, the emperor's own rule was far from secure, however. Paradoxically, the emperor increasingly had less standing in the empire than kings did in their kingdoms.[203] As Bartolus and Baldus had shown, cities could claim to be a *princeps* unto themselves.

Nicholas of Cusa (1401–64), wrote widely on theology, mathematics, and science. His scientific work will be touched upon in chapter 9 below. Here, the emphasis is on his work on the politics and organization of the Catholic Church, *De concordantia catholica*.[204] It has been described as "easily the greatest of fifteenth century political writings, and one of the most interesting of the later middle ages."[205] In this work Cusa argues that the council of the church is more important than the pope, and it has therefore been described as "the greatest and most complex work of conciliar theory."[206] Cusa sees the church as an organic whole:

In my treatise on the Catholic concordance, I believe that it is necessary to examine that union of faithful people that is called the Catholic Church, as well as the parts that together make up the church—i.e.,

its soul and body. Therefore we will consider first the church itself as
a composite whole, then its soul, the holy priesthood, and thirdly its
body, the holy empire.[207]

Cusa therefore believes that "the holy empire itself comes from
God,"[208] but not through the church. Rather, it has a distinct basis. As Karl
Jaspers puts it, "*Sacerdotium* and *imperium* constitute a unity of different
elements."[209] As Cusa cautions, the ancients "were concerned to prohibit
the mixing of the spiritual and temporal concerns among the clergy."[210]
Yet, in so doing, Cusa effectively took away the legitimacy of this polity
compared to others. As Woolf puts it, Dante's *Monarchia* is not the "swan
song" of the empire, and if it had one, then Nicolas of Cusa sang it.[211]

Cusa is of interest for his discussion of the Donation of Constantine,
which leads to his spatial determination of the law, unusual in a theologian. Cusa contends that while Constantine *could* have made the donation, this does not prove that he did. He notes that there is no historical
record of this, and that in distinction there are several instances that show
that the "Roman pontiffs acknowledged the emperors as their overlords"
even after Constantine.[212] Later, though, a transfer did take place:

> For after the cities and places [*civitates et loca*] named in the acts of
> Stephen II became the legal property of St. Peter because of the gift
> of Pepin, the father of Charlemagne, and more land [*loca*] was added
> later because several cities put themselves under the legal jurisdiction
> of St. Peter and their citizens cut their hair in the Roman fashion, there
> was a need for a patrician to defend those lands.[213]

The political power over land comes, though, with a constraint: jurisdiction is tied to the land itself. He speaks, for example, of "the law
that sets limits [*terminorum*] and decrees that no one should cross another's boundaries [*terminus*]," and says that this can sometimes be set
aside because of the "negligence of those in the lower ranks or because of
necessity."[214]

> We should note that his power to command does not extend beyond the
> boundaries of his empire [*terminus imperii*] under him, as is evident in
> the text *Ego Ludovicus*, where although Louis describes himself as em-
> peror, he issues commands only to the inhabitants of the kingdom of
> France and the Lombards who were his *de facto* subjects. And follow-
> ing this we should say that the emperor is said to be lord of the world as

ruler of the empire that the Romans once conquered by their valor . . .
deriving that title from the fact that the Romans had the greater part of
the world under their rule.[215]

Cusa provides a detailed list of these limits. They stretch from the
Caspian Mountains to the gates of Alexander in northern Scythia—the
southern parts of Russia, Ukraine, parts of the Caucasus, Kazakhstan—to
Norway, the lands beyond the Caspian Sea and the Himalayas, China, the
Persian desert, India, Arachosia (Pakistan), Ceylon, and large parts of Africa and Arabia. Cusa's source for this is Ptolemy's geography. Together
the lands so excluded "make up no small part of the world—in fact, almost half of the inhabitable land."[216] No less an authority than Jaspers
had suggested that as a political thinker, Cusa is "antiquated," especially
compared to Marsilius.[217] Yet this is to miss the way that Cusa takes legal arguments into his political theory, developing the claims of the Post-
Glossators, and puts limits to the empire.

Later generations of legal thinkers also developed these claims.[218] The
Spanish theologian and jurist Francisco de Vitoria (1492–1546), for example,
made extensive use of Bartolus in his writings on the indigenous peoples of
North America.[219] Vitoria's broader concerns include the question of spiritual power and civil power. Vitoria suggests that "royal power is not from
the *republica*, but from God himself . . . even though kings are set up by
the *republica*, royal power derives immediately from God . . . the *republica*
does not transfer to the sovereign its *potestas*, but simply its *auctoritas*."[220]
Vitoria therefore makes a distinction between the Latin *potestas*, meaning authority, and strength, capability, *potentia*. Thus, he suggests that "to
ask whether the Church has some 'spiritual power' is equivalent to asking, first, *whether the church has some force or authority in the spiritual
domain*, and then, *whether that force or authority is distinct from civil
power*." His answer is clear: "There must exist in the Church a spiritual
power of some kind, distinct from civil and lay power."[221] This kind of
power is, for him, superior to royal power,[222] which is supreme in civil matters but limited in religious ones: "Royal power comprises all civil power,
since to be king means to reign supreme over all things in the *republica*;
yet the king has no authority over liturgy and spiritual actions; therefore
spiritual power is different from civil power."[223]

Figgis has suggested that writers such as Grotius and Bodin do "not
merely quote Bartolus, but are what they are largely because of him."[224]
Canning argues that a similar debt is found to Baldus, whom he describes
as "a seminal contributor to the juristic mainstream in the development

of European political ideas."[225] Jean Bodin's work on sovereignty will be discussed in chapter 8 below; here some brief remarks on Hugo Grotius as a key figure in the development of international law are in order.[226]

Grotius's major work is *De jure belli ac pacis* (The right of war and peace).[227] It is a book that has been described as bearing "the impression of the military revolution" on every page, where it is useless to try to maintain the old rules and standards of conflict in the face of technological changes.[228] One aspect of the work was to discuss when war was acceptable. Yet anticipating what was later formalized as a distinction between *jus ad bellum* and *jus in bello*, Grotius also sought limits to what was acceptable in war, even though these were extremely wide: it could be lawful to kill a prisoner; to assassinate; to devastate lands and cities even after surrender; to treat civilians, including women and children, as if they were combatants.[229] Grotius's focus is on *imperium* rather than jurisdiction. He suggests that it is exercised over "two subjects, primarily persons, and that alone is sometimes sufficient . . . secondarily the place [*locum*], which is called territory [*territorum*]."[230]

Grotius rehearses the etymological arguments for the meaning of the term, suggesting that the "derivation of the Word *territory* given by Siculus Flaccus, from *terrendis hostibus*, [terrifying the Enemy], seems as probable as that of Varro, from *terendo* [treading upon]; or that of Frontinus, from *terra* [land]; or that of Pomponius the Lawyer, from *terrendi jure* [the right to terrify] which the magistrates have."[231] A more substantive concern is the way *imperium* is acquired:

> But why the *imperium* over any particular place; that is, any part of a territory, that lies, suppose, uninhabited and waste, may not be alienated by a free people, or by a king in concurrence with his people, I see no manner of reason to dispute. Were indeed any part of the people to be transferred, as they have a freedom of will, so have they likewise a right to oppose such an alienation; but the territory, whether wholly, or in part, belongs in common and inseparably to the people; and consequently, is entirely at their disposal.[232]

He continues to suggest that the people as whole take "possession of the whole land [*terras*], both as to the *imperium* and *dominium*, before the lands [*agri*] were parcelled out to private persons."[233] Grotius therefore contends that there is a property right at the basis of *imperium* over *territorium*.

Thus also we see what was the basis of property, which was derived not from a mere internal act of the mind, since one could not possibly guess what others designed to appropriate to themselves, that he might abstain from it; and besides, several might have had a mind to the same thing, at the same time; but it resulted from a certain compact and agreement, either expressly, as by a division; or else tacitly, as by seizure. For as soon as living in common was no longer approved of, all men were supposed, and ought to be supposed to have consented, that each should appropriate to himself, by right of first possession, what could not have been divided . . . the division of lands [ex agrorum divisione] produced a new sort of right.[234]

Grotius does not simply discuss land, but also offers a discussion of rivers and their relation to territory.[235] He poses the same questions of the *Institutes*, but clearly mediated by Bartolus's work. Fundamentally, he asks, if the course of a river changes, what about the territory?[236] In terms of water, Grotius is much more famous for what he said about the sea. One of the chapters of *De Jure Praedae Commentarius* had been separately published as "Mare librum."[237] It produced a response by John Selden, which also drew on the likes of Bartolus and Baldus.[238] The overall contours of this debate are well known: Selden claimed that states can enclose and claim oceans, just as they could appropriate "territory or field,"[239] whereas for Grotius they should remain free and open to all.[240] Selden wanted both *dominium* and *imperium*,[241] and in time Grotius began to recognize that while appropriation was not applicable to the sea, some kind of enclosure was possible, suggesting in *De jure belli ac pacis* that "*imperium* over a part of the sea is acquired, in my opinion, as all other sorts of *imperium*; that is, as we said before, in regard to persons, and in regard to territory."[242] These issues were, of course, in part occasioned by the maritime expansion of the European powers, and disputes between the Dutch, Spanish, Portuguese, and English in the East Indies. These were questions that had begun a couple of centuries previously.

Renaissance and Reconnaissance

While we now know that parts of North America had contacts with Europeans long before Christopher Columbus,[1] his voyage marks a break in the European mentality. As Seaver puts it, this can be conceived as "sailing out of the Middle Ages."[2] Two years after Columbus made landfall, the 1494 Treaty of Tordesillas divided Spanish and Portuguese claims to the parts of the world that European countries supposed were newly discovered. The treaty amended the papal bull *Inter Caetera* from the previous year. Its crucial clause stipulated that

> a boundary or straight line [*una rraya o linea derecha*] be determined and drawn, from pole to pole, on the said ocean [the Atlantic], from the Arctic to the Antarctic pole, north to south. This boundary or line shall be drawn straight, as aforesaid, at a distance of three hundred and seventy leagues west of the Cape Verde Islands, being calculated by degrees, or by any other manner as may be considered the best and readiest, provided the distance shall be no greater than above said.[3]

The dividing line allowed the king of Portugal to establish claims to non-Christian lands the east, the king and queen of Castile those to the west. Broadly the aim was to allow Portugal Africa and Spain the Americas, but it was later discovered that part of South America was actually to the east, hence the creation of Portuguese colonies there, known as Brazil.[4] This demonstrates that latitude was a much more successful marker, until more reliable clocks allowed exact measurement of longitude. What was important about Tordesillas is that it suggests a model that the actual techniques only later caught up with. As historians of cartography have demonstrated, many of the maps of the world in this period were concerned with precisely discovering where this demarcation actually fell on the material earth.[5] What was crucial was that Tordesillas was an attempt to break with the idea that simple occupation led to possession; rather, it divided lands that were not yet known by a calculative measure. The division applied to both "all lands, both islands and mainlands [*asi yslas, como tierra firme*], found and discovered already, or to be found and discovered hereafter."[6] The treaty also made provision for where the line passed through land:

> And should, perchance, the said line and bound from pole to pole, as aforesaid, intersect any island or mainland, at the first point of such intersection of such island or mainland by the said line, some kind of mark or tower shall be erected, and a succession of similar marks shall be erected in a straight line from such mark or tower, in a line identical with the above-mentioned bound. These marks shall separate those portions of such land belonging to each one of the said parties; and the subjects of the said parties shall not dare, on either side, to enter the part of the other, by crossing the said mark or bound in such island or mainland.[7]

As Seed has shown, what established possession varied between different countries. In the case of Portugal, she notes that "while occasionally they planted objects such as stone pillars to indicate the extent of their discoveries, their ability to establish the latitude of a new place provided the central proof of their discovery."[8] This thus produced the need for enhanced techniques of navigation and land measurement, in terms of both achieving landfall and knowing where they were when they arrived. While the compass had been used from at least 1300,[9] there was also the beginning of the reuse of ancient instruments such as the quadrant. The use of the polestar for navigation had worked in the Northern Hemisphere, but beyond this navigators used the astrolabe, which had been developed by the ancient Greeks and used by Islamic scholars in the Middle Ages.[10] In Chaucer's "The Miller's Tale," the student womanizer Nicholas is a would-be astrologer, who has in his belongings an astrolabe:

> His Almageste, and bokes grete and smale,
> His astrelabe, longinge for his art,
> His augrim-stones, layen far apart
> On shelves couched at his beddes heed;
> His presse ycovered with a falding reed.[11]

The *Almagest* is a collection of writings by Ptolemy, on astronomy and mathematics, and makes a perfect literary foil for the astrolabe and the augrim-stones, used for calculations.[12] Chaucer himself wrote a treatise on the astrolabe around 1391.[13] In 1556, Leonard Digges published his *A Boke named Tectonicon*, which promised to show how geometry could be put to work, with diagrams, technical descriptions, and practical exercises.[14] A little later the Elizabethan writer John Dee wrote a preface to a translation of Euclid's *Elements* in 1570, which made a very Euclidean or Aristotelian distinction between different types of mathematics. We are told that number is "a certain mathematicall sume, of *units*," which are indivisible. Magnitude, in contrast, is "a thing *mathematicall*, by participation of some likeness of whose nature, any thing is judged long, broade or thicke." Magnitude is "divisible for ever," and points are ends of lines. Dee tells us that "neither *number*, nor *magnitude*, have any materialite." Arithmetike relates to number, geometrie to magnitude, within science mathematicall.[15] Dee read widely, and refers to Bartolus, Baldus, and Accursius.[16] He also makes a passing reference to "territory or parcel of land."[17]

It is important to stress that for all the technical innovations Digges and Dee had, theoretically they were simply utilizing Euclidean notions. Similarly, Denis Cosgrove has described the working practices of the important Italian Renaissance cartographers Giacomo Gastaldo and Cristoforo Sorte, suggesting that the "compass, cross-staff, quadrant and astrolabe were the daily tools of Gastaldo's and Sorte's trade, Euclidean geometry its theoretical foundation."[18] The Euclidean axioms were important, since there was, as Cosgrove notes, a shift from a merely practical to a renewed interest in speculative geometry. "For sixteenth-century Venice a key to this unity of material and cultural change, of the practical and speculative, lies in the geometry of landscape."[19] By the sixteenth century an astrolabe allowed the construction of lines of latitude on a globe.[20]

For Matthew Edney, "'Empire' is a cartographic construction; modern cartography is the construction of modern imperialism."[21] Given the extensive literature on cartography and conquest that already exists, this chapter does not pretend to add anything in that register. Rather, it encounters a range of thinkers who are known to varying degrees—Machiavelli, Erasmus, More, Luther, Bodin, Botero, Shakespeare, and others—with a different inflection to most work on them. Crucially, the question remains that which has motivated this study from the outset: what is the relation between place and power in their thought? The "discovery," conquest, and division of the New World provide a context within which all the thinkers discussed in this and the subsequent chapter operated. If often absent in their texts in an explicit way, it nonetheless provides the frame within which their arguments were made.[22] It was not a case of a Europe with nation-states with fixed territory as a model that was exported to the rest of the world; rather, the New World proved to be a laboratory where ideas were tried out, concepts forged, and techniques tested and perfected, which were then carried back to Europe.[23] Thus, as Seed suggests, by the middle of the sixteenth century, advances in the use of the astrolabe allowed the fixing of latitude not simply on land, and not simply at noon, but at sea and at any time the sun shone. "For the first time positions throughout the globe were being described by a set of numbers."[24]

The specific focus of this chapter is the period known as the Renaissance, a rebirth of interest in the classical world. As previous chapters have demonstrated, this is not something that can be confined to a specific century, with the resurgence of scholarship in the Carolingian Empire, the translations of Aristotle, and the rediscovery of Roman law all being parts of a much more general story. And the label "Renaissance" is a much later invention, with the first person to use the term being the nineteenth-century French historian Jules Michelet.[25] Yet it was the Swiss historian Jacob Burckhardt who gave it a "definitive portrait" as a fifteenth-century Italian phenomenon.[26] The explosion beyond the borders of Europe was not confined to lands to the west in the Americas. The development of skills in sailing and desires for exploration furthered the links to the east too.[27] What is crucial is the way that the artistic and scholarly advances of the late Middle Ages and the period known as the Renaissance impacted on wider cultural, social, and political trends. To take one example, as Edgerton has shown in detail, the development of linear perspective in art impacted on cartographic practice.[28]

This coincidently linked to the discovery of a manuscript of Ptolemy's *Geography* in Constantinople in 1406, which lacked maps but which the text allowed to be reconstructed, since it provided coordinates.[29] This rediscovery inspired subsequent developments in cartography, in terms of both the focus of cartographic practice and the methods that made it possible.[30] Yet, despite its impact, it was discovered at a time when advances were occurring in related fields, and so it was incorporated into those developments rather than overpowering them.[31]

MACHIAVELLI AND *LO STATO*

Niccolò Machiavelli (1469–1527) was a writer who took much of his inspiration from the writings of ancient Rome.[32] He is most famous for his book

Il Principe, almost always translated as "The Prince," but which means the ruler or the principal rather than the king's son.[33] He also wrote *Discorsi*, discourses on republicanism.[34] References to Cicero, Virgil, and Seneca, implicit or explicit, abound in *Il Principe*; he knew and referred to the Greek historians Polybius and Plutarch; and his *Discorsi* were explicitly framed as reflections on Livy's *Ab urbe condita*. In *Il Principe* historical figures such as the Gracchi and Alexander the Great pepper the narrative alongside Cesare Borgia and the Medici family. Indeed, Machiavelli often makes it appear that the distance between the present day and these historical forbears is actually rather short. He suggests, for example, that a ruler "should read history books, and in them he should study the actions of admirable men."[35] It was clear, in addition, that this was where he felt most at home, telling his patron and the Florentine ambassador to the papacy Francesco Vettori that after a day in the countryside, he entered his study in the evening, taking off his "work clothes, covered in mud and filth, and put on the clothes an ambassador would wear." Then, he suggests, he can enter "the ancient courts of rulers who have long since died."[36]

The apparent split between a work on principalities—that is, polities ruled by autocrats—and one on republics has occasioned an extensive literature on the relation between the two texts, asking if Machiavelli changed his mind between them. This has similarly opened up a debate about when *Il Principe* was composed.[37] It is generally acknowledged that *Il Principe*, apart from its dedication and first chapter, dates from 1513, while the *Discorsi* were written after 1515. The tension has been reconciled by suggestions that what form of government may be appropriate for some instances may not be appropriate for all. Yet take away the key distinction between different kinds of polity and the problems collapse: both works have large numbers of common themes and shared concerns. It is the way of posing them that is challenging: "It has always been as dangerous to propose new ways of thinking and new institutions as it is to seek new oceans and unknown lands [*terre incognite*]."[38] The relation between the dating of these works in the early sixteenth century and the discoveries of the New World in the previous two decades has not been lost.

Yet if *Il Principe* is sometimes seen as a particularly modern work, it makes more sense to see it as the product of a late-medieval writer,[39] an example of and critique of the "mirror for princes" literature common in the Middle Ages, of which those of Aegidius Romanus, Aquinas, and Ptolemy of Lucca have already been discussed.[40] Many of Machiavelli's contemporaries, including Seyssel and Erasmus, had written similar books.[41] As with many of those earlier books, Machiavelli's was a text written for and dedicated to a particular patron. Yet, in this case, these people were not the same. As Wootton argues, it was written for Giuliano de' Medici, rather than his nephew Lorenzo, to whom it was eventually dedicated. For this reason the advice had to be framed in a nonspecific way, because Giuliano was an aspiring, rather than actual, ruler, looking for papal support in gaining a land to rule. This was a land, Wootton suggests, that may have had no preexisting ruler, and "in all probability, territory that had no tradition of urban self-government to overcome."[42]

This reference to "territory" appears unproblematic, and translations of Machiavelli's works abound with the word. Not only that, but Michel Foucault takes Machiavelli as the exemplar of what he characterizes as a medieval model that endures until the sixteenth century: "Sovereignty is not exercised on things, but above all on a territory and consequently on the subjects who inhabit it."[43] Previous chapters have demonstrated that this is not at all accurate as an assessment of the Middle Ages. But to what extent does it hold true of Machiavelli? Foucault contends that Machiavelli's key concern is that of knowing how "a province or a territory acquired through inheritance or by conquest can be held against its internal or external rivals"[44] and of preserving "the surety of the territory or the surety of the sovereign who reigns over the territory."[45] A surety is a bond, and Foucault stresses the importance of the relation. But the Italian word *territorio* is not used in Machiavelli. How, then, does he discuss political control?

One of the key words of Machiavelli's political vocabulary is *stato*. As Price outlines, this is a word that has at least a twofold sense: "a political community existing within certain territorial boundaries as well as the government of such a community."[46] Viroli equally suggests that sometimes Machiavelli "uses the term 'state' in the sense of the territory over which a prince or republic have sovereignty."[47] Skinner suggests that Machiavelli "frequently uses the term *lo stato* in *Il Principe* to denote the lands or territories of princes."[48] Translators often thus render the first meaning as "territory" or "region" as well as "state." This is obviously extremely confusing. Machiavelli's own explanation comes in the first sentence of *Il Principe*: "All states [*stati*]—all dominions [*dominii*] that have held and have authority [*imperio*] over men—have been either republics or principalities [*republiche o principati*]."[49] The first noun of the book is therefore *stato*, and there are two key elements to this sentence. One is that the middle clause defines *stati* rather than being additional to it. States are dominions holding *imperium* over men. The key here is that the object of political rule is men, people. The second is that such polities are divided into two, republics or principalities, those ruled by the many or one. The middle clause thus refers the modern concept of the state to two much older ones—*dominium* and *imperium*. It therefore links relations of property and power to that of men. As *dominium* and *imperium* both have some spatial connotations, there are a whole number of relations set up here.[50] But it is the primacy of men, the populace, that is crucial: they, not territory, are the object of political rule. If not entirely modern, this is an understanding of the temporal extent of rule, entirely distinct and self-standing from any religious justification.

It is important to note that *stato* is very rarely an active subject of a verb: Machiavelli's states do very little.[51] Yet *stato* is not simply an object of analysis and a mode of rule, but can function as an assessment of worth or power. In examining a principality, Machiavelli asks if a ruler has *tanto stato* (sufficient power);[52] and he assesses one situation as a time when rulers "had to make sure none of the Italian powers increased its power [*che veruno di loro occupassi più stato*]."[53] Looking back to history, he suggests that Philip of Macedonia "did not have much power [*non molto stato*], compared to the might [*grandezza*] of Rome and Greece which attacked him."[54]

One of the key issues in Machiavelli's political writings is the notion of acquisition. Yet while translations often find a way of giving the verb *to acquire* an object, Machiavelli often leaves it unspoken. Thus, Machiavelli discusses "republics that acquire,"[55] and speaks of "new acquisitions,"[56] and "the pursuit of expansion."[57] In *Il Principe* he similarly suggests that "it is perfectly natural and normal to desire to acquire; and whenever men do what will succeed towards this end, they will be praised, or at least not condemned."[58] While "territory" might appear to be a legitimate expansion of these phrases, there are many things that can be acquired. Land is merely one. He speaks, for instance, of an instructive comparison to be made of "how much the methods used by the Romans in acquisition [*acquistare*] differed from those used at present to extend their jurisdiction [*ampliano la giurisdizione loro*]."[59] Acquisition can also be of *imperium*, rule or empire: "It is all too easy to acquire empire without acquiring new strength [*acquistare imperio e non forze*], and if you acquire empire without at the same time building up your strength, you are heading for destruction."[60] While this certainly implies a spatial extent, it is by no means exhausted by it. One of the spatial terms that he does use, as Foucault indicated, is that of *provincia*, a province.

It would be relevant here for us to explain the policies pursued by the Roman people when occupying newly acquired provinces [*provincie*] if we had not discussed this question at length in our treatise *Principati*. I will only say this much in passing. The Romans always tried hard when they were acquiring new provinces to have the support of an ally who could serve as a ladder over the defences, or as a gate through the walls, or as an assistant in retaining control once it was acquired. . . . They were never short of such allies to assist them in their undertakings and to help them acquire and hold new provinces.[61]

Machiavelli similarly advises that "anyone who becomes the ruler of a city or of a state . . . should leave nothing as it was in the whole *provincia*."[62] Yet this is sometimes without any opposition being mentioned. Machiavelli discusses, for instance, how new cities can be founded: "Cities are built by free men when a group of people, either under the command of a ruler or acting on their own, are forced to abandon their native land [*paese patrio*] and to seek a new home [*nuova sede*] because of disease, or hunger, or war."[63] At other times existing people will need to be supplanted. This raises the issue of the people as a whole:

If you want to make a populace numerous and well-armed, so that they can conquer a vast empire, then you must accept that you will not be able to get them to do everything you want. If you keep the population small or unarmed so that you can get them to do what you want, then if you do conquer dominions [*acquisti dominio*] you will not be able to hold onto it. . . . So if someone wanted to set up a republic, he would have to ask himself if he wanted it to grow in power and dominion [*di dominio e di potenza*] as Rome did or to remain limited in both.[64]

While growth might be appealing, this therefore raises the potential for civil disorder. But there are other dilemmas. Discussing a new republic, Machiavelli contends that "if it stays within its own boundaries [*termini*], and people see from experience that it is not interested in making conquests, then no one will ever go to war against it out of fear of being attacked by it. This will be all the more true if the constitution or laws of this republic prohibit its expansion [*l'ampliare*]."[65] Yet it still will not be safe, as others will want to conquer it: "It is not possible for a republic to succeed in peacefully enjoying its liberty within a small area [*pochi confini*]."[66] That said, he notes that "the cities of Germany are free to do as they please. They have little rural lands [*hanno poco contado*], and obey the Emperor only when they want."[67]

Expansion is not without other problems. Lands acquired can rebel, though if the rebellion is ended, this cements the control: "Of course it is true that, after a ruler has regained power in a rebel country [*paesi rebellati*], he is much more likely to hang onto it."[68] The Romans in Greece found that they had to destroy "cities in that region [*provincia*] . . . the simple truth is there is no secure way of holding them, short of demolition."[69] Machiavelli sharply distinguishes between different types of acquisitions: "Let me start by saying these states [*questi stati*] that are acquired and annexed to a state that is already securely in the possession of a ruler are either in the same region [*provincia*] as his existing possessions and speak the same language, or they are not."[70] When they are, Machiavelli argues, they are relatively straightforward to command and control. But "states in a region [*stati in una provincia*]" that does not share those characteristics are much harder, and there are various tactics to make this work easier. The rulers themselves can go and live there, or they can send colonies to settle in the area. Alternatively, they can garrison forces there, but this usually consumes all the resources that the new colony produces.[71] "The Romans, in the regions they seized, obeyed these principles admirably. They settled colonies; were friendly towards the weaker rulers, without building up their strength; broke the powerful; and did not allow foreign powers to build up support."[72] He also suggests dividing up lands (*le terre*) in order to exert better control over them and disarming people, except those who supported you, when a new land is added to an existing state.[73]

One of the interesting things to emerge from this discussion is that "state" can apply both to the original state and the lands it acquires. The case of Cesare Borgia, son of Pope Alexander VI, is illustrative of this somewhat vague sense of the term: it does not simply mean "the state," in the modern sense. Machiavelli tells us that the pope "could find no way of making him the lord of any state [*stato*], except part of the states of the church [*stato di Chiesia*]."[74] Even when he was successful in gaining a state, this was later lost, "despite the fact that he used every technique and did all the things a prudent and skilful [*virtuoso*] man ought to do, to entrench himself in those states [*stati*] that the arms and fortunes of others had acquired for him."[75]

Machiavelli was also interested in questions of property in land. For example, he spends some time in the *Discorsi* discussing agrarian reform in ancient Rome. The key parts of the reform were that no citizen should own more than a certain extent of land (*terra*) and that conquests should be divided between all citizens.[76] He is also critical of the land gentry, describing them as "those who live in idleness on the abundant revenue derived from their estates."[77] Indeed, Crick suggests that one of Machiavelli's conditions for republican rule is the town dominating the country, which he claims is based on Machiavelli's contempt not simply for the gentry but also for the peasantry.[78] Strategic control of land is another concern, what might be called the question of terrain. He suggests that a ruler should spend peacetime preparing for war, not simply by keeping his troops well trained but also in terms of his own readiness. He should always be hunting, both to get used to fatigue and also to learn about the terrain (*siti*).

He should take the opportunity to study the nature of the terrain [*la natura de' siti*], how mountains rise, how valleys descend, plains spread, the nature of rivers and marshes. He should spare no effort, and this for two reasons. First, the knowledge of his own land [*sua paese*] will stand him in good stead if he has to defend it against invasion; second, his knowledge and experience of his own terrain [*quelli siti*] will make it easy for him to understand any other terrain [*altra sito*] which he must explore. The hills, the valleys, the plains, the rivers, the marshes of, for example, Tuscany have a good deal in common with those of other regions of Italy. A knowledge of the terrain [*sito*] in one region will make it easy for him to learn about the others. A ruler [*principe*] who lacks this kind of skill does not satisfy the first requirement in a military commander [*capitano*], for it is knowledge of the terrain that enables you to locate the enemy and to get the edge over him when deciding where to camp, how to lead a march, how to draw up the troops on the field of battle [*terre*], and where to build fortifications.[79]

Given the discussion of such issues in both *Il Principe* and the *Discorsi*, it might be expected that his *Dell'arte della guerra*—The art of war—would cover such matters in depth.[80] As with his work on politics, it has been described as not the first modern work on warfare, but perhaps the greatest "medieval compilation."[81] Yet while there is some brief discussion of choosing the site of battle,[82] and some advice on setting up a camp (book 6) and fortifications (book 7), in the main this is a discussion of types of soldiers, and weapons. It does, however, contain perhaps the most succinct summation of the relation between leadership, population, and place in Machiavelli: "More important than the number [*moltitudine*] of soldiers is their *virtù*; more benefit comes at times from terrain [*sito*] than from *virtù*."[83] A similar comment on the lack of attention to geographical aspects could also be made of his history of Florence, a plodding work that is more concerned with internal struggles and power plays.[84]

Shortly before he wrote *Il Principe*, Machiavelli had been interrogated and imprisoned. When he was released, he was confined instead of exiled. The sentence was given in Latin rather than Italian, and stipulated that he was to remain *"in territorio et dominio florentino per unum annum continuum,"* in the lands and dominions of Florence for the span of a year.[85] This was an area of about a day's ride and perhaps a radius of twenty-five miles.[86] What is significant is that the judgment uses the term *territorio*, showing that it was a term that would have made sense to Machiavelli. It was also used by his contemporary Francesco Guicciardini, who talks of a city "that was content with its freedom and its small territory [*territorio*]."[87] Guicciardini took a rather different line to Machiavelli in other respects, asking, "How can one, according to conscience, wage war from a lust to expand one's dominions [*dominio*], in which one commits so many killings, so many sackings, so many violations of women, so many burnings of houses and churches, and an infinite number of other evils?"[88] But the point is important: *territorio* is a word Machiavelli would have known, but he does not use it. Geographical questions are underplayed in his work; territory was not the object of rule.

THE POLITICS OF REFORMATION

Contemporaries of Machiavelli included Erasmus, already mentioned, and Thomas More. Like many of the other humanists of the early sixteenth century, Erasmus wrote advice books for a prince. It is difficult to argue with Skinner's contention that although the writers were distinguished, the writings were derivative.[89] While Erasmus was of fundamental importance in other respects, he is not especially interesting as a political theorist.[90] Three small points can be made. The first is that he recognizes that the vocabulary of political thought bears traces that cannot be simply be remade in a sacred manner: "Always bear in mind that the words 'dominion,' 'imperial authority,' 'kingdom,' 'majesty,' and 'power' [*dominium, imperium, regnum, maiestatem, potentiam*] are pagan [*ethnicorum*] terms, not Christian, the *imperium* of Christians is nothing other than administration, benefaction and guardianship [*custodiam*]."[91] The second is that he offers a reuse of the image of the body politic, suggesting that "the *respublica* is a kind of body composed of different parts, among whose number is the prince himself (even if he is exceptional),"[92] and "what the heart is in the living body the prince is in the *republica*. Since it is the fount of the blood and spirits, it imparts life to the whole body, but if it is impaired, it debilitates every part of the body."[93] The third is that he resorts to an almost classical use of the vocabulary of political extent: "In my judgment it would be most beneficial to the *reipublicae* if the marriage alliances of princes were confined within the boundaries of their kingdom [*intra regni fines continerentur*]; if they must go beyond their frontiers [*limitibus*], they should be united only with near neighbours and then only with those best suited to a pact of friendship."[94]

His friend Thomas More was an important political actor as well as writer. He was schooled in a system where the Bible, Aristotle, and books of law of Justinian were the core teaching.[95] He is perhaps most of interest for the work commonly known as *Utopia*, a word that famously combines the Greek words for no place and happy place. The full title, however, begins *De optimo status reipublicae statu deque nova insula Utopia*—the optimal state of the *respublica* in the new island of Utopia—which clearly situates it within a broad frame of late-Scholastic and Renaissance texts on the best state of a "commonwealth."[96] Like Machiavelli, and many of their contemporaries, he situates his own innovations in relation to the new discoveries of Vespucci.[97] One of the most interesting moments in this text is his description of the geography of Utopia.[98] More suggests that "every city [*ciuitatibus*] has enough country [*agri*] assigned to it so that at least twelve miles of farmland [*minus soli quam duodecim passuum millia una quaeuis*] are available in every direction, though even more where the cities [*urbes*] are farther apart. No city wants to enlarge its boundaries [*nulli urbi cupido promouendorum finium*], for the inhabitants consider themselves cultivators [*agricolas*] rather than landlords [*dominos*]."[99] More also provides an only thinly coded criticism of the enclosure movement.[100]

More had been lord chancellor under Henry VIII, and so was deeply involved in the turbulence of the time. He had been responsible for the condemnation and burning of William Tyndale, who had translated parts of the Bible into English. Tyndale's most political book was *The Obedience of a Christian Man*, dating from 1528, which suggested that religion was subservient to the king.[101] Some years later King Henry claimed that "this is a book for me and all kings to read."[102] A book explicitly dedicated to Henry from this time was Thomas Elyot's *The Boke Named the Governour*, from 1531.[103] It has been described as "one of the most widely-read humanist texts of the 16th century."[104] Henry was apparently sufficiently impressed that Elyot was asked to represent him as ambassador to Emperor Charles V to help with the case for Henry's divorce, and he was sent to the Netherlands to apprehend William Tyndale.[105] Elyot's book is, frankly, largely unremarkable as an example of an advice book to a prince, and Rude suggests that John of Salisbury and Erasmus are close forerunners.[106] In part this is because of the opening description of a state or "weal": "A publike weal is a body lyvyng, compacte or made of sondry astates and degrees of men whiche is diposed by the order of equitie, and governed by the rule and moderation of reason."[107] There are other interesting moments, such as the description of the delights of map reading—"For what a pleasure is it, in one houre, to beholde those realms, cities, sees, and mountaynes, that uneth in an olde mannes life can nat be journaide?"[108]—but Elyot is of more interest as the compiler of a Latin-English edition. The modern reedition of this dictionary describes it as the "first Latin-English dictionary to be based on Renaissance humanist ideals of classical learning." There were Latin-English dictionaries in the fifteenth century, but these were of medieval, not classical, vocabulary.[109] It is of interest here because of its entry for *territorium*: "Territorium, the fyeldes or countraye lyenge within the iurisdiction and boundes of a citie, a territorie."[110]

One much-analyzed work of art dates from this time: Hans Holbein's *Ambassadors*, now hanging in the National Gallery in London, and depicting Jean de Dinteville and Georges de Selve at the court of Henry VIII. Selve represents the French church, Dinteville the nobility. While perhaps the anamorphic skull at the bottom has occasioned the most recent interest as a *memento mori*, the depiction of various instruments and tools is also revealing.[111] Included in the picture are globes, both of heaven and earth; a book of arithmetic held open by a square; compasses, a sundial, and other geometric, navigational, and astronomic devices;[112] a case of flutes; and a hymnbook showing works by Martin Luther. They are generally taken to symbolize the liberal arts of the trivium and quadrivium, but as Jerry Brotton has argued, there are several indications of religious conflict in the objects. The hymnbook puts Luther between the Catholic ambassadors and the English king, where the setting, as the floor indicates, is the sanctuary of Westminster Abbey; there is a largely obscured crucifix in the top-left corner; and the lute has a broken string. The hymnbook is printed, showing the advance of technology, but also the purposes to which it could be put. The arithmetic text is Petrus Apianus's *Eyn newe unnd wolgegründte underweysung aller Kauffmanss Rechnung* (A New and Well-grounded Instruction in all Merchants' Arithmetic), with examples of how profit and loss can be calculated.[113] The first word on the open page is *Dividirt*, "division."[114] The very visible globe has a large number of

Fig. 9. Hans Holbein, *The Ambassadors*, National Gallery, London.

places labeled.[115] It symbolizes not simply the discovery of the New World, with the line of the Treaty of Tordesillas labeled as "Linea Divisioñis Castellanorū et Portugalleñ" and "Brisilici" clearly marked, but also the potential for territorial struggle between the English king and the Catholic nations. It also shows the knowledge of the New World, but also Europe's self-awareness: the continent is labeled "Europa" and it traces voyages of discovery.[116]

Yet it was on the European continent that the key thinkers of the reformation were working. Martin Luther (1483–1546) was foremost among them, with Jean Calvin (1509–64) the key figure of the next generation.[117] Luther exemplifies the belief, summarized by le Goff, of the sixteenth-century theologians that under the Scholastics the Bible itself had been

lost in its exegesis, and that the Reformation had the "justifiable feeling
that it had rediscovered it."[118] Luther's translation of the Bible into the
vernacular, in so doing creating the modern German language in the way
that Dante had done for Italian two centuries before, will not be examined here. Rather, the issue is his political thought. Yet, as Moeller notes,
most German history between 1500 and 1650 has remained the preserve
of theologians, with the consequence that religion is looked at as the only
issue.[119] But to what extent can Luther be read politically? The older attitude is memorably summed up by the Carlyle brothers: "Luther was not a
systematic political thinker . . . indeed he can hardly be described as a political thinker at all."[120] Yet Luther himself claimed on at least two occasions that he had discussed temporal government better than any, except
Augustine, since the apostles.[121] Nonetheless, Luther's views on the topic
are anything but clear, and it has been suggested that none of his work has
been "more misrepresented than his teaching about the nature, extent,
and limits of temporal power."[122] Part of the problem is the lack of explicit
connection of his arguments with those of his near contemporaries: it is
difficult to connect Luther with either Machiavelli or Bodin.[123]

There are several texts of Luther's that might be mined for a political
meaning, including important texts from 1520 on German nobility and
the papacy.[124] The most important political text, however, is *On Secular
or Temporal Authority (Von Weltlicher Oberkeit)*, from 1523.[125] Many of
his political texts were written in German rather than Latin and were expressly directed at a German context. This piece raises some fundamental
questions, not least in the words of its title. *Weltlicher* is derived from
Welt, world. It therefore means "worldly," in opposition to heavenly, as
well as "earthly," as opposed to spiritual. Accordingly, Höpfl has suggested that it carries a somewhat more negative connotation than secular or temporal.[126] Nonetheless, many of Luther's key arguments removed
power from the church. In proposing *sola fide, sola gratia, sola scriptura*,
that it was only faith, grace, and scripture that allowed contact with God,
Luther was taking away the church, and most especially the priest, from
any mediation. In doing so, crucially Luther repudiated any claim by the
church to exercise power over temporal affairs. He argued, for instance,
that "the Donation of Constantine is a great lie by which the pope usurped
[*arrogat*] half of the Roman Empire for himself."[127] Instead, for Luther,
Christians live in two kingdoms—that of Christ and that of the world.
The first is the realm of the church, as this *congregatio fidelium*. In seeing
the church solely as a congregation of the faithful, he was rejecting ideas

of the church's power outside of its spiritual role. The second, accordingly, is not the church's realm but that of temporal authority. Secular power extends over the church too.[128] This means it is crucial that the prince should be a man of God, even though preachers should not advise princes, just as they would not tell a tailor how to make a suit.[129]

Thus, for Luther, the old conflict between the secular kingdom and the priesthood was over. As Figgis puts it, Luther destroyed "the metaphor of the two swords; henceforth there should be but one, wielded by a rightly advised and godly prince."[130] As Luther says in "On Secular Authority," "Our first task is a sound basis for secular law and the sword, so no one will doubt that it is in the world as a result of God's will and ordinance."[131] Luther suggested that "there are three orders [ordines] in this life: the household, the state, and the church [Oeconomiam, Politiam et Ecclesiam]."[132]

> God has appointed three social orders to which he has given the command not to let sins go unpunished. The first is that of the parents, who should maintain strict discipline in their house when ruling the domestics and the children. The second is the Politicus, for the magistrates [Magistratus] bear the sword for the purpose of coercing the obstinate and remiss by means of their power of discipline. The third is that of the church [Ecclesiasticus], which governs by the Word.[133]

Luther uses a range of terms when discussing the second, political, realm, including the Latin respublica, politia, civitas, and Reich, which the English of the time would have described as a "commonwealth."[134]

It is worth a brief look at the remarkable Passional Christi und Antichristi, from 1521.[135] This is a fairly well-known book comprising Lucas Cranach woodcuts, with images of Christ and the Antichrist (depicted as the pope) on facing pages. The left-hand page has a quotation from scripture; the right-hand page a description of the contemporary parlous state of the church, which is acting contrary to the Bible's message. The opening pair, for instance, juxtaposes Christ refusing the crown as an earthly king and the pope behind a gate with cannons and armed soldiers defending himself from armored figures on horseback. Later ones show Jesus expelling the moneylenders from the temple, and the pope selling indulgences; or Jesus washing feet and the pope having his kissed by admirers. The writer of the German text was probably Philipp Melanchthon, a collaborator of Luther.[136] The scripture is in German too. Though this predates the

paſſional Chriſti vnd   Antichriſti.

Antichriſtus.

Chriſtus.

Do Jheſus innen warde/das ſie komnten wurden vnd yhnen tzum konig machen/iſt er abermals vffin Bergk gefloben/er allein.Johan.6.Mein reich iſt nicht von diſſer welt.Joh.18.Die konnige der welt hirſchen yr/vnd die gewalde haben/werden gnedige herrn gonandt/yr aber nicht alſo/ßonder der do groſſer iſt vnther auch/ſall ſich nyddern/als der weniger.Luce.22.

Auß obirkayt die wir ſonder tzweiffell tzum keyßerthuß haben/ vñ auß vnſſer gewalt/ſeynt wir des keyßerthumßs/ ſo ſich das vorledigte ein rechter erbe/ cle.paſtoralis ab ſi.deſen.er re udi.Suma ſummarii.Nichts anders iſt in des Bapſts geyſtliche rechte tzu finden/dan das es ſeynen abgot vnd Antichriſt vbir alle keyßer/konig vñ furſten irheßet/als Petrus vorgeſagt hat. Es werden konnen vnuoꝛſchamſte Biſchoff die die weltlich herſchafft werden voꝛachten.2.Pet.2.   A ij

Fig. 10A and 10B: *Passional Christi und Antichristi*, Wittenberg, 1521.

publication of Luther's Bible translation, it was contemporaneous with the work on the New Testament. It is crucial as an argument against the usurpation of temporal power by the papacy.

What is important for future politics is the way Luther tied the church to specific areas, the notion of a *Landeskirche*. In a not-unrelated development, Michael Gaismair proposed a *Landsordnung* for the Tyrol in the Peasants' War that was a model socialist utopia.[137] Even though he was not politically radical, as Skinner has noted, Luther did not propose political passivity, especially when there was a violation of justice, which was picked up by Calvin and, later, Locke.[138] Calvin's most important political text is *Institutio Christianae Religionis*, initially published in Latin and then translated into French and English.[139] These arguments suited secular rulers, who used them against the universalizing tendencies of either the pope or emperor, and in particular their attempts at supranational jurisdiction. Henry VIII is one example of a less religious attempt at the same time.[140] The 1534 Act of Supremacy in England was an assertion of the principle that *rex est imperator* and the nonrecognition of a superior power.[141] A decade later Charles de Grassaille declared that the king of France was not simply the *imperator*, but also the "vicar of Christ in his kingdom."[142] This was summed up by the formula *une foi, une loi, un roi*—one faith, one law, one king.[143]

Luther's work was not really directed at the establishment of different secular orders, and was not therefore itself a territorial political argument. Nonetheless, the fractures in the church that Luther and thinkers in his wake occasioned led to a particular way of thinking about these issues. Gierke has suggested that the rediscovery of Roman law undermined German community, but it is not clear that this is the case.[144] As chapter 9 will demonstrate, many of the German writers of the next century used the law as a support to their arguments on behalf of individual polities within Germany. Luther, then, though not an especially important political theorist himself, occasioned a situation that the resources of political theory were used to resolve. German political realities, along with the rediscovery of Greek thought and Roman law, are a crucial part of the story. In theological debates, Luther's arguments led to the 1545–63 Council of Trent, which was convened to try to codify doctrine, and rule positions heretical. It is one of the key moments of the Counter-Reformation. In political practice the 1555 Peace of Augsburg was an attempt to mediate the conflicts between princes within the empire who had become Lutherans. While the Augsburg ruling worked in their favor, in a sense it was the end of the radical premise behind Luther, which was that individuals, in an unmediated way, could choose their faith. What Augsburg enforced was the idea that the prince could dictate the faith, rather than the pope or emperor. It was thus a shift from the power of the papacy to the temporal ruler, but it did not produce a democratization of faith. The question of what happened to dissenters, the Jews, and other religions remained unresolved.[145]

BODIN, *RÉPUBLIQUE*, SOVEREIGNTY

A huge range of other writers contributed political works at this time. Foucault highlights Guillaume de La Perrière's *Le miroir politique*, from 1555, because he suggests that it defines politics in a way that excludes a territorial definition: "You will notice that the definition of government in no way refers to territory. One governs things."[146] La Perrière declares that

"Republic or (as older French writers said) public thing [*chose publicque*] is the ordering of a city, on which depends its good or ill [fate]."[147] The English translation renders *Republicque* as "Commonweale." He is more interesting with his definition of "police," which he says "is a word derived from the Greek word *politeia*, which in our language we call *civilitié*. That which the Greeks called politic government [*gouuernement politicque*] the Latins called government of a Republic or civil society."[148] The notion of police here, as much more than a uniformed force for the prevention and detection of crime, with its rich heritage through Adam Ferguson, Adam Smith, and Hegel, has been comprehensively analyzed in the wake of Foucault's work.[149] Overall, though, *Le miroir politique* is a very peculiar book. Much is on family relations—what a husband should do for a wife, and the reverse; with some interesting trees of relations drawn up at various points. These are depicted as actual trees with branches to show how this leads to that, which leads to this, and so on. La Perrière is much exercised by Aristotle's typographies of monarchy, aristocracy, democracy, and so forth, which is a commonplace of the time. Indeed, La Perrière is more of a mark of continuity than break. Overall it is hard to understand Foucault's interest in him, and much easier to sign up to Allen's judgment that *Le miroir politique* is "an odd and very silly book. La Perrière acquired an ill-founded reputation as a poet and man of learning."[150]

Foucault's interest in François Hotman is much easier to understand.[151] Hotman worked on behalf of the Huguenots, and wrote a number of polemical texts.[152] The most important is *Francogallia*, published in 1573 and revised in 1576 and 1586, which looks at the composition of the French nation from Gallish and Frankish elements.[153] Hotman is largely reliant on the Roman historians, but is well read and also draws on the likes of Gregory of Tours and political theorists such as Marsilius. Hotman's case is for a king with limited power within his kingdom: a king who is subject to higher laws. As he suggests, "The king of Francogallia does not have unlimited authority within his kingdom [*non infinitam in suo regno dominationem*] but is circumscribed by well-defined right and specified laws."[154] Hotman is concerned with the introduction of Roman law into a land that had been ruled by Germanic law for seven hundred years.[155] Sixteenth-century France was not the place for an archaic form: public law was a constraint on government, but this should be a product of Frankish law. This comes across most strongly in his book *Antitribonian*, from 1567, a book that was notably written in French.[156] Bartolus and Baldus are occasionally referred to by Hotman, and his work has been described as neo-Bartolism in the attempt to bring law into relation with the present.[157]

Nonetheless, he broadly wants to return to an earlier model of Gaulish kings who held *regium imperium* within their *finibus*,[158] rather than the Frankish aristocracy or the Romans who supplanted and suppressed the Gauls.[159]

Jean Bodin (1530–96) is commonly hailed as the first modern political thinker, and the first theorist of sovereignty. Lewis, for instance, suggests that Marsilius's work has "all the elements of a theory of sovereignty; but the theory itself was lacking. Bodin provided it."[160] For Tooley, Bodin asked a question that, apart from *Defensor Pacis*, was not asked: "What is a state and how it is constructed?"[161] Others see the basis for his ideas in papal proclamations, rendered now in a secular sense;[162] Skinner finds it in germ in Machiavelli and only fully accomplished in Bodin.[163] Yet his French context raises some questions that go back a little further: "The question that he asked . . . was what prerogatives a political authority must hold exclusively if it is not to acknowledge a superior or equal in its territory."[164] The key issue, then, with Bodin, who clearly articulates an understanding of sovereignty, is to what extent this is exercised over territory, and to what extent territory is the object and possibility of sovereignty. Caution is important, for as Parker suggests, "Certainly, he was far more concerned with the restoration of virtue and religion than with the enunciation of a modern, secular view of state power. . . . Bodin can only be transformed into a founder of modern political thought by reading history backwards and extrapolating those elements in his work which have a modern appearance."[165]

His key political text is *Six livres de la république*, from 1576. One issue that any reader of *Les six livres de la république* must contend with is textual. Bodin wrote it originally in French, and it went through several editions in his lifetime.[166] He produced a Latin version in 1586, *De republica libri sex*, again going through several editions, but he did not simply translate the French.[167] Rather, he rewrote the text, taking this as an opportunity to change, amplify, and finesse the argument.[168] A translation of the text was made into English in 1606 by Richard Knolles, but this is neither of the French nor the Latin alone.[169] Rather, as its modern editor has noted, "The translator has carefully worked the multiple strands of Bodin's twin versions into a single, closely textured argument."[170] It is valuable in many ways, but often fuses the prose uncritically, and can be difficult to match passages to either source text. It is perhaps most revealing as an example of early seventeenth-century English political vocabulary. Since that date, there has not been a full English translation. There is an abridgment of the whole work,[171] and four key chapters have been translated more recently.[172]

There are thus enormous textual problems to grapple with.[173] In addition there is a range of marginal glosses in both Latin and French that have only recently begun to be properly analyzed.[174] Mindful of Giesey's admonition that work that does not examine both Bodin's languages is flawed,[175] the reading here tries to negotiate a way through these linguistic issues. In these quotations, where both original languages are provided in brackets, the French precedes the Latin.

Bodin also wrote on a range of other topics, including methodological approaches to history,[176] and the question of witches. But the last cannot be easily dismissed: he wrote it in between the French and Latin editions of the *Six Livres*.[177] The book on witches was originally written in French, the one on history in Latin. Harding thinks this is significant: Bodin wrote *Six Livres* in French originally, like the book on witches, to reach a wider audience and make the argument applicable to the French as a whole, rather than simply to an elite.[178] It has been suggested that there are other differences: Skinner sees *Methodus* as constitutionalist; while in *Six Livres* he is "a virtually unyielding defender of absolutism, demanding the outlawing of all theories of resistance and the acceptance of a strong monarchy as the only means of restoring political unity and peace."[179] Yet there does not seem to be compelling evidence that Bodin changed his mind, and Franklin has contended that, on the contrary, *Six Livres* was "more radical than Bodin knew," because it took the restraints and made them nonbinding.[180]

Three key questions thus arise. How did Bodin define sovereignty? Who or what was that sovereignty exercised over? Do these definitions and limits imply a constitutionalist or absolutist position? Finally, his relation to the Post-Glossators will be interrogated. Bodin provides several definitions of sovereignty. It is important to note that "sovereignty" is the translation of the French *souveraineté*, for which Bodin's Latin uses *majestas*, which is the word from which the English *majesty* comes. This is the first difficulty: "majesty" and "sovereignty" seem to imply different things and, as Leibniz later suggested, may need to be distinguished. Bodin was unconcerned, suggesting that "as for the title 'majesty' it is clear enough that it belongs only to someone who is sovereign."[181] This was not any ruler: "Neither feudal kings nor dukes, marquises, counts, princes may use the title of Majesty, but only Highness, Serenity, or Excellency."[182] The key definition comes in the eighth chapter of the first book.

> Sovereignty [*la souveraineté*] is the absolute and perpetual power vested in a republic [*puissance absolue et perpétuelle d'une Répu-*

blique] which in Latin is termed *majestatem*, the Greeks *akran exousian, kurion arche,* and *kurion politeuma;* and the Italians *segnoria,* a word they use for private persons [*particuliers*] as well as for those who have full control of the state of the Republic [*les affairs d'estat d'une République*], while the Hebrews call it *tomech shévet*—that is, highest power of command. We must now formulate a definition of sovereignty because no jurist or political philosopher has defined it, even though it is the chief point, and the one that needs most to be explained, in a treatise on the republic. Inasmuch as we have said that a republic is a just government of several households and of that which they have in common, with sovereign power, we need to clarify the meaning of sovereign power.[183]

Several things are worth noting here. Bodin suggests that sovereignty is without limit, in terms of either scope or time, but places it in a political unit, the *République*. The 1606 Knolles translation and most since have rendered this as "commonwealth," but this, for Bodin, was a rendering of the term inherited from the classical tradition, at least as far back as Cicero: *republica*. The opening lines of the first book had clarified his understanding of the republic:

> A republic may be defined as the rightly ordered government of a number of families, and of those things which are their common concern, by a sovereign power [*puissance souveraine*].[184]

He later makes it clear that *République* is a translation of the Greek *politeia*.[185] He also subdivides republics in different types, along the lines of a tradition derived from Aristotle: monarchy, aristocracy, and democracy or popular estate (*l'estat est populaire*).[186]

He thus links the term *sovereignty* to Latin, Greek, Italian, and Hebrew terms; yet there is an interesting tension in that he can tie it to all these languages, places, and times, and claim to define it for the first time. Bodin's Latin is importantly different: "*Maiestas est summa in cives ac subditos legibusque soluta potestas,*" which might be rendered as "Majesty is the supreme and absolute legal power over citizens and subjects."[187] He goes on to say that "*maiestas* needs careful definition, because no jurist or political philosopher has in fact attempted to define it, although it is the distinguishing mark of a republic, and an understanding of its nature fundamental to any treatment."[188]

One of the places to look for the definition of sovereignty is the tenth

chapter of book 1, "On the true marks of sovereignty." Yet this is actually
a fairly conventional chapter that sets out a number of things that sovereignty can do, but does not seem to mention its explicit object. As Skinner summarizes, it comprises the "the power to legislate, to make war
and peace, appoint higher magistrates, hear final appeals, grant pardons,
receive homage, coin money, regulate weights and measures and impose
taxes."[189] Other chapters are more revealing, especially in terms of the relation between sovereignty and geography. One of the key passages comes
in the discussion of the requirements of a republic:

> If one turns from the small to the large, it follows by parity of argu-
> ment that the republic should have sufficient territory and appropri-
> ate places for the inhabitants [*territoire suffisant et lieu capable pour
> les habitans*], and sufficiently fertile soil for planting, and well stocked
> enough with beasts to feed and clothe the subjects. It should have a
> mild and equable climate, and an adequate supply of good water for the
> maintenance of their health. If the place itself is not sufficiently cov-
> ered and defensible, it should have buildings and fortification for the
> defence and shelter of the people.[190]

Here "territory" is clearly intended to be a desirable attribute of the
republic, alongside living spaces, fertile soil, and shelter as geographical
concerns. Indeed, Bodin wants to separate out the polity from its mere location or population: "It is neither the town walls [*la ville*] nor its inhabitants [*personnes*] that makes a city [*cité*], but the union of a people under a
sovereign ruler, even if they are only three households."[191] This is because
for Bodin

> the word *cité* is a word of right [*droit*], which signifies not one place
> or region [*point un lieu, ni une place / non locum ac regionem*], as the
> word Town or City; which the Latins call *Urbem, ab Urbo, id est aratio*
> [*Urbem*, from *Urbo*, that is, plowing; the Latin is in the French text],
> for as Varro says, the compass and circuit [*le circuit et pourpris*] of cit-
> ies was marked out by the plough.[192]

A city—that is, a particular kind of political community—is a legal
term, and is an attempt to render the Latin *urbo* into his French. His reference back to Varro perhaps makes it remarkable that he does not tie
this more explicitly to the surrounding *territorium*. Indeed, Bodin rarely
uses this Latin word. The French uses *territoire* more frequently, and the

Knolles translation uses "territorie" in many instances. This is sometimes to render the French *territoire*, but often is the translator's own invention. See, for instance, his claim that Venice "has no very large territory,"[193] when the French is *"qui n'a pas grande estendue de païs,"* which does not have a large extent of countryside.[194] The translator also uses "territorie" in the following passage, when he really mangles the sense: "If the Emperor, or the King of the Romans, left the frontiers of their lands [*païs*], they marched on the lands [*terres*] of the other princes as if strangers [*quasi comme estrangers*]."[195] Knolles often renders the French *terre* with "territorie," such as the "territorie of another prince . . . territorie and protection of the English."[196] In the Latin, Bodin uses *agros* here. Drawing on Plutarch's *Romulus*, he claims that the Romans seized a seventh part of the territories of those they vanquished.[197]

> We read, for instance, that Romulus, founder of Rome and the Roman Republic, divided the whole territory [*territoire / agrum*] into three parts, assigning one third for the upkeep of the Church, a second as the public domain [*domain de la Republique*], and the rest was divided among private individuals. . . . Plutarch . . . tells us that Romulus would set no limits on the territory of Rome [*ne voulut pas borner le territoire de Rome / agrum Romanum ullis finibus terminare voluisse*].[198]

There are also treaties that talk of "such a king, his countries, territories and seignories [*tel Roy, ses païs, terres, et seigneuries*]."[199] Then there are moments when it appears to be the translator's own intervention, with no parallel in either of Bodin's texts: "within the precinct and territory of the province of the magistrates."[200]

It is more interesting when the French and the Latin do not match in the way that might be expected. One of these is the notion of *iurisdictionem praediatoriam*, which the French usually has as some kind of jurisdiction over *territoire*.[201] There are several instances, especially in chapters 5 and 6 of the third book. Bodin talks, for instance, of "the rest of the princes and others having territorial jurisdiction [*depuis les autres Princes ont suyvi chacun en son territoire / qui prædiatoriam iurisdictionem*]."[202]

> Wherefore this jurisdiction which seems to be annexed [*cohærare*] unto the territory or land [*prædiis*] (and yet in truth is not) and is therefore called Prædiatorie, is proper to those are possessed of such lands [*prædia*], by inheritance, or by other lawful right, and that as unto right and lawful owners thereof, in giving fealty and homage unto the sovereign

of the republic, from whom all great commands [*Imperia*] and jurisdictions flow, and is saving also the laws of the imperium [*Imperii legibus*], and the right of the last appeal.[203]

One might ask if the magistrate might forbid a subject to come to the
court, being within the jurisdiction of his territory [*au resort de son territoire / provinciæ fines non excesserit*]? This is not without difficulty,
nevertheless without entering into further dispute, I claim that the magistrate banishing the guilty subject out of the territory of his jurisdiction
[*le territoire de sa jurisdiction / extra fines suæ iurisdictionis*], where
the prince may also then be secretly forbidden to approach the court,
albeit that he cannot expressly forbid him to approach the court.[204]

Bodin immediately cites the Roman jurist Ulpian here, from the *Digest*. What is interesting is the way that he uses the French *territoire* but
when working in Latin goes back to the Caesarean vocabulary of *fines*. He
does this elsewhere when he says that "each Confederate state is a sovereign power with its own distinct magistrates, distinct estates, distinct rev-
enues, distinct domain, distinct territory [*finibus & imperio divisas*]."[205]
The key in his argument is the potential to make the move from magistrates to rulers as a whole. But while the English translation recognizes
where the argument is going, Bodin himself holds back. Bodin declares
that "it is not necessary here to reject the opinion of those who attach jurisdiction to *prædia*, because in so doing we would go beyond the bounds
of this treatise."[206] The French reads *"qui ont attaché la jurisdiction aux
fiefs"*; the Latin phrase is *"qui iurisdictiones prædiis ita cohærere putant."*
A marginal gloss links this argument to Bartolus, Baldus, and Oldradus.
But the English translation renders the key phrase as "which affirme jurisdictions so to cleave unto the territories."[207] Bodin continues to suggest
that "justice so little holds to a *fief* that the sovereign Prince who has sold
or given away a *fief/feudum*, of whatever nature it might be, is not to be
reputed as giving away or selling the jurisdiction."[208]

> And whereas in the Edicts or laws any thing is commanded to be done,
> it is thus understood, that every Magistrate in his own province is to
> be obeyed, for that the magistrate has no power to command outside of
> his own territory or jurisdiction. In ancient times the king pursuant
> or officers, if they were to put in execution the commands of the royal
> magistrates in the territory of such lords as had therein territorial ju-
> risdiction, were first to ask them to leave, until that afterwards it was
> by the most straight decrees of the highest courts forbidden them to

do so, for that therein the sovereign majesty of the king seemed to be something impaired.[209]

Bodin illustrates this by saying that if lords or magistrates pass into the lands of another, they cannot be commanded or corrected by the landowner. It is the "superior magistrate or predominant lord" who rules and determines.[210] But this is to suggest that the sovereign's control over land is of the same standing as any other landowner, which still effectively treats the sovereign as possessing power separate from the area in which it is exercised. In Bodin there is still not a clear separation between land and territory. It is worth underscoring that Bodin does use *territorium* in the Latin, just infrequently. One example is the phrase *"quia nullum est extra territorium magistratui ius imperandi,"* which could be rendered as "the magistrate has no power to command out of his own territory or jurisdiction."[211] What is interesting in passing is that Bodin uses *territoire* in French to render the laws of Rome, concerning provincial governors and their provinces, for which the Latin is *prædiatorium iurisdictionem.*[212]

Bodin clearly states that "sovereignty is, of its very nature, indivisible."[213] What this means is that some form of monarchy is inevitable, and any kind of mixed constitution would be absurd.[214] There is an interesting discussion of the relation between the arithmetic and the geometric and their links to politics on this point.[215] He had earlier tried to construct the harmony of the republic through mathematics, using Plato.[216] He now notes that "geometric or distributive proportion is based on the principle of similarity, arithmetic or commutative proportion on the principle of equality. Harmonic is a fusion of the two which nevertheless does not resemble either."[217] He asks how this could relate to the three types of state or government: "If a single state could thus be compounded of all three, it would surely have to be wholly different from any one of them, just as we can see that the harmonic proportion, which is composed of the arithmetic and geometric, is entirely different from either of these . . . but the mixture of the three basic forms of state [*Républiques*] does not produce a different kind. The combination of royal, aristocratic, and democratic power makes only a democracy [*l'estat populaire*]."[218]

Bodin suggests that this leads to a strongly hierarchical system. At the head is the leader or prince, "whose majesty [*majesté*] does not admit of any division . . . below him are the three estates, which have always been disposed in the same way in all well-ordered republics." The three estates are the clergy, the military—both of which include nobles and commoners—and "the third estate of scholars, merchants, craftsmen, and labour-

ers." Bodin concedes that "aristocratic and popular states also flourish and
maintain a government. But they are not so well united and knit together
as if they had a prince . . . the union of its members depends on unity under a single ruler, on whom the effectiveness of all the rest depends. A sovereign prince is therefore indispensible, for it is his power which informs
all the members of the republic."[219]

What is Bodin's attitude to Roman law? In the preface to his *Methodus*, he describes how "at a time when all things suffered from the crud-
est barbarism, fifteen men appointed by Justinian to codify the laws so
disturbed the sources of legislation that almost nothing pure is dragged
forth from the filth and the mud."[220] Some of the commentators on his
work have seen him as a break from the medieval tradition. Perry Anderson, for example, sees him as the first to "systematically and resolutely . . .
break with the mediaeval conception of authority as the exercise of traditional justice, and to formulate the modern idea of political power as
the sovereign capacity to create new laws, and impose unquestioning
obedience to them."[221] Yet it is easy to be misled in reading Bodin. There
are several incidences where Bodin cites the jurists, including Bartolus
and Baldus, even if only to criticize them. There are multiple references:
Giesey reckons twenty-six to Baldus and fourteen to Bartolus in book 1,
chapter 8, alone.[222] What this means is that even as he criticizes them, his
work builds upon them. Indeed, he describes Bartolus as "one of the greatest of the jurists,"[223] an assessment, Hazeltine suggests, that developed as
he moved from theory to practice.[224] Bodin's education would have been
informed by this tradition. He would have used the same text of the *Corpus Iuris Civilis* with Accursius's *glossa ordinaria* as Bartolus and Baldus
would have used. Shortly after him the gloss would have been abandoned,
and a new edition based on humanistic textual values was used. As Giesey
summarizes, "The medieval legal tradition became moribund soon after
Bodin's time, while the idea of sovereignty which he had propagated had a
remarkably vigorous growth."[225] It is crucial to underline that the latter is
in part dependent on the former.

BOTERO AND *RAGIONE DI STATO*

In his *Essais*, Michel de Montaigne describes how he met a man:

> a gentleman of good appearance who was of the opposing party to ours,
> but I knew nothing of it, for he feigned otherwise. The worst of these
> wars is that the cards are so shuffled [*les cartes sont si meslées*], with

your enemy distinguished from yourself by no apparent mark either of language or of deportment, being brought up in the same laws, manners and customs, so that it is hard to avoid confusion and disorder.[226]

As Hodges notes, *cartes* should be taken to mean both cards and maps: "Rather than delineate territorial confines, the 'maps' in this passage obfuscate identity. . . . Although the 'cartes' may be mixed up, their appearances deceitful, like that of the gentleman Montaigne describes in this passage, the term also signifies a map and may be read as a treatment of precisely what such a signifier points to, its referent, the territory, which may well offer an alternative form of stability in uncertain times."[227]

Skinner has argued that Montaigne endorses reason of state. This has been comprehensively and compellingly challenged by Collins.[228] The individual case is less interesting than the issue itself. This phrase is a shorthand for a whole form of statecraft, sometimes better known by its French equivalent of *Raison d'État*. One of the German representatives to the Westphalia peace negotiations declared that "Reason of state is a wonderful beast, for it chases away all other reasons."[229] As Meinecke puts it, "Raison d'État is the fundamental principle of national conduct, the State's first Law of Motion. It tells the statesman what he must do to preserve the health and strength of the State."[230] The phrase had been used by Guicciardini, possibly for the first time, but the sense is not quite the same: "When I talked of murdering or keeping the Pisans imprisoned, I didn't perhaps talk as a Christian: I talked according to the reason and practice of states [*la ragione ed uso degli stati*]."[231] It is worth noting, with Hexter, that "reason of state" or *Staatsraison* are poor translations of *raison d'état* or *ragione di stato*, because they omit the idea of "right"—the right of the state.[232] Indeed, the idea of acting in the "national interest" might be closer to the sense, even if anachronistic. Yet as Skinner and others have noted, it is Machiavelli's elevation of the conception of prudent action over other virtues that is perhaps the key influence.

The Italian writer Giovanni Botero (1544–1617) is perhaps today the best-known example of this position, in the work *Ragione di stato*, although it is only one of the many works that utilized this term and idea.[233] The first edition of Bodin's work was published in 1589, though he continued to modify the work for the next decade. The opening lines are important in many respects:

State is a stable dominion [*dominio fermo* / strong, firm rule] over people; Reason of State [*Ragione di Stato*] is the knowledge of the means by which such a dominion [*Dominio*] may be founded, preserved and extended [*fondare, conservare, & ampliare*]. Yet, although in the widest sense the term includes all of these, it is concerned most nearly with preservation, and more nearly with extension than with foundation. This is because Reason of State assumes a ruler and a State [*il Prencipe, e lo Stato*] (the one as artificer, the other as his material) whereas they are not assumed—indeed they are preceded—by foundation entirely and in part by extension. But the art of foundation and of extension is the same because the beginnings and the continuations are of the same nature.[234]

This is the final version left by Botero, modified from earlier versions. The opening clause and the parenthetical remark were not in earlier editions;[235] instead, the text begins with the words *Ragione di Stato*. As in Machiavelli, *dominio* has the sense of property, rule, and the object over which rule is exercised, dominions.[236] Descendre, for example, reads *dominio* as meaning both a relation to subjects of domination or power, but also "the territorial and political reality over which this power is exercised, and it has in that case the sense of domain, territory, state."[237] Yet, in a famous analysis, Foucault has suggested that Botero's definition excludes territory. Foucault contends that there is "no territorial definition of the state, it is not a territory, it is not a province or a kingdom, it is only people and a strong domination."[238] In a sense Foucault is correct, because Botero only rarely uses a vocabulary that would admit of a territorial definition, and Descendre is guilty of reading far too much into the texts. But, as the analysis so far has shown, this is hardly surprising: political writings before Botero did not use such concepts either.

Indeed, Botero's spatial imaginary is in fact quite pronounced. This is not so much in *Ragione di Stato* but in two other works. One was a book first published in 1588, *Della cause della grandezza delle citta, On the Cause of the Greatness of Cities*. It again went through many editions. The opening lines of this book are also worth attention:

A city is said to be an assembly of people [*ragunanza d'huomini*], a congregation drawn together to the end that they may thereby the better live at their ease in wealth and plenty. And the greatness of a city is said to be, not the largeness of the site [*spatio del sito*] or the circuit of the walls, but the multitude and number [*moltitudine*] of the inhabitants and their power.[239]

Botero talks of particular cities and whether they were well situated, suggesting that they had "neither great convenience either of territory or traffic [*di territorio, ò di traffico*]."[240] An English translation from 1606 renders this as "*Territory or Trafique*": what is interesting is not so much the use of the term, but its modern spelling, especially when counterposed with "trafique," and the other instances of *territorie* in the translation to render Botero's *territorio*.[241] Botero also provides chapters that analyze the site (*sito*) and the soil (*terreno*) of the city.[242]

The second was *Relazioni universali*, of which four books were published between 1591 and 1598.[243] This was a work that was compiled on behalf of the papacy,[244] providing a synthesis of what was known of the New World, but also important to the development of statistics.[245] It followed the model of books like Sebastian Münster's *Cosmographia*, which provided a detailed description of the sum of geographical knowledge.[246] Münster's book was crucially written in German, and is remarkable for the woodcuts included, perhaps most famously the pictorial representation of Europe as a woman.[247]

This was a time of considerable innovation in cartography, with Abraham Ortelius's *Theatrum Orbis Terrarum* appearing in 1570 and Mercator's atlas in 1595.[248] *Relazioni universali* contains four maps of America, north and south; Africa; Europa; and Asia. They are reasonably accurate by the standards of the time, are likely derived from those of Ortelius, and include latitude and longitude grids. That said, *Relazioni universali* is largely descriptive and not especially interesting. There is, however, one fascinating sentence in the English translation: "The enlarging of Dominion is, the uniting and establishing of divers territories under one sovereigntie and government."[249] This is almost certainly not of Botero's making, but given its date from 1601, it is a remarkable phrasing.

The closest types of phrases in Botero's writings come in his discussion of war. There are different kinds of war: "whether it be waged to secure frontiers [*assicurare i confini*], to increase the dominion [ò *per ampliar l'Imperio*], or simply to win glory and riches, to protect allies or assist friendly powers, or to defend religion and the worship of god."[250] Later in the same book he suggests that they are "waged either defensively or offensively, to acquire from another [*per acquisto dell'autrui*]."[251] Such gains are "acquired a little at a time, but must all be preserved together as a whole."[252] He also contests one of Machiavelli's suggestions about such conquests: "I am surprised that Machiavelli should advise his prince, or tyrant, to transfer himself and his court to conquered lands [*paesi acquistati*]."[253]

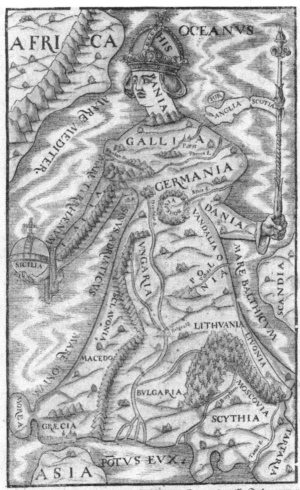

A FRI...CA   OCEANVS

HIS

DANIA

HIB.
ANGLIA   SCOTIA

MARE MEDITER.

GALLIA   PARIS

GERMANIA

DANIA

VANDALIA

VNGARIA

POLONIA

MARE BALTHICVM

SCANDIA

SICILIA

LITHVANIA

LIVONIA

MACEDO...

MOSCOVIA

MOREA

BVLGARIA

GRÆCIA

SCYTHIA

TARTARIA

ASIA   POTVS EVX.

zweyen General Tafeln/vnd in der newen Tafel die allein Europam begreifft. Doch wann man
ansehen will vnd darzu rechnen die grossen Landschafften die gegen Mitnacht gehn/solt wol die
breite Europe vbertreffen die länge. Wie aber Ptolemæus Europam beschrieben hat/ist sein länge
grösser dann die breite. Das ist ein mal gewiß/daß Europa ist ein trefflich fruchtbar vnd wol erba-
wen

Fig. 11: Sebastian Münster, *Cosmographei oder Beschreibung aller
Länder, Herrschafften undfürnemesten Stetten des gantzen Erdbodens*
(Basel, Switzerland: Sebastian Henricpetri, 1588), 55.

Descendre pushes his claims further than Botero's texts will allow, but makes some interesting claims in so doing. He suggests, for example, that the word *popolo* in the book on cities never means simply a fraction of the population, but "it always means the *l'ensemble* of the inhabitants living in a territory—country, town or region."[254] He suggests that Botero provides a "constant enterprise of the *territorialisation* of political questions,"[255] arguing that "it is the first time that a political thinker puts political and geographical questions in relation in a way that is both constant and methodical."[256] In Botero, he suggests, we find the "political technology of territory, the idea that the government of men depends on the political management of space, and the analysis of the effects of spatial structures on government and the obedience of subjects."[257] He claims that he is "the modern thinker of territory—that is of the political nature of space—at the threefold scale of the city, State and world."[258] Overblown though these claims certainly are, there is definitely a profoundly geographical sense.

Joost Lips—or, as he is better known, Justus Lipsius (1547–1606)—is an important figure here. He was deeply influenced by the classical tradition, and has been described as the last Renaissance philosopher.[259] Yet in a sense he predated that tradition, because his most important political work, the *Politica* of 1589,[260] is essentially a patchwork of quotations from Latin and Greek authors or, in Mellor's apt phrase, a "mosaic of maxims."[261] The key figures in this work, and his earlier *Constantia*, from 1584,[262] were Tacitus and Seneca.[263] Lipsius himself singles out Tacitus above all others,[264] and had edited some of his writings. *Politica* alone shows how important the classical heritage was even in the late sixteenth century. The arrangement is interesting. As Lipsius himself declares, "The stones and rafters I have taken from others, but the structure and form of the building are entirely mine. I am the architect, but have collected material from all over."[265] Lipsius supplemented the *Politica* with *Monita et exempla politica*, which was written as a sequel with illustrative historical examples, but this remained unfinished.[266] In *Adversus Dialogistam liber De Una Religione*, from 1590, he replied to attacks on the religious policy of *Politica*.[267] There are no modern editions of either text. He also wrote a detailed commentary on the military aspects of Polybius,[268] and it has been suggested he was an important innovator in military discipline.[269]

Given this collage, it can be difficult to ascertain Lipsius's own views. While the tissue of quotations allows him to make a relatively novel argument, it prevents him from any kind of conceptual advances, simply because he is using the terms inherited from the tradition, albeit with occasional new inflections. Some key aspects can nonetheless be underlined. His thought is essentially an application of Stoic virtues to the contemporary moment: *Constantia* for individual subjects; *Politica* for their leaders.[270] As he says in the opening pages of the second work on its purpose: "It is my aim now to equip those who rule for governing, just as in *De Constantia* I equipped citizens for endurance and obedience."[271] In this aim, Plato and Aristotle, especially the latter, are seen as precursors, but Lipsius notes that they wrote of government generally, whereas his intent is to concentrate on monarchy, *principatum*.[272] Of his contemporaries, Machiavelli is the one who most impresses him, even though he "strays from the road."[273] Lipsius does not mention Bodin in the *Politica*, but he knew his work;[274] Botero's *Della ragione di stato* appeared in the same year as *Politica*.

Senellart claims that the key problematic for Lipsius is "not the abstract relation of a king to his people, nor the domination of a prince over a territory, but the commandment of man by man."[275] Two minor geographical passages are worth noting. He suggests a link between countries and peoples when he notes that "moreover with a certain confusion of nations you French men possess Gaul; you Saxons, Britain; you Normans seize upon Belgica and the lands bordering [*finitima occupante*]";[276] and "he who prevails with this (the sword) shall best settle the borders [*optime de finibus disputat*]."[277] Lipsius cites and translates Plutarch's Greek into Latin here.[278] Kleinschmidt has even suggested that his work "assigned to rulers of territories the task of defining and protecting private ownership in land";[279] where Lipsius more straightforwardly suggests:

> For after men forsook their wild and savage manner of living, and began to build houses and walled towns, to join in society, and to use means offensive and defensive, behold then a certain communion necessarily began among them, and a social participation of diverse things. They parted the earth and bounds [*terram et fines*] between them; they had temples in common; also marketplaces, treasuries, seats of judgment; and principally ceremonies, rites, laws.[280]

One further point is worth noting: that the English translation of his work helped to establish the word *politics* as "the name of the art of governing cities." The English translation of Lipsius dates from 1594; Aristotle's *Politics* was translated four years later.[281]

One other work is important to mention here. This is *Vindiciae, contra Tyrannos*, published in 1579, but possibly written in 1574.[282] Its pseudony-

mous author was "Stephanus Junius Brutus, the Celt," and debates have
continued as to who actually wrote it: possibly the French diplomat Hubert
Languet or Pierre Mournay. It has been described as "the most celebrated
of the anti-monarchist tracts . . . a theological and legal masterpiece."[283] It
is of interest here because it suggests that "where the glory of God and the
kingdom of Christ is concerned, no limits [*limites*], no frontiers [*fines*],
no barriers [*cancelli*], ought to restrict the zeal of pious princes."[284] Yet,
on the other hand, for nonreligious reasons, it advocates the protection of
borders and respect for separate authorities: "If others, in order to procure
subterfuges for their impiety, state that frontiers [*limites*] and jurisdictions
are distinct, and that it is not lawful to thrust a scythe into another's harvest; I too definitely do not consider that under this pretext you may invade foreign borders [*fines*] or seize the jurisdiction of another for yourself,
or remove the harvest of a neighbour to your area, as many do under this
pretence."[285]

KING LEAR: "INTEREST OF TERRITORY, CARES OF STATE"

William Shakespeare's *King Lear* indicates in dramatic form some of the
tensions around politics and land, and is, in the folio edition, one of only
two Shakespeare plays that uses the word *territory*. The word *territories*
is slightly more common, but almost all these are places where *territories*
seems to mean the same thing as *lands*. However, in *Henry VI, Part 2*,
Lord Somerset reports on the situation in France: "That all your interest
in those territories / Is utterly bereft you—all is lost."[286] While this may
appear to be another use of *territories* in a sense of lands, or as a battlefield
fought over and surrendered, the relation of interest shows that it is not
simply property or a strategic sense, but the political control of and stake
in those places. This same phrasing is the one used of *territory* in *King
Lear*. Lear is discussing his plans for the inheritance of his kingdom between his three daughters.

> Meantime, we shall express our darker purpose.
> Give me the map there. Know we have divided
> In three our kingdom; and 'tis our fast intent
> To shake all cares and business from our age,
> Conferring them on younger strengths. . . .
> . . . Tell me, my daughters,
> (Since now we shall divest us both of Rule,

Interest of territory, cares of state)
Which of you shall we say doth love us most?
That we our largest bounty may extend
Where nature doth with merit challenge.[287]

This supposed test of filial obedience provokes strong reactions. The elder daughters, Goneril and Regan, both obsequiously profess their love of their father, the first declaring that her love is "dearer than eyesight, space, and liberty"; the second seeks that Lear "prize me at her worth. . . . Only she comes too short . . . the most precious square of sense possess."[288] The language of geometry, calculation, and economy maps onto the geographies they seek, and they receive lands in return.[289] Somewhat unusually, it appears that they receive their dowries sometime after their marriages. Goneril is rewarded by Lear in the following terms:

Of all these bounds, even from this line to this,
With shadowy forests and with champains rich'd,
With plenteous rivers and wide-skirted meads,
We make thee lady: to thine and Albany's issues
Be this perpetual.[290]

The speech compresses a division of land within a wider realm, of bounds and lines, sets out its characteristics, and establishes a lineage for its inheritance. The way that Lear turns to Cordelia shows this too, describing her as "to whose young love / The vines of France and milk of Burgundy / Strive to be interess'd."[291] Yet Cordelia refuses to play along, and her first response when asked to profess her love is "Nothing, my lord," to which Lear replies, "Nothing will come of nothing." Nothing is, of course, as the void, a spatial category, and one that much exercised the seventeenth-century imagination.[292] She goes on to state that she loves her father "according to my bond; no more nor less."[293] Lear fails to realize that Cordelia has no wish to join her sisters in their cheap flattery, and that she alone probably loves him most. But being honest is no reward: Lear tells her that "thy truth, then, be thy dower." The repercussions of these events quickly lead to developments that can seem unconnected, especially concerning the king's madness, but *King Lear* is a play that is fundamentally structured by this division of land both in the major plot of Lear and his daughters and in the subplot concerning Gloucester and the inheritance of his bastard son Edmund or his legitimate child Edgar.[294]

Lear's reaction to Cordelia is misjudged, as is his belief in his older daughters' love. But Lear is not a foolish king in initially proposing such a division. Rather, as Harry Jaffa has convincingly argued, Lear is struggling with the question of succession and the unity of the kingdom he has created.[295] Albany and Cornwall were the extremities of the kingdom, Albany being the north and the old name for Scotland; Cornwall being an expanse much bigger than the modern county, but a formerly separate kingdom including the southwest and much of Wales. At the time of writing, King James's son was Prince of Wales and Duke of Cornwall.[296] Lear has clearly decided what lands these couples should receive, because they are given their gifts immediately after the speeches of the first two daughters, but before Cordelia's.[297] In other words, Lear is not really comparing the speeches in order to distribute the lands, but using this as a pretense to buy off the two dukes. Indeed, in the folio his opening speech had expressly indicated an intention to do this "that future strife / May be prevented now."[298] Division into three does not necessarily mean each share is equal, and Lear continues to suggest that this is open to question. He rewards Regan's speech with "this ample third of our fair kingdom / No less in space, validity, and pleasure, / Than that conferr'd on Goneril"; but then immediately indicates to Cordelia that she could gain "a third more opulent than your sisters."[299] Cordelia throws the whole procedure off-balance and it is then that Lear acts rashly.

It is therefore important to note that Lear's pronouncement of this test for division was initially intended only for ceremony. Lear's plans are already known, with the apportioning of lands to Cornwall and Albany being the key issue, rather than to the daughters who happen to be married to them. Thus, Lear gives Goneril and Albany some lands close to their existing ones; and the same to Regan and Cornwall, reserving the central portion for Cordelia. Cordelia is being courted by the Duke of Burgundy and the king of France. It may be the principal purpose of Lear's test is to decide who Cordelia gets to marry: if she loves her father most of all, she will surely follow his wishes in this. Jaffa thus claims that Lear is being very strategic in terms of division: "It was an action predestined by the very means required to bring unity to the kingdom. Lear, it appears, delayed the division as long as possible, but he could not put it off indefinitely, any more than he could put off indefinitely his own demise."[300] The intention, he suggests, is that of "living on as king with Cordelia, with Albany and Cornwall acting as his deputies in regions which he could not control without their loyalty anyway."[301] Yet in not going with his plan, Cordelia receives nothing from Lear. Kent's attempts to mediate are swiftly prevented, and Lear apportions Cordelia's share between the first two sisters: "With my two daughters' dowers digest the third."[302] Being without a dowry immediately makes Cordelia a less attractive proposition for Burgundy, who pleads with Lear to reverse things. Lear grants her to France, with lines that are stinging in their rebuke: the dismissal applying as much to Cordelia as to France himself, and welcoming to Burgundy who, unlike France, has shown himself interested in bounty alone. Cordelia leaves, with France, only to return later in the play at the head of the invasion force.

◦◦◦

It seems significant that the word *territory* is not frequently found in Shakespeare's plays, even as late as the early seventeenth century. The word remained uncommon: it does not appear, for example, in the King James Bible, itself based on the earlier work of William Tyndale. We forget the comparatively recent intrusion of the word into our conceptual vocabulary, thinking that the word and the concept can be found throughout history. Yet *territorium* is a very rare word in classical Latin and until the late Middle Ages did not have the sense we might think that it carries. As Leider notes, it may well be that the use of *territory* and other nouns in the opening speeches of *King Lear* suggests a majesty of foreign influence later replaced by words of "native origin."[303] *Land* and *earth* can both be traced to Anglo-Saxon roots and, as chapter 3 showed, feature strongly in *Beowulf*. Several things in terms of Shakespeare's understanding of territory can nonetheless be gleaned from this instance of the word and the wider spatial and land politics of the play. Territory implies a range of political issues: it is controlled, fought over, distributed, divided, gifted, and bought and sold. It is economically important, strategically crucial, and legally significant. King Lear divides Britain into three, just as King James was trying to unite it.

CHAPTER NINE

The Extension of the State

THE CONSOLIDATION OF THE REFORMATION

R ichard Hooker (1554–1600), best known for his book *On the Laws of Ecclesiastical Polity*, has received extensive praise for his importance, principally for his defense of the religious settlement of Queen Elizabeth.[1] *Of the Laws of Ecclesiastical Polity* comprises eight books, only five of which were published in Hooker's lifetime.[2] In the various books, it offers discussions of the foundation of law, the authority of scripture, a defense against Puritanism, and the specific rites of the Church of England.[3] The religious context of Hooker puts him in a curious position in relation to other political thinkers, and he shares little with his contemporaries. But Hooker is a central figure in the transition from the church in England to the Church of England. He uses the term *Church-polity* because, in his words, "it contains both government and also whatsoever besides belongs to the ordering of the Church in public. Neither is anything in this degree more necessary than Church-polity, which is a form of ordering the public spiritual affairs of the Church of God."[4] As Eccleshall puts it, "Hooker placed political society firmly within the context of a rationally ordered universe."[5]

Hooker sometimes uses *territory* in a metaphorical sense: "For once we descend unto probable collections what is convenient for men, we are then in the territory where free and arbitrary determinations, the territory where human laws take place, which laws are after to be considered."[6] But a more fundamental concern is the extent of political and religious power. In book 7, for instance, he asks "how far the power of Bishops hath reached from the beginning in respect of territory or local compass."[7] But as he goes on to suggest, "How far the power which Bishops had did reach, what number of persons was subject unto them at the first, and how large their territories were, it is not the question we have in hand a thing very greatly material to know."[8] And as he later suggests, a city has "some territory belonging unto it."[9] There is therefore a territorially specific definition, but the specific size is unimportant.

> Is it not manifest that in this realm, and so in other the like dominions, where the tenure of lands is altogether grounded on military laws, and held as in fee under princes which are not made heads of the people by force of voluntary election but born the sovereign Lords of those whole and entire territories, which territories their famous progenitors obtaining by way of conquest retained what they would in their own hands and divided the rest to others with reservation of sovereignty and capital interest, the building of Churches and consequently the assigning of either parishes or benefices was a thing impossible without consent of such as were principal owners of land; in which consideration for their more encouragement hereunto they which did so far benefit the Church had by common consent granted (as great equity and reason was) a right for them and their heiress till the worlds end to nominate in those benefices men whose quality the Bishop allowing might admit them thereunto?[10]

Hooker is operating in a situation where the fractures in Christianity have presented a new problem. "While the Church was restrained unto one people it seemed not incommodious to grant their *Kings* the general chiefty of power. Yet now the Church having spread itself over all nations, great inconveniency might thereby grow if every *Christian King* in his several territory should have the like power."[11] His challenge therefore is to work out the relation between the church and state in those territories. But his key opponents are not Catholics, wishing to restore a previous situation, but Puritans advocating further reform; as well as atheists and Catholics, and elements within the church.[12] Hooker's solution is to give rulers not simply jurisdiction over religion in the sense of choosing the confession of their realm, but power over the church "within their own precincts and territories,"[13] or "within the compass of his own territories in such ample sort as the *Kings* of this land may do by the laws thereof."[14] One of Hooker's key aims in book 8 is to prevent potential conflicts between competing jurisdictions, and so he proposes a "universal power" in a territory.[15] This used to be the bishop of Rome, but it has transferred to national assemblies and synods, and in Hooker's case for England to the monarch. Yet Hooker is not simply a slavish apologist. Faulkner suggests that there is an attempt at a reconciliation between religious law and political expediencies.[16]

Johannes Althusius (1563–1638) was the author of a fundamental work of political theory, which went through three editions, in 1603, 1610, and 1614. The first edition is more polemical than later ones.[17] Althusius was writing in the wake of the 1555 Diet of Augsburg, which had allowed Lutherans, but not Calvinists, religious freedom. Skinner has accordingly described the *Politica* as "the most systematic statement of revolutionary Calvinist thought."[18] He was operating in a highly charged environment, with particular politics and determinations. Friedrich notes that his "discussion of the province (*Landschaft*) is . . . fully understandable only in terms of the German territories, each of which had its territorial diet."[19] Althusius was also a lawyer, and made extensive use of Roman law, including Bartolus and Baldus, in constructing his arguments. His citations are extraordinarily detailed, including lawyers, theorists, and biblical texts and theologians as far back as Augustine.[20] As Witte has put it, "Althusius piled citation upon citation, from all manner of seemingly unrelated sources, in demonstration of each simple assertion about what the natural law contained and commanded."[21] He discusses how Althusius must have taken extensive notes from all his reading, dividing these into lists that were then inserted into the manuscript. "It is an utterly fascinating and utterly exhausting display of erudition."[22] These notes are extremely revealing.

Althusius suggests that the *politicus*, the statesman or the political theorist, is "concerned with the fact and sources of sovereignty [*politicus de facto, & capitibus majestatis agit*]."[23] Then, paraphrasing Aristotle, he declares that

> politics is the art of consociating [*consociandi*] men for the purpose of establishing, cultivating, and conserving social life among them. Whence it is called *symbiotiké*. The subject matter of politics is therefore consociation [*consociatio*], in which the *symbiotici* pledge themselves each to the other, by explicit or tacit agreement, to mutual communication of whatever is useful and necessary for the harmonious exercise of social life.[24]

In this passage we can see that the notion of a consociation, a *consociatio*, is an important concept in Althusius's thought. Indeed, it is used alongside a range of other concepts, such as the "*universatis, civitas, reg-*

num, imperium, respublica, communicatio, foedus and *populus,"* to invoke what we might reduce to the state. Friedrich describes these terms as "a rather complex assortment."[25] Althusius goes on to suggest that

> he who takes the rights of sovereignty [*majestatis capita*] away from politics destroys the universal consociation [*universalem consociationem*]. . . . For what would the rector, prince, administrator, and governor of a commonwealth [*gubernator Reip.*] be without the necessary power, without the practice and exercise of sovereignty [*majestatis*].[26]

The *consociatio universalis* is not just any kind of association, but one that includes other associations from families to cities, and smaller-scale political organizations. This polity is self-sufficient, and is provided with *majestas*, which, following Bodin, appears to be understood in a way akin to the French *souveraigneté*. Althusius goes on to stress that "the community [*universitas*] is an consociation formed by fixed laws and composed of many families and collegia living in the same place [*eodem loco*]. It is also called a *civitas*."[27] Here we begin to see the spatial extent of the *consociatio*, which is explored at some length in other parts of the work. Althusius declares that "this community [*universitas*] is either rural [*rustica*] or urban [*urbana*]. A rural community is composed of those who cultivate the fields and exercise rural functions. Such a community is either a hamlet, a village, or a town."[28]

> A city may be either free, municipal, mixed or metropolitan. A free city is so called because it recognizes as its immediate superior the supreme magistrate [the emperor], and is free from the rule [*imperio*] of other princes, dukes, and counts. It is called an imperial city [*civitas imperialis*] in the German polity. . . . And no one doubts that these cities have the rights of princes within their *territorio*. The municipal or provincial city is one that is subject to a territorial lord [*territorii domino*]. . . . It recognises a superior other than the supreme magistrate.
>
> A mixed city is so called because it recognises partly the emperor and partly a duke or count as its superior, and enjoys both imperial and provincial privileges. There are some cities in which dukes or counts have usurped rights, even though the territory [*territorium*] does not actually belong to them. . . .
>
> A metropolis is so called because it is the mother of other cities that it brings forth as colonies.[29]

Each of these polities has subtly different political rights, which Althusius spends some time in adumbrating.

> The rights of the city [or laws, *jura civitatis*], its privileges, statutes, and benefits, which make a city great and celebrated, are also communicated by the citizens . . . it also includes the autonomy of the city, its privileges, right of territory [*jus territorii*], and other public rights that accompany jurisdiction and imperium.[30]

Cities therefore have the *jus territorii* themselves, which is a fundamental determination at odds with other theorists of the time. Althusius suggests that "any city [*civitatis*] that has a distinct and separate rule and territory [*regimen & territorium*] is said to be a province."[31] Nonetheless, a free city "cannot have the personal rights of princes . . . nor exercise jurisdiction beyond their territories [*extra territorium*]."[32] Their jurisdiction is thus exercised within, and limited by, the territory. In the wake of Augsburg, this is unsurprisingly a religious determination:

> Ecclesiastical communion of the realm is the process by which those means that pertain to the public organizing and conserving of the kingdom of Christ are established, undertaken and communicated according to his will throughout the territory [*territorio*] of this universal consociation. . . . Within the boundaries of the realm [*intra fines territorii regni*], this right guides the enjoyment of a pious life by which we acknowledge and worship God in the present world.[33]

It also works in general terms: "All power is limited by definite boundaries and laws, never absolute, infinite, unbridled, arbitrary, and lawless, but every power is bound to laws, right and equity. Likewise, every civil power that is constituted by legitimate means can be terminated and abolished."[34] Then, in a chapter added to later editions of the work, he further clarifies the relation, explicitly stressing how a province can exceed the city, and is a self-sufficient polity.

> We now turn to the province, which contains within its territory many villages, towns, outposts, and cities united under the communion and administration of one law [*jus*]. It is also called a region, district, diocese, or community. . . . I identify the territory of a province as whatever is encompassed by the limits or boundaries within which its laws

are exercised [*Territorium provinciae voco, quod fines & terminus provinciae, intra quos jura illius exercentur, continet*].[35]

This definition of the territory as the area included within the "limits
and boundaries," and over which the laws are exercised, is fundamental. It
is further stressed later in the work, when Althusius suggests that

> even though these heads, prefects, and rectors of provinces recognise
> the supreme magistrate of the realm [*summum regni magistratum*]
> as their superior, from whom their administration and power are con-
> ceded, nevertheless they have rights of sovereignty in their territory
> [*jura majestatis & principis in suo districtu & territorio*], and stand in
> the place of the supreme prince. They prevail as much in their territory
> [*territorio*] as does the emperor or supreme magistrate in the realm [*in
> regno*], except for superiority, pre-eminence, and certain other things
> specifically reserved to the supreme magistrate who does the con-
> stituting. Such is the common judgment of jurists. . . . The head of a
> province therefore has the right of superiority and regal privileges in
> his territory [*jus superioritatis & regalia in suo territorio*], but without
> prejudice to the universal jurisdiction that the supreme prince has. . . .
> This supreme and universal jurisdiction is itself the form and substan-
> tial essence of the sovereignty of the king [*majestatis regiae*] . . . which
> the king by himself cannot abdicate.[36]

There is much to be said of this. It is a reconciliation of the relation between the individual rulers of parts of the empire with the universal power
of the emperor, and, ultimately, the pope. In the language used to work
this, there are echoes of Bartolus of Sassoferrato and Jean Bodin, but also
the theorists of temporal power from previous struggles between kings,
popes, and emperors such as John of Paris, and even a bit of Marsilius of
Padua. However, those advocates of temporal power lacked a vocabulary
to articulate clearly what temporal power was exercised *over*, which Althusius has. He implies the formula that the *rex imperator in regno suo*—
the king is an emperor within his kingdom—but dressed in a different
phrasing: the king has those privileges within the *territorium*. Althusius
articulates a very clear division of powers, with certain things reserved for
the supreme magistrate, but sovereignty (majesty) is reserved to the king.
In suggesting that the king has a sovereignty that he himself cannot ab-
dicate, there is a recognition that the office exceeds the person—elements

of the "king's two bodies" notion. It is therefore a fundamental passage of great importance.

Althusius makes the link between the kingdom, the *regni*, and the *territorium* explicit in several places in his analysis. He declares that the right of the realm

> is the means by which the members, in order to establish good order and the supplying of provisions throughout the territory of the realm [*territorio regni*], are consociated and bound to each other as one people in one body and under one head. This right of the realm [*jus regni*] is also called the right of sovereignty [*jus majestas*].[37]

Then, in a passage not in the English edition, he suggests that "the territory of the realm is the bounded and described place, within which the laws of the realm are exercised."[38] In a twenty-first-century textbook it would be unremarkable; but in the early seventeenth century it really is something of an innovation, especially in the specific terms used.

In making this definition, he refers to several sources. One is the line from the Roman jurist Pomponius in the *Digest* discussed in chapter 7. As previous chapters have outlined, in classical Latin, *territorium* is not straightforwardly rendered as "territory"; it is much smaller scale, and this is an internal unit within the empire, not a separate polity. But this definition is of course crucial for the fourteenth-century jurists rereading Roman law in an entirely different context. Now, with Althusius, it is being brought explicitly into political theory.

Of his predecessors and contemporaries there are several references. One is to Udalricus (Ulrich) Zasius (1461–1536), a lawyer working in the late fifteenth and early sixteenth centuries.[39] Zasius is building on Bartolus of Sassoferrato and Baldus de Ubaldis. Like them, Zasius sees that jurisdiction is in a *territorium*, which is both the object of its rule and the thing that defines its extent.[40] As many before him, he draws on Pomponius's definition.[41] But Zasius—and this seems a fundamental development—ties *supremacy* to territory (*superiotatis* to *territorium*): "*quod ipse dictos actus superiotatis in suo territorio exercuerit.*"[42] This may well be the first time that this relation is expressed in those terms: it certainly becomes crucial to Althusius's definition. Also referenced is Matthias Stephani (1576–1646), author of the *Tractatus de jurisdictione.*[43] Stephani draws on Baldus and Bartolus in suggesting that jurisdiction operates in "the whole territory [*toto territorio*]," and that jurisdiction inheres (*cohæret*) in *territorium*. He discusses the notion of the territorial law (*iure territorii*),[44] and suggests that the notion of *Landes Obrigkeit* is the vulgar equivalent of *iurisdictione territorii*.[45] Crucially, he makes a distinction between the emperor and the princes, with the latter having jurisdiction within their territory.[46] Stephani draws on Zasius, Pomponius, and Siculus Flaccus, *De conditionibus agrum*, so there is a certain internal logic to the references being made at this time. But his key reference appears to be Andreas Knichen, an author also referenced by Althusius.

Knichen (1560–1621) was a German jurist and political adviser. His most important work was the 1600 book *De sublimi et regio territorii iure*.[47] He has rightly been described as "the most respected author on German territorial law";[48] and in a fundamental work of *Begriffsgeschichte*, Willoweit suggests that Bartolus, followed by Knichen, are the key theorists of *territorium* as a notion.[49] Höfert has suggested that "the concept of the territorial state" can be traced back to his work, because it is "an early attempt to systematise territorial state law by using Roman law."[50] There are several important elements to his work. Knichen provides a very brief discussion of the etymology of *territorium*, drawing on Isidore of Seville (though unnamed, he is easily recognizable through the discussion of the notion of a *tauritorium*), Cicero, and Varro in *De lingua Latina*.[51] He also draws on the definition offered by Pomponius from Justinian's *Digest*.[52] Though Knichen is fully conversant in the legal arguments, his relation to Baldus and Bartolus is ambivalent. On the one hand, he regularly cites them in support of his claims, but on the other, he suggests that too much "lamp-oil and labour [*oleum & operam*]" is expended on them, finding arguments in their work that relate to the situation when he is writing.[53] Instead, the basis of claims for *ius superioritatis* should be on the specific legal codes in existence. Knichen is also skeptical of too easy a link being made between the Roman Empire of antiquity and the Holy Roman Empire of his time.[54] Nonetheless, he uses the traditional Latin language of *terminatio, limites, fines*, and *finibus* in relation to *territorium*, to set out both the limits of the *territorium* itself and the extent of jurisdiction.[55]

Knichen explicitly links the notion of *Landeshoheit* with *superioritas territorialis*. He does this in the full title of the book, and it is recurrent throughout his argument. This is important in terms of bringing together distinct lineages of thought, which reoccurs in the treaties of Westphalia in 1648. He also discusses *superioritas* and *iurisdictione* in relation to *Landes Obrigkeit*,[56] and *die Landesfürstliche Obrigkeit* (the lands-princes authority).[57] Thus, there is a relation between *Landes* and *territorium* on the one hand, and *Hoheit/Obrigkeit* and *superioritas/iurisdictione* on the other. While there are lots of conceptual nuances at play here, it is this relation that is crucial.[58] Knichen makes an important distinction between the *majestatem* of the emperor and the Roman Empire and the *plurimum iura territorii* of the princes within it.[59] There are, of course, French and Italian precedents for this, in the idea that a ruler in his territory exercises the power of the emperor in the empire.[60] Knichen, though, articulates the political and legal arguments for why this is the case. Nonetheless, he opposed extending the *Territorialhoheit* to the free and imperial cities, reserving this for the territorial princes. Later editions revised these arguments after objections.

For Knichen, jurisdiction and territory inhere in each other.[61] This is a common claim in the fourteenth-century Post-Glossators, particularly in Baldus, but here it is given a political, as opposed to largely legal, reading. At one point he declares that *territorium* cannot be without *iurisdictione* and the reverse.[62] Jurisdiction is permanently attached to territory. It provides the basis of control over churches, for the administration of justice, and for the raising of taxes and armies.[63] In chapter 5 of the work, he moves to questions of jurisdiction more generally, and stresses that territorial rulers do not have jurisdiction in other territories.[64] It is important to stress that here, as elsewhere, he is discussing elements *within* the empire, not separate states themselves. While Bodin uses the Latin *majestas* and the French *souveraineté* as equivalents, Knichen can be seen as beginning the separation between majesty and sovereignty, which Leibniz would make explicit. Or, at least, Knichen distinguishes between *majestas* and *superioritas*, which seems to be a stage in that separation. While the emperor has majesty, and the princes mirror that in some respects, they have superiority. The majesty of the emperor, though, is integral in the definition of the supremacy of the territorial rulers.

There is very little discussion of Knichen in the secondary literature.[65] Yet the way that he is referred to by his contemporaries and some of the thinkers coming in his wake—Leibniz is a particular example—suggests a great importance. It seems remarkable that there is no modern edition of the text, let alone a translation, and that he has been almost wholly neglected as a thinker. In the notions of *superioritas territorialis* or *Landeshoheit* and *Landes-Obrigkeit*, Knichen is providing a fundamental contribution to the development of political thought, of a type of rule that is circumscribed both spatially and in accordance with the imperial constitution. Willoweit has described it as belonging to "the great achievements of the early publicists."[66]

Francis Bacon (1561–1626) is best known for his works of philosophy and his utopian fable of the New Atlantis, which outlines not the ideal state as much as the founding of a scientific community. He says relatively little in terms of political issues, but the successive versions of his essay "On the True Greatness of Kingdoms" are worth a few words. In the original version, published in the 1612 collection of his *Essays*, he declares that "the greatnes of a state in bulke or territory doth fall under measure; & the greatnes of finances and revenew, doth fall under computation: the population may appear by Musters, and the number of Cities and Towns by Carts and Mappes."[67] Here *carts*, "charts," should be taken to mean plans. Bacon is clear that the original extent of a state is no guarantee of its future: "So are there States that are great in Territory, and yet not apt to conquer or inlarge: and others that have but a small dimention or stemme, and yet apt to be the foundation of great Monarchies."[68] In the 1625 edition of the *Essays*, Bacon expands the text, now under the title "Of the True Greatness of Kingdoms and Estates." The passages reappear, but with more modern spelling and minor revisions: "charts" is now replaced by "cards"; and in the second passage "conquer or inlarge" becomes "enlarge or command."[69] Bacon produced a very close Latin approximation of this argument in his Latin text *De Augmentis*, itself translated into English: "The greatness of an empire as regards its size and territory falls under measure; as regards its revenue under computation. The number of the population may be taken by a census; the number and greatness of cities and towns by maps and surveys."[70]

All this would be of little interest were it not for the way that it relates to arguments made in Bacon's text "On the true Greatness of Britain,"[71] which was written for King James around 1608 but was abandoned, and not published until 1634.[72] Bacon here begins that "the greatness of kingdoms and dominions in bulk and territory doth fall under measure and demonstration that cannot err."[73] But he goes on to lay out a number of negative claims, the first of which is that "in measuring or balancing of greatness, there is commonly too much ascribed to largeness of territory."[74] Of his more positive theses, the first is "that true greatness doth require a fit situation of the place or region"; the second is "that true greatness consisteth essentially in population and breed of men"; and the sixth and final, "that it consisteth in the commandment of the sea."[75] So it is not quantity of territory but its qualities; the population is crucial; and the future for Britain is in the seas. Bacon offers four subsequent theses on territory:

First, *That the territories be compacted, and not dispersed.*

Secondly, *That the region which is the heart and seat of the state, be sufficient to support those parts which are but provinces and additions.*

Thirdly, *That the arms or martial virtue of the state be in some degree answerable to the greatness of dominion.*

And lastly, *That no part or province of the state be utterly unprofitable, but do confer some use or service to the state.*[76]

This is important for one key reason, that the construction of a polity is a balance between the population and the territory.[77]

One final text is worth looking at before we turn to developments of philosophy and mathematics in the seventeenth century, and then proceed to the holy trinity of Filmer, Locke, and Hobbes. This is a text that was long attributed to Walter Ralegh, entitled "Maxims of State," but almost certainly not by him.[78] Olwig has described this as "one of the earliest formulations of the idea of the territorial state,"[79] even though neither term can be taken uncritically.

Policy is an art of government of a commonwealth, and some part of it, according to that state or form of government wherein it is settled for the public good.

State is the frame or set order of a commonwealth, or of the governors that rule the same, especially of the chief and sovereign governor that commandeth the rest.

The state of sovereignty consisteth of five points:

1. The making or annulling of laws.
2. Creating and disposing of magistrates.
3. Power over life and death.
4. Making of war or peace.
5. Highest or last appeal.

Where these five are, either in one or in more, there is the state.[80]

Modern this might be in some respects, but it is also heavily influenced by classical and medieval formulations. Crucially, though, it does not set out a territorial frame. Rather, its interest lies in the stress on *state* as a noun, rather than as an adjective, and the use of sovereignty. The wonderful phrase the "state of sovereignty" also makes an appearance, though the earliest use of the phrase I have found is in the 1606 translation of Bodin.[81]

In these political writings, Ralegh uses the word *territory* a few times, and *territories* a few more, usually only in the sense of lands. It is commonly used to discuss colonial possessions, for instance, such as when he talks of the Seriffe in Barbary as "lord of a small territory."[82] One of the things that a prince with a newly acquired kingdom should do is "to look well to the borders and confining provinces, and if any rule there of great and equal power to himself, to join league with some other borderers, though of less strength, to hinder the attempts (if any should be) by such neighbour prince."[83]

THE GEOMETRY OF THE POLITICAL

In Paris in 1610 Henri IV was assassinated by François Ravaillac. Stephen Toulmin has described this as epoch marking, if not epoch *making*.[84] Indeed, many commentators suggest that this event is important in understanding the outbreak of the Thirty Years' War. Religious toleration appeared to have failed. Some days later, Henri's heart was enshrined at the Jesuit college at La Flèche. A small boy attended the ceremony.[85] Some seven years later, as a twenty-year-old, this Frenchman joined the army of Prince Maurice of Nassau in Holland. In 1619, he went to Germany and attended the coronation of the Emperor Ferdinand II at Frankfurt. Held up by the onset of winter in the Duchy of Neuburg, he remained in a stove-heated room for a day. Making use of the time to think through certain issues he held as important, he began a process of rethinking philosophy from the ground up. The results of this were drafted as *Regulae ad directionem ingenii*, but those works were not published until after his death. Instead, a preliminary study was published in 1637. In 1620 he was in Bohemia, on July 17 he was with the army as it invaded Upper Austria,[86] "and in 1621 in Hungary. Then he abandoned the profession of arms."[87] It is unlikely that he ever was actually involved in conflict, being rather a gentleman observer. In 1628 he settled in Holland, where he lived and worked until 1649. The previous year a peace negotiated in the towns of Münster and Osnabrück had brought the war to an end, which in the standard story laid the basis for the modern European state system. Engaged as a tutor by the queen of Sweden, the thinker proceeded to Stockholm, "where he saw the Queen four or five times in her library, at a very early hour in the morning." However, the story is that the court was preoccupied with the

celebrations of the peace, and he was asked to take part. Declining the opportunity to dance, he did at least write a libretto for the ballet, called *La
Naissance de la paix*.[88] The grateful queen offered him an estate and a pen-
sion, the estate to come from lands recently acquired through the peace.[89]
The Swedish winters proved too much, especially with the tutorials given
at five o'clock in the morning; he contracted pneumonia and died.

The French boy, soldier, philosopher, tutor, and librettist was René Descartes; the 1637 work written in the winter his *Discours de la méthode*;
the treaties the Peace of Westphalia. No direct causal link can be made
between these events—their geographical and temporal proximity seems
mere coincidence. And yet Descartes's view of space outlined in the *Discours*, and elaborated in the *Geometry* as measurable, mappable, strictly
demarcated, and thereby controllable, is precisely that which underpins
the modern notion of political rather than solely geographical borders, the
boundaries of states. Descartes's view of space is as radical a break from
the geometry of Euclid (which, crucially, and despite the common assertion, includes no notion of space) as the modern state is from the Greek
notion of the *polis*.

The point of the biography is partly to correct the standard understanding of Descartes's work as without any context, partly to show the inter-
woven nature of the Thirty Years' War and the most prominent thinker of
the time, and partly to suggest that Descartes should be taken seriously
as a *political* thinker.[90] Despite the fact that the ballet *La Naissance de
la paix* has a disputed authorship,[91] Descartes's life is clearly very closely
related to the events of the time. Descartes is important as indicating a
shift in how we understand the material world, and the fundamental determination of its geography. Indeed, he provides a geographical example
to illuminate his thought processes. This comes in the *Discours*, after recounting the story of how he came to be in the stove-heated room. He
discusses the order and beauty of a city built by a single architect, and
suggests that it is difficult to produce a finely executed product by labor-
ing only on the works of others. He notes that we rarely see people pulling
all the houses down in order to rebuild a city—though Baron Haussmann
came close two centuries later—but that we do see people starting from
scratch with an individual house. He notes that we therefore cannot "reform a state by changing it from the foundations up and overturning it in
order to set it up again," but that we can do so with our own mind.[92] He
returns to this metaphor especially in his reply to the seventh set of objections to his *Meditationes*.[93]

In his division of mind and body, mind is *res cogitans*, matter *res ex-*

tensa. This division, found, for example, in the *Meditationes*,[94] places the notion of extension at the heart of his project. The material world is reduced to a single form, matter in motion, which is fundamentally understood through extension. While Descartes puts all things in doubt initially,[95] a number of arguments or thought experiments, such as the discussion of the changeable properties of wax, convince him that extension is the primary characteristic.[96] As he says in the *Principia Philosophiae*, "There exists something extended in length, breadth and depth and possessing all the properties which we clearly perceive to belong to an extended thing. And it is this extended thing we call 'body' or 'matter.'"[97] These objects exist in a space, a *spatium*, that is similarly extended. Two different things in size and shape can occupy the same *place*, but clearly not the same *space*. And when something moves, it is its place that has changed, not its size or shape.[98] This is crucially important—space, not place, claims for exclusivity.

> Thus we always take a space to be an extension in length, breadth and depth. But with regard to place, we sometimes consider it as internal to the thing which is in the place in question, and sometimes as external to it. Now internal place is exactly the same thing as space; but external place may be taken as being the surface [*superficiem*] immediately surrounding what is in the place.[99]

Descartes follows his *Discours* with three scientific treatises, on dioptrics, meteorology, and geometry. He makes especially bold claims with regard to the third,[100] which he regards as the very model of rigor.[101] Yet it is crucial to understand that Descartes is not simply taking a preexisting understanding of geometry and asserting its importance. Rather, he is transforming how geometry should be understood. In the *Geometrie* itself, he suggests that "all problems in geometry can be simply reduced to such terms that a knowledge of the lengths of certain straight lines is sufficient for their construction."[102] Later in the same work he suggests that "in the method I use all problems which present themselves to geometers reduce to a single type, namely, to the question of finding the values of the roots of an equation."[103] Descartes is suggesting that geometric problems can be understood as a question of quantity, the length of lines, or equations, a problem of number.

This geometry, known as analytic or coordinate geometry, can be traced back to Fermat or Viète's work on algebraic notation in the late sixteenth century, but it reaches a mature form in Descartes's work. This is a break from previous understandings that saw arithmetic and geometry as distinct. Descartes therefore sees geometry as a mode of access to the world rather than a merely mental exercise. The importance of this move is substantial. Descartes thinks that extension is not simply a geometrical but a physical property or ontological determination of the world. The world becomes a geometrical frame within which things are situated, move, and change. As Klein has noted, "Only at this point has the conceptual basis of 'classical' physics, which has since been called 'Euclidean space,' been created. This is the foundation on which Newton will raise the structure of his mathematical science of nature."[104] "Euclidean space" is thus a misnomer, because it takes a term that Euclid does not use, *space*, and applies his mathematical system to it in order to make sense of the world. Euclid's *Elements* was rewritten in algebraic calculus in the seventeenth century, which shows it was compatible with new developments, but only in a form Euclid would not have recognized.[105] As Alexander puts it, it seemed apparent that "if Newtonian mechanics used Euclidean geometry, space must be Euclidean."[106] The original translation of *topos* in Latin was with the word *locus*; only later did the word *spatium* arise to mean not simply extent, as it did in classical Latin, but a container with extent.[107] Euclid, like Plato, sees his geometry as a mathematical system. It is the transfer of this geometric way of thinking, and the three-dimensional space in which it operates, to explain the world through physics that is the fundamental issue in the scientific revolution of the seventeenth century.[108] As Descartes puts it: "All my physics is nothing but geometry."[109]

Descartes thus transfers a geometrical notion, *spatium* as *extensio*, into a way of making sense of the world. But he also effects a revolution in geometry. Descartes says that the "object dealt with by geometricians" is "conceived as a continuous body, or a space indefinitely extended in length, breadth, and height or depth, divisible into various parts that may have various shapes and sizes, and may be moved or transposed in all sorts of ways."[110] But Descartes is unconvinced, because he suggests that "there was nothing at all in these demonstrations which assured me of the existence of their object."[111] Instead, for Descartes, geometry is not adequate as an *abstraction* alone but must be seen as a mode of access to the world itself.[112]

Descartes was not, of course, the only writer framing such issues. Nicolas Copernicus had published *De revolutionibus* in 1543, and his ideas had been developed by thinkers like Giordano Bruno (1548–1600), who had proposed an infinite universe, which also owed something to Lucretius's ideas.[113] Galileo Galilei (1564–1642) had been professor of mathematics and military engineering at Pisa.[114] Galileo was the author of an important work entitled *Le operazioni del compasso geometrico et militare* (1606), which dealt with questions such as aiming cannons, arranging armies, and performing financial calculations.[115] There was also the earlier figure of Nicholas of Cusa, who had been instrumental in the replacement of Aristotelian notions of place with geometrical space.[116] Cusa is interesting—beyond the political-theological work discussed in chapter 7—because of his mathematical works. In his *De staticis experimentis* (Experiments with scales), a key theme is the way that the world of finite objects, which by its very nature is quantitative, can be counted, measured, and weighed.[117] Yet here, and in *De docta ignorantia*, the goal is not knowledge for its own sake, nor even to know the world as an end in itself, but as a means of access to knowledge of God. The finite world is the way to understand the infinite God. Cusa quotes from Wisdom of Solomon 11:21: God has "ordered all things in number, weight and measure."[118]

There are, of course, multiple differences between these writers, and the complexities should not be reduced. Nonetheless, it is not unreasonable to suggest that they are operating within a shared paradigm, where the material world is amenable to scientific, often mathematicized, inquiry. As Husserl has described, there was a two-stage process: Galileo's *geometrization* of nature, and the subsequent *arithmetization* of geometry in which Descartes plays such a fundamental role.[119] As Cassirer puts it, "Descartes' physics is in many regards, both in its explanation of special phenomena and in its general conception of the laws of motion, opposed to Galileo's views. But it is an offspring of the same philosophic spirit."[120] It also seems notable that in the extensive "objections" to Descartes's *Meditationes*, the question of extension is rarely raised, with the focus being more on the thinking subject, knowledge, and God.[121] Antoine Arnauld (1612–94), for instance, accepts the idea of extension but wonders if thinking things are similar, just with the additional potential for thought.[122]

The English philosopher Thomas Hobbes (1588–1679) was one of the respondents to the *Meditations*, which Descartes responded to intemperately, guessing that it was Hobbes. Yet Hobbes too was much taken by geometry, declaring that it was "the onely Science that it hath pleased God hitherto to bestow on mankind."[123] More generally his work was influenced by mechanistic philosophy.[124] His political work will be discussed in much more detail below, but some remarks are helpful here. While Skinner is right to stress Hobbes's conception of "a 'geometry' of politics,"[125] it would be a mistake to think that Hobbes is following Descartes too closely. As Gillespie notes, their geometric methods are distinct.[126] This is important to stress, because Hobbes was likely to have been one of the first English readers to see Descartes's *Discourse, Dioptric, Meteors and Geometry* (1637), which was given to him by Kenelm Digby.[127] Hobbes continues to see a sharp divide between arithmetic and geometry: "For as Arithmeticians teache to adde and substract in *numbers*; so the Geometricians teach the same in *lines, figures* (solid and superficiall,) *angles, proportions, times*, degrees of *swiftnesse, force, power*, and the like."[128] As Christian Huygens notes, Hobbes "seems to reject without cause the use of arithmetical calculation in the investigation of geometrical constructions."[129]

The geometrical method also greatly impressed Baruch Spinoza (1632–77), who structured his entire *Ethics* around propositions, corollaries, demonstrations, and scholia.[130] The particular kind of geometry is open to dispute, though it certainly bears a relation to Euclid's mode of exposition.[131] But the formal nature of the demonstration should not make us think that Spinoza is imposing a method over and above the content. As Montag puts it, "Spinoza rejects the notion of a method prior and therefore external to the process or activity of knowledge itself."[132] Indeed, while Spinoza thinks that extension is *a* fundamental characteristic of matter, he did not think it was the primary characteristic, and he makes a distinction between our perception of extension through the senses and our conception of extension through the understanding.[133] In recent years, Spinoza has come to be seen as an important political philosopher rather than primarily as a metaphysician.[134] Much of this comes in the banner of the "new Spinoza" work undertaken in France and Italy,[135] influenced in part by the readings of Louis Althusser and Gilles Deleuze.

Nonetheless, in terms of his thinking of political space, there is certainly no development. Spinoza reverts to an earlier way of thinking political control over land, insofar as he even discusses it.[136] One of the very few instances comes in his discussion of the dating and authorship of certain books of the Bible. In the process there is a discussion of Ibn Ezra's commentary on Genesis 12:6, where he makes references to Abraham "surveying the land [*terram*] of the Canaanites." The verse says that Abraham traveled through the land, and "at that time the Canaanites were in the land [*terra*]." Because of the "at that time," Ezra apparently claims that "this must have been written after Moses' death, at a time when the Canaanites had been expelled and no longer possessed that region [*regiones*]."[137] Spinoza later returns to the Canaanites and the conquest of their *imperium* and the subsequent division of the land (*terras*) into lots by the twelve tribes.[138] He therefore decides against the use of the term *territorium* to make these concepts more contemporary; like the

Vulgate, he uses *terra*. His work also had a profound influence on Henry de Boulainviller (1658–1722).[139]

The relation between Leibniz and Newton is clouded by two issues. The first is that while it is now generally accepted that they invented calculus independently, Newton and his supporters claimed that Leibniz had plagiarized him. Yet this was an issue that fascinated Leibniz throughout his life, and his work differs in some important ways from Newton's, especially concerning the techniques and notation. Fundamentally, Leibniz was concerned, just as Descartes, with the composition of continuous quantities, such as the infinitesimally small points that together make up a line. He suggested that only geometry "can provide a thread for the Labyrinth of the Composition of the Continuum."[140] His writings on this topic are—like most of his output—extremely extensive.[141] The second is to see their entire relation through the lens of the correspondence that Leibniz conducted with Samuel Clarke, acting as a representative of Newton.[142] This was right at the end of Leibniz's life—the correspondence breaks off after the fifth reply of Clarke, which was sent to Leibniz on October 29, 1716, only two weeks before he died. Yet Newton and Leibniz exchanged some personal correspondence, and shared a number of concerns. The key one is the understanding of the world through numbers. As Leibniz wrote to Newton in 1692, "I would wish that, perfected in geometrical problems, you would continue, as you have begun, to handle nature in mathematical terms."[143]

Yet while Newton and Leibniz shared the mechanical philosophy of Descartes and his followers, there are some important differences. Newton famously argued for an absolute space, a container that was independent of whether there were any bodies in it or not.[144] As Janiak has argued, in articulating this thesis, he had "a Cartesian, and not a Leibnizian, opponent primarily in mind."[145] For Newton, this absolute space was a fundamental determination: "Space is an affection of a being just as a being. No being exists or can exist which is not related to space in some way."[146] What seems remarkable is that by the late seventeenth century, in his *De gravitatione*, a text probably written shortly before 1685, Newton is able to declare that "the terms 'quality,' 'duration,' and 'space' are too well known to be susceptible of definition by other words."[147] He then proceeds through some fundamental definitions:

Definition 1. Place is a part of space which something fills completely.

Definition 2. Body is that which fills place.

Definition 3. Rest is remaining in the same place.

Definition 4. Motion is change of place.[148]

In his *Principia Mathematica*, he clarifies the point further:

Absolute space, of its own nature without reference to anything external, always remains homogeneous and immovable. Relative space is any movable measure or dimension of this absolute space; such a measure or dimension is determined by our senses from the situation of the space with respect to bodies and is popularly used for immovable space, as in the case of space under the earth or in the air or in the heavens, where the dimension is determined from the situation of the space with respect to the earth. Absolute and relative space are the same in species and in magnitude, but they do not always remain the same numerically.[149]

Leibniz, in opposition, stressed that space was entirely relative. The idea of absolute space, separated from anything in it, was a nonsense: "Space without matter is something imaginary."[150] For him, space is not a "certain thing consisting in a supposed pure extension," and "motion is not something absolute, but consists in relation."[151]

Extension or space and the surfaces, lines and points one can conceive in it are only relations of order or orders of coexistence, both for the actually existing thing and for the possible thing one can put in its place.[152]

For Leibniz, extension itself is not the primary characteristic, but "is a relation to something which is extended, something whose diffusion or continuous repetition it implies; it presupposes bodily substance, which involves the capacity for action [*potentiam*] and of resistance and which exists everywhere as corporeal mass, the diffusion of which is contained in extension."[153] While for Leibniz body was extension, it was not substance.[154] As he stresses in his correspondence with Clarke:

I don't say that matter and space are the same thing. I only say, there is no space, where there is no matter; and that space in itself is not an absolute reality. Space and matter differ, as time and motion. However, these things, though different, are inseparable.[155]

As Sallis has shown, Leibniz was not as interested as Descartes in forging philosophy anew, but rather in using modern thought to build on ancient thought.[156] He rejected the idea that the new philosophy demanded a complete break from Aristotle's work.[157] He was clear that alongside the modern mathematics, there must be an understanding of force, for which the Greek notions of *dynamis* and *entelechy* were helpful indications. This is why Leibniz declared that "we must employ the notion of force in addition to that of extension [*la force*]."[158] As Garber puts it, "Though he shared the physics, he did not share the metaphysics on which Descartes grounded his mechanism."[159]

Right at the end of his life, Leibniz proposed some very clear definitions. "Space is the order of co-existing, or the order of existence for all which is contemporaneous."[160] He then goes on to say that *"extension is the magnitude of space. It is false to confound extension, as is commonly done, with extended things, and to view it as substance."*[161] Yet this does not mean he abandons a geometrical model of space: "A *line* originates as the path of the point. A *surface* is the path of a line. The whole *volume of space* or, as is commonly said, the solid [*solidum*] is the path of a surface."[162] Newton's conceptions formed the basis of so much science that followed him it is difficult to see him as anything other than the victor in the dispute. Under his influence, modern science "finally dissolved the cosmic system described in Ptolemy's *Almagest*."[163] Yet Leibniz's work has continued to exercise an influence.[164] Leibniz will be returned to below, for the way this geometry impacted on his understanding of politics.

THE DIVINE RIGHT OF KINGS: HOBBES, FILMER, AND LOCKE

Thomas Hobbes is often looked at as the first modern political philosopher. For Skinner, although he stresses the importance of Machiavelli and Bodin, "it is Hobbes who first speaks, systematically and unapologetically, in the abstract and unmodulated tones of the modern theorist of the state."[165] In addition he notes that Hobbes is one of the first English philosophers to write of "'politics' as the art of governing cities."[166] Yet in many respects Hobbes's arguments remain rooted in previous debates. He was a translator of, among other texts, Thucydides's *History of the Peloponnesian War* and Aristotle's *Rhetoric*, and was widely read in Latin too, including later writers such as Ammianus Marcellinus.[167] His relation to the past is nowhere more apparent than in his discussion of the relation of spiritual and temporal power, for which he uses the terminology and arguments of the two swords. It should be remembered that the full title of his most famous book is *Leviathan; or, The Matter, Forme, & Power of a Commonwealth*

Ecclesiasticall and Civill—that is, the commonwealth is at once political and religious.

Hobbes contends that *"Temporall* and *Spirituall* government, are but two words brought into the world to make men see double and mistake their Lawfull sovereign."[168] He suggests that the distinction "is but words," and that "Power is as really divided, and as dangerously to all purposes, by sharing with another an *Indirect* Power, as a *Direct* one."[169] The fracturing of political power represents divided authority, as Hobbes continually insists, and instead makes a case for the centralization of control in a single source: undivided and unlimited. He sees division as a problem if it is between two rivals for the same kind of power, but also if there is a division between powers, understood in different terms. "From the same mistaking of the present Church for the Kingdom of God, came in the distinction betweene the *Civill* and the *Canon* Laws: The Civil Law being the acts of *Soveraigns* in their own Dominions, and the Canon Law being the Acts of the *Pope* in the same Dominions."[170]

Hobbes is at pains to ensure that even where there is a distinction between powers, there must still be a clear hierarchy. For him the civil, temporal, power must be able to supersede other powers:

> For not withstanding the insignificant distinction of *Temporall*, and *Ghostly*, they are still two Kingdomes, and every Subject is subject to two Masters. For seeing the Ghostly Power challengeth the Right to declare what is Sinne, it challengeth by consequence to declare what is Law, (Sinne being nothing but the transgression of the Law;) and again, the Civill Power challenging to declare what is Law, every Subject must obey two Masters, who both will have their Commands be observed as Law; which is impossible. Or, if it be but one Kingdome, either the *Civill*, which is the Power of the Common-wealth, must be subordinate to the *Ghostly*, and then there is no Soveraignty but the *Ghostly*; or the *Ghostly* must be subordinate to the *Temporall*, and then there is no *Supremacy* but the *Temporall*. When therefore these two Powers oppose one another, the Common-wealth cannot but be in great danger of Civill warre, and Dissolution.[171]

The final sentence might suggest that because the danger to the commonwealth is that of civil war, Hobbes has the English Civil War squarely in mind. This war, fought between 1642 and 1651, undoubtedly provides a crucial context for Hobbes's writings: *Leviathan* was published in 1651. But the division Hobbes talks about here is not between different temporal rulers—king and Parliament, for instance—but between temporal and ghostly, or spiritual, rule. It should be remembered that although Hobbes explicitly says that he wrote the book for an English audience,[172] he had written the book in France, having been in exile there since 1640. As Tuck notes, Hobbes believed the book was as relevant to the problems of France, notably the "Fronde."[173] The Fronde—the word in French means "sling," a weapon used at the time—was a series of revolts that followed the Peace of Westphalia from 1648 to 1653. These struggles were, in part, between the landed aristocracy and the French royalty. A certain reading of Hobbes's concerns can be used to shed light on such disagreements. But for much of the time that Hobbes was in France, it would have been the conflicts known as the Thirty Years' War that would have been foremost in his mind. These were conflicts over the divide between political powers, certainly, but also concerning the role of the pope, and through him the emperor, in politics. Indeed, seeing this context makes sense of the repeated references to the relation between temporal and spiritual power. As he says, the contemporary papacy is "no other than the *Ghost* of the deceased *Romane Empire* sitting crowned on the grave thereof: For so did the Papacy start up on a Sudden out of the Ruines of that Heathen Power."[174]

Hobbes is not impressed by the arguments concerning possession, authorization, and use of the sword. As he suggests in *De Cive*:

> Since the right of the sword [*ius gladij*], is simply the power by right
> to use the sword at his own will, it follows, that the judgment of its
> right use pertains to him; for if the power of judging were in one, and
> the power of execution in another, nothing would be done. For in vain
> would he give judgment, who could not execute his commands; or, if
> he executed them by the power of another, he himself is not said to
> have the power of the sword, but that another, to whom he is only an
> officer. All judgment therefore, in a commonwealth, belongs to the
> possessor of the swords; that is, to him who has the supreme authority
> [*imperium summum*].[175]

Hobbes is at his stinging best when he engages with the contemporary moment. He suggests that the idea that the bishop of Rome had become "Bishop Universall," as successor to Saint Peter, could be "compared not unfitly to the *Kingdome of Fairies*; that is to the old wives *Fables* in England, concerning *Ghosts* and *Spirits*, and the feats they play in the night."[176]

Hobbes pursues this agenda, in part, through a detailed engagement with the writings of Roberto Francesco Romolo Bellarmino (1542–1621), better known as Robert Bellarmine, a cardinal of the Catholic Church and advocate of papal power.[177] Bellarmine's *De summo Pontifice*, part of his *Disputationes*, is the key target.[178] In some important aspects, then, Hobbes was the end, or at least a late stage, of debates, instead of the beginning. In terms of the relation of political power to land, he is in a related position. Hobbes is frequently the inheritor of a terminology forged before him.

> So that the question of the Authority of the Scriptures is reduced to this, *Whether Christian Kings, and the Soveraigne Assemblies in Christian Common-wealths, be absolute in their own Territories, immediately under God; or subject to one Vicar of Christ, constituted over the Universall Church; to bee judged, condemned, deposed, and put to death, as hee shall think expedient, or necessary for the common good.*[179]

For Hobbes, while "the *universal Church* [*Ecclesia universa*] is indeed one *mystical body* [*Corpus mysticum*]," its "head is Christ,"[180] rather than any figure on the earth. In terms of political power it is not ruled by one, but dispersed, each concentrated in one figure. This is also seen in his responses to Bellarmine. He suggests that a plurality of Christian sovereigns had rights of sovereignty in their multiple territories,[181] and that the pope does not have civil power except in the territories he directly controls: "in the Territories of other States" he has neither direct nor indirect civil power.[182] Further, he argues that the clergy take whatever authority they have from the civil sovereign.[183] As he later underlines, "Every Christian Soveraign be the Supreme Pastor of his own Subjects,"[184] a point that he later underlines applies "in their own Dominions."[185]

These are the arguments of the temporal power theorists of the early fourteenth century, yet loosely formulated in terms of the argument that sovereignty is related to territory. Hobbes, though, is, in a sense, still trying to work with an earlier model: his aim is for absolute sovereignty—that is, sovereignty without limits—which is what was previously understood as temporal power but without a counterbalance of spiritual power. Yet in other respects his arguments break new ground. The notion of the empowered sovereign being constituted from the individuals who authorized it is a powerful notion. As Hobbes suggests, "The Multitude so united in one Person, is called a COMMON-WEALTH, in latine CIVITAS. This is the Generation of that great LEVIATHAN."[186] Hobbes is thus taking the name of the monster in the Book of Job to apply at once to the state and the sovereign.[187] It is important to underline that, in Hobbes, *civitas* is translated by both "commonwealth" and "state."

The frontispiece to *Leviathan* has been widely analyzed, and it would appear that there is little is left to say.[188]

But it is worth stressing several points. The sword and the crook or

Fig. 12: Thomas Hobbes, *Leviathan* (London: Andrew Crooke, 1651), frontispiece.
Reproduced by permission of Durham University Library, Cosin T.1.12.

crosier held by the ruler, would, if continued above the illustration, meet as a triangle.[189] Indeed, the style of the artist, Abraham Bosse, is known as *geometrico*. The Latin text comes from Job 41:24, which reads *Non est potestas Super Terram quae Comparatur*: There is no power on earth to compare. The sovereign's body is, as is well known, made up of the bodies of the people, but rendered anonymous, since their backs are turned to the viewer. (A variant of the frontispiece has the body made up of heads with visible faces.)[190] The sovereign, though, is clearly visible, and there is a debate as to whether he is supposed to resemble Charles II, Cromwell, or both, or some other figure.[191] He is surveying, and clearly rules over, a landscape, which has a walled city and surrounding countryside. There are some very visible churches both within and outside the city walls. As Hobbes notes, "The Soveraign of each Country hath Dominion over all that reside therein."[192] The bottom half of the illustration has, in the left column below the sword, the trappings of civil power: castle, crown, cannon, other arms arranged as a "trophy," and an army on a battlefield.[193] The right column below the crosier has the ecclesiastical equivalents: church, miter, and a *fulmen*, a divine thunderbolt or of excommunication,[194] logical divisions, and a canon law court. The two columns clearly relate to the figure, who holds symbols of both sides, but also to the book's subtitle, which stresses the unity. This is also shown in the panel concerning divisions, which has a trident to the left, whose prongs have the word *Syllogis-me* split between them; and then three forks, reading "spiritual" and "temporal"; "directe" and "indirecte"; and "real" and "intentional." At the bottom of this panel there are the two horns of a di-lem-ma, the parts of the word split across the horns. This may be an allusion to the divisions between the Catholic and Protestant churches, or fractures within those churches,[195] but is surely also between secular and spiritual power.

Even though Hobbes has cast a long shadow in the centuries since, he was not the only advocate of centralized power discussed at the time. A somewhat more moderate position was advocated by George Lawson in his *Politica Sacra et Civilis*.[196] In the latter half of the seventeenth century, Robert Filmer probably exerted a stronger influence.[197] He was the principal target of John Locke's *Two Treatises on Government*, for example. Filmer acknowledges a debt to many writers, but Aristotle and, in particular, Bodin are foremost among them, alongside the Bible. One essay, "The Necessity of the Absolute Power of all Kings: and in particular, of the King of England," is comprised entirely of quotations from the 1606 Knolles translation of Bodin.[198] He also wrote a detailed discussion of Aristotle's *Politics*.[199] Nonetheless, there is a striking argument of his own being advanced, which comprises a strong justification for a supreme leader. He discusses, with some concern, the fractures of political authority in England before the Norman Conquest.[200] His main work is *Patriarcha*, which is likely to have been written sometime between 1635 and 1642.[201] The title suggests the content and argument: this is a key source for the argument for patriarchy, the rule of the father. This Filmer derives from Adam, and his lordship over the earth, inherited by his children. Filmer thinks that in "all kingdoms or commonwealths in the world," and this regardless of whether they were legitimately inherited, seized by force, or the rulers elected, "is the only right and natural authority of a supreme father."[202] The point concerning the way they achieved this position is important—Filmer makes no distinction between these different means of gaining power: "There is, and always shall be continued to the end of the world, a natural right of a supreme father over every multitude, although by the secret will of God, many at first do most unjustly obtain the exercise of it."[203]

Filmer contends that this is a right achieved by divine will, not by some choice or transfer of the people. He notes Edward Coke's suggestion that "the first kings of this realme had all the lands of *England* in demesne"[204]—that is, as personal *dominium*. Filmer replies that if this were true, and "if the first kings were chosen by the people (as many think they were), then surely our forefathers were a very bountiful (if not a prodigal) people to give all the lands of the whole kingdom to their kings, with liberty for them to keep what they pleased and to give the remainder to their subjects, clogged and cumbered with a condition to defend their realm. This is but an ill sign of a limited monarchy by original constitution or contract."[205] He also looks back at biblical times:

> Some, perhaps, may think that these princes and dukes of families were but some petty lords under some greater kings, because the number of them are so many that their particular territories could be but small and not worthy the title of kingdoms. But they must consider that at first kings had no such large dominions as they have nowadays. . . . Caesar found more kings in France than there be now provinces there, and at his sailing over into this island he found four kings in our county of Kent. These heaps of kings in each nation are an argument their territories were but small, and strongly confirms our assertion that erection of kingdoms came at first only by distinction of families.[206]

In his response to Filmer, Locke confronts the idea that Adam was given possession of the whole earth and that this gives him sovereignty over the people there.[207] Concerning Abraham, as a descendent of Adam, he asks, "Yet his Estate, his Territories, his Dominions were very narrow and scanty, for he had not the Possession of a Foot of Land, till he bought a Field and a Cave of the Sons of *Heth* to bury *Sarah* in."[208] He therefore does not believe either that contemporary kings are the descendants of kings past, nor that God ever gave the earth to one man. Instead, he contends that the earth was given to men in common. This produces perhaps the key question of his work: if the earth is held in common, how can any individual lay claim to private property in it?

Laslett has rightly suggested that Locke came to the topic of property late. There are few references before the *Two Treatises of Government*, and he claims that "Locke simply had not thought in a systematic way about property before 1679. He had not worked out his justification of ownership in terms of labour."[209] Yet in *Two Treatises* there is a clear articulation of this linkage. It is important, though, to recognize the context in which the *Two Treatises* were written. Their date of publication in 1690 suggests that their immediate political situation was that of the 1688 Glorious Revolution, which saw the Protestant William of Orange replace the deposed James II, brother of Charles II. William was married to James's daughter Mary, who would have become queen on his death, were it not for the birth of a son to James in 1688. Yet this was only one in a succession of political crises: ten years before, there had been rumors of a plot to murder Charles and replace him with James. Locke's sponsor, the Earl of Shaftesbury, had been integral to attempts to pass exclusion bills preventing James, as a Catholic, from becoming king. Laslett's analyses seem to have dated the composition of the *Two Treatises* to this time, rather than immediately before their publication. As Laslett contends, "*Two Treatises* in fact turns out to be a demand for a revolution to be brought about, not the rationalization of a revolution in need of defence."[210] Given this, "*Two Treatises* is an Exclusion Tract, not a Revolution Pamphlet."[211]

While the *First Treatise* is an explicit engagement with Filmer, the *Second Treatise* broadens the questions more generally. It has been described by Skinner as "the classic text of radical Calvinist politics."[212] The key chapter of the *Second Treatise* is the fifth, "Of Property."[213] Locke begins by suggesting that though God has "given the World to Men in common," he has also given them the reason to use it to their benefit. So, while "no body has originally a private Dominion, exclusive of the rest of Mankind," there must be a means to appropriate them in order to be used by any particular man.[214] Locke is therefore collapsing the idea of use and possession. He continues by suggesting that "the Fruit, or Venison, which nourishes the wild *Indian*, who knows no Inclosure, and is still a Tenant in common, must be his, and so his, *i.e.* a part of him, that another can no longer have any right to it, before it can do him any good for the support of his Life."[215]

Locke's solution to this problem is, of course, well known. He begins by suggesting that men have possessions in their own person. The labor of that body and the work of those hands, is, by extension, also the possession of that person. When something held in common is mixed with that labor, the man is legitimate in claiming that new thing as his own, as his property. "It being by him removed from the common state Nature placed it in, it hath by this *labour* something annexed to it, that excludes the common right of other Men. For this *Labour* being the unquestionable Property of the Labourer, no Man but he can have a right to what that is once joined to, at least where there is enough, and as good left in common for others."[216]

Locke then provides examples of basic appropriation, such as gathering acorns or picking apples, suggesting that it was the first step that made them the man's property. It was the labor, not the possession, transport, eating, or keeping of them that made them his. "That *labour* put a distinction between them and common."[217] Locke is clear that this does not require consent from others, and as he extends the examples, this point becomes clear:

> Thus the Grass my Horse has bit; the Turfs my servant has cut; and the Ore I have digg'd in any place where I have a right to them in common with others, become my *Property*, without the assignation or consent of any body. The *labour* that was mine, removing them out of that common state they were in, hath *fixed* my *Property* in them.[218]

Locke is doing something interesting here. For he is not only extending the examples, but he is showing how one kind of property (a horse, for instance) can create another kind (the grass). This is also true, he suggests, when the servant—someone who has sold his time and capacity to labor—produces something that he, as the owner of that servant's labor, can also appropriate. Locke essentially takes the idea of wage labor as unproblematic.[219] These rights are accrued, he suggests, without anybody, or any body, such as a government, providing the legitimation.

Locke recognizes that at the time he is writing, the key question is no longer things on the earth, but the earth itself. This is revealingly described as "that which takes in and carries with it all the rest."[220] In other words, if you own the land, then you own what is on it. Locke contends that "*as much Land* as a man Tills, Plants, Improves, Cultivates, and can use the Product of, so much is his *Property*. He by his Labour does, as it were, inclose it from the Common."[221] This, again, does not require the consent of others. Yet "*appropriation* of any parcel of *Land*" comes with a crucial exception: that there is enough, and as good, left for others. In other words, it is justification that applies when there is no scarcity.[222] Nor does it allow the accumulation of land to the extent that it was wasted.[223] As Locke later notes, "In the beginning all the World was *America*,"[224] and so these generalizations are valid. But today, scarcity forces the need to settle things differently, even if the principles remain. Money is one of the key things that allows what might appear excessive accumulation, since money does not spoil. It is for this reason, among others, that Locke judges that in America, "a King of a large and fruitful Territory there feeds, lodges, and is clad worse than a day Labourer in *England*."[225] Appropriation—that is, the use and possession of land—can produce materially better conditions for all: not in the sense of as much land left for them, but a better living as a result.[226]

Governments, for Locke, are thus established for a range of reasons, but as he explicitly states, "The great and *chief end* therefore, of Mens uniting into Commonwealths, and putting themselves under Government, *is the preservation of their property*."[227] Locke suggests that, in time, it made sense for people to order things more thoroughly:

> But as Families increased, and Industry inlarged their Stocks, their *Possessions inlarged* with the need of them; but yet it was commonly *without any fixed property in the ground* they made use of, till they incorporated, settled themselves together, and built Cities, and then, by consent, they came in time, to set out the *bounds of their distinct Territories*, and agree on limits between them and their Neighbours, and by Laws within themselves, settled the *Properties* of those of the same Society.[228]

The argument of consent to a regime is also linked to property rights. This works in two main registers in terms of geography. Locke suggests that "there being always annexed to the Enjoyment of Land, a Submission to the Government of the Country, of which that Land is a part."[229] In other words, the possession and use of property, protected by the government, implies consent to that government. He is later more explicit, distinguishing between *express consent* to a regime and *tacit consent*. Tacit consent can consist in "every Man, that hath any Possession, or Enjoyment, of any part of the Dominions of any Government, doth thereby give his *tacit Consent*." This may be possession of "Land, to him and his Heirs for ever, or a Lodging only for a Week; or whether it be barely travelling freely on the Highway." From this he makes the argument by situation: "In Effect, it reaches as far as the very being of any one within the Territories of that Government."[230]

Locke was not, of course, working in isolation.[231] His contemporary James Harrington (1611–77) is a curious figure who is best known for his imagined history of *Oceana* published in 1656. Skinner describes it as "arguably the most original and influential of all the English treatises on free states."[232] In that work Harrington treats questions of land and territory in some detail, yet the term *territory* is used in an indiscriminate sense to relate to a whole range of places from ancient Rome to the Middle Ages to his present.[233] Most centrally, Harrington applies *territory* to Oceana as a whole,[234] and describes "lands" as "the parts and parcels of a territory."[235] Yet at other times the term is used as an effective synonym for *land*, and generally it seems unproblematic for him: his concentration is on the particular kinds of politics that unfold within such places.[236]

Locke is therefore an important moment in the consolidation of the idea of territory. While, like Harrington, he is not an innovator in the concept itself, he is important in terms of cementing the relation between political power and territory. To be within the territory is to be subject to the rule, and this is magnified when ownership of land is taken into account.[237] There is therefore equally a reinforcing of the idea that individuals can own land that a sovereign has power over more generally. Indeed, in an earlier text, the "Two Tracts on Government," Locke had justified the imposition of a ruler's own religion on the people, a continuation of the principle of Augsburg, although in "An Essay on Toleration" he argued for a more plural perspective.[238] Finally, it is worth noting that in "The Fundamental Constitutions of Carolina," Locke and his colleagues begin: "Our sovereign lord the king having, out of his royal grace and bounty, granted unto us the province of Carolina, with all the royalties, properties, jurisdictions, and privileges of a county palatine, as large and ample as the county palatine of Durham, with other great privileges; for the better settlement of the government of the said place."[239]

"MASTER OF A TERRITORY"

On the European continent at this time, there were a related series of struggles over political power. At their heart was the status of the Holy Roman Empire, and the same kinds of issues that Althusius and Knichen had grappled with remained. Bodin's equation of sovereignty and majesty was in tension. If the individual rulers within the empire had some measure of political power, to what extent was that sovereignty, and, if it was, how did that relate to the standing of the emperor? One solution advanced was by Theodor Reinking (1590–1664), who, in his work *Tractatus de regimine saeculari et ecclesiastico*, proposed a hierarchical model of power with majesty held by the empire and delegated *ius superioritatis territorialis* to the rulers within it.[240] Drawing on Bartolus, Baldus, Zasius, Knichen, and others, Reinking declares that "the limits of the territory are the limits of jurisdiction [*limites territorii dicuntur limites jurisdictionis*],"[241] and suggests that *superioritatis territorialis* is the equivalent of *der Landes Hohen Obrigkeit*.[242] "Any prince and territorial right [*jura territorii*] has the power in its territory [*in suo territorio*], which the Emperor has in the universe [*Imperator in universo*]. . . . Any Prince is Emperor in his territory [*Quilibet Princeps est Imperator in suo territorio*]."[243] Nonetheless, they did not possess the "legal right of Majesty," which was reserved for the emperor *in universo*.[244] These issues came to a head in the Thirty Years' War, of 1618–48, though so many related conflicts preceded it and continued beyond it that the dating is somewhat arbitrary.[245] It caused enormous devastation across Europe, especially in Germany, though how much is open to dispute.[246] It is important to stress that the war was not, primarily, about religion.[247] The peace congress at Westphalia, conducted between 1643 and 1648, resulted in a treaty between the Spanish and Dutch, and separate settlements between the empire and the French, at Münster, and the empire and the Swedish, at Osnabrück, signed on October 24, 1648. It became known as "the peace of exhaustion."[248] It has become ingrained in the conscience of political theory and international relations as a kind of founding moment of modern states and the international system.[249]

In recent years there has been a consolidated challenge to that prevailing orthodoxy, with a recognition that the treaties of Westphalia say little that is claimed of them, and that to privilege this as a turning point is misleading in a number of ways.[250] Indeed, as Wilson notes, Westphalia is seen by political scientists as positive, but by historians as negative.[251] The most concerted challenge from within international relations itself has come from Benno Teschke, who argues that the settlement of 1648 "expressed and codified the social and geopolitical relations of absolutist sovereignty,"[252] rather than modern sovereignty. For Teschke, the distinction between absolutist and modern sovereignty is crucial,[253] and he claims that "demystifying Westphalia requires retheorizing absolutist sovereignty."[254] The modern notion of sovereignty, supposedly derived from Westphalia, is of a much later date. Teschke claims that "the formation of the modern states-system, based on exclusive territoriality operated by a depersonalized state, must be pushed up to the nineteenth century."[255] Teschke's claims are well taken, and his corrective to the orthodoxy long overdue. However, he is frustratingly vague with his use of *territory* as a term.[256] This is inevitably an outcome of his approach, which makes no references to primary literature at all, with the exception of a few references to the treaties. His claims are often entirely appropriate as general observations, but extremely imprecise in terminology. His greatest worth may be in his challenge to orthodox Marxist accounts, suggesting that "the political organization of the modern world into a territorially divided states-system was not a function of capitalism. . . . Capitalism emerged in a territorially prefigured states-system."[257]

Teschke therefore offers a powerful account that stresses the importance of political-economic elements, but his historical-conceptual inquiry remains impoverished because of a lack of textual fidelity.[258] While these approaches are valuable, therefore, they risk missing what was interesting about Westphalia. It was not nearly as central as the traditional accounts suggest; but it was not as unimportant as the revisionists would have us believe.

In order to understand this, it is worth spending a little time looking at some of the political theory produced in the lead-up to the negotiations. One of the most important interventions was entitled *Dissertatio de ratione status in imperio nostro Romano-Germanico*, purporting to be by Hippolithus à Lapide, and published in 1640. The author was widely believed to be Bogislaw Philipp von Chemnitz (1605–78), the pseudonym taken on because of the controversial nature of the work.[259] The key issue in the text is the relation between the empire as a whole and the individual principalities and free cities that constituted it. In other words, it is an issue of the administration of the empire.[260] Chemnitz argues for constitutional weakness of the emperor as well as outlining the excesses of the Hapsburgs. He stresses the superiority of the estates instead of the majesty of the emperor as the focus of power in the Holy Roman Empire, a claim that would have reconstituted the empire as something closer to a federation.[261] This was what Chemnitz wanted, so his analysis was at once diagnosis and cure. One of the things that was interesting about Chemnitz's claim was the use of the idea of *raison d'état* as pertaining to smaller polities, and therefore for a nonabsolutist purpose.[262] It was used in the negotiations at Westphalia, largely at the wish of the French, and was said to have done more damage to the emperor than some of the battles he lost.[263] It was work informed by the Italian jurists of the fourteenth century, especially in its reading of the legal texts, as well as serving a more polemical and immediate political purpose. Part of this was through an engagement with Bodin. Bodin, as chapter 8 outlined, equated majesty and sovereignty. But what kind of power did the constituent parts of the empire have? Bodin seems to have thought they were sovereign. How, then, could that be reconciled with the *majestas* of the emperor? In Chemnitz's reading, the principalities took on the characteristics of states; the interests of the princes were reason of state. Chemnitz explicitly drew parallels between the shift in political thought and the shift in mathematics, such as in the work of Galileo. Reason of state is the mechanism by which states function, the means to "establish, conserve, and augment a republic."[264]

The Holy Roman Empire was conceived as Christendom, the secular version of the kingdom of God. The emperor, crowned by the pope in the first instance, was intended to have power over the principalities, kingdoms, and cities within a large swath of central Europe. The Reformation had, of course, provided a profound challenge to this. In 1555, the Diet of Augsburg had attempted to find a peace between warring factions.[265] Despite the way that this is usually described, Augsburg was not primarily a religious peace,[266] though it was undoubtedly important. The phrase with which it is most associated is *cuius regio, eius religio*, to whom the region, the religion. This meant that the religion within the empire would accord with the confession of the individual rulers rather than the emperor. It was effectively the religious equivalent of the earlier proposal that the king and emperor were supreme rulers within their domains, yet now it applied to rulers within the bounds of the empire itself.[267] However, this crucial phrase was not included in the text and, as Wilson notes, actually appears in later debates on the settlement after 1586.[268] The settlement was in favor of religious freedom for Lutherans, but not for all Protestants, and so there was an uneasy alliance between the emperor and those polities such as Saxony that had gained what they wanted and did not want to change or challenge things further. Saxony was able to distinguish those Catholics who abided by the Augsburg peace from those who did not.[269]

The Augsburg principle was hard to uphold in practice since it was

dependent on the decisions of individual rulers and could be changed at
any time. It was essentially abandoned at the congress of Westphalia.[270]
Instead, here, the religion of each part of the empire was frozen according
to its situation in 1624, although crucially here the right was extended to
Calvinists too. The peace effectively acted as a constitution of the empire, in a succession that included the 1552 Peace of Passau and the Diet of
Augsburg.[271]

> It is a common misconception to see Westphalia as bringing peace by
> taking religion out of politics. Though it promoted secularization in
> the longer term, it was not a fully secular peace. The Empire remained
> Holy in the sense of Christian. Toleration was extended only to include
> Calvinists. Other dissenters, along with Orthodox Christians, Jews,
> and Muslims, were denied similar constitutional rights.[272]

Yet to judge by the papal reaction, it was far from the wishes of the papacy. Pope Innocent X issued the bull *Zelo domus Dei* (Zeal for the House
of God) in August 1650, but backdated it to November 26, 1648, to reaffirm
earlier verbal protests.[273] The written text claimed that the treaties were
"null and void, invalid, iniquitous, unjust, damnable, reprobate, rejected,
inane, without force or effect, and no one is to observe them, even when
they be ratified by oath."[274]

As Wilson notes, the word *sovereignty* appears only in the English
translation of the Treaty of Münster.[275] So, what did Westphalia grant the
rulers within the empire, and what did it reserve for the emperor? The key
clause is found in both the Münster and Osnabrück treaties.

> And to prevent for the future any differences arising in the political
> state, each and all of the electors, princes and states of the Roman Em-
> pire are so established and confirmed in their ancient rights, preroga-
> tives, freedoms, privileges, free exercise of territorial right [*libero iuris
> territorialis*], in ecclesiastic and political matters, in their domains
> [*ditionibus*], regalia, and possessions by virtue of this present Transac-
> tion: that they never can or ought to be removed by anyone under any
> pretext.[276]

While the treaties did give France some lands from the empire, with
sovereignty and jurisdiction over them, more important, it gave the constituent parts of the empire this "free exercise of territorial right." The
Latin for this phrase was *"libero iuris territorialis,"* or, as it appeared else-

where in the Osnabrück text, *"iure territorii et superioritatis* [territorial right and superiority]" and *"ratione territorii et superoritatis* [reason of territory and superiority]."[277] Importantly, the Latin text describes these rights as belonging to the *Statibus Imperii,* the states within the empire. The German word in the treaties for these territorial rights or superiority was *Landeshoheit.* In a valuable attempt to subvert orthodoxies about the birth of sovereignty and the European states system in 1648, Osiander has cautioned against translating this as "territorial sovereignty," suggesting that the German term *Landeshoheit* is actually "territorial jurisdiction," and that what makes it interesting "is precisely that which makes it different from sovereignty."[278] For Gagliardo, *Landeshoheit* "carried with it nearly all the ingredients or attributes of true sovereignty, but was legally distinct from it, and was everywhere in Germany admitted to be so."[279] Elsewhere in his work, he defines it as "territorial lordship."[280] Oestreich translates *Landeshoheit* as "territorial sovereignty," and stresses that "the dispute over the interpretation of *jus territoriale* determines the history of German liberty until the end of the Empire."[281] The French equivalent of *jure territorii et superioritatis* was *supériorité territoriale.*[282] Crucially, it extended this in both temporal and spiritual registers, since many of the electors were both religious leaders and substantial landowners. The treaties stressed the "free exercise of territorial right in both spiritual and political affairs [*librum iuris territorialis tam in ecclesiasticis quam in politicis exercitium*]."[283]

It is important to note that whether this is right, jurisdiction, or even sovereignty, it is held over territory. This is central to understanding the importance of Westphalia. Quoting an eighteenth-century German jurist, Osiander notes that the autonomy of the estates—free cities and principalities—was limited through the laws of the empire and the constitutional arrangements. What he underplays is that internally—that is, "in their lands and territories"[284]—they were empowered politically. In his conclusion he attempts to suggest that today "there is a clear de facto trend in international politics away from classical sovereignty and toward something closer to *landeshoheit,* territorial jurisdiction under an external legal regime shared by the actors."[285] This is both important (because it shows us that the emergence of territory at Westphalia was not tied to some absolute notion of sovereignty, as is often supposed) and potentially misleading (as it underplays the importance of territory as a concept in itself, distinct from sovereignty).

Gross outlines that some of these aspects are dependent on the negotiations that led to this point, with the different parties making use of the resonances within their own languages, suggesting that "the French draft of the treaties used the expression *souveranitatis iura* for territorial supremacy, but the Emperor succeeded in having the phrase eliminated."[286] It is important to note two key things. First, that the settlement was not simply in terms of the constituent parts of the empire and their relation to the emperor. It also ruled comprehensively in favor of the electors and princes in terms of their relation to other political actors within their territory. Princes, and not individual landowners, had *ius territorialis* or *Landeshoheit*. It was thus an important moment in the assertion of exclusive control within these geographically determined areas. Second, it stressed the right of individual units within the empire to pursue a number of attributes that we would today commonly associate with states. The Westphalia treaties had various clauses that allowed them to have standing armies, raise taxes, make laws and new fortifications, and have an independent negotiation and alliance policy.[287] On August 29, 1645, the *ius belli ac pacis* had been conceded to all independent territorial rulers,[288] which was reinforced in the peace.[289] Yet this was not given without reservation: they were not allowed to make alliances against the emperor or empire. This is indeed the key point: the treaties codified and reinforced an already existing state of affairs rather than distributing a wider set of rights. The elements within the empire were not yet states, because these rights came with their status as constituent parts of the empire.[290] Nonetheless, taken together, these two points can be seen as crucial stages in the assertion of the state as laying claim to the monopoly of physical violence, both within the polity and as the means by which it would exceed its borders.[291] In practice, the empire could neither act within the parts of the empire, nor raise a consolidated force, even if in theory it retained those rights. Effectively it put an end to the empire as much more than simply a geographical term, and ended any chance of German unification under the emperor.[292]

One of the writers trying to make sense of where this left the empire was Samuel Pufendorf (1632–94). Although Pufendorf says relatively little about the political control of land,[293] he had written a text in 1667 entitled *De statu imperii Germanici*, in which he begins to grapple with the question of territory in relation to the empire.[294] Pufendorf had proposed the notion of *de systematibus civitatum*, which international relations theorist Martin Wight has translated as "states system." Pufendorf notes that contemporary Germany is an "irregular body and like some misshapen monster" in the sense that it is a peculiarity that is neither a limited kingdom nor a "system of several states knit and united in a League, but something that fluctuates between these two."[295] Pufendorf notes how tenaciously princes cling to "rights of territorial superiority [*juriam superioritatis territorialis*]."[296] Wight has suggested that he is aiming for a system of "several states that are so connected as to seem to constitute one body but whose members retain sovereignty."[297] The Holy Roman Empire would be reconstituted as a collection of states that were independent but under the nominal authority of the emperor. As Nardin notes, this notion of a states system would extend beyond the empire to Europe as a whole.[298] Yet this is to get ahead of the story.

One of those who responded to Pufendorf was Leibniz, both in writings and in a brief correspondence.[299] It is often forgotten that, as well as his work on mathematics and philosophy, Leibniz was a political adviser and historian.[300] Leibniz was employed by the elector of Mainz, and was sent to Paris on his behalf in 1672. When he returned, he was employed by the electors of Hanover, and remained there for the rest of his life (1676–1716), serving a number of heads, the last of whom became the British king George I. The last period, in Hanover, is looked at as his mature philosophy.[301] Early in his career he had sent somewhat sycophantic letters to Hobbes praising his "writings on political theory" and especially *De Cive*,[302] though he would soon become much more critical. Leibniz was trained as a Scholastic and tried to reconcile this with the new philosophy of Descartes and his successors, rather than simply trying to overthrow it. He shows a familiarity with the writings of Bodin, Knichen, Althusius, Filmer, Locke, and Pufendorf, of merely his near contemporaries in political theory.[303] In addition, although his master's degree was in philosophy, he had a bachelor's degree and a doctorate in law, the latter for a thesis on difficult legal cases,[304] which dealt with issues relating to the conflict of the laws. He wrote a preface to a large compilation of legal texts,[305] and there and elsewhere frequently cited Justinian's *Digest*, admiring the rigor of the writings, even suggesting that "there are no authors whose style is more akin to the geometers than the old Roman jurists in the *Digest*."[306]

Leibniz was also an innovator in applied branches of mathematics. Ian Hacking has described him, for instance, as "the first philosopher of probability," who "anticipated, often in great detail, many of our modern probabilistic conceptions."[307] This carried over into his political work, especially in terms of the emerging science of statistics.[308] In some of his correspondence, Leibniz claimed to have solved the problem of calculating longitude at sea.[309] In terms of safe passage and colonial exploration, this was extremely important. He was interested in geography in a much broader sense too, as evidenced by his prehistory of the earth, *Protogaea*.[310]

In his work "Entwurff gewißer Staats-Tafeln" (The design of state-boards),
he recommended the use of land and sea maps as part of state practice. As
Tang puts it, "One element of the sovereign exercise of power is to keep
constant watch over the land as the natural substrate of the state."[311] This
relation between state power and geography is perhaps Leibniz's most important political innovation. Indeed, it appears to be significant enough to
challenge Friedrich's claim that in political theory, "it would seem that
the extraordinary imaginative originality which characterizes his work as
a metaphysician and mathematician is lacking. . . . Recurrent claims to
the contrary have not succeeded in establishing Leibniz as a thinker of the
first rank on law and politics; no basically novel insight can be attributed
to him."[312] Even the editor of the English edition of his political writings
agrees that he pales in comparison to Hobbes, Spinoza, or Locke.[313] Yet, in
Foucault's phrase, Leibniz is a "general theoretician of force as much from
the historical-political point of view as from that of physical science."[314]

Leibniz thought there were six kinds of natural communities. First, between man and wife; second, between parents and children; third, between
master and slave; and fourth, the household, which is composed of the previous three types. These are for the purpose of meeting daily needs.[315] It is
with the fifth and sixth that he begins to develop some ideas beyond Aristotelian notions, dependent, in part, on his reading of Althusius.[316]

> The fifth natural community is the civil community [*bürgerliche Ge-
> meinschaft*]. If it is small, it is called a city; a province [*Landschaft*]
> is a society of different cities, and a kingdom or a large dominion is a
> society of different provinces—all to attain happiness for to be secure
> in it—whose members sometimes live together in a city sometimes
> spread out over the land [*Land*]. Its purpose is temporal welfare [*zeitli-
> che Wohlfahrt*].
>
> The sixth natural society is the Church of God, which would
> probably have existed among men even without revelation, and been
> preserved and spread by pious and holy men. Its purpose is eternal
> happiness.[317]

While this certainly is dependent on some medieval notions, in terms
of hierarchy and arrangement, there are two more modern elements: first,
the idea of the *bürgerliche Gemeinschaft*, which, while equated with the
polity generally, anticipates some of Hegel's formulations of the *bürgerliche Geschellschaft*; second, the spatial elements of this determination.

In 1677, shortly after he had arrived back in Germany from Paris, Leibniz was asked by the Duke of Hanover to clarify the position of the rulers within the empire, in preparation for the peace congress of Nijmegen.[318] That congress, attempting to end Louis XIV's war against Holland, was calling ambassadors from across Europe. But the rulers of some smaller polities within the empire were not invited, on the grounds that the emperor and the electors would represent them.[319] The issue at stake—in the wake of Westphalia—was effectively that of sovereignty. The notion of territorial right or superiority was proclaimed by the treaties, but the old feudal structure of the empire was still evident. If the rulers of the small principalities were under a higher authority, what kind of power did they have? The Hanoverian position was thus complicated: they wanted both to be independent and to become an elector to the empire—a position that seemed irreconcilable. On the one hand, they were claiming territorial independence; on the other, submitting to imperial power.[320] Leibniz wrote two important texts at this time. The first, drafted between June and October 1677, was entitled "De Jure Suprematus ac Legationis Principum Germaniae," which was published under the pseudonym of "Caesarinus Fürstenerius"—a playful name that stresses the equivalence of the emperor and prince, or "Prince as Emperor."[321] According to Aiton, it "reveals remarkable insight into the nature of government and the problems of applying political and legal theories in a complex situation."[322] The second was a French dialogue on related matters entitled *Entrétiens de Philarete et d'Eugène*, between a representative of a prince who was not an elector, and an ambassador of an elector.[323] While the Latin piece was the more formal document, it quickly outgrew his original aim of an intervention, and the dialogue was therefore written to convey the ideas in a more direct manner. Indeed, the dialogue was provided to the delegates at the congress.[324] The dialogue went through a number of editions, "each successively adapted to constantly changing political conditions."[325]

Leibniz takes his task as one of "explaining the concept of *suprematu*," which he suggests is to enter into "a thorny and little-cultivated province" for such an "important and common a concept."[326] He suggests that this is in part because the focus has too often been on ancient ideas rather than their contemporary manifestations: "Among vulgar jurists [*jurisperitis*] this does not surprise me; for them, all wisdom appears collected in the corpus of Roman law alone."[327] He recognizes that Grotius did some valuable work on this, drawing on history but making it relevant today, but that his work is not especially useful. In sum, Leibniz suggests that "whoever wants to speak of supremacy, commonly called *sovereignty* [*Suprematu . . . la Souveraineté*], lacks the aid of good writers."[328]

Conversant in Latin, French, and German, Leibniz is revealing because
he recognizes the politics of language. He begins with some definitions of
key terms. Much of this derives from Aristotle. But the discussion of spatial terms develops some important points:

> A dominion [*ditio*] is an area of inhabited land served by a common
> administration. A larger dominion is called a region [*Regio*], and when
> a region is part of another still large dominion it is said to be a province
> [*Provincia*]. Territory is a name common to a *civitas* or a dominion or
> a tract of land [*terrae tractui*]. But in addition to its fundamental mean-
> ing, it also expresses the aggregate of laws and rights, so that just as
> inheritance and patrimony involve the whole of the things and rights
> in some family or dwelling, so territory signifies the whole of laws
> and rights which can come to obtain in an inhabited portion of the
> earth.[329]

What is important here is the way that Leibniz identifies territory, *territorium*, with both an extent of land and a polity, the *civitas*, and the way
that it is linked to a particular legal regime. For Leibniz, there is then the
question of how different levels of legal-political power relate to this area.

> Hence there arises what the German jurists call territorial superiority
> [*Superioritatem terriorialem*—i.e., *Landeshoheit*], or the high right of
> territory [*sublime territorii jus*]. But the Italians preceded them, and
> Baldus used to say that superiority inheres in territory, like mist to
> a swamp. In this right, moreover, in addition to jurisdiction and the
> mild power of coercion, there is also contained the right of military
> might. The more closely they are mixed together, the more accurately
> it is necessary to distinguish them. For the Lord of the jurisdiction and
> the Lord of the territory are two different things. I call jurisdiction the
> right of deciding cases or of handing down judgements and of coercing
> obstinate private persons. I say that this right of coercing (which the
> ancient jurists called *Imperium*) lies in being able, when necessary, to
> use force on stubborn people.[330]

Leibniz sees jurisdiction as operating at quite a low level, and military
power as operating at a higher level, because rather than individuals, it allows the power of "keeping the whole dominion in its duty." So the "lord
of some village or burg can have all jurisdiction," but this does not mean

that he can do everything. In such instances he would need to call upon the "Lord of the territory."[331]

> He who considers these things with care will see that territorial superiority consists in the highest right of forcing or coercing, which differs as much from the simple faculty of coercing as does public from private in Roman law. . . . This right, in turn, belongs not only to the princes of the Empire, but also to the counts. For a long time there was doubt concerning the free cities, but recently, especially since the peace of Münster [Pace imprimis Monasteriensi], the question seems to have been resolved. And what we call territorial superiority seems to be identical to what the French call *la souveraineté*, in a slightly looser sense.[332]

Leibniz therefore makes a crucial distinction between majesty, as the power to demand obedience and loyalty, without being commanded themselves, and sovereignty, which he sees as being stressed in the treaties of Westphalia, as concerned with territory. In stressing the high right of territory (*sublime territorii jus*) as more than mere territorial supremacy, Leibniz echoes Knichen's formulation.[333] But he goes further in drawing a direct relation between *suprematus*, understood as high right of territory, and the notion of *la souveraineté*.[334] He therefore breaks with the understanding of sovereignty that Western thought had taken from Bodin, seeing that there can be multiple levels of political power.[335] The sovereign is someone who is "powerful enough to make himself considerable in time of peace and in time of war, by treaties, arms and alliances."[336] It is true that at times Leibniz suggests that there are differences between sovereignty and territorial superiority, but this is because there are "degrees of *seigneurie*," of lordship. "There is a lord [*Seigneur*] of the jurisdiction, Lord of the Territory and Sovereign, that is a free Prince, or Republic."[337] This helpfully outlines the post-Westphalia position of the empire—external authority but internal noninterference in the estates.[338] He is attacking Hobbes's view of sovereignty for its absolutism, and in recognizing that sovereignty has to be divided, clarifying its meaning. Leibniz is suggesting that the princes within the empire—even those who are not electors— are as powerful "in their territories as the Emperor in the Empire."[339] In Riley's felicitous phrase, he sees sovereignty as a "comparative rather than a superlative standard."[340]

Leibniz is not, therefore, trying to undervalue majesty, but trying to conceptually distinguish it from sovereignty. Majesty, for him, held out the possibility of a universalism he wanted in terms of the reassertion of Christendom. Leibniz wanted a single body of Christian states, a union, with the emperor as temporal head and pope as spiritual: the emperor as "the secular arm of the Universal Church."[341] In some respects his proposal bears comparison to Dante's *Monarchia*,[342] though Leibniz had a more positive view of the papacy. While this might appear surprising for a Lutheran, he seems to take as unproblematic the idea that confession at a local level could be reconciled with universalism. For Stewart his attempt to reconcile Protestants and Catholics in a *respublica Christiana* means that his politics can be summed up simply as theocracy.[343] Yet as a contemporary of Louis XIV of France, he had a powerful challenge. Louis's own political strategy was aiming at a similar goal, but Leibniz wrote his polemic "Mars Christianissimus" against him.[344] Even though the Turks were at the gates of Vienna, Leibniz saw Louis as the greater threat, and as the reason for the empire's weakness in the east.[345] As Stewart has shown, these broader writings demonstrate that Leibniz was not simply working in the interests of Hanover, but was impelled by a grander political vision for the west.[346]

It is sovereignty, then, rather than majesty, that is diminished in Leibniz's work, but in such a way that it becomes a more appropriate indicator of political actualities. For Leibniz sovereignty was internal competence, and external recognition. It did not imply that all polities were equal, and there could still be a hierarchical model of power. As Riley notes, it "did not exclude ultimate allegiance to a universal 'Christian republic.'"[347] Leibniz goes on to stress that his argument does not mean that the power of the territorial rulers is absolute, but that there can be a higher authority to appeal to. It is a reconciliation of "a plurality of sovereignties with the unity of the Republic of the Empire."[348] Equally, as the examples of the empire, Switzerland, and the United Provinces show, "several territories, moreover, can unite in one body, retaining their singular territorial superiority."[349] If the minor German princes are as much sovereigns as the kings of Europe, then, on the one hand, sovereignty does not mean as much;[350] on the other, though, he is providing a much more modern notion of sovereignty, because it is the political control of territory, of whatever size, that is crucial. Leibniz defines the role of rulers in a way that removes absolutism: "Sovereign or potentate is that Lord or State who is master of a territory [*Souverain ou Potentat est ce Seigneur ou cet Estat qui est maistre d'un territoire*]."[351]

Leibniz therefore described himself as offering "the first true definition of sovereignty, as something distinct from Majesty,"[352] but this is a notion explicitly tied to territory. Leibniz is intriguing as a figure because he is an Aristotelian by training who derives some notions from the temporal power theorists; he is informed by Roman law and scholarship on it, but sees a need to develop this legal thought in relation to contemporary concerns; and is deeply informed by the German theorists of political practice. He thus brings together three key strands of the story. Leibniz's suggestion that the sovereign is he "who is master of a territory"[353] is a fundamental moment in the development of Western political thought.[354] While it might appear that Hobbes's absolute sovereignty and Newton's absolute space define modern politics and geography, Leibniz's relational views of both are closer to how politics was actually practiced.

Territory as a Political Technology

The idea of a territory as a bounded space under the control of a group of people, usually a state, is therefore historically produced. Other ways of organizing the relation between place and power have existed, were combined in diverse ways, labeled with multiple terms, argued for and against, and understood differently. Some of these ideas were reappropriated, rearranged, and revised by later thinkers. Others were abandoned along the way. Nonetheless, the notion of space that emerges in the scientific revolution is defined by *extension*. Territory can be understood as the political counterpart to this notion of calculating space, and can therefore be thought of as the *extension of the state's power*. Equally the state in this modern form extends across Europe and from there across the globe. Therefore, from around this time we are justified in talking of *the extension of the state*—in this plural sense.

If the modern concept of territory is established by this time, this is not to suggest that future developments are unimportant. Far from it. Yet we should understand in what ways they are important. There are, of course, fundamental changes to particular territories, and debates about its understanding, how other political-theoretical concepts such as justice and rights apply to it, but the concept seems to be in place by then. This may partly explain the relatively unproblematic way in which the term is used and implicitly understood in mainstream political and geographical discussions.

Nonetheless, the historical-conceptual analysis offered here should not simply be used to support that mainstream view of territory. Territory should be understood as a political technology, or perhaps better as a bundle of political technologies. Territory is not simply land, in the political-economic sense of rights of use, appropriation, and possession attached to a place; nor is it a narrowly political-strategic question that is closer to a notion of terrain. Territory comprises techniques for measuring land and controlling terrain. Measure and control—the technical and the legal—need to be thought alongside land and terrain. What is crucial in this designation is the attempt to keep the question of territory open. Understanding territory as a political technology is not to define territory once and for all; rather, it is to indicate the issues at stake in grasping how it was understood in different historical and geographical contexts.

How this idea was put into practice, with historical and geographical specificity, would take several other books. There are, of course, a good number of studies of the histories of specific territories and geographies of state formation. Philosophers, theologians, jurists, geometers, historians, explorers, surveyors, and cartographers all play their part. Yet several things are worth attention. In what remains of this book, two broad areas of analysis of territory from this point on will be briefly sketched: the nation and the technical. Much valuable work has already done on these questions, and it is to be hoped that *The Birth of Territory* will provide a historical and theoretical background to those studies.

಄

Given the historical parameters of this study, the concept of the nation and the ideology of nationalism are outside its bounds. Yet it is perhaps worth underscoring that the relation between the nation and the state takes place within the spatial framework that the concept of territory produces. As Fulbrook puts it, "Historically, the formation of *states* with a centralised government administering and controlling a clearly defined geographical territory preceded the articulation of ideas of the *nation*."[1] The qualification to Fulbrook's point is that it was the idea of the state and territory that preceded the nation; in practice it was much more complicated and geographically variegated.

It is clear that the treaties of Westphalia and the others from the second half of the seventeenth century did not introduce a uniform, and universally recognized, system. States, such as France, whose territory was already well established, embarked on projects of nation building within those existing borders. Breuilly notes that one of the issues behind the revolutionary wars of the late eighteenth century was the sovereignty of various enclaves within France that had some allegiance to the Holy Roman Empire. "The modern conception of France as a tightly bounded space within which the French state was sovereign was opposed to an older conception of power as varying bundles of privileges related to different groups and territories."[2] Similar things happened within England somewhat earlier and the Scandinavian countries at a similar time.[3]

Other national groups sought to create a state to represent them within defined geographical areas.[4] These would include those like the Italian and German unification projects in the nineteenth century, as well as a host of independence movements across the world in the twentieth and twenty-first centuries. While the boundaries of states in Europe continue to be an issue today—1945 was important in securing Western borders, but 1989 opened up a whole range of issues in Central and Eastern Europe—in this earlier period many of the borders were still porous and ill defined, and sovereignty was overlapping.[5] Germany had many internal boundary disputes to solve (whether part of a state was in the confederation or not): its external boundaries were more or less secure depending on whom that boundary was with. For example, its boundary with France—"the most modern, boundary-conscious European state"—was "fixed with political-administrative precision"; whereas its southern border was simply a line drawn on the map of Austria. In the north, with the disputed province of Schleswig-Holstein, "the 'boundary' dispute arose via the question of 'national sovereignty.'" The way this dispute worked out only served to reinforce the notion that nation-state was a territorially sovereign state. "Boundaries came to matter more in this political conception."[6] Breuilly notes that only with the Weimar Republic did Germany actually become a *state*—under Bismarck it had been a *Reich*, an *empire*: "The tragedy was that this state was also the product of defeat—its boundaries were seen as artificial and its constitution as imposed."[7]

⟡

Much has been written about the importance of cartography in state projects. Escolar suggests that the techniques of this rejuvenated cartography were used for "bureaucratic and administrative management and territorial control of state power in the states of Western Europe" in the sixteenth century.[8] They were prepared to invest heavily in this: as Harley notes, "The state became—and has remained—a principal patron of cartographic activity in many countries."[9] While Kain and Baignet suggest that "by definition, state mapping can be practised only after the establishment of the state,"[10] this is in danger of missing the way that, in order to establish *what* is actually controlled, mapping becomes both a requirement and a tool of power. Christian Jacob's important study on the relation between sovereignty, empire, and cartography is indicative: "The power of maps, however, is also a tool for power: ruling over a province, a nation, a kingdom, an empire, protecting or conquering a territory; imposing upon it the rationality of an administrative grid, a political project of reform or of development."[11] Given the benefit states gained from accurate maps, it is no surprise that the key sponsors of advances in cartographic techniques were states. As Harley puts it, "At the very time maps were being transformed by mathematical techniques, they were also being appropriated as an intellectual weapon of the state system."[12]

Other techniques, such as the ability to more accurately measure longitude, had important political-cartographic implications.[13] As James Scott has argued, this was about making the state legible: "The premodern state was, in many crucial respects, partially blind; it knew precious little about its subjects, their wealth, their landholdings and yields, their location, their very identity. It lacked anything like a detailed 'map' of its terrain and its people."[14]

These projects have been studied in some detail in, among other places, Denmark,[15] India,[16] and Mexico.[17] But it is in France that the most extensive early project took place.[18] Following the 1659 Treaty of the Pyrenees, which put an end to the conflict between Spain and France, a joint commission was established to set the exact boundary between the two states, an event that is looked at as inaugurating "the first official boundary in the modern sense."[19] There was also the work done on the border fortifications of the country by Sébastien le Prestre de Vauban, and the cartographic work done by four generations of the Cassini family.[20] The resultant map has been described as "the first original map based on a trigonometric topographic land survey."[21] Yet even though various iterations of this project were completed in the eighteenth century, the revolution took this further. While the Cassini map had intended to use geometric and calculative techniques to make sense of the existing landscape, the revolution attempted to impose the grid over the top, with the rectangular *départments* reordering the geopolitical landscape.[22] As Breuilly notes, just as the revolutionary calendar attempted to secure a more rigorous understanding of time, so too did the internal restructuring of France attempt to undermine "traditional, legitimate understandings of space." But reason and nature were in alliance, as boundaries were often established "upon criteria such as the catchment area of a river."[23]

The mapping and control of territory is, in large part, dependent on such techniques. Only with these kinds of abilities could modern boundaries be established as more than a simple line staked out on the ground.

For mountainous regions, for deserts or tundra, or particularly for the abstract division of unknown places in the colonized world, such techniques were crucial. They are made possible through a calculative grasp of the material world, what Lefebvre calls abstract notions of space, or indeed abstract space. One of Lefebvre's comments on abstract space is relevant here: "As a product of violence and war, it is political; instituted by a state, it is institutional."[24] There is an inherent violence to these techniques. In the famous title of Yves Lacoste's 1976 book, "Geography is, above all, making war."[25] Baudrillard's line of the map preceding the territory has been picked up by James Corner, Geoff King, and John Pickles, among others. For Corner, this is always the case, because "space only becomes territory through acts of bounding and making visible."[26] While Corner recognizes that Baudrillard is going one stage further, the claim is still central.[27]

The key is, of course, what kind of map is required, or what kind of cartographic techniques are needed for the production of territory. Nonetheless, techniques that related to territory were not confined to the cartographic.[28] While it is sometimes suggested that the Western model of the state and its territory was exported to the rest of the world,[29] there is perhaps more truth in seeing the way that in the colonial theater many of the techniques could be perfected in a purer form.[30] Earlier chapters showed how some of these ideas of surveying, division of virgin lands, and so forth, colonial practices for the management of populations, led, or were partnered by, developments in legal and technical practices. One of the most widely studied large-scale cartographic, and thereby territorial, projects is the rectangular land survey in the United States, begun under President Thomas Jefferson, but with earlier antecedents.[31] One of the most interesting of these can be found in the measuring instruments developed by English mathematician and astronomer Edmund Gunter (1581–1626), who had been a professor at Gresham College. Among other mathematical achievements, he introduced the terms *cosine* and *cotangent*. His most famous study was *Use of the Sector, Crosse-Staffe, and Other Instruments*.[32] Among these instruments were Gunter's line or scale, which was an early slide rule; aids for maritime navigation, including a quadrant; and perhaps most important, the deceptively simple Gunter's chain. This was sixty-six feet long (twenty-two yards), with one hundred links, and originally made from either iron or brass, and therefore liable to heat-induced errors. It could be used to measure landscapes, because the length of eighty chains was exactly one mile.[33] The chain gave both the unit of length of a "link" and that of a "chain," which is the length of a modern cricket pitch.

Paul Alliès suggests that administration in a general sense is crucial to what he calls "the invention of territory." For Alliès it is sufficiently important that "we can say that administration produced territory," that "administration is therefore a constitutive moment of territory" or integral to "the production of territory."[34] Yet this works the other way too, since he also suggests that "territory was this wonderful invention whereby bourgeois power would tame social interactions and their spatial movement."[35] Equally the relation between these kinds of techniques and the economic has been studied in some detail, particularly in three key studies by Pierre Dockés, Frank Swetz, and Richard W. Hadden.[36] The last of these—which develops ideas found in the work of Jacob Klein and others—is in a sense the most interesting. Hadden's argument is that mathematical examples can be tied to trade and capitalism. His argument is in part based on a shift in the emphasis concerning property in land, but this is a development that clearly predates the early modern era. Rather, Hadden is most useful in recognizing how the advances in mathematical techniques of that period found an immediate "market" in early capitalism. It is of course crucial to recognize that the establish of national markets helped to constitute and consolidate the spaces in which they operated, even if, as Marx suggests, ultimately capital seeks to move beyond "every spatial barrier . . . to conquer the whole world for its market."[37]

In the late seventeenth century the idea of political arithmetic was developed by writers such as John Graunt and William Petty. Petty had been a student of Hobbes,[38] and he developed a means of analyzing human behavior, especially collectively, through "number, weight or measure," and aspired to the same rigor as science.[39] Graunt, especially in his *Natural and Political Observations Made upon the Bills of Mortality* (1682), had similarly used numerical techniques.[40] Political arithmetic was one of the forerunners of the notion of statistics, which etymologically means the study of states. Alongside Ian Hacking's pioneering work on probability, there are a range of other important works that trace this particular relation of calculation and the political.[41]

In the extensive literature on the state, the territorial dimension has often been neglected or assumed as unproblematic. This is despite the stress on its importance in Max Weber's famous definition of the state:

> The state is that human community, which within a certain area or territory [*Gebietes*]—this "area" belongs to the feature—has a (successful) *monopoly of legitimate physical violence*.[42]

However, the territorial part of this definition—in distinction to community, legitimacy, and violence—has been rather neglected.[43] Yet just as the sovereign state was only one of the potential ways to organize politically,[44] alternative ways of spatial ordering have existed. Territory is a word, concept, and practice, and the complicated relation between these three terms can only be grasped with historical, geographical, and conceptual specificity. In his work on the history of political thought, Quentin Skinner rightly separates the concept and the word, but suggests that when a concept exists, a word or a vocabulary will be developed to discuss it, and that, as a general rule, "The possession of a concept will at least *standardly* be signalled by the employment of a corresponding term."[45] In a sense Skinner is right, although the word *territory* is derived from a much older word, *territorium*, which did not always have the same meaning. On the other hand, concepts that appear much closer to what is now labeled "territory" were previously known by other terms. And the practices of making, controlling, and defending exist in complicated relations to these words and concepts. Territory, then, as word, concept, and practice, is a historical, geographical, and political question.

☙❧

At the beginning of this book, Rousseau's discourse on inequality was quoted. In the light of the story outlined here, we can read Rousseau's suggestion in a new context. In terms of the question of the state of territory, it is clear that there have been many "crimes, wars, murders . . . miseries and horrors" resulting from the division and ordering of the world. Yet Rousseau comes too late: the genie is out of the bottle. He does not simply come too late in terms of the particular ordering of states and their spaces—these would continue to be fought over for centuries, and continue today—but *conceptually* too late. He is writing at a time, in the mid-eighteenth century, when politics was fundamentally conceived as operating with discrete, bounded spaces under the control of a group of people, usually the state. Where those boundaries were was still open to question, of course, and what political structures should operate within the area was widely debated, as it was in Rousseau's own writings. But the effective structure was now widely assumed: it had become the static background behind the action of political struggles.

Indeed, this is found in Rousseau's own writings. Rousseau declares that

a body politic can be measured in two ways, by the extent of its territory [*l'étendue du territoire*] and by the number of its people, and an appropriate ratio has to obtain between these two measures for the State
to be given its genuine size: The men make up the State, and the land
[*terrain*] feeds the men; thus the ratio requires that there be enough
land [*terre*] to support its inhabitants, and as many inhabitants as the
land [*terre*] can feed.[46]

He notes the extreme cases where people are "unevenly distributed
across the territory [*territoire*] and crowded in one place while others get
depopulated."[47] He similarly talks of "the resources provided by a large
territory [*territoire*]."[48]

Rousseau recognizes the dual aspect of land property and state territory. Individuals can lay claim to particular sites, which can be within the
larger territory of the polity. In this respect we should understand "the soil
as both public territory and the patrimony of private individuals."[49] The
sovereign and the private individual can therefore have different rights to
the same land.[50] Rousseau recognizes that there is a link between these
processes, suggesting that "it is intelligible how individuals' combined
and contiguous pieces of land [*terre*] become the public territory [*territoire*], and how the right of sovereignty, extending from subjects to the
land they occupy, becomes at once real and personal."[51]

Yet while the emergence of the idea may have shifted from people to
land, it now works the other way round. Rousseau notes that while they
previously called themselves kings of peoples, now "present-day monarchs
more shrewdly call themselves Kings of France, of Spain, of England,
etc. By thus holding the land [*terrain*], they are quite sure of holding the
inhabitants."[52] Rousseau thus rehearses an argument that was anticipated
in, among others, Locke. This is made explicit when he suggests that "once
the State is instituted, consent consists in residence; to dwell in the territory is to submit to sovereignty [*habiter le territoire c'est se soumettre à la
souveraineté*]."[53]

To be in the territory is to be subject to sovereignty; you are subject
to sovereignty while in the territory, and not beyond; and territory is the
space within which sovereignty is exercised: it is the spatial extent of sovereignty. Sovereignty, then, is exercised over territory: territory is that
over which sovereignty is exercised. In explicitly endorsing Leibniz's definition, Rousseau proves himself to be a thinker of his time, of our time,
where politics, state, and space come together in the concept of territory.

Similar claims could be made about Montesquieu's *L'esprit des lois*, the political writings of Rousseau's contemporary David Hume,[54] or the giant of eighteenth-century thought, Immanuel Kant.[55] The birth of territory is a long and complicated story, as this study has attempted to show. Territory is a *historical* question: produced, mutable, and fluid. It is *geographical*, not simply because it is one of the ways of ordering the world, but also because it is profoundly uneven in its development. It is a word, a concept, and a practice, where the relation between these can only be grasped genealogically. It is a *political* question, but in a broad sense: economic, strategic, legal, and technical. By this time, though, it had reached maturity. Whether it is now into its old age is a topic for another place, but reports of its demise are likely to have been exaggerated.

INTRODUCTION

1. Jean-Jacques Rousseau, *Discours sur l'origine et les fondements de l'inégalité*, II, 1. Jean-Jacques Rousseau, *Oeuvres complètes*, vol. 3, *Du contrat social: Écrits politiques*, ed. Bernard Gagnebin and Marcel Raymond (Paris: Gallimard, 1964); Jean-Jacques Rousseau, "Discourse on the Origin and Foundation of Inequality," in Jean-Jacques Rousseau, *The Discourses and Other Early Political Writings*, ed. Victor Gourevitch (Cambridge: Cambridge University Press, 1997).

2. This was the way it was used in Adam Ferguson, *An Essay on the History of Civil Society*, ed. Fania Oz-Salzberger (Cambridge: Cambridge University Press, 1995). See G. W. F. Hegel, *Elements of the Philosophy of Right*, ed. Allen W. Wood, trans. H. B. Nisbet (Cambridge: Cambridge University Press, 1991).

3. Rousseau, *Discours sur l'origine et les fondements de l'inégalité*, II, 1.

4. Rousseau, *Discours sur l'origine et les fondements de l'inégalité*, II, 24.

5. A fuller attempt to show how an understanding of territory can illuminate contemporary events is made in my *Terror and Territory: The Spatial Extent of Sovereignty* (Minneapolis: University of Minnesota Press, 2009).

6. Blaise Pascal, *Pensées* (Paris: Bookking, 1995), 113–14; and Blaise Pascal, *Pensées*, trans. A. J. Krailsheimer (London: Penguin, 1995), 16.

7. See, for example, Peter Sahlins, *Boundaries: The Making of France and Spain in the Pyrenees* (Berkeley: University of California Press, 1989); Thongchai Winichakul, *Siam Mapped: A History of the Geo-Body of a Nation* (Honolulu: University of Hawaii Press, 1994); Anssi Paasi, *Territories, Boundaries and Consciousness: The Changing Geographies of the Finnish-Russian Border* (London: John Wiley and Sons, 1996); and Christer Jönsson, Sven Tägil, and Gunnar Törnqvist, *Organizing European Space* (London: Sage, 2000).

8. David Storey, *Territories: The Claiming of Space* (London: Routledge, 2011); and David Delaney, *Territory: A Short Introduction* (Oxford: Blackwell, 2005). See also Malcolm Anderson, *Frontiers: Territory and State Formation in the Modern World* (Cambridge: Polity, 1996).

9. This is a common complaint. See, for example, Jean Gottmann, *The Signifi-cance of Territory* (Charlottesville: University Press of Virginia, 1973), ix; John Ruggie, "Territoriality and Beyond: Problematizing Modernity in International Relations," *International Organization* 47 (1993): 174; and Claude Raffestin, *Pour une géographie du pouvoir* (Paris: Libraires Techniques, 1980), 143. For a recent attempt to develop a science, rather than history, of territory, see Andrea Mubi Brighenti, "On Territorology: Towards a General Science of Territory," *Theory, Culture & Society* 27 (2010): 52–72.

10. Anthony Giddens, *A Contemporary Critique of Historical Materialism*, vol. 1, *Power, Property and the State* (London: Macmillan, 1981), 5–6, 11; see Anthony Giddens, *The Nation-State and Violence*, vol. 2 of *A Contemporary Critique of Historical Materialism* (Berkeley: University of California Press, 1987), 120.

11. Joe Painter, "Territory-Networks," presented to Association of American Geographers Annual Meeting, Chicago, March 7–11, 2006, http://dro.dur.ac.uk/8537/1/8537 .pdf, 3.

12. John Agnew, "The Territorial Trap: The Geographical Assumptions of International Relations Theory," in *Mastering Space: Hegemony, Territory and International Political Economy*, by John Agnew and Stuart Corbridge, 78–100 (London: Routledge, 1995); and John Agnew, "Sovereignty Regimes: Territoriality and State Authority in Contemporary World Politics," *Annals of the Association of American Geographers* 95 (2005): 437–61.

13. John Agnew, *Hegemony: The New Shape of Global Power* (Philadelphia: Temple University Press, 2005), 42.

14. Agnew, *Hegemony*, 161.

15. For a related inquiry, see Alexander B. Murphy, "The Sovereign State System as Political-Territorial Ideal: Historical and Contemporary Considerations," in *State Sovereignty as a Social Construct*, ed. Thomas Biersteker and Cindi Weber, 81–120 (Cambridge: Cambridge University Press, 1996).

16. See, for example, Philip L. Wagner, *The Human Use of the Earth* (London: Free Press of Glencoe, 1960); Robert Ardrey, *The Territorial Imperative: A Personal Inquiry into the Animal Origins of Property and Nations* (London: Collins, 1967); and Torsten Malmberg, *Human Territoriality: Survey of Behavioural Territories in Man with Pre-liminary Analysis and Discussion of Meaning* (The Hague: Mouton Publishers, 1980).

17. See also Edward W. Soja, "The Political Organization of Space," Commission on College Geography Resource Paper No. 8, Association of American Geographers, Washington, DC. Soja's account is much broader than this, and it remains one of the best single pieces written about territory.

18. Robert D. Sack, "Human Territoriality: A Theory," *Annals of the Association of American Geographers* 73 (1983): 55–74; and Robert D. Sack, *Human Territoriality: Its Theory and History* (Cambridge: Cambridge University Press, 1986).

19. Sack, *Human Territoriality*, 1.

20. Sack, *Human Territoriality*, 19.

21. Robert D. Sack, *Homo Geographicus: A Framework for Action, Awareness and Moral Concern* (Baltimore: Johns Hopkins University Press, 1997), 272n1.

22. Sack, *Human Territoriality*, 83–85. A related criticism might be offered of his *Conceptions of Space in Social Thought: A Geographical Perspective* (London: Mac-

millan, 1980). For discussions, see Ronald J. Johnston, "Territoriality and the State," in *Geography, History, and Social Sciences*, ed. Georg B. Benko and Ulf Strohmayer, 213–25 (Dordrecht: Kluwer, 1995); Ronald J. Johnston, "Out of the 'Moribund Backwater': Territory and Territoriality in Political Geography," *Political Geography* 20 (2001): 677–93; and Hans Vollard, "The Logic of Political Territoriality," *Geopolitics* 14 (2009): 687–706.

23. Edward W. Soja, *Postmodern Geographies: The Reassertion of Space in Critical Social Theory* (London: Verso, 1989), 150n9.

24. Claude Raffestin, "Repères pour une théorie de la territorialité humaine," in *Réseaux territoriaux*, ed. Gabriel Dupuy, Georges Amar, and Bernard Barraque, 263–79 (Caen: Paradigme, 1988), 264.

25. Raffestin, *Pour une géographie du pouvoir*, 46; see also Claude Raffestin, "Could Foucault Have Revolutionized Geography?" trans. Gerald Moore, in *Space, Knowledge and Power: Foucault and Geography*, ed. Jeremy Crampton and Stuart Elden, 129–37 (Aldershot, UK: Ashgate, 2007).

26. Raffestin, *Pour une géographie du pouvoir*, 17.

27. Raffestin, *Pour une géographie du pouvoir*, 129. In a more recent piece, Raffestin equates his understanding of territory with Lefebvre's idea of social space. Claude Raffestin, "Space, Territory and Territoriality," trans. Samuel A. Butler, *Environment and Planning D: Society and Space* 30 (2012): 126.

28. A similar criticism might be leveled against Rhys Jones, *Peoples/States/Territories: The Political Geographies of British State Transformation* (Oxford: Blackwell, 2007), esp. 3, 34. Jones is similarly good on the particular practices of state territorial formation but tends to collapse territory into territoriality, which loses the conceptual precision and analytic purchase of the former term.

29. Claude Raffestin, "Elements for a Theory of the Frontier," trans. Jeanne Ferguson, *Diogenes* 34 (1986): 1–18.

30. Painter, "Territory-Networks," 6.

31. Jean Gottmann, *La politique des états et leur géographie* (Paris: Armand Colin, 1951), 71.

32. See, for example, Gottmann, *La politique des états et leur géographie*, 72–73.

33. Saskia Sassen, *Territory, Authority, Rights: From Medieval to Global Assemblages* (Princeton, NJ: Princeton University Press, 2006); and Jeremy Larkins, *From Hierarchy to Anarchy: Territory and Politics before Westphalia* (London: Palgrave, 2010).

34. Edward S. Casey, *The Fate of Place: A Philosophical History* (Berkeley: University of California Press, 1997).

35. See, for example, Clarence J. Glacken, *Traces on the Rhodian Shore: Nature and Culture from Ancient Times to the End of the Eighteenth Century* (Berkeley: University of California Press, 1967); Denis Cosgrove, *Social Formation and Symbolic Landscape*, with a new introduction (Madison: University of Wisconsin Press, 1998); and Kenneth Olwig, *Landscape, Nature and the Body Politic: From Britain's Renaissance to America's New World* (Madison: University of Wisconsin Press, 2002).

36. Linda M. Bishai, *Forgetting Ourselves: Secession and the (Im)possibility of Territorial Identity* (Lanham, MD: Lexington Books, 2006), 59. For a study of the history of a word, rather than a concept, see Stéphane Bealuac, *The Power of Language in the*

Making of International Law: The Word Sovereignty *in Bodin and Vattel and the Myth of* Westphalia (Leiden, Netherlands: Martinus Nijhoff, 2004), esp. 8.

37. Reinhart Koselleck, *The Practice of Conceptual History: Timing History, Spacing Concepts* (Stanford, CA: Stanford University Press, 2002); and Reinhart Koselleck, *Begriffsgeschichten* (Frankfurt am Main: Suhrkamp, 2006).

38. Reinhard Koselleck, *Futures Past: On the Semantics of Historical Time*, trans. Keith Tribe (New York: Columbia University Press, 2004), 88. Tribe's introduction provides a helpful discussion of Koselleck's work.

39. For exceptions, see Otto Brunner, *Land and Lordship: Structures of Governance in Medieval Austria*, trans. Howard Kaminsky and James Van Horn Melton (Philadelphia: University of Pennsylvania Press, 1992); and Armin Wolf, *Gesetzgebung in Europa, 1100–1500, Zur Enstehung der Territorialstaaten* (Munich: C. H. Beck'sche Verlagsbuchhandlung, 1996).

40. Paul Alliès, *L'invention du territoire* (Grenoble: Presses Universitaires de Grenoble, 1980), 9.

41. See Quentin Skinner, *The Foundations of Modern Political Thought*, 2 vols. (Cambridge: Cambridge University Press, 1978); Quentin Skinner, *Visions of Politics*, 3 vols. (Cambridge: Cambridge University Press, 2002); J. G. A. Pocock, *Politics, Language and Time: Essays on Political Thought and History* (London: Methuen, 1972); and J. G. A. Pocock, *Political Thought and History: Essays on Theory and Method* (Cambridge: Cambridge University Press, 2009). For a discussion, see Kari Palonen, *Quentin Skinner: History, Politics, Rhetoric* (Cambridge: Polity, 2003); and the essays in Richard Rorty, J. B. Schneewind, and Quentin Skinner (eds.), *Philosophy in History: Essays on the Historiography of Philosophy* (Cambridge: Cambridge University Press, 1984).

42. Quentin Skinner, "A Reply to My Critics," in *Meaning and Context: Quentin Skinner and His Critics*, ed. James Tully, 231–88 (Princeton, NJ: Princeton University Press, 1988). See David Wootton, preface to *Divine Right and Democracy: An Anthology of Political Writing in Stuart England*, ed. David Wootton, 9–19 (Harmondsworth, UK: Penguin, 1986).

43. See my *Mapping the Present: Heidegger, Foucault and the Project of a Spatial History* (London: Continuum, 2001). The analysis offered here is also informed by my other, more explicitly theoretical, work. The state/space connection was a key concern in *Understanding Henri Lefebvre: Theory and the Possible* (London: Continuum, 2004); the relation between space and calculation was the focus of *Speaking against Number: Heidegger, Language and the Politics of Calculation* (Edinburgh: Edinburgh University Press, 2006).

44. For fuller treatments of Foucault himself, see Stuart Elden, "Governmentality, Calculation, Territory," *Environment and Planning D: Society and Space* 25 (2007): 562–80; and Stuart Elden, "How Should We Do the History of Territory?," *Territory, Politics, Governance* 1 (2013): 5–20.

45. On the relation between *Begriffsgeschichte* and the Cambridge school, see Michael Freeden, *Ideologies and Political Theory: A Conceptual Approach* (Oxford: Clarendon Press, 1996), chap. 3.

46. See Dominick LaCapra, *Rethinking Intellectual History: Texts, Contexts, Language* (Ithaca, NY: Cornell University Press, 1983), 36 and chap. 1 generally.

47. See Elden, *Terror and Territory*.

48. See Karl Marx, *Capital: A Critique of Political Economy*, vol. 3, trans. David Fernbach (Harmondsworth, UK: Penguin, 1981); and Karl Marx, *Grundrisse: Foundations of the Critique of Political Economy (Rough Draft)*, trans. Martin Nicolaus (Harmondsworth, UK: Penguin, 1973).

49. Perry Anderson, *Passages from Antiquity to Feudalism* (London: NLB, 1974); and Perry Anderson, *Lineages of the Absolutist State* (London: NLB, 1974).

50. This is one of the critiques Henri Lefebvre levels against mainstream Marxism. See Henri Lefebvre, *La production de l'espace* (Paris: Anthropos, 1974); Henri Lefebvre, *The Production of Space*, trans. Donald Nicholson-Smith (Oxford: Blackwell, 1991); Henri Lefebvre, *De l'État*, 4 vols. (Paris: UGE, 1975–78); Henri Lefebvre, *State, Space, World: Selected Essays*, ed. Neil Brenner and Stuart Elden, trans. Gerald Moore, Neil Brenner, and Stuart Elden (Minneapolis: University of Minnesota Press, 2009); and for a discussion, Neil Brenner and Stuart Elden, "Henri Lefebvre on State, Space, Territory," *International Political Sociology* 3 (2009): 353–77.

51. Michael Mann, *The Sources of Social Power*, vol. 1, *A History of Power from the Beginning to AD 1760* (Cambridge: Cambridge University Press, 1986); and Michael Mann, *The Sources of Social Power*, vol. 2, *The Rise of Classes and Nation States, 1760–1914* (Cambridge: Cambridge University Press, 1993).

52. Michel Foucault, "Questions à Michel Foucault sur la géographie," in *Dits et écrits, 1954–1988*, vol. 3, ed. Daniel Defert and François Ewald (Paris: Gallimard, 1994), 32; and Michel Foucault, "Questions on Geography," trans. Colin Gordon, in *Space, Knowledge and Power*, ed. Jeremy Crampton and Stuart Elden (Aldershot, UK: Ashgate, 2007), 176.

53. Foucault, "Questions à Michel Foucault," 33; and Foucault, "Questions on Geography," 177.

54. See Elden, *Mapping the Present*.

55. See, for instance, the arguments of Dipesh Chakrabarty, *Provincializing Europe: Postcolonial Thought and Historical Difference*, new ed. (Princeton, NJ: Princeton University Press, 2008).

56. For two useful discussions, see Michael Freeden, "What Should the 'Political' in Political Theory Explore?," *Journal of Political Philosophy* 13 (2005): 113–34; and Michael Freeden, "Thinking Politically and Thinking about Politics: Language, Interpretation and Ideology," in *Political Theory: Methods and Approaches*, ed. David Leopold and Marc Stears, 196–215 (Oxford: Oxford University Press, 2008).

57. David Harvey, "The Urban Process under Capitalism: A Framework for Analysis," *International Journal of Urban and Regional Research* 2 (1978): 101–31.

58. See, for example, Colin McFarlane, "The City as Assemblage: Dwelling and Urban Space," *Environment and Planning D: Society and Space* 29 (2011): 649–71.

59. See, for example, Neil Brenner, David J. Madden, and David Wachsmuth, "Assemblage Urbanism and the Challenges of Critical Urban Theory," *City* 15 (2011): 225–40.

CHAPTER I

1. Michel Foucault, "Nietzsche, la généalogie, l'histoire," in *Dits et écrits, 1954–1988,* ed. Daniel Defert and François Ewald, 4 vols. (Paris: Gallimard, 1994), 2:137, 2:140; and Michel Foucault, "Nietzsche, Genealogy, History," in *The Foucault Reader,* ed. Paul Rabinow (New York: Pantheon Books, 1984), 77, 80.

2. Martin Bernal, *Black Athena: The Afroasiatic Roots of Classical Civilisation,* vol. 1, *The Fabrication of Ancient Greece, 1785–1985* (London: Vintage, 1987).

3. For a helpful discussion, see Adam T. Smith, *The Political Landscape: Constellations of Authority in Early Complex Polities* (Berkeley: University of California Press, 2003).

4. Nicole Loraux, *Né de la terre: Mythe et politique à Athènes* (Paris: Éditions de Seuil, 1996), 9.

5. Isocrates, *Panegyricus,* in *Isocrates,* Greek-English ed., trans. George Norlin, 3 vols. (London: William Heinemann, 1928), 1:24. See Isocrates, *Panathenaicus,* 2:124–25; and Isocrates, *Peace,* 2:49.

6. For Pericles's funeral oration, see Thucydides, *De Bello Peloponnesiaco,* ed. Franciscus Goeller (Leipzig, Germany: In Libraria Caroli Cnobloch, 1836); and Thucydides, *History of the Peloponnesian War,* trans. Rex Warner (Harmondsworth, UK: Penguin, 1954), bk. 2, 34–46.

7. See M. S. Lane, *Method and Politics in Plato's* Statesman (Cambridge: Cambridge University Press, 1998), 106.

8. Plato, *Menexenus,* in *Timaeus, Critias, Cleitophon, Menexenus, Epistles,* Greek-English ed., trans. R. G. Bury (Cambridge, MA: Harvard University Press, 1966), 237b–c.

9. Plato, *Politeia,* in *Platonis Opera,* ed. Ioannes Burnet, vol. 4 (Oxford: Clarendon, 1902); and Plato, *Republic,* trans. Robin Waterfield (Oxford: Oxford University Press, 1993), 414b–c.

10. Plato, *Republic,* 414c. See 389b–c and Julia Annas, *An Introduction to Plato's* Republic (Oxford: Oxford University Press, 1981), 107–8, on lies generally.

11. See Euripides, *Phoenician Maidens,* in *Euripides in Four Volumes,* trans. Arthur S. Way, Greek-English ed., 4 vols. (London: William Heinemann, 1912), 3:638; and Pausanias, *Description of Greece,* Greek-English ed., trans. W. H. S. Jones, 4 vols. (London: William Heinemann, 1935), vol. 4, IX.12.2.

12. Plato, *Republic,* 414d–e.

13. Bruce Rosenstock, "Athena's Cloak: Plato's Critique of the Democratic City in the *Republic,*" *Political Theory* 22 (August 1994): 369–70.

14. Julia Annas, introduction to Plato, *Statesman,* ed. Julia Annas and Robin Waterfield, ix–xxiv (Cambridge: Cambridge University Press, 1995), xi.

15. Plato, *Politicus,* in *Platonis Opera,* vol. 1; and Plato, *Statesman,* 269b.

16. Plato, *Statesman,* 271a–b.

17. Plato, *Statesman,* 272d–e. On this myth in the *Statesman,* see Harvey Ronald Scodel, *Diaresis and Myth in Plato's* Statesman, *Hypomnemata* 85 (Göttingen, Germany: Vanenhoeck and Ruprecht, 1987), 79; Lane, *Method and Politics,* 105–7; and Pierre Vidal-Naquet, "Plato's Myth of the Statesman, the Ambiguities of the Golden

Age and of History," in *The Black Hunter: Forms of Thought and Forms of Society in the Greek World*, trans. Andrew Szegedy-Marzak (Baltimore: Johns Hopkins University Press, 1986), chap. 14.

18. Lane, *Method and Politics*, 106.

19. For explicit references to these, see also Plato, *Sophist*, 247c; *Laws*, 727e; *Timaeus*, 23d–e; and *Critias*, 109c–d. See Plato, *Theaetetus, Sophist*, Greek-English ed., trans. Harold North Fowler (Cambridge, MA: Harvard University Press, 1921).

20. Nicole Loraux, *L'invention d'Athènes: Histoire de l'oraison funèbre dans la "cité classique"* (Paris: Éditions de l'école des Hautes Etudes en Sciences Sociales, Mouton Editeur, 1981), 2.

21. Aristotle, *Rhetoric*, 1360b; compare *Politics*, 1269a4. The standard edition of Aristotle is the one edited by Immanuel Bekker, *Aristotelis Opera* (Berlin: W. de Gruyter, 1831). Its pagination and line numbers are retained in all scholarly editions and translations. The best modern editions of the Greek are discussed in Jonathan Barnes, ed., *The Cambridge Companion to Aristotle* (Cambridge: Cambridge University Press, 1995). Quotations are based on the translations by George A. Kennedy, *On Rhetoric: A Theory of Civic Discourse* (New York: Oxford University Press, 1991); *Politics*, by C. D. C. Reeve (Indianapolis: Hackett, 1996); and the *Nicomachean Ethics*, by Roger Crisp (Cambridge: Cambridge University Press, 2000).

22. The sources here are Apollodorus, *The Library*, Greek-English ed., trans. James George Frazer, 2 vols. (London: William Heinemann, 1921), vol. 2, III, 14, 6; Hygini, *Fabulae*, ed. Peter K. Marshall (Stuttgart: B. G. Teubner, 1993); *The Myths of Hyginus*, trans. and ed. Mary Grant, University of Kansas Publications in Humanistic Studies, No. 34 (Lawrence: University of Kansas Press, 1960), 166; and *Eratosthenis Catasterismorum Reliquiae*, ed. Carolus Robert (Berolini: Weidmannos, 1963), XIII; and Benjamin Powell, *Athenian Mythology: Erichthonius and the Three Daughters of Cecrops* (1906; repr., Chicago: Aves Publishers, 1976) (citations refer to the 1976 edition).

23. Euripides, *Ion*, trans. K. H. Lee (Warminster, UK: Aris and Phillips, 1997), line 20, 269–70. See also Marc Huys, *The Tale of the Hero Who Was Exposed at Birth in Euripidean Tragedy: A Study of Motifs* (Louvain, Belgium: Leuven University Press, 1995), 381.

24. Nicole Loraux, *Les enfants d'athéna: Idées athéniennes sur la citoyenneté et la division des sexes* (Paris: François Maspero, 1981), 46n42; and Robert Graves, *The Greek Myths*, 2 vols. (New York: George Braziller, 1957), 1:99.

25. K. H. Lee, "Commentary," in Euripides, *Ion*, 163.

26. Euripides, *Ion*, 589–90.

27. Aristophanes, *The Wasps*, in *Aristophanes*, Greek-English ed., trans. Benjamin Bickley Rogers, 3 vols. (London: William Heinemann, 1924), 1076.

28. John Peradotto, "Oedipus and Erichthonius: Some Observations on Paradigmatic and Syntagmatic Order," *Arethusa* 10 (1977): 85–102; and Robert Parker, "Myths of Early Athens," in *Interpretations of Greek Mythology*, ed. Jan Bremmer, 187–214 (London: Croom Helm, 1987).

29. Homer, *Iliad*, Greek-English ed., trans. A. T. Murray, 2 vols. (London: William Heinemann, 1924), 2:546–49.

30. Loraux, *Les enfants d'athéna*, 37. In Sophocles, *Ajax*, ed. A. F. Garvie, Greek-English ed. (Warminster, UK: Aris and Phillips, 1998), 201, the crew of Ajax's ship are said to be "men of the race descended from the earthbound [*kthonion*] line of Erechthe's."

31. Loraux, *L'invention d'Athènes*, 39.

32. Loraux, *L'invention d'Athènes*, 150.

33. Plutarch, *Alexandros*, 26, in *Lives*, Greek-English ed., trans. Bernadotte Perrin, 11 vols. (Cambridge, MA: Harvard University Press, 1914–26), vol. 7.

34. Arrian, *Anabasis Alexandri*, III, 2, Greek-English ed., trans. P. A. Brunt, 2 vols. (Cambridge, MA: Harvard University Press, 1976).

35. Plutarch, *Alexandros*, 26. See also Pseudo-Callisthenes, *Die Griechische Alexanderroman: Rezension* Γ, ed. Ursula von Lauenstein (Meisenheim am Glan, Germany: Anton Hain, 1962); and *The Greek Alexander Romance*, trans. Richard Stoneman (London: Penguin, 1991), 1:32. I am grateful to Veronica della Dora for pointing me to this example.

36. Sources used here are Apollodorus, *Library*, I, III.4.1–2; Graves, *Greek Myths*, 1:196; Euripides, *Bacchanals*, in *Euripides in Four Volumes*, 3:264, 3:538–40, 3:1026; Euripides, *Phoenician Maidens*, 638–75; and E. R. Dodds, "Commentary," in Euripides, *Bacchae*, ed. E. R. Dodds, 2nd ed. (Oxford: Clarendon, 1960), 144.

37. Euripides, *Phoenician Maidens*, 934–35.

38. Euripides, *Phoenician Maidens*, 939–40.

39. On the survivors' names, see W. K. C. Guthrie, *In the Beginning: Some Greek Views on the Origins of Life and the Early State of Man* (Ithaca, NY: Cornell University Press, 1957), 21–22.

40. Plato, *Laws*, 663e8–664a2.

41. Arlene W. Saxonhouse, "Myths and the Origins of Cities," in *Greek Tragedy and Political Theory*, ed. J. Peter Euben, 252–73 (Berkeley: University of California Press, 1986), 255–56.

42. Saxonhouse, "Myths and the Origins of Cities," 256–57.

43. Saxonhouse, "Myths and the Origins of Cities," 259.

44. Lee, introduction to Euripides, *Ion*, 35.

45. For a discussion, see Rosenstock, "Athena's Cloak," esp. 365–66.

46. Euben, introduction to *Greek Tragedy and Political Theory*, 37.

47. Hereafter the play is cited by line number of the Greek, which is usually found in the margins of translations. Texts used are Sophocles, *Antigone*, ed. Mark Griffith (Cambridge: Cambridge University Press, 1999), for the Greek; and primarily the translation by Robert Fagles in Sophocles, *The Three Theban Plays* (New York: Viking Penguin, 1982), though several other versions have also been consulted.

48. Sophocles, *Antigone*, 36.

49. Sophocles, *Antigone*, 79.

50. Sophocles, *Antigone*, 545.

51. Sophocles, *Antigone*, 776.

52. Perhaps the most famous other example of this is the desecration of Hector's body by Achilles. Hector's dying words are a request for an honorable burial—one immediately denied by Achilles. Homer, *Iliad*, XXII, 337–60.

53. See Sophocles, *Antigone*, 1016–18, 1040–45.

54. Sophocles, *Antigone*, 36.

55. Sophocles, *Antigone*, 1065–70

56. Sophocles, *Antigone*, 1100–1101.

57. See Sophocles, *Antigone*, 1103–4.

58. For an analysis of this death, see Nicole Loraux, *Façons tragiques de tuer une femme* (Paris: Hachette, 1985), 61–62.

59. Sophocles, *Antigone*, 2–3.

60. Sophocles, *Antigone*, 50–68.

61. Sophocles, *Antigone*, 855–56.

62. Sophocles, *Antigone*, 860–62, 584–85.

63. Sophocles, *Antigone*, 174.

64. Richard Emil Braun, introduction to Sophocles, *Antigone*, trans. Richard Emil Braun (New York: Oxford University Press, 1973), 5, 8.

65. Sophocles, *Antigone*, 660–62.

66. Sophocles, *Antigone*, 531–32.

67. Braun, introduction, 8.

68. Sophocles, *Oedipus at Colonus*, 406.

69. See Lowell Edmunds, *Theatrical Space and Historical Place in Sophocles'* Oedipus at Colonus (Lanham, MD: Rowman and Littlefield, 1996).

70. Judith Butler, *Antigone's Claim: Kinship between Life and Death* (New York: Columbia University Press, 2000), 60–61.

71. Butler, *Antigone's Claim*, 77; see 79.

72. Butler, *Antigone's Claim*, 57.

73. That said, it is not entirely clear in the play. Initial remarks suggest that it was at the seventh gate that the brothers died (141–47), we are not told of the body being moved, and yet at the end of the play Creon has to go "to the furthest part of the field" (1196–97). Mark Griffith, introduction to Sophocles, *Antigone*, 31, notes that some critics suggest that it is Creon's refusal to allow the body to "be removed for burial *outside* the borders" that makes it a step beyond the norm.

74. Jean-Pierre Vernant, "Ambiguïté et renversement: Sur la structure énigmatique d'"Œdipe Roi,'" in *Œdipe et ses mythes*, ed. Jean-Pierre Vernant and Pierre Vidal-Naquet, 23–53 (Paris: La Découverte, 1988), 23–24. See William Blake Tyrell and Larry J. Bennett, *Recapturing Sophocles'* Antigone (Lanham, MD: Rowman and Littlefield, 1998), 69.

75. Sophocles, *Antigone*, 110, 133.

76. Sophocles, *Antigone*, 209–10.

77. Sophocles, *Antigone*, 1098.

78. Sophocles, *Antigone*, 278–79.

79. Sophocles, *Antigone*, 508–9.

80. Sophocles, *Antigone*, 506.

81. Sophocles, *Antigone*, 693–94.

82. Sophocles, *Antigone*, 370–71.

83. Sophocles, *Antigone*, 175–85.

84. Sophocles, *Antigone*, 682.

85. Sophocles, *Antigone*, 725.

86. Sophocles, *Antigone*, 726–39.

87. See John D. B. Hamilton, "Antigone: Kinship, Justice, and the *Polis*," in *Myth and the Polis*, ed. Dori C. Pozzi and John M. Wickersham, 86–98 (Ithaca, NY: Cornell University Press, 1991).

88. Sophocles, *Antigone*, 842–43, 936.

89. Sophocles, *Antigone*, 807.

90. Sophocles, *Antigone*, 1122, 199.

91. Sophocles, *Antigone*, 155.

92. See Sophocles, *Antigone*, 187.

93. For an extended treatment, see Edmunds, *Theatrical Space and Historical Place*, 101ff.

94. See M. I. Finley, *Politics in the Ancient World* (Cambridge: Cambridge University Press, 1983), 1–2.

95. See, for example, Homer, *Iliad*, III, 50, where it means people, and XVI, 437, where it means the land of Lycia. See Raphael Sealey, *A History of the Greek City States, ca. 700–338 B.C.* (Berkeley: University of California Press, 1976), 91; and Charles W. Fornara and Loren J. Samons II, *Athens from Cleisthenes to Pericles* (Berkeley: University of California Press, 1991), 48.

96. We should therefore note the importance of the *demes* to *demokratia*, but not think that is the principal sense of the term. For this point more generally, see Sealey, *History of the Greek City States*, 301; and David Whitehead, *The Demes of Attica, 508/7–ca. 250 B.C.: A Political and Social Study* (Princeton, NJ: Princeton University Press, 1986), 37. On the term *deme*, see P. J. Rhodes, trans., glossary, in Aristotle, *The Athenian Constitution* (Harmondsworth, UK: Penguin, 1984), 179.

97. See Herodotus, *Histories*, trans. A. D. Godley, Greek-English ed., 4 vols. (London: William Heinemann, 1925), 5:66, 5:69. For a discussion, see Sealey, *History of the Greek City States*, 150; and H. T. Wade-Gery, "Studies in the Structure of Attic Society: The Laws of Kleisthenes," *Classical Quarterly* 27 (Jan. 1933): 17–29.

98. Aristotle, *Constitution of Athens*, ed. John Edwin Sandys (London: Macmillan, 1912); and Aristotle, *The Athenian Constitution*, trans. P. J. Rhodes (Harmondsworth, UK: Penguin, 1984), 21.4.

99. Aristotle, *Athenian Constitution*, 21.4–5. See C. Hignett, *A History of the Athenian Constitution to the End of the Fifth Century B.C.* (Oxford: Clarendon, 1952), 129.

100. Aristotle, *Athenian Constitution*, 21.3.

101. Herodotus, *Histories*, 5:66.

102. Aristotle, *Politics*, 1319b19–27. See Aristotle, *Athenian Constitution*, 21.2; and Herodotus, *Histories*, 6:131. A. Andrewes, "Kleisthenes' Reform Bill," *Classical Quarterly*, n.s., 27 (1977): 241, suggests that it is unlikely that mixing people up was as much of a goal as Aristotle suggests.

103. Herodotus, *Histories*, 5:69.

104. Whitehead, *Demes of Attica*, 18–19.

105. John S. Traill, *The Political Organisation of Attica: A Study of the Demes, Trittyes, and Phylai, and Their Representation in the Athenian Council*, supplement, *Hesperia* 14 (Princeton, NJ: American School of Classical Studies at Athens, 1975),

96–97. These figures are accepted by P. J. Rhodes, *A Commentary on the Aristotelian Athenaion Politeia* (Oxford: Clarendon Press, 1981), 252.

106. Strabo, *The Geography of Strabo*, trans. Horace Leonard Jones, Greek-English ed., 8 vols. (London: William Heinemann, 1924), vol. 4, 9:1, 9:16.

107. Traill, *Political Organisation of Attica*, 97.

108. Pierre Lévêque and Pierre Vidal-Naquet, *Clisthène l'Athénien: Essai sur la représentation de l'espace et du temps dans la pensée politique grecque de la fin du VIᵉ siècle à la mort de Platon* (Paris: Annales Litteraires de l'Université de Besançon, 1964), 15.

109. Whitehead, *Demes of Attica*, 27; for a survey of earlier scholarship, see xviii–xx.

110. C. W. J. Eliot, *Coastal Demes of Attika: A Study of the Policy of Kleisthenes*, *Phoenix: Journal of the Classical Association of Canada*, supplementary vol. 5 (Toronto: University of Toronto Press, 1962), 4.

111. Eliot, *Coastal Demes*, 3.

112. Eliot, *Coastal Demes*, 3.

113. Lévêque and Vidal-Naquet, *Clisthène l'Athénien*, 73–74.

114. On *genos* and *phratria*, see Sealey, *History of the Greek City States*, 22.

115. Pierre Vidal-Naquet, "1993 Preface," in Pierre Lévêque and Pierre Vidal-Naquet, *Cleisthenes the Athenian: An Essay on the Representation of Space and Time in Greek Political Thought from the End of the Sixth Century to the Death of Plato*, trans. David Ames Curtis (Atlantic Highlands, NJ: Humanities Press, 1996), xxxii; and Lévêque and Vidal-Naquet, *Clisthène l'Athénien*, 92–93.

116. See D. M. Lewis, "Lévêque (P.) and Vidal-Naquet (P.), *Clisthène l'Athénien*," *Journal of Hellenic Studies* 85 (1965): 222–23.

117. Lévêque and Vidal-Naquet, *Clisthène l'Athénien*, 107; see 10.

118. Lévêque and Vidal-Naquet, *Clisthène l'Athénien*, 78.

119. Lévêque and Vidal-Naquet, *Clisthène l'Athénien*, 123. On Anaximander, see Charles H. Kahn, *Anaximander and the Origins of Greek Cosmology* (New York: Columbia University Press, 1960); and Paul Seligman, *The Apeiron of Anaximander: A Study in the Origin and Function of Metaphysical Ideas* (London: Athlone Press, 1962).

120. Lévêque and Vidal-Naquet, *Clisthène l'Athénien*, 102. On this question generally, see M. R. Wright, *Cosmology in Antiquity* (London: Routledge, 1995).

121. Lévêque and Vidal-Naquet, *Clisthène l'Athénien*, 109.

122. Roger J. P. Kain and Elizabeth Bagnet, *The Cadastral Map in the Service of the State: A History of Property Mapping* (Chicago: University of Chicago Press, 1992), 1.

123. O. A. W. Dilke, *The Roman Land Surveyors: An Introduction to the Agrimensores* (Newton Abbot, UK: David and Charles, 1971), 22. See also Joseph C. Carter, Stephen M. Thompson, and Jessica Trelogan, "Dividing the Chora," in *Chora und Polis*, ed. Frank Kolb, 127–45 (Munich: R. Oldenbourg, 2004).

124. Wesley E. Thompson, "The Deme in Kleisthenes' Reforms," *Symbolae Osloenses* 46 (1971): 72; see Andrewes, "Kleisthenes' Reform Bill," 243.

125. S. D. Lambert, *The Phratries of Attica* (Ann Arbor: University of Michigan Press, 1993), 7.

126. Whitehead, *Demes of Attica*, 67.

127. Thompson, "Deme in Kleisthenes' Reforms," 76; and Andrewes, "Kleisthenes' Reform Bill," 245.

128. D. M. Lewis, "Cleisthenes and Attica," *Historia* 12 (Jan. 1963): 30.

129. Lewis, "Lévêque (P.) and Vidal-Naquet (P.)," 223.

130. D. M. Lewis, "C. W. J. Eliot, *Coastal Demes of Attica*," *Gnomon: Kritische Zeitschrift für die Gesamte Klassische Altertumswissenschaft* 35, no. 6 (September 1963): 724.

131. Lewis, "C. W. J. Eliot, *Coastal Demes of Attica*," 724. See Lambert, *Phratries of Attica*, 7.

132. Whitehead, *Demes of Attica*, xxi.

133. Merle K. Langdon, "The Territorial Basis of the Attic Demes," *Symbolae Osloenses* 60 (1985): 7.

134. Langdon, "Territorial Basis of the Attic Demes," 11. But see Andrewes, "Kleisthenes' Reform Bill," 244.

135. Langdon, "Territorial Basis of the Attic Demes," 12.

136. Aristophanes, *Aves*, 997, scholium in Felix Jacoby, *Die Fragmente der Griechischen Historiker* (Leiden, Netherlands: E. J. Brill, 1964), vol. 3 B, no. 375, which gives the fragment as "*horismoi tes poleos.*" See Langdon, "Territorial Basis of the Attic Demes," 13.

137. Langdon, "Territorial Basis of the Attic Demes," 13.

138. Moses I. Finley, *Studies in Land and Credit in Ancient Athens, 500–200 B.C.: The Horos Inscriptions* (New Brunswick, NJ: Rutgers University Press, 1951), 4.

139. Finley, *Studies in Land and Credit*, 5–6.

140. Finley, *Studies in Land and Credit*, 3–4.

141. Strabo, *Geography of Strabo*, 1:1.4.7.

142. Though the meaning of this passage is of course disputed. See Thompson, "Deme in Kleisthenes' Reforms"; and Langdon, "Territorial Basis of the Attic Demes," 13.

143. Eliot, *Coastal Demes*, 147.

144. Lewis, "C. W. J. Eliot, *Coastal Demes of Attica*," 724.

145. The best example is found in John S. Traill, *Demos and Trittys: Epigraphical and Topographical Studies in the Organisation of Attica* (Toronto: Athenians, 1986).

146. Rhodes, *Commentary on the Aristotelian* Athenaion Politeia, 763.

147. Andrewes, "Kleisthenes' Reform Bill," 245.

148. Aristotle, *Athenian Constitution*, 21.4–5.

149. These two variants are offered by Rhodes, *Commentary on the Aristotelian* Athenaion Politeia, 258.

150. Aristotle, *Athenian Constitution*, 21.5. On the deme names, see Lewis, "Cleisthenes and Attica," 26–27.

151. Rhodes, *Commentary on the Aristotelian* Athenaion Politeia, 258.

152. Langdon, "Territorial Basis of the Attic Demes," 7.

153. Rhodes, *Commentary on the Aristotelian* Athenaion Politeia, 251–52.

154. Sealey, *History of the Greek City States*, 151; on the reforms generally, see 150–55.

155. P. J. Bicknell, "Kleisthenes as Politician: An Exploration," in *Studies in*

Athenian Politics and Genealogy, Historia Einzelschriften Heft 19, 1–53 (Wiesbaden, Germany: Franz Steiner, 1972), 18; see Eliot, *Coastal Demes*, 3–4.

156. Rhodes, *Commentary on the Aristotelian* Athenaion Politeia, 252; Lambert, *Phratries of Attica*, 8; and Hignett, *History of the Athenian Constitution*, 136.

157. Whitehead, *Demes of Attica*, 352; and W. K. Lacey, *Family in Classic Greece* (Ithaca, NY: Cornell University Press, 1968), 90–91.

158. Rhodes, *Commentary on the Aristotelian* Athenaion Politeia, 253; Hignett, *History of the Athenian Constitution*, 129; and Lambert, *Phratries of Attica*, 2.

159. G. R. Stanton, "The Tribal Reform of Kleisthenes the Alkeonid," *Chiron* 14 (1984): 1–41.

160. Pierre Lévêque, "The *Da-* Root: Repartition and Democracy," in Lévêque and Vidal-Naquet, *Cleisthenes the Athenian*.

161. Sylvie Vilatte, *Espace et temps: La cité aristotélicienne de la* Politique (Paris: Annales Littéraires de l'Université de Besançon, 1995), 67.

162. For the Greek text, see *Platonis Opera*, vol. 5. The main translation used is that of Trevor J. Saunders, Plato, *The Laws* (Harmondsworth, UK: Penguin, 1970); though this has been compared to Plato, *Works*, vol. 9, pt. 1, English-Greek ed., trans. R. G. Bury (London: William Heinemann, 1926); and Plato, *The Laws of Plato*, trans. Thomas L. Pangle (Chicago: University of Chicago Press, 1980).

163. Annas, introduction to Plato, *Statesman*, xii.

164. Saunders, introduction to Plato, *Laws*, 39.

165. Plato, *Laws*, 702c–e. For a discussion, see Loraux, *Né de la terre*, 185; and A.J. Graham, *Colony and Mother City in Ancient Greece* (New York: Barnes and Noble, 1964), 4.

166. Sealey, *History of the Greek City States*, 30–31.

167. Graham, *Colony and Mother City*, xvii.

168. Jean Gottmann, *The Significance of Territory* (Charlottesville: University Press of Virginia, 1973), 17–18.

169. Plato, *Laws*, 704a–e, 737e.

170. Gottmann, *Significance of Territory*, 19.

171. Leo Strauss, *The Arguments and Actions of Plato's* Laws (Chicago: University of Chicago Press, 1975), 54.

172. Plato, *Laws*, 704b–c.

173. 1 stade = 193 yards = 176 meters. See Rhodes, "Glossary and Subject Index," in Aristotle, *Athenian Constitution*, 174.

174. See Plato, *Republic*, 2.370e, which notes that it is practically impossible to build a *polis* without the need for imports.

175. Glenn R. Morrow, *Plato's Cretan City: A Historical Interpretation of the* Laws (Princeton, NJ: Princeton University Press, 1960), 30; see 95.

176. For a discussion, see John Sallis, *Chorology: On Beginning in Plato's* Timaeus (Bloomington: Indiana University Press, 1999); John Sallis, "The Politics of the χώρα," in *The Ancients and the Moderns*, ed. Reginald Lilly, 59–71 (Bloomington: Indiana University Press, 1996); and John Sallis, "Traces of the *Chōra*," in *Retracing the Platonic Text*, ed. John Russon and John Sallis, 57–69 (Evanston: Northwestern University Press, 2000).

177. Sallis, "Politics of the χώρα," 69.

178. Luce Irigaray, *Speculum de l'autre femme* (Paris: Éditions de Minuit, 1974); Julia Kristeva, *La Révolution du langage poétique: L'avant-garde à la fin du XIXᵉ siècle: Lautréamont et Mallarmé* (Paris: Éditions du Seuil, 1974); Judith Butler, *Bodies That Matter: On the Discursive Limits of Sex* (London: Routledge, 1993), chap. 1; and Jacques Derrida, *Khōra* (Paris: Galilée, 1993).

179. Sallis, "Politics of the χώρα," 70.

180. Strauss, *Arguments and Actions of Plato's Laws*, 54.

181. Morrow, *Plato's Cretan City*, 30, 95.

182. G. Charles Picard, "L'administration territoriale de Carthage," in *Mélanges d'archáeologie et d'histoire offerts à André Piganiol*, ed. Raymond Chevallier, 3 vols. (Paris: SEVPEN, 1966), 3:1257–65, 3:1258–59.

183. Plato, *Laws*, 737. On this issue generally, see Philip Brook Manville, *The Origins of Citizenship in Ancient Athens* (Princeton, NJ: Princeton University Press, 1986), chap. 5.

184. Plato, *Laws*, 736.

185. Plato, *Laws*, 737a.

186. Plato, *Laws*, 736e.

187. Plato, *Laws*, 736b.

188. Lacey, *Family in Classic Greece*, 179.

189. Plato, *Laws*, 737e–738a.

190. Plato, *Laws*, 738c–d.

191. Plato, *Laws*, 739d–740a.

192. Plato, *Laws*, 740a–c.

193. Morrow, *Plato's Cretan City*, 118.

194. For a list of vocabulary derived from *polis*, and its usual English translations, see David Keyt and Fred D. Miller Jr., introduction to *A Companion to Aristotle's Politics*, ed. David Keyt and Fred D. Miller Jr. (Oxford: Blackwell, 1991), 2.

195. See Plato, *Laws*, 814c; and Morrow, *Plato's Cretan City*, 113.

196. Plato, *Laws*, 741b–c. On this, see M. I. Finley, "The Alienability of Land in Ancient Greece," *Eirene* 7 (1968): 25–32.

197. Plato, *Laws*, 842e–843b.

198. Plato, *Laws*, 843c–e.

199. See Plato, *Laws*, 757a–b.

200. Plato, *Laws*, 744c–d.

201. Plato, *Laws*, 744d–745a.

202. Morrow, *Plato's Cretan City*, 104; and Plato, *Republic*, 416d–417b.

203. Paul Cartledge, "Greek Political Thought: The Historical Context," in *The Cambridge History of Greek and Roman Thought*, ed. Christopher Rowe and Malcolm Schofield (Cambridge: Cambridge University Press, 2005), 11–59, 14.

204. Nicole Loraux, *Mothers in Mourning*, trans. Corinne Pache (Ithaca, NY: Cornell University Press, 1998), 67.

205. Morrow, *Plato's Cretan City*, 121–22.

206. Plato, *Laws*, 848c–d.

207. See the note by Saunders, in Plato, *Laws*, 218n27.

208. Plato, *Laws*, 738d. Wesley E. Thompson, "The Demes in Plato's *Laws*," *Eranos: Acta Philologica Suecana* 63, fasc. 3-4 (1965): 134; see also Whitehead, *Demes of Attica*, 50.

209. Plato, *Laws*, 746d-747a; see 771.

210. Plato, *Laws*, 760b-d.

211. Plato, *Laws*, 760e-761c.

212. Plato, *Laws*, 778c.

213. Plato, *Laws*, 778e-779c.

214. Aristotle, *Nicomachean Ethics*, 1160a10-30; and Aristotle, *Politics*, 1252a1-7. Compare Plato, *Laws*, 680-81. See Lacey, *Family in Classic Greece*, 21; Sarah B. Pomeroy, *Families in Classical and Hellenistic Greece: Representations and Realities* (New York: Oxford University Press, 1997), 36-37; and William J. Booth, "Politics and the Household: A Commentary of Aristotle's *Politics* Book One," *History of Political Thought* 2 (Summer 1981): 203-26.

215. Aristotle, *Politics*, 1252b27-31.

216. Aristotle, *Nicomachean Ethics*, 1160a10-30; and Aristotle, *Politics*, 1252a1-7, 1280b38-1281a1.

217. Aristotle, *Politics*, 1252b27-31, 1253a1-3.

218. Aristotle, *Nicomachean Ethics*, 1170b31-1171a1.

219. The metaphor of the ship of state is commonly found in Greek authors. See, for example Plato, *Laws*, 758a, 945c; and Sophocles, *Antigone*, 162.

220. Aristotle, *Politics*, 1326a39-b6.

221. Aristotle, *Politics*, 1326b22-24.

222. Aristotle, *Politics*, 1326b26.

223. Aristotle, *Politics*, 1265a19-20.

224. Aristotle, *Politics*, 1326b39-1327a10.

225. Aristotle, *Politics*, 1330a34-b17.

226. Aristotle, *Politics*, 1267b22-33.

227. Lévêque and Vidal-Naquet, *Clisthène l'Athénien*, 123-25.

228. Aristotle, *Politics*, 1330b22-30.

229. Aristotle, *Politics*, 1330b31-1331a15.

230. Aristotle, *Politics*, 1329b36-1330a2.

231. Aristotle, *Politics*, 1330a9-20.

232. Aristotle, *Politics*, 1260b40-1261a1.

233. Aristotle, *Politics*, 1276a15-23.

234. Aristotle, *Politics*, 1276a24-32; see also 1280b12-15, 1326a5-10.

235. Aristotle, *Politics*, 1280b29-36; see 1274b32-1275b20.

236. Aristotle, *Politics*, 1276a40-b4.

237. Aristotle, *Eudemian Ethics*, 1242a22-23.

238. Aristotle, *Eudemian Ethics*, 1242a40-b2.

239. Aristotle, *Politics*, 1274b32-1275b20.

240. Vilatte, *Espace et temps*, 33.

241. See Aristotle, *Metaphysics*, in Bekker, ed., *Aristotelis Opera*; Aristotle, *Metaphysics*, translated in *The Works of Aristotle*, vol. 8, *Metaphysica*, ed. W. D. Ross (Oxford: Clarendon Press, 1930), 1016b29-31.

242. Aristotle, *Physics*, in Bekker, ed., *Aristotelis Opera*; Aristotle, *Physics*, trans. Robin Waterfield (Oxford: Oxford University Press, 1996), 226b18–227a34. For a really helpful discussion, see Helen S. Lang, *The Order of Nature in Aristotle's Physics: Place and the Elements* (Cambridge: Cambridge University Press, 1998). Lang rightly contends that "there is no such thing as 'physical space' in Aristotle" (230). Despite its title, Keimpe Algra, *Concepts of Space in Greek Thought* (Leiden, Netherlands: E. J. Brill, 1995), is a historically and textually nuanced study.

243. Aristotle, *Physics*, 231a24.

244. See, for example, Plato, *Republic*, 527a. See Edward A. Maziarz and Thomas Greenwood, *Greek Mathematical Philosophy* (New York: Frederick Ungar, 1968), 7. The notion of geometry as a land-measuring device by the Egyptians is described in Herodotus, *Histories*, II, 109.

245. See Mario Bunge, "Le lieu et l'espace," in *Penser avec Aristotle: Études réunies sous la direction de M. A. Sinaceur* (Toulouse, France: Érès, 1991), 483–88; Vilatte, *Espace et temps*; Maziarz and Greenwood, *Greek Mathematical Philosophy*; Thomas Heath, *A History of Greek Mathematics*, 2 vols. (New York: Dover, 1981); Martin Heidegger, *Platon: Sophistes* (Frankfurt am Main, Germany: Vittorio Klostermann, 1992), §15; Jacob Klein, *Greek Mathematical Thought and the Origin of Algebra* (New York: Dover, 1992); and David Rapport Lachterman, *The Ethics of Geometry: A Genealogy of Modernity* (New York: Routledge, 1989).

246. Aristotle, *Politics*, 1296b17–20.

247. Aristotle, *Politics*, 1301b29–32.

248. Aristotle, *Politics*, 1301a25–30, 1280a7–34.

249. Aristotle, *Politics*, 1318a27–40.

250. Aristotle, *Politics*, 1131a10–1132a32.

251. François de Polignac, *La naissance de la cité grecque: Cultes, espace et société VIIIᵉ–VIIᵉ siècles avant J.C.* (Paris: Éditions la Découverte, 1984), 8'.

252. de Polignac, *La naissance de la cité grecque*, 5, 66.

253. Loraux, *Né de la terre*, 49–50; see Loraux, *L'invention d'Athènes*, 450n35.

254. Loraux, *Né de la terre*, 50; see de Polignac, *La naissance de la cité grecque*, 16.

255. C. D. C. Reeve, introduction to Aristotle, *Politics*, lvii; see 245. Reeve refers to *Politics*, 1328b2–23.

256. See Martin Heidegger, *Einführung in die Metaphysik, Gesamtausgabe Band 40* (Frankfurt am Main, Germany: Vittorio Klostermann, 1983); Martin Heidegger, *Hölderlins Hymne "Der Ister," Gesamtausgabe Band 53* (Frankfurt am Main, Germany: Vittorio Klostermann, 1984); *Parmenides, Gesamtausgabe Band 54* (Frankfurt am Main, Germany: Vittorio Klostermann, 1982).

257. Cartledge, "Greek Political Thought," 17–18.

258. See Manville, *Origins of Citizenship*, 36–38.

259. Cornelius Castoriadis, "The Greek Πόλις and the Creation of Democracy," in Lilly, *Ancients and the Moderns*, 46.

260. Manville, *Origins of Citizenship*, 42, 93.

261. Sealey, *History of the Greek City States*, 19.

262. de Polignac, *La naissance de la cité grecque*, 57.

263. de Polignac, *La naissance de la cité grecque*, 64.

264. Manville, *Origins of Citizenship*, 36. See Sealey, *History of the Greek City States*, 19.

265. Chester G. Starr, *The Origins of Greek Civilization, 1100–650 BC* (London: Jonathan Cape, 1962), 338.

266. Chester G. Starr, *Individual and Community: The Rise of the Polis, 800–500 B.C.* (Oxford: Oxford University Press, 1986), 71. See, though, Herodotus, *Histories*, IX, 13.

267. Plato, *Laws*, 704. See Morrow, *Plato's Cretan City*, 30, 95; and Strauss, *Arguments and Actions of Plato's Laws*, 54.

268. Plato, *Laws*, 707–8.

269. Manville, *Origins of Citizenship*, 5–6; and Reeve, introduction to Aristotle, *Politics*, 245.

270. Manville, *Origins of Citizenship*, 53. See A. C. Bradley, "Aristotle's Conception of the State," in *Companion to Aristotle's* Politics, ed. Keyt and Miller, 14–15.

271. Nicole Loraux, *The Divided City: On Memory and Forgetting in Ancient Athens*, trans. Corinne Pache with Jeff Fort (New York: Zone Books, 2002), 100.

272. Castoriadis, "Greek Πόλις and the Creation of Democracy," 46.

273. Thucydides, *History of the Peloponnesian War*, VII, 77. The gender implications of this are the focus of Nicole Loraux, *Les expériences de Tirésias: Le féminin et l'homme grec* (Paris: Gallimard, 1989).

274. Thucydides, *History of the Peloponnesian War*, I, 143. On this passage, see Oddone Longo, "Atene fra Polis e Territorio: In Margine a Tucidide I 143, 5," *Studia Italiani Filologia Classica*, n.s., 46 (1974): 5–21.

275. Aeschylus, *Persians*, edit. and trans. Edith Hall, Greek-English ed. (Warminster, UK: Aris and Phillips, 1996), line 347.

276. John E. Stambaugh, "The Idea of the City: Three Views of Athens," *Classical Journal* 69 (April–May 1974): 309–10, 311–12.

277. de Polignac, *La naissance de la cité grecque*, 156–57.

278. de Polignac, *La naissance de la cité grecque*, 42, 77.

279. de Polignac, *La naissance de la cité grecque*, 46.

280. de Polignac, *La naissance de la cité grecque*, 104–5.

281. de Polignac, *La naissance de la cité grecque*, 107.

282. François de Polignac, "Preface to the English Edition," *Cults, Territory, and the Origins of the Greek City-State*, trans. Janet Lloyd (Chicago: University of Chicago Press, 1995), xiii–xiv.

283. Thucydides, *History of the Peloponnesian War*, IV, 118.

CHAPTER 2

1. Tacitus, *Annales*, I, 1. For Tacitus I have used *The Histories*, trans. Clifford H. Moore, and *The Annals*, trans. John Jackson, Latin-English ed., 4 vols. (London: William Heinemann, 1937); *Histories*, trans. Kenneth Wellesley (Harmondsworth, UK: Penguin, 1964); *The Annals of Imperial Rome*, trans. Michael Grant, rev. ed., (Harmondsworth, UK: Penguin, 1975); and *The Annals: The Reigns of Tiberius, Claudius and Nero*, trans. J. C. Yardley (Oxford: Oxford University Press, 2008).

2. For the earlier period, an exceptional guide is T. J. Cornell, *The Beginnings of Rome: Italy and Rome from the Bronze Age to the Punic Wars (c. 1000–264 BC)* (London: Routledge, 1995); see also Jean-Michel David, *The Roman Conquest of Italy*, trans. Antonia Nevill (London: Routledge, 1997).

3. Frank E. Adcock, *Roman Political Ideas and Practice* (Ann Arbor: University of Michigan Press, 1959), 19.

4. M. I. Finley, *Politics in the Ancient World* (Cambridge: Cambridge University Press, 1983), 1–2.

5. F. R. Cowell, *Cicero and the Roman Republic* (Harmondsworth, UK: Penguin, 1948), 135–37.

6. See Lily Ross Taylor, *The Voting Districts of the Roman Republic: The Thirty-Five Urban and Rural Tribes* (Rome: American Academy in Rome, 1960); and Fergus Millar, *The Crowd in Rome in the Late Republic* (Ann Arbor: University of Michigan Press, 1998).

7. Cowell, *Cicero and the Roman Republic*, 159–60; see Arthur Keaveney, *Rome and the Unification of Italy* (London: Croom Helm, 1987).

8. K. J. Beloch, *Römische Geschichte bis zum Beginn der Punischen Kriege* (Berlin: Walter de Gruyter, 1926), 620–21.

9. David Shotter, *Rome and Her Empire* (London: Longman, 2003), 70, 75, 82.

10. William V. Harris, *War and Imperialism in Republican Rome* (Oxford: Clarendon Press, 1979), 60.

11. Andrew Lintott, *Imperium Romanum: Politics and Administration* (London: Routledge, 1993), 129.

12. See Stephen L. Dyson, *Community and Society in Roman Italy* (Baltimore: Johns Hopkins University Press, 1992), 1; and Steven K. Drummond and Lynn H. Nelson, *The Western Frontiers of Imperial Rome* (Armonk, NY: M. E. Sharpe, 1994), 238.

13. P. A. Brunt, "The Army and the Land in the Roman Revolution," *Journal of Roman Studies* 52 (1962): 69–86. See also Keith Hopkins, *Conquerors and Slaves* (Cambridge: Cambridge University Press, 1978), 6, 36.

14. Plutarch, *Romulus*, XVII, in *Lives*, Greek-English ed., trans. Bernadotte Perrin, 11 vols. (Cambridge, MA: Harvard University Press, 1914–26), vol. 1.

15. O. A. W. Dilke, *The Roman Land Surveyors: An Introduction to the* Agrimensores (Newton Abbot, UK: David & Charles, 1971), 183.

16. Hopkins, *Conquerors and Slaves*, 49–50.

17. There are several good biographies. See, in particular, Gérard Walter, *Caesar*, trans. Emma Craufurd, 2 vols. (London: Cassell, 1953); Matthias Gelzer, *Caesar: Politician and Statesman*, trans. Peter Needham (Oxford: Basil Blackwell, 1968); Christian Meier, *Caesar*, trans. David McLintock (London: HarperCollins, 1995); and Adrian Goldsworthy, *Caesar: Life of a Colossus* (New Haven, CT: Yale University Press, 2007). On his posthumous reception, see Maria Wyke, *Caesar: A Life in Western Culture* (London: Granta Books, 2007). On his written style, see F. E. Adcock, *Caesar as Man of Letters* (Cambridge: Cambridge University Press, 1956); Leo Raditsa, "Julius Caesar and His Writings," *Aufstieg und Niedergang der Römischen Welt* 1 (1973): 417–64; and William W. Batstone and Cynthia Damon, *Caesar's Civil War* (Oxford: Oxford University

Press, 2006). On the war in Gaul, see also Barry Cunliffe, *Greeks, Romans and Barbarians: Spheres of Interaction* (London: B. T. Batsford, 1988), chap. 6.

18. Caesar, *De bello gallico*, in *The Gallic Wars*, English-Latin ed., trans. H. J. Edwards (London: William Heinemann, 1917). I have generally used the translation in *The Conquest of Gaul*, trans. S. A. Handford, revised by Jane Gardner (Harmondsworth, UK: Penguin, 1982).

19. Caesar, *De bello civili*, in *The Civil Wars*, English-Latin ed., trans. A. G. Peskett (London: William Heinemann, 1914). I have generally used the translation in *The Civil War, Together with the Alexandrian War, the African War, and the Spanish War by Other Hands*, trans. Jane F. Gardner (Harmondsworth, UK: Penguin, 1967). For the Latin for the other texts in this volume, which continue the story, see "Caesar," *Alexandrian, African and Spanish Wars*, English-Latin ed., trans. A. G. Way (London: William Heinemann, 1955).

20. Adcock, *Caesar as Man of Letters*, 5. The previous view—in Gelzer and Meier, for instance—was that the book was compiled from the reports after his return to Rome, and made a whole. For a discussion, see T. P. Wiseman, "The Publication of *De Bello Gallico*," in *Julius Caesar as Artful Reporter: The War Commentaries as Political Instruments*, ed. Kathryn Welch and Anton Powell, 1–9 (London: Duckworth with Classical Press of Wales, 1998); and Adrian Goldsworthy, "'Instinctive Genius': The Depiction of Caesar the General," same volume, 193–219.

21. See J. P. V. D. Balsdon, "The Veracity of Caesar," *Greece & Rome*, 2nd ser., 4 (March 1957): 19–28; and J. Cuff, "Caesar the Soldier," *Greece & Rome*, 2nd ser., 4 (March 1957): 29–35.

22. M. Cary, *The Geographic Background of Greek and Roman History* (Oxford: Clarendon, 1949), 315.

23. Caesar, *De bello gallico*, I, 1.

24. C. R. Whittaker, *Rome and Its Frontiers: The Dynamics of Empire* (London: Routledge, 2004), 144.

25. Honorius, "Cosmographia Iulii Caesaris," in *Geographi Latini minores*, ed. Alexander Riese (1878), 21–23, trans. in O. A. W. Dilke, *Greek and Roman Maps* (London: Thames and Hudson, 1985), 183–84. See Claude Nicolet, *Space, Geography, and Politics in the Early Roman Empire*, trans. Hélène Leclerc (Ann Arbor: University of Michigan Press, 1991), 96. The original is *L'Inventaire du Monde: Géographie et politique aux origins de l'Empire romain* (Paris: Fayard, 1988).

26. Nicolet, *Space, Geography, and Politics*, 97.

27. Whittaker, *Rome and Its Frontiers*, 144.

28. Elizabeth Rawson, *Intellectual Life in the Late Roman Republic* (London: Duckworth, 1985), 260.

29. See, for example, the maps in Caesar, *Conquest of Gaul*, 266–67; and Caesar, *Seven Commentaries on the Gallic War, with an Eighth Commentary by Aulus Hirtius*, trans. Carolyn Hammond (Oxford: Oxford University Press, 1996), 1–li.

30. Gavin A. Sundwall, "Ammianus Geographicus," *American Journal of Philology* 117 (1996): 619.

31. C. R. Whittaker, *Frontiers of the Roman Empire: A Social and Economic Study* (Baltimore: Johns Hopkins University Press, 1994), 67.

32. Whittaker, *Rome and Its Frontiers*, 67.

33. Adcock, *Caesar as Man of Letters*, 63, see 16; and Jane F. Gardner, introduction to Caesar, *Civil War*, 26–27.

34. I have made extensive use of H. Merguet, *Lexikon zu den Schriften des Cäsars und seiner Fortsetzer* (Jena, Germany: Gustav Fischer, 1886); and esp. E. G. Sihler, *A Complete Lexicon of the Latinity of Caesar's Gallic War* (Amsterdam: B. R. Grüner, 1968).

35. Caesar, *De bello gallico*, V, 42; VII, 73; VII, 85. *Terra* is only rarely used to describe the land of a people (*De bello gallico*, I, 30). It also sometimes means the earth as world: *orbis terrarum* (*De bello gallico*, IV, 7; VI, 14; VII, 29). Caesar does not use the word *limes*.

36. Caesar, *De bello gallico*, II, 18; see also V, 9; VII, 36; VII, 69; VII, 73; Caesar, *De bello civili*, I, 47–48; and "Caesar," *Spanish War*, 38.

37. For example, Caesar, *De bello gallico*, I, 7; I, 52.

38. Caesar, *De bello gallico*, II, 17; see I, 43, etc.

39. Caesar, *De bello gallico*, VII, 23.

40. See, for example, Virgil, *Aeneid*, V, 316, 321, 325, 327. I have used the Latin text in *Works*, trans. H. Rushton Fairclough, revised by G. P. Goold, Latin-English ed., 2 vols. (Cambridge, MA: Harvard University Press, 1999–2000); and largely followed the translation by Frederick Ahl (Oxford: Oxford University Press, 2007). It is used in the singular in Juvenal, *Satire*, VI, 582, in *Juvenal and Persius*, ed. and trans. Susanna Morton Braund, Latin-English ed. (Cambridge, MA: Harvard University Press, 2004); and Statius, *Thebaid*, VI, 594, trans. D. R. Shackleton-Bailey, Latin-English ed. (Cambridge, MA: Harvard University Press, 2004).

41. Caesar, *De bello gallico*, III, 29; see I, 38; VII, 69.

42. Caesar, *De bello gallico*, III, 20.

43. Caesar, *De bello gallico*, I, 9–10; see II, 16; IV, 10; and many other instances.

44. Caesar, *De bello gallico*, I, 1. There are countless references of this kind.

45. Caesar, *De bello gallico*, II, 15.

46. Caesar, *De bello gallico*, VI, 10; see VII, 66.

47. Caesar, *De bello gallico*, VI, 44.

48. Caesar, *De bello gallico*, V, 56; IV, 1; VII, 33; and Hirtius, *De bello gallico*, VIII, 31.

49. Daphne Nash, "Territory and State Formation in Central Gaul," in *Social Organisation and Settlement*, ed. David Green, Colin Haselgrove, and Matthew Spriggs, BAR International Series (Supplementary) 47 (ii) (Oxford, 1978), 455–75, esp. 461, 463, 473n3. The term is regularly used in *De bello gallico*.

50. Caesar, *De bello gallico*, VI, 13.

51. Caesar, *De bello gallico*, I, 2; see I, 5; I, 31; and *De bello civili*, I, 39; and also IV, 3; VI, 12, for *agrum* in the sense of neighboring lands.

52. Cicero, *Pro M. Marcello Oratio*, in *Works*, vol. 14, trans. N. H. Watts (London: William Heinemann, 1972), 28.

53. Valerie A. Maxfield, "Mainland Europe," in *The Roman World*, ed. John Wacher, 2 vols., 1:139–97 (London: Routledge, 1987), 1:140. See Caesar, *De bello gallico*, IV, 4; IV, 16.

54. Susan Mattern, *Rome and the Enemy: Imperial Strategy in the Principate* (Berkeley: University of California Press, 1999), 33, see 76. See Caesar, *De bello gallico*, I, 47; and G. Walser, *Caesar und die Germanen: Studien zur politischen Tendenz römischer Feldzugsberichte* (Wiesbaden, Germany: F. Steiner, 1956).

55. C. M. Wells, *German Policy of Augustus: An Examination of the Archaeological Evidence* (Oxford: Clarendon Press, 1972), 29.

56. Wells, *German Policy of Augustus*, 30. See also E. A. Thompson, *The Early Germans* (Oxford: Clarendon Press, 1965).

57. Walter Goffart, *Barbarians and Romans, A.D. 418–584: The Techniques of Accommodation* (Princeton, NJ: Princeton University Press, 1980), 13; see Walter Goffart, *Barbarian Tribes: The Migration Age and the Later Roman Empire* (Philadelphia: University of Pennsylvania Press, 2006); and Tacitus, *Germania*, trans. Herbert W. Benario, English-Latin ed. (Warminster, UK: Aris and Philips, 1999).

58. Wells, *German Policy of Augustus*, 30.

59. Caesar, *De bello gallico*, IV, 16.

60. Caesar, *De bello gallico*, IV, 19.

61. Justine Davis Randers-Pehrson, *Barbarians and Romans: The Birth Struggle of Europe, A.D. 400–700* (London: Croom Helm, 1983), 5.

62. Tacitus, *Germania*, 28.

63. Gaius Suetonius Tranquillus, *Divus Julius*, ed. H. E. Butler and M. Cary, revised by G. B. Townend (Bristol, UK: Bristol Classical Press, 1982); and Gaius Suetonius Tranquillus, *The Twelve Caesars*, trans. Robert Graves, revised by Michael Grant (Harmondsworth, UK: Penguin, 1979), 25.

64. A. N. Sherwin-White, "Caesar as Imperialist," *Greece & Rome*, 2nd ser., 4 (1957): 40–41. The reference to Pompeius can be found in Plutarch, *Pompeius*, XXXIII, in *Lives*, vol. 1, where the Greek word is *horos*.

65. Plutarch, *Caesar*, XX, XXXII, in *Lives*, vol. 7.

66. Whittaker, *Rome and Its Frontiers*, 63; and Mary T. Boatwright, Daniel J. Gargola, and Richard J. A. Talbert, *The Romans: From Village to Empire* (New York: Oxford University Press, 2004), 246.

67. Robin Seager, *Pompey the Great: A Political Biography*, 2nd ed. (Oxford: Blackwell, 2002), 171. On this generally, see Adrian Goldsworthy, *Caesar's Civil War, 49–44 BC* (Oxford: Osprey, 2002); the discussion in his *Caesar*; and Batstone and Damon, *Caesar's Civil War*.

68. Erich S. Gruen, *The Last Generation of the Roman Republic* (Berkeley: University of California Press, 1974), 365.

69. Suetonius, *Divus Julius*, 32–33. In Plutarch's Greek, the wording is "let the die be cast [*hanerriphtho kubos*]": imperative, not indicative (*Caesar*, XXXII). This may be a quotation from Menander. Appian calls it the proverb "*ho kubos hanerriphtho*." See Appian, *The Civil Wars*, II, 35, in *Roman History*, trans. Horace White, Greek-English ed., 4 vols. (Cambridge, MA: Harvard University Press, 1912), vols. 3–4. I have used the translation of John Carter (Harmondsworth, UK: Penguin, 1996). According to the notes to the Butler and Cary edition of Suetonius, Erasmus conjectured *esto* for *est* to make Suetonius fit those accounts (85). This is the reading suggested by Lewis and Short. See Goldsworthy, *Caesar*, 378.

70. Whittaker, *Rome and Its Frontiers*, 63.

71. Caesar, *De bello civili*, III, 57.

72. Gelzer, *Caesar*, 273.

73. Caesar, *De bello civili*, I, 9.

74. Inscription in Taranto, cited by E. Badian, "Christian Meier, *Caesar*," *Gnomon: Kritische Zeitschrift für die Gesamte Klassicsche Altertumswissenschaft* 62 (1990): 22–39.

75. Colin Wells, *The Roman Empire*, 2nd ed. (London: Fontana, 1992), 12; see Boatwright, Gargola, and Talbert, *Romans*, 265.

76. Gruen, *Last Generation of the Roman Republic*, 497. See Andrew Lintott, *Violence in Republican Rome* (Oxford: Oxford University Press, 1999).

77. On his life, see David Stockton, *Cicero: A Political Biography* (Oxford: Oxford University Press, 1971); Elizabeth Rawson, *Cicero: A Portrait* (London: Allen Lane, 1975); and W. K. Lacey, *Cicero and the End of the Roman Republic* (London: Hodder and Stoughton, 1978). There is much revealing information in the letters he wrote to Atticus: Cicero, *Ad Atticum*, in *Epistole a Attico*, ed. Carlo di Spigno, Latin-Italian ed., 2 vols. (Turin: Unione Tipografico-Editrice, 1998).

78. On this aspect of his career, see Jonathan Powell and Jeremy Paterson, eds., *Cicero: The Advocate* (Oxford: Oxford University Press, 2004).

79. Michael Grant, introduction to Cicero, *On Government*, trans. Michael Grant (Harmondsworth, UK: Penguin, 1993), 7.

80. Ronald Syme, *Ten Studies in Tacitus* (Oxford: Clarendon Press, 1970), 119.

81. Syme, *Ten Studies in Tacitus*, 119. See Janet Coleman, *A History of Political Thought: From Ancient Greece to Early Christianity* (Oxford: Blackwell, 2000), chap. 5.

82. Grant, introduction to Cicero, *On Government*, 6.

83. Cicero, *Ad Atticum*, VIII, 11 (February 49 BCE).

84. Donald Earl, *The Moral and Political Tradition of Rome* (London: Thames and Hudson, 1967), 59, who also notes that the use of the word *dominatio* by Cicero is itself highly critical.

85. Miriam Griffin, introduction to Cicero, *On Duties*, ed. M. T. Griffin and E. M. Atkins (Cambridge: Cambridge University Press, 1991), xii–xiii, notes that *De Officiis* was written after the event of Caesar's assassination to justify what Cicero saw as tyrannicide.

86. Ronald Syme, *The Roman Revolution* (Oxford: Oxford University Press, 1939), 201. The reference is to Tacitus, *Annales*, III, 28.

87. Cicero, *Epistulae ad Familiares*, in *The Letters to His Friends*, trans. W. Glynn Williams, Latin-English ed., 3 vols. (London: William Heinemann, 1927–29), X, 28; see XII, 4.

88. Cicero, *The Philippic Orations of Cicero*, Latin text, edited with English notes by Rev. John Richard King (Oxford: Clarendon Press, 1868). On these see Hartvig Frisch, *Cicero's Fight for the Republic: The Historical Background of Cicero's Philippics* (Copenhagen: Gyldendalske Boghandel, 1946). See Demosthenes, *Orations*, trans. J. H. Vince, English-Greek ed., vol. 1 (Cambridge, MA: Harvard University Press, 1930). In Plutarch's *Lives* (vol. 7), which presents parallel biographies, Demosthenes and Cicero appear together.

89. Neal Wood, *Cicero's Social and Political Thought* (Berkeley: University of California Press, 1988), 8.

90. Cicero, *De officiis*, ed. M. Winterbottom (Oxford: Clarendon, 1994); *On Duties*. On this see Andrew R. Dyck, *A Commentary on Cicero, De Officiis* (Ann Arbor: University of Michigan Press, 1996); and A. A. Long, "Cicero's Politics in *De officiis*," in *From Epicurus to Epictetus: Studies in Hellenistic and Roman Philosophy* (Oxford: Oxford University Press, 2006), chap. 15.

91. Cicero, *On the Ideal Orator (De Oratore)*, ed. and trans. James M. May and Jakob Wisse (New York: Oxford University Press, 2001).

92. For the Latin texts, I have used *De re publica, De legibus*, trans. Clinton Walker Keyes, English-Latin ed. (Cambridge, MA: Harvard University Press, 1928); except for the former, where it is available in *De re publica: Selections*, ed. James E. G. Zetzel (Cambridge: Cambridge University Press, 1995). For the English, see *On the Commonwealth and On the Laws*, ed. and trans. James E. G. Zetzel (Cambridge: Cambridge University Press, 1999), which I have based translations on, although I have also consulted *The Republic* and *The Laws*, trans. Niall Rudd (Oxford: Oxford University Press, 1998).

93. Zetzel, introduction to Cicero, *On the Commonwealth and On the Laws*, xvii; see Griffin, introduction to Cicero, *On Duties*, xxvii.

94. Polybius, *Historion*, VI.

95. A. E. Douglas, *Cicero*, Greece and Rome, New Surveys in the Classics, No. 2 (Oxford: Clarendon Press, 1968), 28.

96. Douglas, *Cicero*, 28.

97. Rawson, *Cicero*, 148.

98. Lacey, *Cicero and the End of the Roman Republic*, 12. On this topic generally, see Roland Poncelet, *Cicéron traducteur de Platon: L'expression de la pensée complexe en latin classique* (Paris: E. de Boccard, 1957); D. M. Jones, "Cicero as Translator," *Bulletin of the Institute of Classical Studies*, no. 6 (1959): 22–34; and J. G. F. Powell, "Cicero's Translations from Greek," in *Cicero the Philosopher*, ed. J. G. F. Powell, 273–300 (Oxford: Clarendon Press, 1995).

99. D. H. Berry, introduction to Cicero, *Defence Speeches*, trans. D. H. Berry (Oxford: Oxford University Press, 2000), xxi.

100. Jones, "Cicero as Translator," 33. This is the view, broadly speaking, of Poncelet, *Cicéron traducteur de Platon*, although Jones is a little more sympathetic.

101. M. L. Clarke, *The Roman Mind: Studies in the History of Thought from Cicero to Marcus Aurelius* (London: Cohen and West, 1956), 44.

102. Zetzel, introduction to Cicero, *On the Commonwealth and On the Laws*, xix. For an argument that it was in part designed to counter the views of Lucretius, see Wood, *Cicero's Social and Political Thought*, 63. See Lucretius, *On the Nature of the Universe*, trans. Ronald Melville (Oxford: Oxford University Press, 1997); and James H. Nichols Jr., *Epicurean Political Philosophy: The De rerum natura of Lucretius* (Ithaca, NY: Cornell University Press, 1972).

103. Cicero, *De officiis*, I, xi, 35; II, i, 3.

104. Cicero, *De officiis*, II, viii, 29; II, xiii, 45; III, i, 4; III, xxi, 83. This characterization and references come from Griffin, introduction to Cicero, *On Duties*, xiii.

105. Malcolm Schofield, "Cicero's Definition of *Res Publica*," in *Cicero the Philosopher*, ed. Powell, 66–67.

106. Stockton, *Cicero*, 306.

107. Griffin, introduction to Cicero, *On Duties*, xxvii.

108. Cicero, *De legibus*, I, v, 15; III, ii, 4. On this, see Andrew R. Dyck, *A Commentary on Cicero, De Legibus* (Ann Arbor: University of Michigan Press, 2004).

109. P. L. Schmidt, "The Original Version of the *De re publica* and the *De legibus*," in *Cicero's Republic*, ed. J. G. F. Powell and J. A. North, 7–16 (London: Institute of Classical Studies, 2001).

110. Cicero, *De re publica*, I, xxv, 39–40. The missing text, enclosed in {}, is filled in from the Keyes edition. See *De re publica*, I, xxxii, 48, and the discussion at II, xxxiii, 44–45. On this definition, see Zetzel, "Commentary," in *De re publica: Selections*, 127–29; Griffin and Atkins, "Notes on Translation," in *On Duties*, xliv; and Schofield, "Cicero's Definition of *Res Publica*," in *Cicero the Philosopher*, ed. Powell, 63–83.

111. See J. A. S. Evans, "'Res Publica Restituta': A Modern Illusion?" in *Polis and Imperium: Studies in Honour of Edward Togo Salmon*, ed. J. A. S. Evans, 280–84 (Toronto: Hakkert, 1974).

112. Wood, *Cicero's Social and Political Thought*, 125. That said, Cicero does use the phrase *"status civitatis"* (*De re publica*, I, xx, 33); and *"statum rei publicae"* (*In Catilinam*, I.3, in *Catilinarians*, ed. Andrew R. Dyck [Cambridge: Cambridge University Press, 2008]) or *"rei publicae statum"* (*De lege agraria*, in *The Speeches*, Latin-English ed., trans. John Henry Freese [London: William Heinemann, 1930], II, iii, 8) to mean the state or condition of the *civitas* or *res publica*.

113. Wood, *Cicero's Social and Political Thought*, 120.

114. Cicero, *De officiis*, I, xvi, 50; see 51–53.

115. Cicero, *De officiis*, I, xvii, 54–55. For a discussion, see Dyck, *Commentary on Cicero, De Officiis*, 172–74.

116. Cicero, *Ad Atticum*, I, 16 (from May 61 BCE).

117. Cicero, *Ad Atticum*, II, 1 (from June 60 BCE).

118. This is borne out by the various lexicons in existence. I have consulted H. Merguet, *Lexikon zu den Reden des Cicero*, 3 vols. (Jena, Germany: Gustav Fischer, 1877–84); H. Merguet, *Lexikon zu den Philosophiscen Schriften des Cicero*, 3 vols. (Jena, Germany: Gustav Fischer, 1887–94); H. Merguet, *Handlexicon zu Cicero* (Leipzig, Germany: Theodor Weicher, 1905); William Abbott Oldfather, Howard Vernon Canter, and Kenneth Morgan Abbott, *Index Verborum Ciceronis Epistularum* (Urbana: University of Illinois Press, 1938); and Kenneth Morgan Abbott, William Abbott Oldfather, and Howard Vernon Canter, *Index Verborum Ciceronis Rhetorica* (Urbana: University of Illinois Press, 1964). There is equally no discussion in J. Helloegouarc'h, *Le vocabulaire latin des relations et des parties politiques sous la république*, 2nd ed. (Paris: Société d'Édition "Les Belles Lettres," 1972).

119. Cicero, *Oratio Philippica*, II, xl, 102–3. In *Philippic Orations*, 94, there is no commentary on this passage, and *territorium* does not appear in the index. In *Philippics I–II*, ed. John T. Ramsey (Cambridge: Cambridge University Press, 2003), 312, there is a brief note on the political, but not conceptual, context.

120. Indeed, the translation in Cicero, *Second Philippic Oration*, ed. W. K. Lacey,

(Warminster, UK: Aris and Phillips, 1986), does render *territorium* simply as "lands." In his lexicons, Merguet suggests *Gebiet*, or "region."

121. Varro, *De Lingua Latina*, in *On the Latin Language*, trans. Roland G. Kent, Latin-English ed., 2 vols. (London: William Heinemann, 1938), V, iv, 21.

122. Drummond and Nelson, *Western Frontiers of Imperial Rome*, 89, 246. See Robert K. Sherk, "Roman Geographical Exploration and Military Maps," *Aufstieg und Niedergang der Römischen Welt* 2 (1974): 553–55; J. J. Wilkes, *Dalmatia* (London: Routledge and Kegan Paul, 1969), 109, 111; and M. Rostovtzeff, *The Social and Economic History of the Roman Empire*, 2nd ed., 2 vols. (Oxford: Clarendon Press, 1957), 1:244–46, which note that inscriptions substantiate this.

123. Dyson, *Community and Society in Roman Italy*, 23; and Wilkes, *Dalmatia*, 391. For discussions, see Rostovtzeff, *Social and Economic History of the Roman Empire*, 1:232; and Karlheinz Dietz and Thomas Fischer, *Die Römer in Regensberg* (Regensberg, Germany: Friedrich Pustet, 1996), 142–44, on the use of the phrase *territorium contributum* for the military.

124. Seneca, *De Consolatione ad Marciam*, XVII, in *Moral Essays*, trans. John W. Basore, Latin-English ed., 3 vols. (London: William Heinemann, 1932), 2:2–97. See *Concordantiae Senecanae*, ed. R. Busa and A. Zampolli, 2 vols. (Hildesheim, Germany: Georg Olms, 1975).

125. Pliny the Elder, *Naturalis Historia / Natural History*, Latin-English ed., trans. H. Rackham, W. H. S. Jones, and D. E. Eichholz, 10 vols. (London: William Heinemann, 1938–63), XXIX, xxxiv, 106.

126. Cicero, *De re publica*, II, iv, 8.

127. Cicero, *De re publica*, III, ix, 15. This story is also found in Plutarch, *Moralia*, 210e, trans. Frank Cole Babbitt, Greek-Latin ed., 14 vols. (London: William Heinemann, 1931).

128. For a helpful discussion of land, which notes that there were no real developments in agricultural methods, see A. H. M. Jones, *The Later Roman Empire, 284–602: A Social, Economic, and Administrative Survey*, paperback ed., 2 vols. (1964; repr., Oxford: Basil Blackwell, 1986), vol. 2, chap. 20 (citations refer to the 1986 edition).

129. Jonathan Powell and Niall Rudd, "Explanatory Notes," in Cicero, *Republic* and *The Laws*, 187. See also Wood, *Cicero's Social and Political Thought*, 35.

130. Cicero, *De re publica*, II, xiv, 26.

131. Cicero, *De re publica*, II, xviii, 33.

132. Cicero, *Cato maior de senectute*, ed. J. G. F. Powell (Cambridge: Cambridge University Press, 1988), IV, 11; see *In Catilinam*, II, 5.

133. Cicero, *De re publica*, III, xv, 24.

134. Cicero, *Pro Murena*, 22, in Cicero, *Works*, 29 vols., vol. 10, trans. C. Macdonald (Cambridge, MA: Harvard University Press, 1977).

135. The extensions to Cicero's rather sparse prose are provided by Grant's translation in *On Government*, 120.

136. Cicero, *De lege agraria*, II, xxi, 56. On this, see E. G. Hardy, "The Policy of the Rullan Proposal in 63 B.C.," *Journal of Philology* 32, no. 64 (1913): 228–60; and Evan T. Sage, "Cicero and the Agrarian Proposals of 63 B.C.," *Classical Journal* 16 (1921): 230–36. Cicero's long-standing opposition to agrarian reform is discussed in Thomas N.

Mitchell, *Cicero: The Senior Statesman* (New Haven, CT: Yale University Press, 1991). On previous attempts at land reform, see the accounts of the Gracchi in Plutarch, *Lives*, vol. 10; Appian, *Civil Wars*, I, 9–17; and H. H. Scullard, *From the Gracchi to Nero: A History of Rome from 133 B.C. to A.D. 68*, 5th ed. (London: Methuen, 1982), chap. 2.

137. See "Lex Agraria," in *Roman Statutes*, ed. M. H. Crawford, 2 vols., 1:113–80 (London: Institute of Classical Studies, 1996).

138. Berry, introduction to Cicero, *Defence Speeches*, xvi.

139. Cicero, *De lege agraria*, II, xxv, 67.

140. See, for example, Cicero, *De lege agraria*, I, iv, 10–11; II, xxv, 66; II, xxix–xxxi, 79–85; II, xxxii, 88; III, ii, 8; III, iv, 15–16. On the word *ager*, see Varro, *De Lingua Latina*, V, vi, 34.

141. Cicero, *De legibus*, I, xxi, 55, citing Tabula, VII, 2–5. For a reconstruction of the text of the tables, see E. H. Warmington, ed., *Remains of Old Latin*, 4 vols. (London: William Heinemann, 1935), 3:424–515; and Crawford, *Roman Statutes*, 2:555–721; and for a discussion, see André Magdelain, *La loi à Rome: Histoire d'un concept* (Paris: Société d'Édition les Belles Lettres, 1978); Cornell, *Beginnings of Rome*, chap. 11. For Livy, I have used *Ab urbe condita*, trans. B. O. Foster, Frank Gardner Moore, Evan T. Sage, and Alfred C. Schlesinger, 14 vols. (London: William Heinemann, 1919–59). I have made use of the Penguin translations: *The Early History of Rome: Books I–V of The History of Rome from Its Foundation*, trans. Aubrey de Sélincourt (Harmondsworth, UK: Penguin, 1971); *Rome and Italy: Books VI–X of The History of Rome from Its Foundation*, trans. Betty Radice (Harmondsworth, UK: Penguin, 1982); *The War with Hannibal: Books XXI–XXX of The History of Rome from Its Foundation*, trans. Aubrey de Sélincourt (Harmondsworth, UK: Penguin, 1965); and *Rome and the Mediterranean: Books XXXI–XLV of The History of Rome from Its Foundation*, trans. Henry Bettenson (Harmondsworth, UK: Penguin, 1976).

142. Clinton Walker Keyes, note to Cicero, *De legibus*, 360n1; see Dyck, *Commentary on Cicero, De Legibus*, 214–15.

143. Cicero, *De re publica*, VI, xix, 20.

144. Cicero, *Pro Archia Poeta*, in *The Speeches*, Latin-English ed., trans. N. H. Watts (London: William Heinemann, 1923), X, 23.

145. Cicero, *Pro L. Cornelio Balbo Oratio*, in *The Speeches*, Latin-English ed., trans. R. Gardner (London: William Heinemann, 1958), 64. See also *Pro Milone*, in *Works*, 14:98: "*fines imperi populi Romani.*"

146. See, for instance, Cicero, *Pro Sexto Roscio Amerino*, in *Speeches*, 1930, XVI, 47.

147. Cicero, *Pro M. Marcello Oratio*, 5; 8.

148. Cicero, *De re publica*, II, iii, 6; II, v, 10.

149. Cicero, *De re publica*, II, vi, 11. There is also the description of Syracuse: *In Verram*, in *The Verrine Orations*, trans. L. H. G. Greenwood, Latin-English ed. (London: William Heinemann, 1949), II, v, 97–98.

150. Cicero, *De re publica*, II, xi, 21–22.

151. Mattern, *Rome and the Enemy*, 27.

152. All of these histories have, however, been drawn on for points of detail when appropriate. See also F. W. Walbank, "The Geography of Polybius," *Classica et Mediaevalia* 9 (1948): 155–82.

153. Tacitus, *Agricola, Germania, Dialogus*, trans. M. Hutton and W. Peterson Latin-English ed. (Cambridge, MA: Harvard University Press, 1970). I have also consulted the Benario edition cited above, and largely used the translations in Tacitus, *The Agricola and the Germania*, trans. H. Mattingly and revised by S. A. Handford (Harmondsworth, UK: Penguin, 1970). The likely full title is provided in "Introduction to *Germania*," in Tacitus, *Agricola, Germania, Dialogus*, 119.

154. As Ronald Syme, *Tacitus*, 2 vols. (Oxford: Oxford University Press, 1958), 1:253n1, suggests, both titles are fairly suspect, and *Ab Excessu Divi Augusti* may have been the title or subtitle of the latter.

155. Syme, *Tacitus*, 1:126.

156. Cary, *Geographic Background of Greek and Roman History*, 315.

157. Jürgen Rapsch, Dietmar Najock, and Adam Nowosad, *Concordantia in Corpus Sallustianum* (Hildesheim, Germany: Olms-Weidmann, 1991); David W. Packard, *A Concordance to Livy*, 4 vols. (Cambridge, MA: Harvard University Press, 1968); A. Gerber and A. Greef, *Lexicon Taciteum*, 2 vols. (Leipzig, Germany: B. G. Teubneri, 1903); and D. R. Blackman and G. G. Betts, eds., *Concordantia Tacitea: A Concordance to Tacitus*, 2 vols. (Hildesheim, Germany: Olms-Weidmann, 1986).

158. See D. C. Earl, *The Political Thought of Sallust* (Cambridge: Cambridge University Press, 1961); and for a general assessment, Ronald Syme, *Sallust* (Berkeley: University of California Press, 1964).

159. Syme, *Ten Studies in Tacitus*, 120.

160. Sallust, *Bellum Iugurthinum*, in Sallust, *Works*, trans. J. C. Rolfe, English-Latin ed. (Cambridge, MA: Harvard University Press, 1921), XVII.

161. Sallust, *Bellum Iugurthinum*, XVII.

162. Strabo, *The Geography of Strabo*, trans. Horace Leonard Jones, Greek-English ed., 8 vols. (London: William Heinemann, 1924), XVII, 3, 1; Herodotus, *Histories*, trans. A. D. Godley, Greek-English ed., 4 vols. (London: William Heinemann, 1925), 2:16; and Varro, *De Lingua Latina*, V, v, 31. See Rolfe's note in Sallust, *Works*, 170. Pliny the Elder suggests a threefold division: *Naturalis Historia*, III, i, 3. Pliny is, however, generally very weak on geography. See John F. Healey, introduction to Pliny the Elder, *Natural History: A Selection*, trans. John F. Healy, ix–xl (Harmondsworth, UK: Penguin, 2004), xxi.

163. S. A. Handford, "Life and Writings of Sallust," in Sallust, *The Jugurthine War and The Conspiracy of Catiline*, trans. S. A. Handford (Harmondsworth, UK: Penguin, 1963), 10.

164. Sallust, *Bellum Iugurthinum*, LXXIX.

165. Sallust, *Bellum Iugurthinum*, LXXIX.

166. See, for example, Livy, *Ab urbe condita*, VII, 23; IX, 3; XXXIII, 6.

167. Livy, *Ab urbe condita*, IV, 48.

168. Livy, *Ab urbe condita*, IV, 47.

169. Livy, *Ab urbe condita*, VI, 16.

170. Livy, *Ab urbe condita*, VI, 37–38.

171. See, for example, Livy, *Ab urbe condita*, XXXIV, 48.

172. Livy, *Ab urbe condita*, XXIX, 5.

173. Livy, *Ab urbe condita*, XXVII, 31.

174. Livy, *Ab urbe condita*, XXVII, 32.

175. For example, Livy, *Ab urbe condita*, XXXVIII, 48.

176. Livy, *Ab urbe condita*, XXXVIII, 48; see XXXVII, 54.

177. Livy, *Ab urbe condita*, II, 49.

178. Livy, *Ab urbe condita*, XXVII, 5.

179. Livy, *Ab urbe condita*, XXII, 14; see XXXV, 48.

180. Livy, *Ab urbe condita*, I, 33.

181. Livy, *Ab urbe condita*, XXXVIII, 40.

182. Livy, *Ab urbe condita*, XLV, 29.

183. Livy, *Ab urbe condita*, I, 6–7.

184. Plutarch, *Romulus*, IX–X. The claim it was by a companion comes from Ovid, *Fasti*, IV, 843, trans. James George Frazer, Latin-English ed., 2nd ed. (Cambridge, MA: Harvard University Press, 1989), whose story also slightly differs in other respects. For a discussion, see Joseph Rykwert, *The Idea of a Town: The Anthropology of Urban Form in Rome, Italy and the Ancient World* (London: Faber and Faber, 1976), 27–29, 44–45.

185. Livy, *Ab urbe condita*, I, 9–13; and Cicero *De re publica*, II, vii, 12. See Melissa M. Matthes, *The Rape of Lucretia and the Founding of Republics: Readings in Livy, Machiavelli, and Rousseau* (University Park: Pennsylvania State University Press, 2000), esp. 42–43, 49; and Robin May Schott, "Sexual Violence, Sacrifice, and Narratives of Political Origins," in *Birth, Death, and Femininity: Philosophies of Embodiment*, ed. Robin May Schott, 25–48 (Bloomington: Indiana University Press, 2010).

186. Livy, *Ab urbe condita*, XXXIV, 62.

187. Syme, *Ten Studies in Tacitus*, 18.

188. F. R. D. Goodyear, *Tacitus*, Greece and Rome, New Surveys in the Classics, No. 4 (Oxford: Clarendon Press, 1970), 36–37; generally on his style see Syme, *Tacitus*, vol. 1, chap. 26, and "Appendix F: Style and Words," 2:711–45; and Benedetto Fontana, "Tacitus on Empire and Republic," *History of Political Thought* 14 (1993): 27–40.

189. Tacitus, *Historiae*, I, 70.

190. Tacitus, *Annales*, I, 38; see XII, 33.

191. Tacitus, *Annales*, I, 60.

192. Tacitus, *Historiae*, III, 41.

193. Tacitus, *Annales*, IV, 32.

194. Tacitus, *Annales*, IV, 32.

195. Tacitus, *Annales*, IV, 33.

196. Tacitus, *Historiae*, III, 15.

197. Tacitus, *Annales*, XII, 31–32. See also XIII, 55, in which he notes that "then the Ampsivarii occupied the land [*agros*]."

198. Tacitus, *Annales*, II, 5–6.

199. Tacitus, *Annales*, II, 56.

200. Tacitus, *Annales*, II, 14.

201. Tacitus, *Historiae*, III, 9.

202. Tacitus, *Historiae*, III, 8.

203. Tacitus, *Germania*, 35.

204. Tacitus, *Annales*, IV, 5.

205. Tacitus, *Annales*, I, 51.

206. Tacitus, *Annales*, III, 62.

207. Tacitus, *Historiae*, III, 60.

208. Tacitus, *Annales*, II, 64.

209. Tacitus, *Annales*, XI, 24.

210. Tacitus, *Annales*, XII, 23; see also Seneca, *De Brevitate Vitae*, 13 (in *Moral Essays*, vol. 2), which suggests that it was the expansion of Italian, not provincial, lands (*agro*).

211. Plutarch, *Romulus*, XI; and Livy, *Ab urbe condita*, I, 44. See Thomas Wiedemann, *Cicero and the End of the Roman Republic* (London: British Classical Press, 1994), 92.

212. Aulus Gellius, *Noctes Atticae*, in *Attic Nights*, trans. John C. Rolfe, Latin-English ed., 3 vols. (London: William Heinemann, 1927), XIII, xiv.

213. J. S. Richardson, "*Imperium Romanum*: Empire and the Language of Power," *Journal of Roman Studies* 18 (1991): 3.

214. On Augustus's expansion of the *pomerium*, see Pat Southern, *Augustus* (London: Routledge, 1998), 169; and Ronald Syme, *Historia Augusta Papers* (Oxford: Clarendon Press, 1983), esp. 137.

215. As well as Cicero, see Plutarch, *Romulus*, XI; and Virgil, *Aeneid*, V, 755–56. For a discussion, see Rykwert, *Idea of a Town*, 65–66. Rykwert, following Plutarch, notes that the plow was lifted when the area of the gates was reached, so they were not sacred, and allowed impure goods to enter or leave the city, thus giving the word *porta* from *portare*, "to carry."

216. Tacitus, *Annales*, XII, 24. On these places, see Lawrence Richardson Jr., *A New Topographical Dictionary of Ancient Rome* (Baltimore: Johns Hopkins University Press, 1992).

217. Andrew Wallace-Hadrill, *Augustan Rome* (London: Bristol Classical Press, 1993), 23.

218. Boatwright, Gargola, and Talbert, *Romans*, 62.

219. Tacitus, *Germania*, 29, 32.

220. Tacitus, *Agricola*, 23.

221. Tacitus, *Historiae*, III, 5.

222. Tacitus, *Annales*, I, 60.

223. Tacitus, *Agricola*, 33: *finem Britanniae*.

224. Tacitus, *Annales*, I, 60.

225. Tacitus, *Historiae*, IV, 64.

226. Tacitus, *Historiae*, V, 1.

227. Tacitus, *Historiae*, IV, 37.

228. Tacitus, *Annales*, XIV, 23.

229. Tacitus, *Annales*, I, 9.

230. Tacitus, *Germania*, 45.

231. Livy, *Ab urbe condita*, XXII, 12; XXII, 15; XXXI, 24; XXXII, 13; XXXIV, 28; XLI, 14.

232. Tacitus, *Annales*, I, 50.

233. Tacitus, *Annales*, II, 7.

234. Tacitus, *Germania*, 29.

235. Whittaker, *Frontiers of the Roman Empire*, 46. The Penguin translation renders *limes* as "frontier line of defence."

236. Tacitus, *Agricola*, 41. On this passage, see Benjamin Isaac, *The Limits of Empire: The Roman Army in the East*, rev. ed. (Oxford: Clarendon Press, 1992), 397; his "The Meaning of the Terms *Limes* and *Limitanei*," *Journal of Roman Studies* 78 (1988): 128; and Lintott, *Imperium Romanum*, 42.

237. Werner Eck, *The Age of Augustus*, trans. Deborah Lucas Schneider (Oxford: Blackwell, 2003), 15.

238. David Shotter, *Augustus Caesar* (London: Routledge, 1991), 136; and Syme, *Roman Revolution*, 217.

239. Shotter, *Rome and Her Empire*, 137.

240. Cassius Dio, *Roman History*, trans. Herbert Baldwin Foster, Greek-English ed., 9 vols. (London: William Heinemann, 1924), L, 1.

241. *Res Gestae Divi Augusti: The Achievements of the Divine Augustus*, with an introduction and commentary by P. A. Brunt and J. M. Moore, English-Latin ed. (Oxford: Oxford University Press, 1967), 34. For this text, I have also consulted the version in Velleius Paterculus, *Compendium of Roman History*, and *Res Gestae Divi Augusti*, trans. Frederick W. Shipley, English-Latin ed. (London: William Heinemann, 1924).

242. Ronald Syme, *The Augustan Aristocracy* (Oxford: Clarendon, 1986), 2; Shotter, *Rome and Her Empire*, 147.

243. Tacitus, *Annales*, I, 9.

244. Meier, *Caesar*, 496.

245. Southern, *Augustus*, 21; Wells, *Roman Empire*, 12; and Eck, *Age of Augustus*, 10.

246. Gardner, introduction to Caesar, *Civil War*, 24.

247. H. Galsterer, "A Man, a Book, and a Method: Sir Ronald Syme's *Roman Revolution after Fifty Years*," in *Between Republic and Empire: Interpretations of Augustus and His Principate*, ed. Kurt A. Raaflaub and Mark Toher, 1–20 (Berkeley: University of California Press, 1990), 15.

248. Cassius Dio, *Roman History*, LII, 41.

249. Wallace-Hadrill, *Augustan Rome*, 85–86.

250. Chester G. Starr, *Civilization and the Caesars: The Intellectual Revolution in the Roman Empire* (Ithaca, NY: Cornell University Press, 1954), 3. This book is a study of the second period.

251. "Key to Technical Terms," in Tacitus, *Annals of Imperial Rome*, 401.

252. Suetonius, *Divus Augustus*, ed. John M. Carter (Bristol: Bristol Classical Press, 1982), II, 7; translated in *The Twelve Caesars*. See Shotter, *Augustus Caesar*, 27; Ronald Syme, "Imperator Caesar: A Study in Nomenclature," *Historia* 7 (1958): 172–88; and Paul Zanker, *The Power of Images in the Age of Augustus*, trans. Alan Shapiro (Ann Arbor: University of Michigan Press, 1990), 98.

253. Fergus Millar, "State and Subject: The Impact of Monarchy," in *Caesar Augustus: Seven Aspects*, ed. Fergus Millar and Erich Segal, 37–60 (Oxford: Clarendon, 1984), 37.

254. Edward Gibbon, *The History of the Decline and Fall of the Roman Empire*, ed. David Womersley, 3 vols. (Harmondsworth, UK: Penguin, 1985), 1:95. For a helpful

discussion of Gibbon and his motives, see Karen O'Brien, *Narratives of Enlightenment: Cosmopolitanism from Voltaire to Gibbon* (Cambridge: Cambridge University Press, 1997), chap. 6. For an exhaustive study of Gibbon's role in eighteenth-century European culture, see J. G. A. Pocock, *Barbarism and Religion*, 4 vols. (Cambridge: Cambridge University Press, 1999–2005).

255. Livy, *Ab urbe condita*, XXVII, 19. For its use in the republic, see R. Combès, *Imperator: Recherches sur l'emploi et la signification du titre d'Imperator dans la Rome républicaine* (Paris: Presses Universitaires de France, 1966).

256. *Res Gestae Divi Augusti*, 4.

257. For an exhaustive treatment, see Fergus Millar, *The Emperor in the Roman World (31 BC–AD 337)* (London: Duckworth, 1977).

258. Wells, *Roman Empire*, 7. See *Res Gestae Divi Augusti*, 30.

259. Eck, *Age of Augustus*, 124.

260. Shotter, *Augustus Caesar*, 24. On these issues, see also André Magdelain, *Auctoritas Principis* (Paris: Société d'Édition "Les Belles Lettres," 1947); and W. H. Gross, "The Propaganda of an Unpopular Ideology," in *The Age of Augustus: Interdisciplinary Conference Held at Brown University, April 30–May 2, 1982*, ed. Rolf Winkes, 29–45 (Louvain-la-Neuve, Belgium: Archaeologica Transatlantica, 1985).

261. A thorough study is John Richardson, *The Language of Empire: Rome and the Idea of Empire from the Third Century BC to the Second Century AD* (Cambridge: Cambridge University Press, 2008). He is unfortunately only occasionally as attentive to the nuances of words relating to territory as those of *provincia* and *imperium*. See, for example, 27–30, 74–79.

262. For a helpful discussion, see Carter, introduction to Suetonius, *Divus Augustus*, 10–11.

263. Justinian, *Digest*, I, iv, 1. *The Digest of Justinian*, ed. Theodor Mommsen with Paul Krueger, translation ed. Alan Watson, Latin-English ed., 4 vols. (Philadelphia: University of Pennsylvania Press, 1985).

264. Southern, *Augustus*, 1; and Nicholas Purcell, "Maps, Lists, Money, Order and Power," *Journal of Roman Studies* 80 (1990): 178. For a discussion, see A. H. M. Jones, "The *Imperium* of Augustus," *Journal of Roman Studies* 41, nos. 1–2 (1951): 112–19.

265. Richardson, "*Imperium Romanum*," 1; and Lintott, *Imperium Romanum*, 22. The term originates with Sallust; see Richardson, *Language of Empire*, 98–99.

266. Richardson, *Language of Empire*, 145.

267. Michael Grant, *From* Imperium *to* Auctoritas: *A Historical Study of Aes Coinage in the Roman Empire, 49 B.C.–A.D. 14* (Cambridge: Cambridge University Press, 1946).

268. Claude Nicolet, "L'Empire romain," *Ktèma* 8 (1983): 166.

269. Southern, *Augustus*, 104–5; see Magdelain, *Auctoritas Principis*; and Grant, *From* Imperium *to* Auctoritas.

270. Juvenal, *Satire*, X, 81. On gifts to the community and city, see Paul Veyne, *Le pain et le cirque: Sociologie historique d'un pluralisme politique* (Paris: Éditions du Seuil, 1976); and Paul Veyne, *Bread and Circuses: Historical Sociology and Political Pluralism*, abridged by Oswyn Murray, trans. Brian Pearce (London: Allen Lane, 1990).

271. *Res Gestae Divi Augusti*, prelude. This alone renders the assessment of Sue-

tonius that "Augustus never wantonly invaded any country, and felt no temptation to increase the *imperium* or enhance his military glory" (*Divus Augustus*, II, 21) highly dubious. See E. S. Gruen, "The Imperial Policy of Augustus," in *Between Republic and Empire: Interpretations of Augustus and His Principate*, ed. Kurt A. Raaflaub and Mark Toher, 395–416 (Berkeley: University of California Press, 1990), 410.

272. For a discussion, see Paul Rehak, *Imperium and Cosmos: Augustus and the Northern Campus Martius* (Madison: University of Wisconsin Press, 2006), 54–58.

273. For their unreliability, see Zvi Yavetz, "The *Res Gestae* and Augustus' Public Image," in *Caesar Augustus*, ed. Millar and Segal, 1–36, 22, who notes that despite this being the scholarly consensus, they have nonetheless set the terms of scholarly engagement.

274. P. A. Brunt, *Roman Imperial Themes* (Oxford: Clarendon, 1990), 433. It is worth noting that Roman leaders could only have a "triumph" if they gained new lands, not merely recapturing ones previously lost. For a general discussion, see Mary Beard, *The Roman Triumph* (Cambridge, MA: Harvard University Press, 2007).

275. *Res Gestae Divi Augusti*, 26, 30.

276. Wells, *Roman Empire*, 77; Eck, *Age of Augustus*, 94; and Nicolaus of Damascus, *Life of Augustus*, ed. Jane Bellemore, Greek-English ed. (Bristol, UK: Bristol Classical Press, 1984), 2.

277. Virgil, *Aeneid*, I, 278–79. On the imperial themes of his work, see Philip R. Hardie, *Virgil's Aeneid: Cosmos and Imperium* (Oxford: Clarendon Press, 1986); Richard F. Thomas, *Virgil and the Augustan Reception* (Cambridge: Cambridge University Press, 2001); and Eve Adler, *Vergil's Empire: Political Thought in the Aeneid* (Lanham: Rowman and Littlefield, 2003). On his language, see Monroe Nichols Wetmore, ed., *Index Verborum Vergilianus* (Hildesheim, Germany: Georg Olms, 1979).

278. Virgil, *Aeneid*, I, 286–87; see VI, 791–95; and more generally his *Georgics*, in *Works*.

279. Virgil, *Aeneid*, I, 236.

280. Horace, *Odes*, IV, 15, 13–16, in *The Odes and Epodes*, trans. C. E. Bennett, Latin-English ed. (Cambridge, MA: Harvard University Press, 1988).

281. Ovid, *Fasti*, II, xxiii, 683–84. This work is a calendar of Roman festivals, and here comes in the context of his praise of Terminus, the god of boundaries, *finis*. The date is February 23. On Terminus, see Rykwert, *Idea of a Town*, 106–17; and Dyson, *Community and Society in Roman Italy*, 122–23. Ovid makes it clear that this means a division of lands into separate realms: "This is your land [*ager*], and that is his" (II, xxiii, 678). See Thomas Habinek, "Ovid and Empire," in *The Cambridge Companion to Ovid*, ed. Philip Hardie, 46–61 (Cambridge: Cambridge University Press, 2002); and esp. Richardson, *Language of Empire*, 132–34, which suggests that he provides the first clear indication of *imperium* as a "territorial empire" (133). On the more general politics of poetry at this time, see Michèle Lowrie, *Writing, Performance, and Authority in Augustan Rome* (Oxford: Oxford University Press, 2009).

282. P. Aelius Aristides, *Orationes*, XXVI, 81, in *Quae supersunt omnia*, ed. Bruno Keil (Berolini: Weidmannos, 1898); and P. Aelius Aristides, *The Complete Works*, trans. Charles A. Behr, 2 vols. (Leiden, Netherlands: E. J. Brill, 1981–86), 2:90.

283. Aristides, *Orationes*, XXVI, 28 (2:79). The quotation is from Homer, *Iliad*,

trans. A. T. Murray, Greek-English ed., 2 vols. (London: William Heinemann, 1924), IX, 476.

284. Aristides, *Orationes*, XXVI, 61 (2:86).

285. Lintott, *Imperium Romanum*, 12; see 2.

286. Southern, *Augustus*, 163–64. There were thus elements of ignorance, willful disregard, and assumptive arrogance in the Roman position. See Brunt, *Roman Imperial Themes*, 433; and more generally Joseph Vogt, *Orbis: Ausgewählte Schriften zur Geschichte des Altertums* (Freiburg, Germany: Herder, 1960), chap. 7.

287. Boatwright, Gargola, and Talbert, *Romans*, 255.

288. Beard, *Roman Triumph*, 20.

289. Lintott, *Imperium Romanum*, 2; see L. Wilkinson, *The Roman Experience* (New York: Alfred A. Knopf, 1974), 77–78. See, for example, the one struck in 70 BCE after the Social War, in Wiedemann, *Cicero and the End of the Roman Republic*, 30.

290. Reproduced in Hopkins, *Conquerors and Slaves*, facing 223; see also Antony R. Birley, *Hadrian: The Restless Emperor* (London: Routledge, 1997), 242, 258.

291. Reproduced in Boatwright, Gargola, and Talbert, *Romans*, 331.

292. Ann L. Kuttner, *Dynasty and Empire in the Age of Augustus: The Case of the Boscoreale Cups* (Berkeley: University of California Press, 1995), 86–87. On his imperialism generally, see Wells, *German Policy of Augustus*, chap. 1.

293. Rehak, *Imperium and Cosmos*, 92.

294. Nicolet, *Space, Geography, and Politics*, figs. 6, 7, and 8.

295. Kuttner, *Dynasty and Empire in the Age of Augustus*, 90.

296. This is a phrase with a history. Livy has Hannibal characterize Rome as *caput orbis terrarum*—the capital of the known world (*Ab urbe condita*, XXI, 30). For a general discussion of this theme in Greek and Roman thought, see James S. Romm, *The Edges of the Earth in Ancient Thought: Geography, Exploration, and Fiction* (Princeton, NJ: Princeton University Press, 1992).

297. See E. Badian, *Roman Imperialism in the Late Republic* (Ithaca, NY: Cornell University Press, 1968); and Harris, *War and Imperialism in Republican Rome*.

298. Wells, *German Policy of Augustus*, 5; and Shotter, *Augustus Caesar*, 53.

299. Wells, *German Policy of Augustus*, 249.

300. E. A. Thompson, *Romans and Barbarians: The Decline of the Western Empire* (Madison: University of Wisconsin Press, 1982), 7. See Gruen, "Imperial Policy of Augustus," 407–8. Historical sources include Tacitus, *Annales*, I, 57–61; Velleius Paterculus, *Compendium of Roman History*, II, cxvii–cxix; and Cassius Dio, *Roman History*, LVI, 18–22.

301. Suetonius, *Divus Augustus*, II, 23.

302. Tacitus, *Annales*, I, 3.

303. Tacitus, *Annales*, I, 11; see Cassius Dio, *Roman History*, LVI, 33; and Drummond and Nelson, *Western Frontiers of Imperial Rome*, 20–21. For a discussion, see Josiah Ober, "Tiberius and the Political Testament of Augustus," *Historia: Zeitschrift für Alte Geschichte* 31 (1982): 306–28.

304. Gruen, "Imperial Policy of Augustus," 409–10.

305. Wells, *Roman Empire*, 78; and Isaac, *Limits of Empire*, 397.

306. Wells, *German Policy of Augustus*, 245; see Maxfield, "Mainland Europe," 137.

307. Drummond and Nelson, *Western Frontiers of Imperial Rome*, 21.

308. Wells, *German Policy of Augustus*, 246.

309. See Wells, *Roman Empire*, 69, 74–75, 125, 246, 248; Gruen, "Imperial Policy of Augustus," 408; and Eck, *Age of Augustus*, 102–3.

310. Gibbon, *History of the Decline and Fall of the Roman Empire*, 1:32.

311. For a general survey, see Derek Williams, *The Reach of Rome: A History of the Roman Imperial Frontier, 1st–5th Centuries AD* (London: Constable, 1996).

312. Though see Stephen L. Dyson, *The Creation of the Roman Frontier* (Princeton, NJ: Princeton University Press, 1985), which argues from the perspective of the republic. For some cautions, see Isaac, *Limits of Empire*, 373.

313. See, for example, Polybius, *Historion*, III, 54, 2, trans. W. R. Paton, Greek-English ed., 6 vols. (London: William Heinemann, 1922–27), where they are described as an *acropolis*; and Herodian, *Regnum post Marcum*, trans. C. R. Whittaker, Greek-English ed. (London: William Heinemann, 1969), II, 11, 8, which describes them as a barricade. These references are provided by Whittaker, *Frontiers of the Roman Empire*, 26. For a later period, see Neil Christie, "The Alps as Frontier (A.D. 168–774)," *Journal of Roman Archaeology* 4 (1991): 410–30.

314. Cicero, *De Provinciis Consularibus*, in *Speeches*, XIV, 34.

315. Livy, *Ab urbe condita*, V, 34.

316. Livy, *Ab urbe condita*, XXI, 30.

317. Livy, *Ab urbe condita*, XXI, 35.

318. Whittaker, *Frontiers of the Roman Empire*, 26; and Henry Francis Pelham, "The Roman Frontier System," in *Essays by Henry Francis Pelham*, ed. F. Haverfield (Oxford: Clarendon, 1911), 164–65.

319. Isaac, *Limits of Empire*, 397. On the law concerning property boundaries, see Alan Watson, *The Law of Property in the Late Republic* (Oxford: Clarendon Press, 1968), chap. 5.

320. See Theodore Mommsen, *The Provinces of the Roman Empire: The European Provinces*, ed. T. Robert S. Broughton (Chicago: University of Chicago Press, 1968).

321. Lintott, *Imperium Romanum*, 23.

322. Lintott, *Imperium Romanum*, 23–24; see Isaac, *Limits of Empire*, 397.

323. Lintott, *Imperium Romanum*, 27.

324. Dilke, *Roman Land Surveyors*, 99.

325. Dilke, *Roman Land Surveyors*, 15.

326. Dilke, *Roman Land Surveyors*, 98. Generally, see Max Weber, *The Agrarian Sociology of Ancient Civilizations*, trans. R. I. Frank (London: NLB, 1976), chaps. 6 and 7.

327. Roger J. P. Kain and Elizabeth Bagnet, *The Cadastral Map in the Service of the State: A History of Property Mapping* (Chicago: University of Chicago Press, 1992), 1.

328. Polybius, *Historion*, VI, 41–42. He contrasts it to a Greek camp. See Gibbon, *History of the Decline and Fall of the Roman Empire*, 1:44–45.

329. "Balbi ad Celsum exposition et ratio omnium formarum," in F. Bluhme, K. Lachmann, and A. Rudorff, eds., *Die Schriften der römischen Feldmesser*, 2 vols. (Berlin: Georg Reimer, 1848), 1:92; translated in Craige B. Champion, ed., *Roman Imperialism: Readings and Sources* (Oxford: Blackwell, 2004), 305. See generally Sherk, "Roman Geographical Exploration and Military Maps."

330. *Corpus Agrimensorum Romanorum*, ed. C. Thulin (Leipzig, Germany: B. G. Teubner, 1913). A selection can be found in Brian Campbell, *The Writings of the Roman Land Surveyors*, Journal of Roman Studies Monograph No. 9 (London: Society for the Promotion of Roman Studies, 2000), which provides the Latin text, translation, and commentary.

331. Dilke, *Roman Land Surveyors*, 21; and O. A. W. Dilke, *Mathematics and Measurement* (London: British Museum, 1987), 7.

332. Dilke, *Roman Land Surveyors*, 51.

333. Plautus, *Poenulus*, Prologue, 48–99, in M. Accii Plauti, *Comoediae*, ed. J. F. Gronovii, 5 vols. (London: A. J. Valpy, 1829), 3:1258; see Campbell, *Writings of the Roman Land Surveyors*, xlvi.

334. Dilke, *Roman Land Surveyors*, 87.

335. On Sextus Julius Frontinus in this regard, see Dilke, *Roman Land Surveyors*, 40–41. Dilke notes that his "knowledge of land survey was extensive, so that we may suspect he had been a land commissioner" (41). For his wider work, see Frontinus, *The Stratagems* and *The Aqueducts of Rome*, trans. Charles E. Bennett, Latin-English ed. (London: William Heinemann, 1925); and *De Aquaeductu urbis Romae*, ed. R. H. Rodgers (Cambridge: Cambridge University Press, 2004).

336. Frontinus, *De agrorum qualitate*, in Campbell, *Writings of the Roman Land Surveyors*, 2–5, 2, 3.

337. Campbell, *Writings of the Roman Land Surveyors*, xxviii.

338. Frontinus, *De controversiis*, in Campbell, *Writings of the Roman Land Surveyors*, 4–9, 4, 5. On these generally, see Dilke, *Roman Land Surveyors*, 106–8. On territorial jurisdiction more specifically, see also Agennius Urbicus, *De controversiis agrorum*, in Campbell, *Writings of the Roman Land Surveyors*, 17–49. This was probably written in the late fourth or early fifth century.

339. Frontinus, *De controversiis*, 6, 7.

340. Frontinus, *De controversiis*, 6, 7. It is possible that this is a gloss of later date.

341. Jenaro Costas Rodríguez, *Frontini Index* (Hildesheim, Germany: Olms-Weidmann, 1985).

342. Frontinus, *Strategematon*, in *Stratagems* and *Aqueducts of Rome*.

343. Campbell, *Writings of the Roman Land Surveyors*, xxxi.

344. Siculus Flaccus, *De condicionibus agrorum*, in Campbell, *Writings of the Roman Land Surveyors*, 102–33, 104, 105. There is also a helpful Latin-French edition, trans. M. Claval-Lévêque et al. (Naples: Jovere, 1993).

345. Siculus Flaccus, *De condicionibus agrorum*, 104, 105.

346. Siculus Flaccus, *De condicionibus agrorum*, 104, 105.

347. Siculus Flaccus, *De condicionibus agrorum*, 120, 121.

348. Siculus Flaccus, *De condicionibus agrorum*, 104, 105.

349. Siculus Flaccus, *De condicionibus agrorum*, 130, 131.

350. Siculus Flaccus, *De condicionibus agrorum*, 104, 105.

351. Brian Campbell, "Shaping the Rural Environment: Surveyors in Ancient Rome," *Journal of Roman Studies* 86 (1996): 78; and Campbell, *Writings of the Roman Land Surveyors*, xxxvii.

352. Dilke, *Roman Land Surveyors*, 44.

353. Campbell, *Writings of the Roman Land Surveyors*, xxxvii.

354. Olwen Brogan, "The Roman *Limes* in Germany," *Archaeological Journal* 92 (1935): 1; Whittaker, *Frontiers of the Roman Empire*, 200; and Edward N. Luttwak, *The Grand Strategy of the Roman Empire from the First Century A.D. to the Third* (Baltimore: Johns Hopkins University Press, 1976), 19. On military handbooks, see Brian Campbell, "Teach Yourself How to Be a General," *Journal of Roman Studies* 77 (1987): 13–29.

355. *Historia Augusta*, Marcus Aurelius, trans. David Magie, Latin-English ed., 3 vols. (London: William Heinemann, 1921), 22–23.

356. André Piganiol, "La notion de limes," *Quintus Congressus Internationalis Limitis Romani Studiosorum*, 119–22 (Zagreb: Jugoslavenka Akademija Znanosti, 1963), 122.

357. David Shotter, *Nero* (London: Routledge, 1997), 25, attributes his downfall in part to his lack of interest in military matters.

358. Pelham, "Roman Frontier System," 166; and Shotter, *Rome and Her Empire*, 304.

359. Shotter, *Rome and Her Empire*, 305.

360. Frontinus, *Strategematon*, I, iii, 10. *Limite* has the simple sense of a road in *Strategematon*, I, v, 10.

361. Pelham, "Roman Frontier System," 168; and Lintott, *Imperium Romanum*, 42.

362. Birley, *Hadrian*, 1, 147, 200–201.

363. Birley, *Hadrian*, 128, 134.

364. Wells, *Roman Empire*, 203; Birley, *Hadrian*, 1, 78; and Shotter, *Rome and Her Empire*, 323.

365. Birley, *Hadrian*, 96–97; see Suetonius, *Divus Augustus*, II, 21; and Gruen, "Imperial Policy of Augustus," 410n74.

366. Fergus Millar, *The Roman Empire and Its Neighbours* (London: Duckworth, 1981), 104.

367. See Harris, *War and Imperialism in Republican Rome*, 131–33.

368. Luttwak, *Grand Strategy of the Roman Empire*; see Arther Ferrill, "The Grand Strategy of the Roman Empire," in *Grand Strategies in War and Peace*, ed. Paul Kennedy, 71–85 (New Haven, CT: Yale University Press, 1991).

369. Luttwak, *Grand Strategy of the Roman Empire*, 60.

370. Fergus Millar, "Emperors, Frontiers and Foreign Relations, 31 B.C. to A.D. 378," *Britannia* 13 (1982): 2.

371. Champion, *Roman Imperialism*, 278; and J. C. Mann, "Power, Force and the Frontiers of the Empire," *Journal of Roman Studies* 69 (1979): 179.

372. Whittaker, *Frontiers of the Roman Empire*, 66–67; see Mattern, *Rome and the Enemy*, 109–10; and Isaac, *Limits of Empire*, 417.

373. Isaac, *Limits of Empire*, 387.

374. Michael J. Shapiro, *Violent Cartographies: Mapping Cultures of War* (Minneapolis: University of Minnesota Press, 1997), xii–xiii; Whittaker, *Frontiers of the Roman Empire*, 6–19; and Isaac, *Limits of Empire*, 416–17.

375. Whittaker, *Rome and Its Frontiers*, 28. I am more persuaded by the account

of Isaac. See, in particular, Isaac, *Limits of Empire*, 408–9; and Isaac, "Meaning of the Terms *Limes* and *Limitanei*."

376. H. Schonberger, "The Roman Frontier in Germany: An Archaeological Survey," *Journal of Roman Studies* 59, no. 2 (1969): 161.

377. Owen Lattimore, "The Frontier in History," in *Studies in Frontier History: Collected Papers, 1928–1958*, 469–91 (London: Oxford University Press, 1962), 487.

378. Whittaker, *Frontiers of the Roman Empire*, 201; see 43.

379. See Wells, *German Policy of Augustus*, 248.

380. Maxfield, "Mainland Europe," 137; see Paul Lemerle, "Invasions et migrations dans les Balkans deplus la fin de l'époque romaine jusqu'au VIIIe siècle," *Revue Historique* 211 (1954): 265–308; and A. D. Lee, *Information and Frontiers: Roman Foreign Relations in Late Antiquity* (Cambridge: Cambridge University Press, 1993).

381. For a general discussion, see David Braund, "River Frontiers in the Environmental Psychology of the Roman World," in *The Roman Army in the East*, ed. David L. Kennedy, Supplementary Series No. 18 (Ann Arbor, MI: Journal of Roman Archaeology, 1996): 43–47.

382. Decimus Magnus Ausonius, *Gratiarum actio*, II, 7, in *Decimi Magni Ausonii Opera*, ed. R. P. H. Green (Oxford: Clarendon Press, 1999). See Isaac, "Meaning of the Terms *Limes* and *Limitanei*," 133.

383. Fronto, *Principia Historiae*, 4, in *The Correspondence of Marcus Cornelius Fronto*, Latin-Greek-English ed., 2 vols. (London: William Heinemann, 1920), 2:200, 2:201. See, though, Seneca, *Naturales Quaestiones*, VI, 7, trans. Thomas H. Corcoran, Latin-English ed., 2 vols. (London: William Heinemann, 1972), where he sees them as more linear boundaries.

384. For a general survey, see David J. Breeze and Brian Dobson, *Hadrian's Wall*, 4th ed. (Harmondsworth, UK: Penguin, 2000); and Brian Dobson, "The Function of Hadrian's Wall," *Archaeologica Aeliana*, 5th ser., 14 (1986): 1–30.

385. *Historia Augusta*, Hadrianus, 11; and *Lives of the Later Caesars*, trans. Antony Birley (Harmondsworth, UK: Penguin, 1976).

386. *Historia Augusta*, Hadrianus, 12.

387. Birley, *Hadrian*, 128.

388. Isaac, "Meaning of the Terms *Limes* and *Limitanei*," 131.

389. Whittaker, *Rome and Its Frontiers*, 192; Whittaker, *Frontiers of the Roman Empire*, 47–48; and Luttwak, *Grand Strategy of the Roman Empire*, 88.

390. Millar, *Roman Empire and Its Neighbours*, 111.

391. Breeze and Dobson, *Hadrian's Wall*, xvii and chap. 4.

392. For a survey, see William S. Hanson and Gordon S. Maxwell, *Rome's North West Frontier: The Antonine Wall* (Edinburgh: Edinburgh University Press, 1983); and George MacDonald, "The Building of the Antonine Wall: A Fresh Study of the Inscriptions," *Journal of Roman Studies* 11 (1921): 1–24.

393. *Historia Augusta*, Antonius Pius, V.

394. Shotter, *Rome and Her Empire*, 345; and Hanson and Maxwell, *Rome's North West Frontier*, chap. 5.

395. For a discussion, see J. C. Mann, "Hadrian's Wall: The Last Phases," in *The End of Roman Britain*, ed. P. J. Casey, 144–51 (Oxford: BAR British Series 71, 1979).

396. Dilke, *Roman Land Surveyors*, 199; and Birley, *Hadrian*, 116–17.

397. Shotter, *Rome and Her Empire*, 328.

398. David J. Breeze, *The Northern Frontiers of Roman Britain* (London: Batsford, 1982), 84.

399. See David Divine, *The North-West Frontier of Rome: A Military Study of Hadrian's Wall* (London: Macdonald, 1969).

400. Procopius, *Anecdota*, XXIV, 12, in *Procopius*, trans. H. B. Dewing, Greek-English ed., 7 vols. (London: William Heinemann, 1954), vol. 6.

401. Procopius, *Anecdota*, XXIV, 12–13; see Isaac, "Meaning of the Terms *Limes* and *Limitanei*," 136. In *Suidae Lexicon*, ed. Ada Adler, 5 vols. (Leipzig, Germany: B. G. Teubner, 1928–38), vol. 1, pt. 2, 432, the Greek term *eschatia* is used to mean the zones at the edge of the empire, which Diocletian felt a need "to strengthen with sufficient forces and build forts." In "Meaning of the Terms *Limes* and *Limitanei*," Isaac has suggested that "this is the only extant definition of what the word *limes* might mean" (135).

402. Millar, *Roman Empire and Its Neighbours*, 105.

403. Breeze, *Northern Frontiers of Roman Britain*, 161; and John Cecil Mann, "The Frontiers of the Principate," *Aufstieg und Niedergang der Römischen Welt* 2 (1974): 508.

404. Birley, *Hadrian*, 116, 133–34.

405. R. G. Collingwood, "The Fosse," *Journal of Roman Studies* 14 (1924): 256.

406. David J. Breeze, "Britain," in *The Roman World*, ed. John Wacher, 2 vols., 1:198–222 (London: Routledge, 1987), 1:199; see Dilke, *Roman Land Surveyors*, 197–98.

407. V. E. Nash-Williams, *The Roman Frontiers in Wales*, rev. under the direction of Michael G. Jarrett, 2nd ed. (Cardiff: University of Wales Press, 1969).

408. Whittaker, *Frontiers of the Roman Empire*, 59; and Millar, *Roman Empire and Its Neighbours*, 105. More generally see S. Thomas Parker, *Romans and Saracens: A History of the Arabian Frontier* (Boston: American Schools of Oriental Research, 1986).

409. Cary, *Geographic Background of Greek and Roman History*, 173.

410. For the earlier view, see, for example, A. Poidebard, *Le trace de Rome dans le desert de Syrie: Le limes de Trajan à la conquète Arabe, Recherches aériennes (1925–1932)* (Paris: Librarie Orientaliste Paul Geuthner, 1934); Sir George MacDonald, "Rome in the Middle East," *Antiquity* 82 (1934): 373–80; and Jean Baradez, *Fossatum Africae: Récherches aériennes sur l'organisation des confines sahareiens à l'époque Romaine* (Paris: Arts et metiers graphiques, 1949). For more recent accounts, see David Kennedy and Derrick Riley, *Rome's Desert Frontier from the Air* (Austin: University of Texas Press, 1990); E. Dabrowa, "The Frontier in Syria in the First Century A.D.," in *The Defence of the Roman and Byzantine East*, ed. Philip Freeman and David Kennedy, 2 vols., 1:93–108 (Oxford: BAR International Series 297, 1986); and David Cherry, *Frontier and Society in Roman North Africa* (Oxford: Clarendon, 1998).

411. Mattern, *Rome and the Enemy*, 113; and Whittaker, *Rome and Its Frontiers*, 192.

412. Mattern, *Rome and the Enemy*, 114.

413. Ammianus Marcellinus, *Rerum Gestarum*, XXVI, 4, 5, trans. John C. Rolfe, Latin-English ed., 3 vols. (London: William Heinemann, 1935). There is a slightly ab-

breviated more recent translation by Walter Hamilton: *The Later Roman Empire (A.D. 354–378)* (Harmondsworth, UK: Penguin, 1986). See R. C. Blockley, *Ammianus Marcellinus: A Study of His Historiography and Political Thought* (Brussels: Latomus, 1975); Robin Seager, *Ammianus Marcellinus: Seven Studies in His Language and Thought* (Columbia: University of Missouri, 1986); and John Matthews, *The Roman Empire of Ammianus* (London: Duckworth, 1989).

414. Ammianus Marcellinus, *Rerum Gestarum*, XX, 1.

415. Ammianus Marcellinus, *Rerum Gestarum*, XVI, 2, 12.

416. Ammianus Marcellinus, *Rerum Gestarum*, XVI, 12, 59.

417. Ammianus Marcellinus, *Rerum Gestarum*, XXIX, 1, 14.

418. See, for example, Palladius Rutilius Taurus, *Opus Agriculturae*, IV (Mensis Martius), x, 16, in *Opus Agriculturae, De veterinaria Medicina, De insitione*, ed. Robert H. Rodgers (Leipzig, Germany: B. G. Teubner), 1975, which talks of *territorio Neapolitano* as being amenable to growing a particular tree; and Servius Maurus Honoratus, *Servii Grammatici qui feruntur in Vergilii carmina commentarii*, ed. Georg Thilo and Hermann Hagen, 3 vols. (Leipzig, 1881–1902), on *Aeneid*, V, 755; VII, 661. The former suggests "territorium dictum est quasi terriborium, tritum bubus [bovus] et aratro [*territorium* is so called as if *terribovium*, broken by a plow and oxen]." This will be utilized by Isidore of Seville, with a key variant: *tauritorium* for *terribovium*.

419. *De Rebus Bellicus*, ed. M. W. C. Hassall and Robert Ireland (Oxford: BAR International Series 63, 1979).

420. Javier Arac, "Frontiers of the Late Roman Empire: Perceptions and Realities," in *The Transformation of Frontiers: From Late Antiquity to the Carolingians*, ed. Walter Pohl, Ian Wood, and Helmut Reimitz, 5–13 (Leiden, Netherlands: Brill, 2001), 5.

421. *De Rebus Bellicus*, preface, 10.

422. *De Rebus Bellicus*, V, 6–7. For a discussion, see Stephen Johnson, "Frontier Policy in the Anonymous," in *De Rebus Bellicus*, pt. I, 67–75.

423. *De Rebus Bellicus*, VI, 1–3.

424. *De Rebus Bellicus*, XX, 1. On this aspect, see E. A. Thompson, *A Roman Reformer and Inventor* (New York: Arno Press, 1979), 72–73.

425. Mann, "Power, Force and the Frontiers of the Empire," 179.

426. For a helpful discussion, see W. S. Hanson, "The Nature and Function of Roman Frontiers," in *Barbarians and Romans in North-West Europe: From the Later Republic to Late Antiquity*, ed. John C. Barrett, Andrew Fitzpatrick, and Lesley Macinnes (Oxford: BAR International Series 471, 1989).

427. For an account that concentrates more on people than fortifications, see Hugh Elton, *Frontiers of the Roman Empire* (Bloomington: Indiana University Press, 1996).

428. Mattern, *Rome and the Enemy*, 113; and Cherry, *Frontier and Society in Roman North Africa*, 33.

429. Friedrich Kratochwil, "Of Systems, Boundaries and Territoriality: An Inquiry into the Formation of the State System," *World Politics* 39 (October 1986): 35–36.

430. For a detailed study, see Sir Mortimer Wheeler, *Rome beyond the Imperial Frontiers* (London: G. Bell and Sons, 1954).

431. For some of the recognition, see Lattimore, "Frontier in History," 469–70;

Shapiro, *Violent Cartographies*, xii–xiii; and Kratochwil, "Of Systems, Boundaries and Territoriality," 35.

432. Mattern, *Rome and the Enemy*, 114.

433. Lintott, *Imperium Romanum*, 192–93.

434. Cunliffe, *Greeks, Romans and Barbarians*, 3.

435. Cunliffe, *Greeks, Romans and Barbarians*, 79.

436. Perry Anderson, *Passages from Antiquity to Feudalism* (London: NLB, 1974), 88.

437. Luttwak, *Grand Strategy of the Roman Empire*, 81. See also, more generally, Benet Salway, "Sea and River Travel in the Roman Itinerary Literature," in *Space in the Roman World: Its Perception and Presentation*, ed. Richard Talbert and Kai Brodersen, 43–96 (Antike Kultur und Geschichte, Münster, Germany: Lit Verlag, 2004).

438. Tacitus, *Historiae*, III, 21.

439. Pliny the Elder, *Naturalis Historia*, VII, xx, 84.

440. Suetonius, *Divus Julius*, I, 57.

441. Wells, *Roman Empire*, 139.

442. Peter Brown, *The World of Late Antiquity: From Marcus Aurelius to Muhammed* (London: Thames and Hudson, 1971), 13. For a discussion, see Tønnes Bekker-Nielsen, "*Terra Incognita*: The Subjective Geography of the Roman Empire," in *Studies in Ancient History and Numismatics Presented to Rudi Thomsen*, ed. Askel Damsgaard-Madsen, Erik Christiansen, and Erik Hallager, 148–61 (Århus, Denmark: Aarhus University Press, 1988).

443. A. H. M. Jones, *The Decline of the Ancient World* (London: Longman, 1966), 312.

444. Drummond and Nelson, *Western Frontiers of Imperial Rome*, 15.

445. Peter Brown, *Religion and Society in the Age of Saint Augustine* (London: Faber and Faber, 1972), 16.

446. Brown, *Religion and Society in the Age of Saint Augustine*, 15.

447. Lintott, *Imperium Romanum*, 120.

448. Millar, "Emperors, Frontiers and Foreign Relations," 21–22.

449. Appian, *Civil Wars*, V, 65; see A. N. Sherwin-White, *Roman Foreign Policy in the East, 168 B.C. to A.D. 1* (London: Duckworth, 1984).

450. Syme, *Roman Revolution*, 290. On the future situation, see Timothy D. Barnes, *The New Empire of Diocletian and Constantine* (Cambridge, MA: Harvard University Press, 1982).

451. Wells, *Roman Empire*, 135.

452. Jones, *Decline of the Ancient World*, 50. For a much fuller discussion of this period, see his more substantial study, *Later Roman Empire*.

453. Robert Folz, *The Concept of Empire in Western Europe from the Fifth to the Fourteenth Century*, trans. Sheila Ann Ogilvie (London: Edward Arnold, 1969), 4.

454. Jones, *Later Roman Empire*, 2:1068.

455. Gibbon, *History of the Decline and Fall of the Roman Empire*, 1:926.

456. See introduction to *De Rebus Bellicus*, I, which suggests that though Caesar had set the course of history back four hundred years by holding the Rhine frontier, its eventual fall was inevitable.

CHAPTER 3

1. For a general discussion, see Raymond Williams, *Keywords: A Vocabulary of Culture and Society*, 2nd ed. (London: Fontana, 1983), 207–8.

2. Denys Hay, *The Medieval Centuries* (London: Methuen, 1953), 163.

3. Flavio Biondo, *Forliviensis Historiarum ab inclinatione Romanorum* (Basileae: Frobeniana, 1531).

4. Hay, *Medieval Centuries*, 163.

5. Jacques le Goff, *The Medieval Imagination*, trans. Arthur Goldhammer (Chicago: University of Chicago Press, 1988), 19.

6. Brian Tierney, *Religion, Law and the Growth of Constitutional Thought, 1150–1650* (Cambridge: Cambridge University Press, 1982), 1.

7. Lynette Olson, *The Early Middle Ages: The Birth of Europe* (London: Palgrave, 2007), 198.

8. André Ségal, "Périodisation et didactique: Le 'moyen âge' comme obstacle à l'intelligence des origines de l'Occident," in *Périodes: La construction du temps historique* (Paris: Éditions de l'École des Hautes Études en Sciences Sociales, 1991), 105–14; see Elizabeth A. R. Brown, "On 1500," in *The Medieval World*, ed. Peter Linehan and Janet L. Nelson, 691–710 (London: Routledge, 2001). On the relation of medieval thought to contemporary theory, see Bruce Holsinger, *The Premodern Condition: Medievalism and the Making of Theory* (Chicago: University of Chicago Press), 2005; Jeffrey J. Cohen, *Medieval Identity Machines* (Minneapolis: University of Minnesota Press, 2003); and Andrew Cole and D. Vance Smith, eds., *The Legitimacy of the Middle Ages: On the Unwritten History of Theory* (Durham, NC: Duke University Press, 2010).

9. Cary J. Nederman, "The Politics of Mind and Word, Image and Text: Retrieval and Renewal in Medieval Political Theory," *Political Theory* 25 (October 1997): 717. This claim is pursued in his *Lineages of European Political Thought: Explorations along the Medieval/Modern Divide from John of Salisbury to Hegel* (Washington, DC: Catholic University of America Press, 2009).

10. David Woodward, "Medieval *Mappaemundi*," in *The History of Cartography*, vol. 1, *Cartography in Prehistoric, Ancient, and Medieval Europe and the Mediterranean*, ed. J. B. Harley and David Woodward, 286–370 (Chicago: University of Chicago Press, 1987), 342.

11. John Dagenais and Margaret R. Greer, "Decolonizing the Middle Ages: Introduction," *Journal of Medieval and Early Modern Studies* 30 (2000): 431.

12. See Janet L. Abu-Loghod, *Before European Hegemony: The World System, A.D. 1250–1350* (Oxford: Oxford University Press, 1989).

13. Cary J. Nederman, *Worlds of Difference: European Discourses of Toleration, c. 1100–c. 1550* (University Park: Pennsylvania State University Press, 2000), 15.

14. Kathleen Biddick, *The Shock of Medievalism* (Durham, NC: Duke University Press, 1998), 1–2.

15. Bruce Holsinger, *Neomedievalism, Neoconservatism, and the War on Terror* (Chicago: Prickly Paradigm Press, 2007).

16. Philip G. Cerny, "Neomedievalism, Civil War and the New Security Dilemma: Globalisation as Durable Disorder," *Civil Wars* 1 (1999): 36–64.

17. James Anderson, "The Shifting Stage of Politics: New Medieval and Postmodern Territorialities?," *Environment and Planning D: Society and Space* 14 (1996): 133–53.

18. Alain Minc, *Le nouveau Moyen Âge* (Paris: Gallimard, 1993).

19. See Edward W. Said, *Orientalism* (New York: Vintage, 1978); and Suzanne Conklin Akbari, "From Due East to True North: Orientalism and Orientation," in *The Postcolonial Middle Ages*, ed. Jeffery Jerome Cohen, 19–34 (London: Palgrave, 2000). On this exchange more generally, see the other essays in the Cohen volume and Ananya Jahanara Kabir and Deanne Williams, eds., *Postcolonial Approaches to the European Middle Ages: Translating Cultures* (Cambridge: Cambridge University Press, 2005).

20. John B. Morrall, *Political Thought in Medieval Times*, 3rd ed. (London: Hutchinson University Library, 1971), 10–11.

21. George H. T. Kimble, *Geography in the Middle Ages* (London: Methuen, 1938), 68.

22. Walter Ullmann, *Medieval Political Thought* (Harmondsworth, UK: Peregrine, 1975), 229. For a set of essays that complicates any medieval versus modern distinction, see Peter Haidu, *The Subject Medieval/Modern: Text and Governance in the Middle Ages* (Stanford, CA: Stanford University Press, 2004).

23. Hugh Fraser Stewart, *Boethius: An Essay* (1891; repr., New York: Burt Franklin, 1974), 257 (citations refer to the 1974 edition).

24. Hay, *Medieval Centuries*, 8.

25. Hay, *Medieval Centuries*, 3.

26. Sylvia Tomasch, "Introduction: Medieval Geographical Desire," in *Text and Territory: Geographical Imagination in the European Middle Ages*, ed. Sylvia Tomasch and Sealy Gilles, 1–12 (Philadelphia: University of Pennsylvania Press, 1998), 5. On related themes, see also Paul Zumthor, *La mesure du monde: Représentation de l'espace au Moyen age* (Paris: Seuil, 1993); Barbara A. Hanawat and Michal Kobialka, eds., *Medieval Practices of Space* (Minneapolis: University of Minnesota Press, 2000); Andrew Spicer and Sarah Hamilton, eds., *Defining the Holy: Sacred Space in Medieval and Early Modern Europe* (Aldershot, UK: Ashgate, 2005); Beat Kümin, ed., *Political Space in Pre-industrial Europe* (Aldershot, UK: Ashgate, 2009); and Keith D. Lilley, ed., *Mapping Medieval Geographies: Geographical Encounters in the Latin West and Beyond, 300–1600* (Cambridge: Cambridge University Press, 2013).

27. F. C. Copleston, *A History of Medieval Philosophy* (London: Methuen, 1972), 50. For a unique contemporary analysis, see William E. Connolly, *The Augustinian Imperative: A Reflection on the Politics of Morality* (Newbury Park, CA: Sage, 1993).

28. Saint Augustine, *Confessionum Libri XIII*, in *Corpus Christianorum: Series Latina*, vols. 47–48 (Turnholti: Typographi Brepols, 1955); trans. R. S. Pine-Coffin as *Confessions* (Harmondsworth, UK: Penguin, 1961), I, 13.

29. Peter Brown, *Augustine of Hippo: A Biography* (Berkeley: University of California Press, 1967), 36; see 271. This is an invaluable study. See also his companion book of essays *Religion and Society in the Age of Saint Augustine* (London: Faber and Faber, 1972).

30. Kimble, *Geography in the Middle Ages*, 13.

31. Marcia L. Colish, *Medieval Foundations of the Western Intellectual Tradition, 400–1400* (New Haven, CT: Yale University Press, 1997), 35.

32. Saint Augustine, *De Civitate Dei*, in *Corpus Christianorum*, vol. 27; trans. Henry Bettinson as *Concerning the City of God against the Pagans*, ed. David Knowles (Harmondsworth, UK: Penguin, 1972). I have also consulted the translation by R. W. Dyson under the same title (Cambridge: Cambridge University Press, 1998).

33. Maurice Keen, *A History of Medieval Europe* (London: Penguin, 1991), 19.

34. Charles Howard McIlwain, *The Growth of Political Thought in the West: From the Greeks to the End of the Middle Ages* (London: Macmillan, 1932), 159. On Saint Augustine's political thought, see John Neville Figgis, *The Political Aspects of S. Augustine's "City of God"* (London: Longmans, Green, 1921); and Paul Weithman, "Augustine's Political Philosophy," in *The Cambridge Companion to Augustine*, ed. Eleonore Stump and Norman Kretzmann, 234–52 (Cambridge: Cambridge University Press, 2001).

35. Colish, *Medieval Foundations of the Western Intellectual Tradition*, 35.

36. Paulus Orosius, *The Seven Books of History against the Pagans*, trans. Roy J. Defferari (Washington, DC: Catholic University of America Press, 1964). The Latin text is available in *Histoires (contre les païens)*, ed. and trans. Marie-Pierre Arnaud-Lindet, Latin-French ed., 3 vols. (Paris: Belles Lettres, 1980).

37. Saint Augustine, *The Retractions*, II, 43, in *Patrologiæ Cursus Completus, Series Latina*, ed. J. P. Migne, 221 vols. (Paris: Garnier, 1844–78), vol. 32, col. 648; trans. Sister Mary Inez Bogan in *The Fathers of the Church*, vol. 60 (Washington, DC: Catholic University of America Press, 1968), 209 (where it is noted as II, 69). See R. W. Dyson, *The Pilgrim City: Social and Political Ideas in the Writings of St. Augustine of Hippo* (Woodbridge, UK: Boydell Press, 2001), 51.

38. Jerome, "Epistolorum 123," in Saint Jérôme, *Lettres*, ed. Jérôme Labourt, Latin-French ed., 8 vols. (Paris: Société d'Édition "Les Belles Lettres," 1961), 7:93; and *Jerome: Letters and Select Works*, trans. W. H. Fremantle, *Select Library of Nicene and Post-Nicene Fathers*, ser. 2, vol. 6 (Edinburgh: T. and T. Clark, 1892), 237. The reference to Lucan is Annaeus Lucanus, *De Bello Civili [Pharsalia]*, ed. D. R. Shackleton Bailey (Stutgardiae: Teubner, 1988); and Lucan, *Civil War*, trans. S. H. Braund (Oxford: Clarendon Press, 1992), V, 274.

39. Colish, *Medieval Foundations of the Western Intellectual Tradition*, 35.

40. Augustine, *De Genesi ad litteram*, *Patrologiae Cursus Completus, Series Latina*, vol. 34, "On the Literal Interpretation of Genesis: An Unfinished Book," in *Saint Augustine on Genesis*, trans. Roland J. Teske (Washington, DC: Catholic University of America Press, 1991). See Brown, *Augustine of Hippo*, 312; and R. W. Dyson, *St Augustine of Hippo: The Christian Transformation of Political Philosophy* (London: Continuum, 2005), 32.

41. David Knowles, introduction to Augustine, *De Civitate Dei*, xv; echoing Figgis, *Political Aspects of S. Augustine's "City of God,"* 3.

42. For a sampling of these writers, see John Gregory, ed., *The Neoplatonists* (London: Kyle Cathie, 1991).

43. Dyson, *Pilgrim City*, 210.

44. Figgis, *Political Aspects of S. Augustine's "City of God,"* 51–52.

45. P. R. I. Brown, "Saint Augustine," in *Trends in Medieval Political Thought*, ed. Beryl Smalley, 1–21 (Oxford: Basil Blackwell, 1965), 1. This essay is reprinted in Brown, *Religion and Society in the Age of Saint Augustine*, 25–45.

46. See Psalms 46:4; 48:1, 2, 8; 87:3—all cited in Augustine, *De Civitate Dei*, XI, 1.

47. Andreas Osiander, *Before the State: Systematic Political Change in the West from the Greeks to the French Revolution* (Oxford: Oxford University Press, 2007), xix.

48. See, for example, Augustine, *De Civitate Dei*, XII, 1; XV, 1; XVI, 17. For a helpful discussion of *civitas*, and its relation to other key terms, including *societas, res publica,* and *populus*, see R. T. Marshall, *Studies in the Political and Socio-Religious Terminology of the* De civitate Dei (Washington, DC: Catholic University of America Press, 1952).

49. Augustine, *De Civitate Dei*, X, 32 and XI, 1. See R. A. Markus, *Saeculum: History and Society in the Theology of St. Augustine* (Cambridge: Cambridge University Press, 1970), 71, 101, 133.

50. Augustine, "In Psalmum LXIV," in *Patrologiae Cursus Completus, Series Latina*, vol. 36, col. 773; *Expositions on the Book of Psalms*, ed. A. Cleveland Coxe, *Select Library of Nicene and Post-Nicene Fathers*, ser. 1, VIII (Edinburgh: T. and T. Clark, 1892), 268 (where it is numbered Psalm LXV).

51. Augustine, "Epistularum CCXX," in *Epistulae*, ed. Al. Goldbacher, 5 vols. (Vindobonae: F. Tempsky, 1911), 4:437; "Letter 220 to Boniface," in *Political Writings*, ed. E. M. Atkins and R. J. Dodaro (Cambridge: Cambridge University Press, 2001), 222.

52. See Dyson, *St Augustine of Hippo*, 4.

53. See, for example, Etienne Gilson, *The Christian Philosophy of Saint Augustine*, trans. L. E. M. Lynch (London: Victor Gollancz, 1961), 181, where he distinguishes between the "kingdom of God" of the church and Augustine's "city of God."

54. Dyson, *Pilgrim City*, 56–57, 179.

55. Dyson, *Pilgrim City*, 10–11.

56. Augustine, *De Civitate Dei*, XV, 19.

57. Augustine, *De Civitate Dei*, XV, 18.

58. Markus, *Saeculum*, 103; and 223.

59. Augustine, *De Civitate Dei*, I, preface.

60. Dyson, *Pilgrim City*, 180; and Gilson, *Christian Philosophy of Saint Augustine*, 180.

61. Augustine, *De Civitate Dei*, XIV, 1. See Ephesians 2:19 and Philippians 3:20.

62. Markus, *Saeculum*, 59.

63. Augustine, *De Civitate Dei*, XI, 1.

64. Augustine, *De Civitate Dei*, II, 21; see Augustine, "Epistularum CXXXVIII," in *Epistulae*, III, 135; and Augustine, "Letter 138 to Marcellinus," in *Political Writings*, 35.

65. Augustine, *De Civitate Dei*, I, preface.

66. Augustine, *De Civitate Dei*, I, preface.

67. Augustine, *De Civitate Dei*, II, 19.

68. Augustine, *De Civitate Dei*, II, 21. The quotation is from Cicero, *De re publica*, I, 25; see 39. The first reference is to a part that is not extant.

69. Cicero, *De re publica*, I xxv. 39. See Markus, *Saeculum*, 64, and the discussion in chap. 2.

70. Jeremy Duquesnay Adams, *The* Populus *of Augustine and Jerome: A Study in the Patristic Sense of Community* (New Haven, CT: Yale University Press, 1971), 2.

71. Adams, Populus *of Augustine and Jerome*, 2.

72. Adams, Populus *of Augustine and Jerome*, 2.

73. Augustine, *De Civitate Dei*, V, 11.

74. Adams, Populus *of Augustine and Jerome*, 109–11.

75. Augustine, *De Civitate Dei*, XIX, 21. For a discussion of some of the issues that arise from this, see P. J. Burnell, "The Status of Politics in St. Augustine's *City of God*," *History of Political Thought* 13 (1992): 13–29.

76. Herbert A. Deane, *The Political and Social Ideas of St. Augustine* (New York: Columbia University Press, 1963), 118–20; see Markus, *Saeculum*, 64.

77. Markus, *Saeculum*, 64–65.

78. Augustine, *De Civitate Dei*, XIX, 24.

79. Augustine, *De Civitate Dei*, XIX, 24.

80. Augustine, *De Civitate Dei*, XIX, 24.

81. Augustine, *De Civitate Dei*, XIX, 24.

82. Augustine, *De Civitate Dei*, IV, 4. Compare the story in Cicero, *De re publica*, II, xiv, 24.

83. Augustine, *De Civitate Dei*, IV, 7.

84. Adams, Populus *of Augustine and Jerome*, 112.

85. Adams, Populus *of Augustine and Jerome*, 113–15.

86. Gilson, *Christian Philosophy of Saint Augustine*, 171.

87. Augustine, *De Civitate Dei*, XV, 21.

88. Brown, *Augustine of Hippo*, 324.

89. Augustine, *De Civitate Dei*, XIV, 28; see II Corinthians 10:17.

90. Gilson, *Christian Philosophy of Saint Augustine*, 172.

91. Knowles, introduction to Augustine, *De Civitate Dei*, xvii.

92. Adams, Populus *of Augustine and Jerome*, 139, provides references for what he calls "territorial association" in relation to *populus* in a range of passages of the *City of God*, but "place" seems closer to the sense.

93. Augustine, *De Civitate Dei*, XVIII, 24; see 2.

94. Augustine, *De Civitate Dei*, XVII, 2.

95. Augustine, *De Civitate Dei*, XVII, 2.

96. Augustine, *De Civitate Dei*, XVIII, 11.

97. Augustine, *De Civitate Dei*, XVIII, 12; see chap. 1 above.

98. Augustine, *De Civitate Dei*, XII, 24; see Genesis 2:7.

99. Adams, Populus *of Augustine and Jerome*, 139.

100. Markus, *Saeculum*, 52.

101. Augustine, "Epistularum LXVI"" and "Epistularum CLXXXV," in *Epistulae*, 2:236, 3:4; Augustine, "Letter 66 to Crispin" and "Letter 185 to Boniface," in *Political Writings*, 133, 175.

102. Augustine, *De Civitate Dei*, IV, 29.

103. Augustine, *De Civitate Dei*, V, 12. This chapter also suggests that while kingdom (*regnum*) comes from *regnare* (to reign), king (*rex*) comes from *regere* (to rule).

104. Augustine, *De Civitate Dei*, IV, 29. See also the note in the Knowles ed., 171n101.

105. Augustine, *De Civitate Dei*, XXI, 4. Cf. editor's note on 969n6: "this is obscure," which gives the example of charcoal as a foundation of a temple.

106. Justinius, *Epitoma historiarum philippicarum Pompei Trogi* (Paris: Plon Fratres, 1845), I, 1. This is cited in Augustine, *De Civitate Dei*, IV, 6. It is picked up in John of Paris, *De regia potestate et papali*, 3.

107. Augustine, "Epistularum CXXXVIII," in *Epistulae*, 3:135; and Augustine, "Letter 138 to Marcellinus," in *Political Writings*, 35.

108. Augustine *De excidio urbis Romae*, VI, in *Patrologiae Cursus Completus, Series Latina*, vol. 40, col. 721; see Markus, *Saeculum*, 61.

109. Markus, *Saeculum*, 149.

110. Augustine, *Confessionum*, VII, 1.

111. Augustine, *Confessionum*, VII, 1.

112. Augustine, *De Civitate Dei*, XI, 6.

113. Augustine, *Confessionum*, XI, 23.

114. Augustine, *Confessionum*, XI, 26.

115. Augustine, *De Civitate Dei*, XI, 5.

116. Augustine, *Confessionum*, VII, 1.

117. Augustine, *Confessionum*, XII, 27.

118. Augustine, *Confessionum*, XIII, 9.

119. Deane, *Political and Social Ideas of St. Augustine*, 69; and Gilson, *Christian Philosophy of Saint Augustine*, 179.

120. Arthur Monaghan, *Consent, Coercion, and Limit: The Medieval Origins of Parliamentary Democracy* (Kingston and Montreal: McGill–Queen's University Press, 1987), 29; though see Deane, *Political and Social Ideas of St. Augustine*, 172.

121. See Dyson, *St Augustine of Hippo*, 142–46.

122. Boethius, *The Theological Tractates* and *The Consolation of Philosophy*, trans. H. F. Stewart, E. K. Rand, and S. J. Tester, Latin-English ed. (Cambridge, MA: Harvard University Press, 1973).

123. Stewart, *Boethius*, 81. For a thorough discussion of the heritage of ideas from antiquity, see Richard Sorabji, *Time, Creation and the Continuum: Theories in Antiquity and the Early Middle Ages* (London: Duckworth, 1983); and Richard Sorabji, *Matter, Space and Motion: Theories in Antiquity and Their Sequel* (London: Duckworth, 1988).

124. Stewart, *Boethius*, 118.

125. V. E. Watts, introduction to Boethius, *The Consolation of Philosophy*, trans. V. E. Watts (Harmondsworth, UK: Penguin, 1969), 7.

126. Boethius, *In Librum Aristoteles Peri Hermeneias*, II, 3, ed. Carolus Meiser (Lipsaie: B. G. Teubner, 1880), 79–80.

127. Howard Rollin Patch, *The Tradition of Boethius: A Study of His Importance in Medieval Culture* (1935; repr., New York: Russell and Russell, 1970), 32 (citations refer to the 1970 edition).

128. Jonathan Barnes, "Boethius and the Study of Logic," in *Boethius: His Life, Thought and Influence*, ed. Margaret Gibson, 73–89 (Oxford: Basil Blackwell, 1981), 74; Copleston, *History of Medieval Philosophy*, 54; and Watts, introduction to Boethius, *Consolation of Philosophy*, 12. A modern edition is Porphyry, *Introduction*, trans. Jonathan Barnes (Oxford: Clarendon Press, 2003).

129. Stewart, *Boethius*, 26; and Patch, *Tradition of Boethius*, 4. See David Pingree, "Boethius' Geometry and Astronomy," in Gibson, *Boethius*, 155–61.

130. Patch, *Tradition of Boethius*, 31.

131. Ernst H. Kantorowicz, "Plato in the Middle Ages," in *Selected Studies* (Locust Valley, NY: J. J. Augustin, 1965), 184–85.

132. R. W. Southern, *The Making of the Middle Ages* (London: Hutchinson, 1953), 167. A similar claim is made in Hans von Campenhausen, *The Fathers of the Latin Church*, trans. Manfred Hoffman (London: A & C Black, 1964), 289.

133. This can be found in *The Riverside Chaucer*, ed. Larry Benson, 3rd ed. (Oxford: Oxford University Press, 1987), 473–585. On the debt, see Stewart, *Boethius*, 214; and Bernard L. Jefferson, *Chaucer and the Consolation of Philosophy of Boethius* (New York: Haskell House, 1965).

134. Copleston, *History of Medieval Philosophy*, 55; see Hay, *Medieval Centuries*, 54, for a description.

135. Colish, *Medieval Foundations of the Western Intellectual Tradition*, 43.

136. Colish, *Medieval Foundations of the Western Intellectual Tradition*, 44.

137. Isidore of Seville, *Etymologiae*, ed. W. M. Lindsay (Oxford: Oxford University Press, 1911); *The Etymologies of Isidore of Seville*, trans. Steven A. Barney, W. J. Lewis, J. A. Beach, and Oliver Berghof (Cambridge: Cambridge University Press, 2006).

138. Joseph Canning, *A History of Medieval Political Thought, 300–1450* (London: Routledge, 1996), 20. The most comprehensive study is Jacques Fontaine, *Isidore de Seville et la culture classique dans l'espagne Wisigothique*, 2 vols. (Paris: Études Augustiniennes, 1959). On the context of his time, see Roger Collins, *Visigothic Spain, 409–711* (Oxford: Blackwell, 2004).

139. See Isidore of Seville, *De Ecclesiasticis Officiis*, trans. Rev. Thomas L. Knoebel (New York: Newman Press, 2008).

140. Isidore of Seville, *Historia de regibus Gothorum, Vandalorum et Suevorum*, in *Monumenta Germaniae Historica Auctorum Antiquissimorum Tomus XI, Chronicum Minorum* (Berlin: Weidmann, 1894), 2:241–303; and Isidore of Seville, *History of the Kings of the Goths, Vandals, and Suevi*, trans. Guido Donini and Gordon B. Ford Jr. (Leiden, Netherlands: E. J. Brill, 1966), prologue, 1.

141. L. D. Reynolds and N. G. Wilson, *Scribes and Scholars: A Guide to the Transmission of Greek and Latin Literature*, 2nd ed. (Oxford: Clarendon Press, 1974), 74.

142. Southern, *Making of the Middle Ages*, 183.

143. Cristina Farronato, *Eco's Chaosmos: From the Middle Ages to Postmodernity* (Toronto: University of Toronto Press, 2003), 107; a discussion that draws on Umberto Eco, "Riflessioni sulle tecniche di citazione nel medioevo, in Ideologie e pratiche del reimpiego nell'alto medioevo," *Settimane di Studi del Centro Italiano di Studi sull'Alto Medioevo* 46 (1999): 461.

144. Ernest Brehaut, *An Encyclopedist of the Dark Ages: Isidore of Seville* (1912; repr., New York: Burt Franklin, 1964), 48 (citations refer to the 1964 edition).

145. Fontaine, *Isidore de Seville*, 2:849–51; Brehaut, *Encyclopedist of the Dark Ages*, 35; and John Henderson, *The Medieval World of Isidore of Seville: Truth from Words* (Cambridge: Cambridge University Press, 2007), 6.

146. Hay, *Medieval Centuries*, 18.

147. See Alfred Hiatt, "Mapping the Ends of Empire," in Kabir and Williams, *Postcolonial Approaches to the European Middle Ages*, 48–76; on these maps generally, see Leo Bagrow, *History of Cartography*, revised and enlarged edition by R. A. Skelton (London: C. A. Watts, 1964), chap. 3.

148. On this, see Akbari, "From Due East to True North."

149. The case for his believing in a flat earth is found in Brehaut, *Encyclopedist of the Dark Ages*, 50–54; a much more plausible account against is found in Wesley M. Stevens, "The Figure of the Earth in Isidore's 'De natura rerum,'" *Isis* 71 (1980): 268–77. On his cartographic work, see John Williams, "Isidore, Orosius and the Beatus Map," *Imago Mundi* 49 (1997): 7–32; Evelyn Edson, *Mapping Time and Space: How Medieval Mapmakers Viewed Their World* (London: British Library, 1997), 36–51; Naomi Reed Kline, *Maps of Medieval Thought: The Hereford Paradigm* (Woodbridge: Boydell Press, 2001), 13–18; Woodward, "Medieval *Mappaemundi*," 301–2; and, on the geographical more generally, Natalia Lozovsky, *"The Earth Is Our Book": Geographical Knowledge in the Latin West, ca. 400–1000* (Ann Arbor: University of Michigan Press, 2000), 103–13.

150. See Rudolf Simek, "The Shape of the Earth in the Middle Ages and Medieval *Mappaemundi*," in *The Hereford World Map: Medieval World Maps and Their Context*, ed. P. D. A. Harvey, 293–303 (London: British Library, 2006); Evelyn Edson, *The World Map, 1300–1492: The Persistence of Tradition and Transformation* (Baltimore: Johns Hopkins University Press, 2007), 206; Jeffery Burton Russell, *Inventing the Flat Earth: Columbus and Modern Historians* (Westport, CT: Praeger, 1991); and Suzanne Conklin Akbari, *Idols in the East: European Representations of Islam and the Orient, 1100–1450* (Ithaca, NY: Cornell University Press, 2009), chap. 1.

151. *Semeiança del Mundo: A Medieval Description of the World*, ed. William E. Bull and Harry F. Williams (Berkeley: University of California Press, 1959).

152. Isidore of Seville, *Etymologiae*, XIV, v, 20–21.

153. Isidore of Seville, *Etymologiae*, XV, xv, 1. My account here is indebted to Henderson, *Medieval World of Isidore of Seville*, 178–80. On these issues more generally, see Ronald Edward Zupko, "Weights and Measures, Western European," in *Dictionary of the Middle Ages*, ed. Joseph R. Strayer, 13 vols. (New York: Charles Scribner's Sons, 1989), 12:582–91; and A. W. Richeson, *English Land Measuring to 1800: Instruments and Practices* (Cambridge, MA: Society for the History of Technology / MIT Press, 1966), chap. 2.

154. Henderson, *Medieval World of Isidore of Seville*, 164.

155. Isidore of Seville, *Etymologiae*, XVIII, xxxvii, 1. This is as it was used by Virgil.

156. Isidore of Seville, *Etymologiae*, XVI, v, 22; see Servius Maurus Honoratus, *Servii Grammatici qui feruntur in Vergilii carmina commentarii*, ed. Georg Thilo and Hermann Hagen, 3 vols. (Leipzig, 1881–1902); on *Aeneid*, V, 755.

157. Henderson, *Medieval World of Isidore of Seville*, 179.

158. Isidore of Seville, *Etymologiae*, XV, xiv, 1–3.

159. Isidore of Seville, *Etymologiae*, III, x, 1.

160. Brehaut, *Encyclopedist of the Dark Ages*, 131. For a detailed discussion, see Fontaine, *Isidore de Seville*, 1:393–412.

161. J. M. Wallace-Hadrill, *Early Germanic Kingship in England and on the Continent* (Oxford: Clarendon Press, 1971), 53; trading on Isidore of Seville, *Sententiarum Libri Tres*, in *Opera Omnia, Patrologiae Cursus Completus, Series Latina*, vol. 83, col. 723.

162. Brehaut, *Encyclopedist of the Dark Ages*, 164–65.

163. See Barbara H. Rosenwein, *Negotiating Space: Power, Restraint, and Privileges of Immunity in Early Medieval Europe* (Ithaca, NY: Cornell University Press, 1999), 215. On these sources, which also extend to archaeological evidence, see Edward James, *The Franks* (Oxford: Basil Blackwell, 1988), chap. 1.

164. Markus, *Saeculum*, 162–63.

165. Walter Goffart, *The Narrators of Barbarian History (A.D. 550–800): Jordanes, Gregory of Tours, Bede, and Paul the Deacon* (Princeton, NJ: Princeton University Press, 1988), 13. For a valuable general discussion, see Paul Fouracre, "Space, Culture and Kingdoms in Early Medieval Europe," in *The Medieval World*, ed. Peter Linehan and Janet L. Nelson, 366–80 (London: Routledge, 2001).

166. For a detailed discussion, see A. H. Merrills, *History and Geography in Late Antiquity* (Cambridge: Cambridge University Press, 2005).

167. Isidore of Seville, *Historia de regibus Gothorum*, 47.

168. Isidore of Seville, *Historia de regibus Gothorum*, 23.

169. It is essential to note that the importance of cities in the Roman Empire is replaced, at this time, with diverse and multiple sites of power. For a collection of essays on this theme, see Mayke de Jong and Frans Theunis with Carine van Rhijn, eds., *Topographies of Power in the Early Middle Ages* (Leiden, Netherlands: Brill, 2001).

170. *The Chronicle of Hydatius and the Consularia Constantinopolitana: Two Contemporary Accounts of the Final Years of the Roman Empire*, ed. and trans. R. W. Burgess, Latin-English ed. (Oxford: Clarendon Press, 1993), XVII, pp. 82–83.

171. Walter Goffart, *Barbarian Tribes: The Migration Age and the Later Roman Empire* (Philadelphia: University of Pennsylvania Press, 2006), 101.

172. Gregory of Tours, *Libri Historiarum X*, in *Gregorii Turonensis Opera*, ed. W. Arndt and Br. Krusch (Hanover, Germany: Impensis bibliopolii Hahniani, 1884–85); trans. Lewis Thorpe as *The History of the Franks* (Harmondsworth, UK: Penguin, 1974).

173. Olson, *Early Middle Ages*, 29–31. For useful maps of the barbarian movements and kingdoms, see David Ditchburn, Simon MacLean, and Angus Mackay, eds., *Atlas of Medieval Europe* (London: Routledge, 2007), 8, 10.

174. Copleston, *History of Medieval Philosophy*, 58

175. J. M. Wallace-Hadrill, *The Long-Haired Kings and Other Studies in Frankish History* (London: Methuen, 1962), 163.

176. James, *Franks*, 162; generally Malcolm Todd, *The Early Germans* (Oxford: Blackwell, 1992).

177. Wallace-Hadrill, *Long-Haired Kings*, 153; see Herwig Wolfram, *History of the Goths*, trans. Thomas J. Dunlap (Berkeley: University of California Press, 1988), 94–95.

178. Ullmann, *Medieval Political Thought*, 87.

179. Lennart Ejerfeldt, "Myths of the State in the West European Middle Ages," in *The Myth of the State*, ed. Haralds Biezais, 160–69 (Stockholm: Almqvist and Wiksell, 1972), 163.

180. Wallace-Hadrill, *Long-Haired Kings*, 212.

181. For a discussion, see Walter Pohl, Ian Wood, and Helmut Reimitz, eds., *The Transformation of Frontiers: From Late Antiquity to the Carolingians* (Leiden, Netherlands: Brill, 2001).

182. Wallace-Hadrill, *Long-Haired Kings*, 1. There is only a very brief mention of the term in Max Bonnett, *Le Latin de Grégoire de Tours* (1890; repr., Hildesheim, Germany: Georg Olms, 1968), 245 (citations refer to the 1968 edition).

183. Gregory of Tours, *Libri Historiarum*, III, 35; see IV, 42.

184. Gregory of Tours, *Libri Historiarum*, IV, 42; see IV, 44.

185. Gregory of Tours, *Libri Historiarum*, V, 14.

186. Gregory of Tours, *Libri Historiarum*, VI, 14.

187. Gregory of Tours, *Libri Historiarum*, VI, 21.

188. Gregory of Tours, *Libri Historiarum*, VIII, 15.

189. Gregory of Tours, *Libri Historiarum*, V, 3.

190. Gregory of Tours, *Libri Historiarum*, VIII, 45; see X, 3.

191. Gregory of Tours, *Libri Historiarum*, II, 35; see II, 37, for "*territurium Turonicum.*"

192. Gregory of Tours, *Libri Historiarum*, X, 8.

193. Gregory of Tours, *Libri Historiarum*, IX, 5; see IX, 24. As Fernand Braudel points out (*The Mediterranean in the Ancient World*, trans. Siân Reynolds [London: Penguin, 2001], 265), there was nonetheless a fundamental difference between the ancient world city and the medieval one, where the former extended into agricultural areas, while the latter made a strict distinction between those *intra muros* and those without.

194. Gregory of Tours, *Libri Historiarum*, III, 15.

195. Gregory of Tours, *Libri Historiarum*, IX, 18; see also VIII, 43; IX, 19.

196. Gregory of Tours, *Libri Historiarum*, IX, 32.

197. Gregory of Tours, *Libri Historiarum*, II, 32; Thorpe renders *regnum* as "territory."

198. Iordanes, *De origine actibusque Getarum*, ed. Francesco Giunta and Antonino Grillone (Rome: Nella Sede Dell'Istituto, 1991); and Iordanes, *The Gothic History of Jordanes*, trans. Charles Christopher Mierow (Princeton, NJ: Princeton University Press, 1915).

199. Saxo Grammaticus, *Gesta Danorum/Danmarkshistorien*, ed. Karsten Friis-Jensen, trans. Peter Zeeburg, Latin-Danish ed., 2 vols. (Copenhagen: Gads Forlag, 2005); and *The History of the Danes, Books I–IX*, ed. Hilda Ellis Davidson, trans. Peter Fisher, 2 vols. (Cambridge: D. S. Brewer, 1979–80), vol. 1. The second volume is a commentary by Fisher and Davidson. See also Karsten Friis-Jensen, ed., *Saxo Grammaticus: A Medieval Author between Norse and Latin Culture* (Copenhagen: Museum Tusculanum Press, 1981).

200. Pauli, *Historia gentis Langobardorum*, In Usum Scholarum ex Monumentis Germaniae Historicis Recusa (Hanover, Germany: Hahnsche, 1878); trans. William Dudley Foulke as Paul the Deacon, *History of the Langobards* (Philadelphia: Department of History, University of Pennsylvania, 1907), III, XXXII. Foulke's translation has "the territories of the Langobards will be up to this place."

201. Of course, the Eastern or Byzantine Empire endured for another thousand years.

202. Goffart, *Barbarian Tribes*, 55.

203. See also Walter Goffart, *Barbarians and Romans, A.D. 418–584: The Techniques of Accommodation* (Princeton, NJ: Princeton University Press, 1980); Justin Davis Randers-Pehrson, *Barbarians and Romans: The Birth Struggle of Europe, A.D. 400–700* (London: Croom Helm, 1983); E. A. Thompson, *Romans and Barbarians: The Decline of the Western Empire* (Madison: University of Wisconsin Press, 1982); Herwig Wolfram, *The Roman Empire and Its Germanic Peoples*, trans. Thomas Dunlap (Berkeley: University of California Press, 1997); and John C. Barrett, Andrew Fitzpatrick, and Lesley Macinnes, eds., *Barbarians and Romans in North-West Europe: From the Later Republic to Late Antiquity* (Oxford: BAR International Series 471, 1989).

204. Olson, *Early Middle Ages*, 32. On this more generally, see Henri Pirenne, *Medieval Cities: Their Origins and the Revival of Trade*, trans. Frank D. Halsey (Princeton, NJ: Princeton University Press, 1925); Giovanno Tabacco, *The Struggle for Power in Medieval Italy: Structures of Political Rule*, trans. Rosalind Brown Jenson (Cambridge: Cambridge University Press, 1989); and Keith D. Lilley, *Urban Life in the Middle Ages, 1000–1450* (London: Palgrave, 2002).

205. For a range of studies, see Angus MacKay, *Spain in the Middle Ages: From Frontier to Empire, 1000–1500* (London: Macmillan, 1977); Robert Bartlett and Angus MacKay, eds., *Medieval Frontier Societies* (Oxford: Clarendon Press, 1989); and David Abulafia and Nora Berend, eds., *Medieval Frontiers: Concepts and Practices* (Aldershot, UK: Ashgate, 2002).

206. Monaghan, *Consent, Coercion, and Limit*, 54; see P. D. King, *Law and Society in the Visigothic Kingdom* (Cambridge: Cambridge University Press, 1972), chap. 2.

207. See P. D. King, "The Alleged Territoriality of Visigothic Law," in *Authority and Power: Studies on Medieval Law and Government Presented to Walter Ullmann on his Seventieth Birthday*, ed. Brian Tierney and Peter Linehan, 1–11 (Cambridge: Cambridge University Press, 1980); P. D. King, "King Chindasvind and the First Territorial Law-Code of the Visigothic Kingdom," in *Visigothic Spain: New Approaches*, ed. Edward James, 131–57 (Oxford: Clarendon Press, 1980); and Collins, *Visigothic Spain*, 225–27.

208. Goffart, *Narrators of Barbarian History*, 235.

209. Colish, *Medieval Foundations of the Western Intellectual Tradition*, 65.

210. See, for example, *Bede's Ecclesiastical History of the English People*, ed. Bertram Colgrave and R. A. B. Mynors, Latin-English ed. (Oxford: Clarendon Press, 1969), I, 1. See Bede, *A History of the English Church and People*, trans. Leo Sherley-Price (Harmondsworth, UK: Penguin, 1955). On the geographical elements, see Lozovsky, *"Earth Is Our Book,"* 86–94.

211. On Easter, see Bede, *Historia ecclesiastica*, V, 21; on dating, see, for example, I, 2, when Julius Caesar's becoming consul is dated "in year 693 from Rome, and sixty years before our Lord." For a discussion, see Edson, *Mapping Time and Space*, chap. 4 and 97.

212. Bede, *Historia ecclesiastica*, V, 24.

213. See Robert Brown, "An Inquiry into the Origin of the Name 'Sunderland'; and

as to the Birthplace of the Venerable Bede," in Society of Antiquaries of Newcastle upon Tyne, *Archaeologica Aeliana; or, Miscellaneous Tracts Relating to Antiquity*, vol. 4, 273–83 (Newcastle upon Tyne: Society of Antiquaries, 1855), 277, 280. The caution comes from Bertram Colgrove, "Historical Introduction," in *Bede's Ecclesiastical History of the English People*, xix. See also the discussion in Peter Hunter Blair, *The World of Bede* (Cambridge: Cambridge University Press, 1990), 3–4, where he translates *territorium* as "an estate" and suggests *sundorland* is "land set apart for a special purpose," and is not a specific place.

214. Bede, *Historia ecclesiastica*, II, 3; III, 3; III, 24; III, 26.

215. Bede, *History of the English Church and People*, editor's note, 334; see *Bede's Ecclesiastical History of the English People*, 72–73n3. For a fuller discussion, see F. M. Stenton, *Anglo-Saxon England*, 2nd ed. (Oxford: Clarendon, 1971), 278–79; Diana Wood, *Medieval Economic Thought* (Cambridge: Cambridge University Press, 2002), chap. 4; and Robert A. Dodgshon, *The European Past: Social Evolution and Spatial Order* (Houndmills, UK: Macmillan, 1987), 158–59.

216. Bede, *Historia ecclesiastica*, I, 25.

217. Bede, *Historia ecclesiastica*, IV, 13; see also III, 24.

218. Bede, *Historia ecclesiastica*, IV, 13.

219. Bede, *Historia ecclesiastica*, IV, 26.

220. L. O. Aranye Fradenburg, "Pro Patria Mori," in *Imagining a Medieval English Nation*, ed. Kathy Lavezzo, 3–38 (Minneapolis: University of Minnesota Press, 2004), 31.

221. Bede, *Historia ecclesiastica*, III, 21.

222. See Mary A. Parker, *Beowulf and Christianity* (New York: Peter Lang, 1987).

223. Nicholas Howe, *Writing the Map of Anglo-Saxon England: Essays in Cultural Geography* (New Haven, CT: Yale University Press, 2008), 188.

224. Citations from *Beowulf* are given by line number. I have used the Old English text presented in George Jack, ed., *Beowulf: A Student Edition* (Oxford: Oxford University Press, 1994). The translations are usually based on the one by Michael Alexander (Harmondsworth, UK: Penguin, 1973), though I have regularly departed from his renderings. In this I have often made use of Jack's line glosses; and Michael Swanton, ed., *Beowulf* (Manchester, UK: Manchester University Press, 1997), which is bilingual, with helpful notes. Bruce Mitchell and Fred C. Robinson, eds., *Beowulf: An Edition with Relevant Shorter Texts* (Malden, MA: Blackwell, 1998), has a detailed glossary, which has also proved very useful.

225. Ritchie Girvan, *Beowulf and the Seventh Century: Language and Content* (London: Methuen, 1971); Fred C. Robinson, *The Tomb of Beowulf and Other Essays on Old English* (Oxford: Blackwell, 1993). An attempt to deny this, and to suggest that the Geats, like Beowulf and the monsters, are fantasies, is found in Jane Acomb Leake, *The Geats of* Beowulf: *A Study in the Geographical Mythology of the Middle Ages* (Madison: University of Wisconsin Press, 1967).

226. See, for example, the map entitled "The Geography of Beowulf," in Mitchell and Robinson, *Beowulf*, xiii.

227. In this it differs from Gillian R. Overing and Marijane Osborn, *Landscape of Desire: Partial Stories of the Medieval Scandinavian World* (Minneapolis: University of

Minnesota Press, 1994); and Margaret Gelling, "The Landscape of Beowulf," *Anglo-Saxon England* 31 (2002): 7–11.

228. Such attempts are found in R. T. Farrell, "Beowulf, Swedes and Geats," *Saga Book of the Viking Society* 18 (1970–73): 220–96; and B. Raw, "Royal Power and Royal Symbols in *Beowulf*," in *The Age of Sutton Hoo: The Seventh Century in Northwestern Europe*, ed. M. O. H. Carver, 167–74 (Woodbridge, UK: Boydell Press, 1998).

229. For a reading of these particular places, see Stuart Elden, "Place Symbolism and Land Politics in *Beowulf*," *Cultural Geographies* 16 (2009): 447–63. That piece provides extensive references to other discussions.

230. *Maxims*, I (A), in *The Exeter Anthology of Old English Poetry: An Edition of Exeter Dean and Chapter Ms 3501*, vol. 1, *Texts*, ed. Bernard J. Muir, 2nd ed. (Exeter, UK: Exeter University Press, 2000), 250. There is a translation by James Earl, "Maxims I, Part I," *Neophilologus* 67 (1983): 277–83. See James W. Earl, *Thinking about Beowulf* (Stanford, CA: Stanford University Press, 1994), 65.

231. *Beowulf*, 1725–34.

232. *Beowulf*, 73–74.

233. *Beowulf*, 1176–80.

234. *Beowulf*, 1881.

235. *Beowulf*, 2145–62.

236. *Beowulf*, 2195–99.

237. *Beowulf*, 2207–8.

238. *Beowulf*, 2389–90.

239. *Beowulf*, 2989–90.

240. *Beowulf*, 2469–71.

241. *Beowulf*, 2492–93.

242. *Beowulf*, 2606–8.

243. *Beowulf*, 2884–90.

244. See *Beowulf*, 2910–15.

245. *Beowulf*, 221, 1913.

246. *Beowulf*, 1623.

247. *Beowulf*, 209.

248. *Beowulf*, 1357.

249. *Beowulf*, 2836.

250. *Beowulf*, 2310.

251. *Beowulf*, 311.

252. *Beowulf*, 2062.

253. *Beowulf*, 31.

254. *Beowulf*, 242, 253.

255. *Beowulf*, 1904.

256. *Beowulf*, 580.

257. *Beowulf*, 521, 2915.

258. Howe, *Writing the Map*, 45–46.

259. Howe, *Writing the Map*, 188.

260. Howe, *Writing the Map*, 188–89; see Fabienne Michelet, *Creation, Migration,*

and Conquest: Imaginary Geography and Sense of Space in Old English Literature
(Oxford: Oxford University Press, 2006), 74.

261. Howe, *Writing the Map*, 189

262. *Beowulf*, 910–13.

263. *Beowulf*, 1390.

264. *Beowulf*, 2027, 3080.

265. *Beowulf*, 2211, 144.

266. *Beowulf*, 2210; see Howe, *Writing the Map*, 190–91.

267. *Beowulf*, 2733–36.

268. Mitchell and Robinson, "The Geatish-Swedish Wars," in Mitchell and Robinson, *Beowulf*, 181.

269. *Beowulf*, 1205–7.

270. *Beowulf*, 2359–60.

271. Jack, introduction to *Beowulf*, 12; and Swanton, *Beowulf*, 197. The references are to the story of Chochilaicus, in Gregory of Tours, *Historia Francorum* III, 3, and the *Liber historiae Francorum*, ed. Bruno Krusch, in *Monumenta Germanica Historica: Scriptores Rerum Merovingicarum* (Hanover, Germany: Impensis Bibliopolii Hahniani, 1888), vol. 2; trans. Bernard S. Bachrach (Lawrence, KS: Coronado, 1973), XIX. For a discussion, see John McNamara, "Beowulf and Hygelac: Problems for Fiction in History," *Rice University Studies* 62 (Spring 1976): 55–63; and Richard A. Gerberding, *The Rise of the Carolingians and the Liber Historiae Francorum* (Oxford: Clarendon Press, 1987), 39–40.

272. Mitchell and Robinson, introduction to Mitchell and Robinson, *Beowulf*, 29; see Jack, introduction to *Beowulf*, 12; and Mitchell and Robinson, "Geatish-Swedish Wars," 181–82.

273. *Beowulf*, 2472–74.

274. *Beowulf*, 2910–13.

275. *Beowulf*, 2999–3003.

276. This is a key theme of Michelet, *Creation, Migration, and Conquest*. See esp. 91, 107.

277. The vocabulary of *lond*, *londe*, and *launde* continues in Chaucer, although he also uses words such as *contree*. The key text is "The Knight's Tale." See Geoffrey Chaucer, *The Canterbury Tales*, ed. Jill Mann (London: Penguin, 2005). See also Suzanne Conklin Akbari, "Orientation and Nation in Chaucer's *Canterbury Tales*," and Sylvia Tomasch, "*Mappae Mundi* and 'The Knight's Tale': The Geography of Power, the Technology of Control," in *Chaucer's Cultural Geography*, ed. Kathryn L. Lynch, 102–34, 192–224 (New York: Routledge, 2002).

278. This account thus differs from the claim of Michelet, *Creation, Migration, and Conquest*, 74–75, that "to secure a territory and to prosper is a recurrent concern of the *Beowulf* poet, thus testifying to the importance of spatial control and of land possession" (see also 109, 114). In her otherwise remarkable account, Michelet uses the term *territory* in a way that lacks precision and textual reference.

CHAPTER 4

1. See the Vatican's own descriptions of the room and paintings, at "Raphael's Rooms," Vatican City State, http://www.vaticanstate.va/EN/Monuments/The_Vatican _Museums/Raphael_s_Rooms.htm.

2. The donation of Constantine was derived in part from the *Legendi sancti Silvestri*, Saint Sylvester's Legend. See John Holland Smith, *Constantine the Great* (London: Hamish Hamilton, 1971), chap. 17; Joseph Canning, *A History of Medieval Thought, 300–1450* (London: Routledge, 1996), 73; and Walter Ullmann, *A Short History of the Papacy in the Middle Ages*, 2nd ed. (London: Routledge, 2003), 36–37. On the donation generally, see John J. Ign von Döllinger, *Fables Respecting the Popes of the Middle Ages: A Contribution to Ecclesiastical History*, trans. Alfred Plummer (London: Rivingtons, 1871), chap. 5.

3. Paulus Hinschius, ed., *Decreteles Pseudo-Isidorianae et Capitula Angilramni*, 2 vols. (Leipzig, Germany: Bernardi Tauchnitz, 1863), 1:249–54. On these more generally, see Walter Ullmann, *The Growth of Papal Government in the Middle Ages: A Study in the Ideological Relation of Clerical to Lay Power*, 2nd ed. (London: Methuen, 1962), 180–84.

4. R. W. Carlyle and A. J. Carlyle, *A History of Mediævel Political Theory in the West*, 6 vols. (Edinburgh: William Blackwell and Sons, 1903–36), 1:288.

5. Dante, *Inferno* XV, 115–17, in Dante Aligheri, *The Divine Comedy*, trans. with a commentary by Charles S. Singleton, Italian-English ed., 3 vols. (Princeton, NJ: Princeton University Press, 1970–75).

6. Charles Howard McIlwain, *The Growth of Political Thought in the West: From the Greeks to the End of the Middle Ages* (London: Macmillan, 1932), 271. Yet other historical accounts, such as Gregory of Tours, do not mention it, thus helping to establish a date before which it did not exist.

7. Confusingly, this is Pope Stephen III in some accounts, since a previous pope named Stephen had died before being consecrated, and so is counted by some and not others.

8. Lynette Olson, *The Early Middle Ages: The Birth of Europe* (London: Palgrave, 2007), 40. On these lands, see Mgr. L. Duchesne, *The Beginnings of the Temporal Sovereignty of the Popes, A.D. 754–1073*, trans. Arnold Harris Mathew (London: Kegan Paul, 1908); Peter Partner, *The Lands of St. Peter: The Papal State in the Medieval Ages and the Early Renaissance* (London: Eyre Methuen, 1972); and Thomas F. X. Noble, *The Republic of St. Peter: The Birth of the Papal State, 680–825* (Philadelphia: University of Pennsylvania Press, 1984).

9. "The Donation of Constantine," in *The Treatise of Lorenzo Valla on the Donation of Constantine*, ed. and trans. Christopher B. Coleman, Latin-English ed. (Toronto: Renaissance Society of America / University of Toronto Press, 1993), chap. 13. Valla's text appears as "Discourse of Lorenzo Valla" in this edition. A more recent edition is Lorenzo Valla, *The Donation of Constantine*, trans. G. W. Bowerstock (Cambridge, MA: Harvard University Press, 2007).

10. "Donation of Constantine," XIV, §6. Valla ("Discourse of Lorenzo Valla," 126, 127) is critical of this, suggesting, "Are not provinces and cities, 'places'? And when you

have said provinces you add cities, as though the latter would not be understood with the former." A general discussion can be found in Christopher B. Coleman, *Constantine the Great and Christianity: Three Phases: The Historical, the Legendary, and the Spurious* (New York: Columbia University Press, 1914), esp. pt. 3; and F. Zinkeisen, "The Donation of Constantine," *English Historical Review* 9 (1894): 625–32.

11. "Donation of Constantine," XIV, §7.

12. Honorius of Augsburg, *Summa Gloria*, in *Monumenta Germanaie Historica: Libelli de Lite Imperatorum et Pontificum Saeculis XI et XII* (Hanover, Germany: Impensis Bibliopolii Hahniani, 1897), 3:63–80, sec. 17; see Carlyle and Carlyle, *History of Mediævel Political Theory*, 4:289–91; and Ullmann, *Growth of Papal Government*, 414–19.

13. John B. Morrall, *Political Thought in Medieval Times*, 3rd ed. (London: Hutchinson University Library, 1971), 25, 28.

14. Walter Ullmann, *Medieval Political Thought* (Harmondsworth, UK: Peregrine, 1975), 137.

15. Carlyle and Carlyle, *History of Mediævel Political Theory*, 4:291.

16. Innocent IV, letter to Friedrich II, 1245, in Eduard Winkelmann, ed., *Acta imperii inedita*, 2 vols. (Innsbruck: Wagner'schen Universitäts-Buchhandlung, 1885), 2:698.

17. See, for example, "Letter of Gregory IX to Frederick II," October 1236, in Brian Tierney, *The Crisis of Church and State, 1050–1300* (Toronto: University of Toronto Press, 1988), 143–44. For a useful discussion, see Robert Black, "The Donation of Constantine: A New Source for the Concept of the Renaissance?," in *Language and Images of Renaissance Italy*, ed. Alison Brown, 51–85 (Oxford: Clarendon Press, 1955).

18. Walter Ullmann, *Medieval Papalism: The Political Theories of the Medieval Canonists* (London: Methuen, 1949), 108.

19. Ullmann, *Medieval Papalism*, 163.

20. Tierney, *Crisis of Church and State*, 142.

21. Black, "Donation of Constantine," 70; and Christopher B. Coleman, "Introductory," in *Treatise of Lorenzo Valla on the Donation of Constantine*, 3. Nicholas of Cusa, *De concordantia catholica libri tres*, in *Opera Omnia*, ed. Gerhardus Kallen (Hamburg: Felicis Meiner, 1959–68), vol. 14; and *The Catholic Concordance*, ed. and trans. Paul E. Sigmund (Cambridge: Cambridge University Press, 1991), III, II, 299. On Cusa, see chap. 7.

22. "Discourse of Lorenzo Valla," 28, 29.

23. "Discourse of Lorenzo Valla," 32, 33.

24. "Discourse of Lorenzo Valla," 34, 35.

25. "Discourse of Lorenzo Valla," 44, 45, 46, 47, 132, 133, 162, 163, 170, 171; see Salvatore I. Camporeale, "Lorenzo Valla's Oratio on the Pseudo-Donation of Constantine: Dissent and Innovation in Early Renaissance Humanism," *Journal of the History of Ideas* 57 (1996): 20.

26. "Discourse of Lorenzo Valla," 54, 55.

27. John 18:36, quoted in "Discourse of Lorenzo Valla," 54, 55.

28. "Discourse of Lorenzo Valla," 60, 61, quoting Matthew 22:21.

29. "Discourse of Lorenzo Valla," 82, 83.

30. Ronald K. Delph, "Valla Grammaticus, Agostino Steuco, and the Donation of Constantine," *Journal of the History of Ideas* 57 (1996): 55.

31. See Camporeale, "Lorenzo Valla's Oratio on the Pseudo-Donation of Constantine," 15.

32. "Donation of Constantine," XIV.

33. "Discourse of Lorenzo Valla," 94, 95.

34. "Discourse of Lorenzo Valla," 94, 95.

35. "Discourse of Lorenzo Valla," 84, 85.

36. See Riccardo Fubini, "Humanism and Truth: Valla Writes against the Donation of Constantine," *Journal of the History of Ideas* 57 (1996): 80; drawing upon Wolfram Setz, "Einleitung" to Lorenzo Valla, *De falso credita et ementita Constantini donatione*, ed. Wolfram Setz, *Quellen zur Geistesgeschichte des Mittelalters* (Weimar: Monumenta Germaniae Historica, 1976), 10:15.

37. Delph, "Valla Grammaticus, Agostino Steuco," 58.

38. Delph, "Valla Grammaticus, Agostino Steuco," 57.

39. See, for example, Carlo Ginzburg, "Lorenzo Valla on the 'Donation of Constantine,'" in *History, Rhetoric and Proof: The Menachim Stern Jerusalem Lectures*, 54–70 (Hanover, NH: University Press of New England, 1999), 54. On the general context at the time Valla was writing, see Anthony Black, *Monarchy and Community: Political Ideas in the Later Conciliar Controversy, 1430–1450* (Cambridge: Cambridge University Press, 1970).

40. "Discourse of Lorenzo Valla," 20, 21.

41. Denys Hay, *The Medieval Centuries* (London: Methuen, 1953), 5.

42. Robert Folz, *The Concept of Empire in Western Europe from the Fifth to the Fourteenth Century*, trans. Sheila Ann Ogilvie (London: Edward Arnold, 1969), 3.

43. Folz, *Concept of Empire in Western Europe*, 3.

44. Lewis Thorpe, introduction to *Two Lives of Charlemagne* (London: Penguin, 1969), 4–5, 190n12.

45. Rosamond McKitterick, *The Frankish Kingdoms under the Carolingians, 751–987* (London: Longman, 1983), 48.

46. Einhard, *Vita Caroli Magni*, ed. Georgius Heinricus Pertz (Hanover, Germany: Impensis Bibliopolii Aulici Hahniani, 1839); trans. Lewis Thorpe as "The Life of Charlemagne," in *Two Lives of Charlemagne*, III, 28.

47. Tierney, *Crisis of Church and State*, 18. See Ernst Kantorowicz, *Laudes Regiae: A Study in Liturgical Acclamation and Mediaeval Ruler Worship* (Berkeley: University of California Press, 1958).

48. Marcia L. Colish, *Medieval Foundations of the Western Intellectual Tradition, 400–1400* (New Haven, CT: Yale University Press, 1997), 337; and Folz, *Concept of Empire in Western Europe*, 85.

49. Colish, *Medieval Foundations of the Western Intellectual Tradition*, 336.

50. Folz, *Concept of Empire in Western Europe*, 21. On Charlemagne, see Paul Edward Dutton, *Charlemagne's Mustache and Other Cultural Clusters of a Dark Age* (London: Palgrave Macmillan, 2004).

51. Morrall, *Political Thought in Medieval Times*, 24.

52. McKitterick, *Frankish Kingdoms*, 72.

53. Einhard, *Vita Caroli*, III, 28.

54. Ullmann, *Short History of the Papacy*, 83–84.

55. Ullmann, *Growth of Papal Government*, 61.

56. Canning, *History of Medieval Political Thought*, 68.

57. McKitterick, *Frankish Kingdoms*, 131.

58. The translation is taken from Canning, *History of Medieval Political Thought*, 68.

59. Ullmann, *Growth of Papal Government*, 98, 114.

60. McKitterick, *Frankish Kingdoms*, 131; drawing upon Peter Classen, "Romanum gubernans imperium: Zur vorgeschichte der Kaisertitulatur Karls des Großen," *Deutsches Archiv für die Erforschung des Mittelalters names der Monumenta Germaniae Historica* 9 (1951): 103–21.

61. McKitterick, *Frankish Kingdoms*, 72.

62. Ullmann, *Growth of Papal Government*, 112; see Ernst Kantorowicz, *Frederick the Second, 1194–1250* (London: Constable, 1931), 441; and Walter Ullmann, *The Carolingian Renaissance and the Idea of Kingship: The Birkbeck Lectures, 1968–9* (London: Methuen, 1969), 139.

63. Ullmann, *Growth of Papal Government*, 112–13.

64. J. M. Wallace-Hadrill, "The *Via Regia* of the Carolingian Age," in *Trends in Medieval Political Thought*, ed. Beryl Smalley, 22–41 (Oxford: Basil Blackwell, 1965), 23.

65. Colish, *Medieval Foundations of the Western Intellectual Tradition*, 72.

66. Folz, *Concept of Empire in Western Europe*, 85.

67. Peter Armour, "Dante and Popular Sovereignty," in *Dante and Governance*, ed. John Woodhouse, 27–45 (Oxford: Clarendon Press, 1997), 31.

68. Ullmann, *Medieval Political Thought*, 67.

69. Ullmann, *Medieval Political Thought*, 70.

70. Colish, *Medieval Foundations of the Western Intellectual Tradition*, 72.

71. Henri Pirenne, *Mohammed and Charlemagne*, trans. Bernard Miall (London: George Allen and Unwin, 1939), 284–85; and Olson, *Early Middle Ages*, 66–67.

72. Pirenne, *Mohammed and Charlemagne*, 236; for a powerful case for one, see Ullmann, *Carolingian Renaissance and the Idea of Kingship*; and McKitterick, *Frankish Kingdoms*, chaps. 6 and 8.

73. See Henri Pirenne, *Economic and Social History of Medieval Europe*, trans. I. E. Clegg (London: Kegan Paul, 1936); and Pirenne, *Mohammed and Charlemagne*, 17. More generally, see Pirenne, *A History of Europe from the Invasions to the XVI Century*, trans. Bernard Miall (London: George Allen and Unwin, 1939). For critical discussions, see Anne Riising, "The Fate of Henri Pirenne's Thesis on the Consequences of Islamic Expansion," *Classica et Medievalia* 13 (1952): 87–130; Alfred F. Havighurst, ed., *The Pirenne Thesis: Analysis, Criticism and Revision* (Boston: D. C. Heath, 1958); and Richard Hodges and David Whitehouse, eds., *Mohammed, Charlemagne and the Origins of Europe: Archaeology and the Pirenne Thesis* (Ithaca, NY: Cornell University Press, 1983); and, more generally, Michael McCormick, *Origins of the European Economy: Communications and Commerce, A.D. 300–900* (Cambridge: Cambridge University Press, 2001); and Adriaan Verhulst, *The Carolingian Economy* (Cambridge: Cambridge University Press, 2002).

74. Pirenne, *Mohammed and Charlemagne*, 234.

75. Pirenne, *Mohammed and Charlemagne*, 285.

76. R. Van Deem, "The Pirenne Thesis and Fifth-Century Gaul," in *Fifth-Century Gaul: A Crisis of Identity?*, ed. John Drinkwater and Hugh Elton, 321–33 (Cambridge: Cambridge University Press, 1992), 325.

77. Pirenne, *Mohammed and Charlemagne*, 147.

78. Ullmann, *Growth of Papal Government*, 272.

79. Olson, *Early Middle Ages*, 197.

80. Jacques le Goff, *Medieval Civilization, 400–1500*, trans. Julia Barrow (Oxford: Basil Blackwell, 1988), 140.

81. Jacques le Goff, *La civilisation de l'occident médiéval* (Paris: Flammarion, 1997), 121; and le Goff, *Medieval Civilization, 400–1500*, 145.

82. An early example is Notker der Stammler, *Taten Kaiser Karls des Grossen* (Munich: Monumenta Germaniae Historica, 1980); trans. Lewis Thorpe as Notker the Stammerer, "The Life of Charlemagne," in *Two Lives of Charlemagne*.

83. *La Chanson de Roland: Manuscrit d'Oxford*, ed. Raoul Mortier (Paris: La geste francor, 1940); and *The Song of Roland*, trans. Dorothy L. Sayers (Harmondsworth, UK: Penguin, 1957). See Peter Haidu, *The Subject of Violence: The* Song of Roland *and the Birth of the State* (Bloomington: Indiana University Press, 1993).

84. Quoted in Einhard, *Vita Caroli*, III, 31.

85. Einhard, *Vita Caroli*, II, 7.

86. "Division Regnorum," in *Monumenta Germaniae Historica Legum*, sec. 2, *Capitularia Regum Francorum*, ed. Alfredus Boretius (Hanover, Germany: Impensis Bibliopolii Hahniani, 1883), text no. 45, 126–30; and "Charlemagne's Division of the Kingdoms," in *Carolingian Civilisation: A Reader*, ed. Paul Edward Dutton, 129–33 (Peterborough, Ontario: Broadview, 1993).

87. Folz, *Concept of Empire in Western Europe*, 98–99.

88. David Nicholas, *The Evolution of the Medieval World: Society, Government and Thought in Europe, 312–1500* (London: Longman, 1992), 123.

89. Le Goff, *Medieval Civilization, 400–1500*, 265.

90. Le Goff, *Medieval Civilization, 400–1500*, 266.

91. McIlwain, *Growth of Political Thought*, 230.

92. "Mandatum ad Ottonem Frisingensem Datum," in *Monumenta Germaniae Historica Legem IV: Constitutiones et acta publica: Imperatorum et regum, 911–1197*, ed. Ludewicus Weiland (Hanover, Germany: Impensis Bibliopolii Hahniani, 1843), 1:224; see Carlyle and Carlyle, *History of Mediævel Political Theory*, 3:173.

93. R. N. Swanson, *The Twelfth-Century Renaissance* (Manchester, UK: Manchester University Press, 1998), 89.

94. See Charles Christopher Mierow, "Bishop Otto of Freising: Historian and Man," *Transactions and Proceedings of the American Philological Association* 80 (1949): 393–402.

95. Otto of Freising, *Gesta Friderici I Imperatoris: Monumenta Germaniae historica. Scriptores rerum germanicarum in usum scholarum*, vol. 46 (Hanover, Germany: Hahn, 1884); and *The Deeds of Frederick Barbarossa*, trans. Charles Christopher Mierow (New York: Columbia University Press, 1953).

96. Walter Ullmann, *Law and Politics in the Middle Ages: An Introduction to the Sources of Medieval Political Ideas* (Cambridge: Cambridge University Press, 1975), 95; and Nicholas, *Evolution of the Medieval World*, 123.

97. Hay, *Medieval Centuries*, 103; and Folz, *Concept of Empire in Western Europe*, 25.

98. Morrall, *Political Thought in Medieval Times*, 79.

99. Samuel von Pufendorf, *Der statu imperii Germanici*, in *Die Verfassung des deutschen Reiches*, ed. Horst Denzer, Latin-German ed. (Stuttgart: Insel Verlag, 1994), chap. 6, p. 9. See the discussion in chap. 9.

100. Voltaire, *Essai sur les moeurs et l'esprit des nations*, in *Oeuvres Completes*, 70 vols. (Paris: La Société Litterarire-Typographique, 1785–89), 17:267.

101. Benjamin Arnold, *Princes and Territories in Medieval Germany* (Cambridge: Cambridge University Press, 1991), 280; and McIlwain, *Growth of Political Thought*, 160. On the later stages of the Holy Roman Empire, see John G. Gagliardo, *Reich and Nation: The Holy Roman Empire as Idea and Reality* (Bloomington: Indiana University Press, 1980).

102. Arthur Stephen McGrade, *The Political Thought of William of Ockham: Personal and Institutional Principles* (Cambridge: Cambridge University Press, 1974), 150.

103. Keen, *History of Medieval Europe*, 162; see Kantorowicz, *Frederick the Second*.

104. McIlwain, *Growth of Political Thought*, 179.

105. Le Goff, *Medieval Civilization, 400–1500*, 3.

106. F. R. Cowell, *Cicero and the Roman Republic* (Harmondsworth, UK: Penguin, 1948), 323; see J. Oliver Thomson, *History of Ancient Geography* (Cambridge: Cambridge University Press, 1948), 320; and C. R. Whittaker, *Frontiers of the Roman Empire: A Social and Economic Study* (Baltimore: Johns Hopkins University Press, 1994), 12.

107. O. A. W. Dilke, *Greek and Roman Maps* (London: Thames and Hudson, 1985), 167.

108. A comprehensive atlas of the ancient world, but using modern technology, is Richard J. A. Talbert, ed., *Barrington Atlas of the Greek and Roman World* (Princeton, NJ: Princeton University Press, 2000).

109. This is the characterization of Gavin A. Sundwall, "Ammianus Geographicus," *American Journal of Philology* 117 (1996): 619n2. C. R. Whittaker, *Rome and Its Frontiers: The Dynamics of Empire* (London: Routledge, 2004), 74, suggests that works like Evelyn Edson, *Mapping Time and Space: How Medieval Mapmakers Viewed Their World*, 36–51 (London: British Library, 1997), and David Woodward, "Reality, Symbolism, Time, and Space in Medieval World Maps," *Annals of the Association of American Geographers* 75 (1985): 510–21, are in danger of thinking that previous cartographers were "like us but more stupid."

110. O. A. W. Dilke, "Maps in the Service of the State: Roman Cartography to the End of the Augustan Era," "Roman Large-Scale Mapping in the Early Empire," and "Itineraries and Geographical Maps in the Early and Late Roman Empires," in *The History of Cartography*, vol. 1, *Cartography in Prehistoric, Ancient, and Medieval Europe and the Mediterranean*, ed. J. B. Harley and David Woodward, 201–11, 212–33, 234–57 (Chicago: University of Chicago Press, 1987); O. A. W. Dilke, *The Roman Land Sur-*

veyors: An Introduction to the Agrimensores (Newton Abbot, UK: David and Charles, 1971); and Dilke, *Greek and Roman Maps.*

111. A. D. Lee, *Information and Frontiers: Roman Foreign Relations in Late Antiquity* (Cambridge: Cambridge University Press, 1993), 86.

112. Claude Nicolet, *Space, Geography, and Politics in the Early Roman Empire,* trans. Hélène Leclerc (Ann Arbor: University of Michigan Press, 1991), 1.

113. Nicolet, *Space, Geography, and Politics,* 2.

114. Nicolet, *Space, Geography, and Politics,* 2.

115. Nicholas Purcell, "Maps, Lists, Money, Order and Power," *Journal of Roman Studies* 80 (1990): 180.

116. Pietro Janni, *La mappa e il periplo: Cartografia antica e spazio odologico* (Rome: Giorgio Bretschneider, 1984); Tønnes Bekker-Nielsen, "*Terra Incognita*: The Subjective Geography of the Roman Empire," in *Studies in Ancient History and Numismatics Presented to Rudi Thomsen,* ed. Askel Damsgaard-Madsen, Erik Christiansen, and Erik Hallager (Århus, Denmark: Aarhus University Press, 1988), 148–61; Richard J. A. Talbert, "Greek and Roman Maps," *Journal of Roman Studies* 77 (1987): 210–12; Richard J. A. Talbert, "Rome's Empire and Beyond: The Spatial Aspect," in *Gouvernants et gouvernés dans l'Imperium Romanum (IIIe av. J.-C.-Ier ap. J.-C.),* ed. E. Hermon, Cahiers des Études Anciennes, 26 (Quebec: Société des Études anciennes du Quebec, 1990), 215–23; and Richard J. A. Talbert, "L'inventaire du monde: Geographie et politique aux origins de l'Empire romain," *American Historical Review* 94 (December 1989): 1351.

117. Susan Mattern, *Rome and the Enemy: Imperial Strategy in the Principate* (Berkeley: University of California Press, 1999), 26.

118. See Arthur H. Robinson and Barbara Bartz Petchenik, *The Nature of Maps: Essays toward Understanding Maps and Mapping* (Chicago: University of Chicago Press, 1976), 123.

119. Kai Brodersen, *Terra Cognita: Studien zur römischen Raumerfassung* (Hildesheim, Germany: Georg Olms, 1995); see Sundwall, "Ammianus Geographicus," 619n2; and Kai Brodersen, "Mapping (in) the Ancient World," *Journal of Roman Studies* 94 (2004): 183–90.

120. Janni, *La mappa e il periplo;* see Lee, *Information and Frontiers,* 86–87.

121. Nicholas Purcell, "The Creation of Provincial Landscape: The Roman Impact on Cisalpine Gaul," in *The Early Roman Empire in the West,* ed. Thomas Blagg and Martin Millett, 7–29 (Oxford: Oxbow Books, 2002), 12; and Lee, *Information and Frontiers,* 86.

122. Lee, *Information and Frontiers,* 86.

123. See Sundwall, "Ammianus Geographicus," 635n36; Purcell, "Maps, Lists, Money, Order and Power," 178; and James S. Romm, *The Edges of the Earth in Ancient Thought: Geography, Exploration, and Fiction* (Princeton, NJ: Princeton University Press, 1992), 29.

124. Dilke, *Roman Land Surveyors,* 109.

125. Kai Brodersen, "The Presentation of Geographical Knowledge for Travel and Transport in the Roman World: *Itineraria non tantum adnotata sed etiam picta,*" in

Travel and Geography in the Roman Empire, ed. Colin Adams and Ray Laurence, 7–21 (London: Routledge, 2001), 12.

126. Sundwall, "Ammianus Geographicus," 619; see Talbert, "Greek and Roman Maps," 211.

127. Talbert, "Rome's Empire and Beyond."

128. Thomson, *History of Ancient Geography,* 219.

129. Brodersen, *Terra Cognita;* see also his "Presentation of Geographical Knowledge."

130. Brodersen, "Presentation of Geographical Knowledge," 9.

131. See Benjamin Isaac, *The Limits of Empire: The Roman Army in the East,* rev. ed. (Oxford: Clarendon Press, 1992), 401–2; Fergus Millar, "Emperors, Frontiers and Foreign Relations, 31 B.C. to A.D. 378," *Britannia* 13 (1982): 1–23; Lee, *Information and Frontiers,* 85; David Cherry, *Frontier and Society in Roman North Africa* (Oxford: Clarendon, 1998), 32; and Robert K. Sherk, "Roman Geographical Exploration and Military Maps," *Aufstieg und Niedergang der Römischen Welt* 2 (1974): 534–62.

132. Purcell, "Creation of Provincial Landscape," 9.

133. Nicolet, *Space, Geography, and Politics,* 95.

134. Sundwall, "Ammianus Geographicus," 621, traces this to Louis XIV's minister Jean-Baptiste Colbert.

135. Nicolet, *Space, Geography, and Politics,* 189.

136. Dilke, *Greek and Roman Maps,* 42.

137. Nicolet, *Space, Geography, and Politics,* 101.

138. Nicolet, *Space, Geography, and Politics,* 9, 16.

139. Nicolet, *Space, Geography, and Politics,* 9, 171.

140. Nicolet, *Space, Geography, and Politics,* 17. Talbert, "L'inventaire du monde," 1351, suggests that his arguments are, in contrast, "exaggerated."

141. Nicolet, *Space, Geography, and Politics,* 101.

142. Nicolet, *Space, Geography, and Politics,* 8. That said, there is a debate as to whether it was actually composed later, under Tiberius. For various discussions, see Dilke, *Greek and Roman Maps,* 62–65; Ronald Syme, *Anatolica: Studies in Strabo,* ed. Antony Birley (Oxford: Clarendon Press, 1995); Christian Jacob, "Mapping in the Mind: The Earth from Ancient Alexandria," in *Mappings,* ed. Denis Cosgrove, 24–49 (London: Reaktion, 1999); Daniela Dueck, *Strabo of Amasia: A Greek Man of Letters in Augustan Rome* (London: Routledge, 2000); and Daniela Dueck, Hugh Lindsay, and Sarah Pothecary, eds., *Strabo's Cultural Geography: The Making of a* Kolossourgia (Cambridge: Cambridge University Press, 2005). On forerunners to Strabo, also including Herodotus, see F. W. Walbank, "The Geography of Polybius," *Classica et Mediaevalia* 9 (1948): 155–82. Dueck notes that, compared to Ptolemy, Strabo offers "more of a tourist-guide with political orientation" (*Strabo of Amasia,* 125).

143. Pliny the Elder, *Naturalis Historia,* III, 17 (*Naturalis Historia/Natural History,* Latin-English ed., trans. H. Rackham, W. H. S. Jones, and D. E. Eichholz, 10 vols. [London: William Heinemann, 1938–63]). See Dilke, *Greek and Roman Maps,* 52; on Strabo, see 62–65; and Thomson, *History of Ancient Geography,* 332.

144. Nicolet, *Space, Geography, and Politics,* 101. For a detailed listing of Pliny's references, see Dilke, *Greek and Roman Maps,* 44–52.

145. Dilke, *Greek and Roman Maps*, 42.

146. Brodersen, *Terra Cognita*, 277–78; Brodersen, "Presentation of Geographical Knowledge," 20n8; see Edson, *Mapping Time and Space*, 11.

147. Brodersen, *Terra Cognita*, 275–77.

148. Benet Salway, "Travel, *Itineraria* and *Tabellaria*," in *Travel and Geography in the Roman Empire*, ed. Adams and Laurence, 29.

149. Benet Salway, "The Nature and Genesis of the Peutinger Map," *Imago Mundi* 57 (June 2005): 128.

150. Salway, "Nature and Genesis of the Peutinger Map," 128. On this see Herbert Bloch, "A New Edition of the Marble Plan of Ancient Rome," *Journal of Roman Studies* 51 (1961): 143–52.

151. Livy, *Ab urbe condita*, XLI, 28.

152. Brodersen, "Presentation of Geographical Knowledge," 20n9.

153. Otto Cuntz, ed., *Itineraria Romana*, vol. 1, *Itineraria Antonini Augusti et Burdigalense* (Stuttgart, Germany: B. G. Teubneri, 1990). See Brodersen, "Presentation of Geographical Knowledge," 13; Brodersen, *Terra Cognita*, 165–90; and Salway, "Travel, *Itineraria* and *Tabellaria*."

154. Bekker-Nielsen, "*Terra Incognita*," 152.

155. R. Moynihan, "Geographical Mythology and Roman Imperial Ideology," in *Age of Augustus*, ed. Winkes, 151.

156. Konrad Miller, *Die Peutingersche Tafel*, expanded ed. (Stuttgart, Germany: Strecker und Schröder, 1916); and Ekkehard Weber, ed., *Tabula Peuteringeriana: Codex Vindobonesis 324* (Graz, Austria: Akademie Druck-u. Verlagsanstanlt, 1976).

157. Salway, "Nature and Genesis of the Peutinger Map," 119.

158. Richard Talbert, "Peutinger's Roman Map: The Physical Landscape Framework," in *Wahrnehmung und Fassung geographischer Räume in der Antike*, ed. Michael Rathmann, 221–30 (Mainz am Rhein, Germany: Philipp von Zabern, 2007), 221.

159. Dilke, *Roman Land Surveyors*, 109.

160. Dilke, *Roman Land Surveyors*, 110–11; see O. A. W. Dilke, *Mathematics and Measurement* (London: British Museum, 1987), 38; and Salway, "Nature and Genesis of the Peutinger Map," 128. Salway critically assesses the account of the map found in J. Wilhelm Kubitschek, "Eine römische Straßenkarte," *Jahreshefte des Österreichischen archäologischen Instituts in Wien*, no. 5 (1902): 20–96.

161. See Thomson, *History of Ancient Geography*, 379; and Dilke, *Mathematics and Measurement*, 38.

162. Salway, "Nature and Genesis of the Peutinger Map," 120.

163. For a thorough discussion, see Richard J. A. Talbert, *Rome's World: The Peutinger Map Reconsidered* (Cambridge: Cambridge University Press, 2010); Richard J. A. Talbert, "Cartography and Taste in Peutinger's Roman Map," in *Space in the Roman World: Its Perception and Presentation*, Antike Kultur und Geschichte, ed. Richard Talbert and Kai Brodersen, 113–41 (Münster, Germany: Lit Verlag, 2004); and his "Travel and Geography in the Roman Empire (Review)," *American Journal of Philology* 123 (2002): 529–34.

164. Claudius Ptolemy, *The Geography*, trans. Edward Luther Stevenson (New York: Dover, 1991); and *Ptolemy's Geography: An Annotated Translation of the Theoretical*

Chapters, ed. and trans. J. Lennart Berggren and Alexander Jones (Princeton, NJ: Princeton University Press, 2000).

165. Salway, "Nature and Genesis of the Peutinger Map," 131n1. See, more generally, Alfred Stückelberger, "Ptolemy and the Problem of Scientific Presentation of Space," in *Space in the Roman World,* ed. Talbert and Brodersen, 27–40.

166. On this map, see P. D. A. Harvey, *Mappa Mundi: The Hereford World Map,* 2nd ed. (Hereford: Hereford Cathedral, 2002); P. D. A. Harvey, ed., *The Hereford World Map: Medieval World Maps and Their Context* (London: British Library, 2006); Scott D. Westrem, *The Hereford Map: A Transcription and Translation of the Legends with Commentary* (Turnhout, Belgium: Brepols, 2001); Kathy Lavezzo, *Angels on the Edge of the World: Geography, Literature, and the English Community, 1000–1534* (Ithaca, NY: Cornell University Press, 2006); and John Block Friedmann, *The Monstrous Races in Medieval Art and Thought* (Syracuse, NY: Syracuse University Press, 2000), chap. 3. For a detailed discussion of the role of Jerusalem, see Edson, *Mapping Time and Space;* and, more generally, Natalia Lozovsky, *"The Earth Is Our Book": Geographical Knowledge in the Latin West, ca. 400–1000* (Ann Arbor: University of Michigan Press, 2000), 20.

167. P. D. A. Harvey, "Local and Regional Cartography in Medieval Europe," in *History of Cartography,* vol. 1, ed. Harley and Woodward, 464; and P. D. A. Harvey, *Medieval Maps* (Toronto: University of Toronto Press, 1991), 7.

168. See Peter Barber, "Medieval Maps of the World," in *Hereford World Map,* ed. Harvey, 1–44.

169. J. B. Harley and David Woodward, preface to *History of Cartography,* vol. 1, ed. Harley and Woodward, xvi.

170. Edson, *Mapping Time and Space,* viii.

171. John Kirtland Wright, *Geographical Lore of the Time of the Crusades: A Study in the History of Medieval Science and Tradition in Western Europe* (New York: American Geographical Society, 1925), 246.

172. Wright, *Geographical Lore of the Time of the Crusades,* 2.

173. See Alessandro Scafi, *Mapping Paradise: A History of Heaven on Earth* (Chicago: University of Chicago Press, 2006).

174. Colish, *Medieval Foundations of the Western Intellectual Tradition,* 325.

175. Hay, *Medieval Centuries,* 86. On this see F. D. Hunt, *Holy Land Pilgrimage in the Later Roman Empire, AD 312–460* (Oxford: Clarendon Press, 1982); Jonathan Sumption, *Pilgrimage: An Image of Mediaeval Religion* (Totowa, NJ: Rowman and Littlefield, 1975); and C. Raymond Beazley, *The Dawn of Modern Geography,* 3 vols. (1897–1906; repr., New York: Peter Smith, 1949) (citations refer to the 1949 edition).

176. Ullmann, *Short History of the Papacy,* 166. Gregory's address can be found in *Patrologiae Cursus Completus, Series Latina,* vol. 148, col. 329; and in Oliver J. Thatcher and Edgar Holmes McNeal, eds., *A Source Book for Medieval History* (New York: Scribners, 1905), 512–13.

177. See the anonymous account of the First Crusade in the *Gesta Francorum,* Latin-English ed., in *The Deeds of the Franks and the Other Pilgrims to Jerusalem,* ed. Rosalind Hill (Oxford: Clarendon Press, 1972). On the Crusades, see Peter Lock, *The Routledge Companion to the Crusades* (London: Routledge, 2006); and Conor Kustick, *The Social Structure of the First Crusade* (Leiden, Netherlands: Brill, 2008).

178. Jonathan Riley-Smith, *The First Crusade and the Idea of Crusading*, 2nd ed. (London: Continuum, 2009), 21.

179. See, for example, Joshua 15–21 and Numbers 34.

180. The only exception seems to be Second Esdras, from the Apocrypha, which only exists in its Latin translation, not the original Hebrew. For a discussion of land politics, see W. D. Davies, *The Gospel and the Land: Early Christianity and Jewish Territorial Doctrine* (Berkeley: University of California Press, 1974).

181. Michel Foucault, *Sécurité, Territoire, Population: Cours au Collège de France (1977–1978)*, ed. Michel Senellart (Paris: Seuil/Gallimard, 2004).

182. See, for example, Rachel Havrelock, *River Jordan: The Mythology of a Dividing Line* (Chicago: University of Chicago Press, 2011). Havrelock notes, for instance, that David Ben-Gurion held "a weekly study group on the book of Joshua . . . attended by biblical scholars, politicians and military officials" (14).

183. See "Urban II's Sermon," in *Medieval Worlds: A Sourcebook*, ed. Roberta Anderson and Dominic Aidan Bellenger, 88–92 (London: Routledge, 2003).

184. Wim Blockmans and Peter Hoppenbrouwers, *Introduction to Medieval Europe, 300–1550*, trans. Isola van den Hoven (London: Routledge, 2007), 196.

185. Janet L. Abu-Loghod, *Before European Hegemony: The World System, A.D. 1250–1350* (Oxford: Oxford University Press, 1989), 47.

186. See George H. T. Kimble, *Geography in the Middle Ages* (London: Methuen, 1938), 69.

187. Walther I. Brandt, introduction to Pierre Dubois, *The Recovery of the Holy Land*, trans. Walther I. Brandt, 3–65 (New York: Columbia University Press, 1956), 3, 5.

188. Pierre Dubois, *De recuperatione Terre Sancte: Dalla "Respublica Christiana" ai primi nazionalismi e alla politica antimediterranea*, ed. Angelo Diotti (Firenze, Italy: Leo S. Olschki, 1977); and Dubois, *Recovery of the Holy Land*, I, 99.

189. Innocent IV, *Super Libros Quinque Decretalium*, III, xxxiiii, 8 (Frankfurt am Main, Germany, 1570), fol. 430v.

190. Edward Gibbon, *History of the Decline and Fall of the Roman Empire*, ed. David Womersley, 3 vols. (Harmondsworth, UK: Penguin, 1985), 3:728.

191. For discussions, see J. J. Saunders, *The History of the Monghol Conquests* (London: Routledge and Kegan Paul, 1971); and Peter Jackson, *The Mongols and the West, 1221–1410* (Harlow, UK: Pearson, 2005).

192. Keen, *History of Medieval Europe*, 190.

193. Ferdinand Braudel, *A History of Civilisations*, trans. Richard Mayne (Harmondsworth, UK: Penguin, 1995), 304; see Norman Davies, *Europe: A History* (Oxford: Oxford University Press, 1996), chap. 1.

194. Keen, *History of Medieval Europe*, 190. See Jack Weatherford, *Genghis Khan and the Making of the Modern World* (New York: Crown Publishers, 2004).

195. Hay, *Medieval Centuries*, 103.

196. "Sententia de iure statuum terrae," May 1, 1231, in *Monumenta Germanaie Historica: Constitutiones et acta publica imperatorum et regum*, vol. 2, *1198–1272*, ed. Ludewicus Weiland (Hanover, Germany: Impensis Bibliopolii Hahniani, 1896), 420.

197. Arnold, *Princes and Territories*, 217.

198. On Westphalia, and this relation, see chap. 9.

199. Arnold, *Princes and Territories*, 232–33. On 39 he suggests "territorial lord-ship" as a translation of *Landeshoheit*, and trades on Robert Sack's analysis of territo-riality (p. 67), but the term then loses any specificity and conceptual precision. On the distinction between manorial lordship and territorial lordship, see B. H. Slicher van Bath, *The Agrarian History of Western Europe, A.D. 500–1850*, trans. Olive Ordish (New York: St. Martin's Press, 1964), 49–53; and Georges Duby, *Rural Economy and Country Life in the Medieval West*, trans. Cynthia Postan (Columbia: University of South Carolina Press, 1968).

200. See, for example, F. L. Ganshof, *Feudalism*, trans. Philip Grierson (London: Longmans, Green, 1952); Marc Bloch, *Feudal Society*, trans. L. A. Manyon (London: Routledge and Kegan Paul, 1961); J. S. Critchley, *Feudalism* (London: George Allen and Unwin, 1978); and Georges Duby, *The Three Orders: Feudal Society Imagined*, trans. Arthur Goldhammer (Chicago: University of Chicago Press, 1980).

201. See, for example, Perry Anderson, *Passages from Antiquity to Feudalism* (London: NLB, 1974); Perry Anderson, *Lineages of the Absolutist State* (London: NLB, 1974); Barry Hindess and Paul Q. Hirst, *Pre-capitalist Modes of Production* (London: Routledge and Kegan Paul, 1975); and John Haldon, *The State and the Tributary Mode of Production* (London: Verso, 1993). For its use to describe premedieval societies, see Max Weber, *The Agrarian Sociology of Ancient Civilizations*, trans. R. I. Frank (London: NLB, 1976).

202. See Francis Oakley, *Empty Bottles of Gentilism: Kingship and the Divine in Late Antiquity and the Early Middle Ages (to 1050)* (New Haven, CT: Yale University Press, 2010), 185.

203. Le Goff, *Medieval Civilization, 400–1500*, 91–92.

204. Joseph R. Strayer and Rushton Coulborn, "The Idea of Feudalism," in *Feudal-ism in History*, ed. Rushton Coulborn, 3–11 (Princeton, NJ: Princeton University Press, 1956), 3.

205. Carlyle and Carlyle, *History of Mediæval Political Theory*, 3:19. The entirety of this volume is an attempt to situate the political theory in the context of feudalism and its laws.

206. Nicolas Brussel, *Nouvel examen de l'usage general des fiefs* (Paris: Claude Prud'homme and Claude Robustel, 1727), 1:2; see McIlwain, *Growth of Political Thought*, 181.

207. Strayer and Coulborn, "Idea of Feudalism," 4–5.

208. This is convincingly proposed by Oakley, *Empty Bottles of Gentilism*, 187–88.

209. See the essays in Wendy Davies and Paul Fouracre, eds., *Property and Power in the Early Middle Ages* (Cambridge: Cambridge University Press, 2002).

210. R. W. Southern, *The Making of the Middle Ages* (London: Hutchinson, 1953), 99.

211. Wood, *Medieval Economic Thought*, 225.

212. Hay, *Medieval Centuries*, 78.

213. Olson, *Early Middle Ages*, 179.

214. Nicholas, *Evolution of the Medieval World*, 218. On pre–Norman Conquest models in England, see H. Cabot Lodge, "The Anglo-Saxon Land Law," in *Essays in Anglo-Saxon Law*, ed. Henry Adams, H. Cabot Lodge, Ernest Young, and J. Laurence

Laughlin, 55–119 (Boston: Little, Brown and Company, 1905); Della Hooke, "Territorial Organisation in the Anglo-Saxon West Midlands: Central Places, Central Areas," in *Central Places, Archaeology and History*, ed. Eric Grant, 79–93 (Sheffield, UK: Department of Archaeology and Prehistory, University of Sheffield, 1986); Della Hooke, *The Landscape of Anglo-Saxon England* (London: Leicester University Press, 1998); G. R. J. Jones, "Cults, Saxons and Scandinavians," in *An Historical Geography of England and Wales*, ed. R. A. Dodgshon and R. A. Butlin, 2nd ed., 45–68 (London: Academic Press, 1990).

215. See John Hudson, *Land, Law, and Lordship in Anglo-Norman England* (Oxford: Clarendon Press, 1994); John Hudson, *The Formation of the English Common Law: Law and Society in England from the Norman Conquest to Magna Carta* (London: Longman, 1996); and George Garnett, *Conquered England: Kingship, Succession, and Tenure, 1066–1166* (Oxford: Oxford University Press, 2007).

216. Della Hooke, "Introduction: Later Anglo-Saxon England," in *Anglo-Saxon Settlements*, ed. Della Hooke, 1–8 (Oxford: Basil Blackwell, 1988), 3. See Bruce M. S. Campbell, "People and Land in the Middle Ages, 1066–1500," in *Historical Geography of England and Wales*, ed. Dodgshon and Butlin, 69–121.

217. See H. C. Darby, *Domesday England* (Cambridge: Cambridge University Press, 1977); and Tom Williamson, *Shaping Medieval Landscapes: Settlement, Society, Environment* (Macclesfield, UK: Windgather, 2004).

218. Lynn White Jr., *Medieval Technology and Social Change* (London: Oxford University Press, 1962), 37. For some serious criticisms, see P. H. Sawyer and R. H. Hilton, "Technical Determinism: The Stirrup and the Plough," *Past & Present* 4 (1963): 90–100.

219. Le Goff, *Medieval Civilization, 400–1500*, 222.

220. Maurice Keen, *A History of Medieval Europe* (London: Penguin, 1991), 48.

221. Nicholas, *Evolution of the Medieval World*, 125.

222. Diana Wood, *Medieval Economic Thought* (Cambridge: Cambridge University Press, 2002), 95; see generally A. W. Richeson, *English Land Measuring to 1800: Instruments and Practices* (Cambridge, MA: Society for the History of Technology and MIT Press, 1966), chap. 2; and G. G. Coulton, *The Medieval Village* (Cambridge: Cambridge University Press, 1931), 44–48.

223. Coulton, *Medieval Village*, 48.

224. Le Goff, *Medieval Civilization, 400–1500*, 177.

225. Nicholas, *Evolution of the Medieval World*, 219.

226. Wood, *Medieval Economic Thought*, 95.

227. Southern, *Making of the Middle Ages*, 179.

228. Harald Kleinschmidt, *Understanding the Middle Ages: The Transformation of Ideas and Attitudes in the Medieval World* (Woodbridge, UK: Boydell Press, 2000), 51.

229. For a valuable discussion of the period 1350–1520, arguing for a much longer shift between feudalism and capitalism, see Christopher Dyer, *An Age of Transition? Economy and Society in England in the Later Middle Ages* (Oxford: Clarendon Press, 2005).

230. On the latter, see Georges Duby, *France in the Middle Ages: From Hugh Capet to Joan of Arc*, trans. Juliet Vale (Oxford: Blackwell, 1991).

231. Harold Dexter Hazeltine, "Roman and Canon Law in the Middle Ages," in *The*

Cambridge Medieval History, vol. 5, *Contest of Empire and Papacy*, ed. J. R. Tanner, C. W. Previté-Orton, and Z. N. Brooke, 697–764 (Cambridge: Cambridge University Press, 1957), 728; and Peter Stein, *Roman Law in European History* (Cambridge: Cambridge University Press, 1999), 39.

232. "Magna Carta," in *Medieval Worlds: A Sourcebook*, ed. Roberta Anderson and Dominic Aidan Bellenger, 152–60 (London: Routledge, 2003).

CHAPTER 5

1. For a helpful discussion of his influence, see Amnon Linder, "The Knowledge of John of Salisbury in the Late Middle Ages," *Studi Medievali*, ser. 3, 18 (1977): 315–66.

2. Cary J. Nederman, "Editor's Introduction," in John of Salisbury, *Policraticus: On the Frivolities of Courtiers and the Footprints of Philosophers*, ed. and trans. Cary J. Nederman, xv–xxviii (Cambridge: Cambridge University Press, 1990), xx. See Raymond Klibansky, *The Continuity of Platonic Translation during the Middle Ages: Outline of a Corpus Platonicum Medii Aevi* (London: Warburg Institute, 1939); and George H. T. Kimble, *Geography in the Middle Ages* (London: Methuen, 1938), 74.

3. Nederman, "Editor's Introduction," in John of Salisbury, *Policraticus*, xx.

4. John of Salisbury, *Episcopi Carnotensis Metalogicon*, ed. Clement C. J. Webb (Oxford: Clarendon, 1929); and *The Metalogicon of John of Salisbury: A Twelfth-Century Defence of the Verbal and Logical Arts of the Trivium*, trans. Daniel D. McGarry (Gloucester: Peter Smith, 1971).

5. Hans Lieberschütz, *Mediaeval Humanism in the Life and Writings of John of Salisbury* (London: Warburg Institute, 1950), 34.

6. The dates are disputed. See Gilles Constable, "The Alleged Disgrace of John of Salisbury in 1159," *English Historical Review* 6970 (1954): 65–76, who dates his first exile to 1156.

7. Kate Langdon Forham, "Salisburian Stakes: The Uses of 'Tyranny' in John of Salisbury's *Policraticus*," *History of Political Thought* 11 (Autumn 1990): 397.

8. Jacques le Goff, *Medieval Civilization, 400–1500*, trans. Julia Barrow (Oxford: Basil Blackwell, 1988), 273.

9. Good general introductions include Lieberschütz, *Mediaeval Humanism*; and Clement C. J. Webb, *John of Salisbury* (London: Methuen, 1932).

10. John Dickinson, "Introduction: The Place of the Policraticus in the Development of Political Thought," in *The Statesman's Book of John of Salisbury*, trans. John Dickinson (New York: Russell and Russell, 1927), xvii. John of Salisbury, *Episcopi carnotensis Policratici*, ed. Clement C. J. Webb, 2 vols. (Oxford: Clarendon, 1909). There is no complete English translation, but an almost complete version can be patched together from John of Salisbury, *Frivolities of Courtiers and Footprints of Philosophers*, trans. Joseph B. Pike (New York: Octagon Books, 1972) (books 1, 2, and 3 and excerpts from 7 and 8); and *Statesman's Book of John of Salisbury* (books 4, 5, and 6 and excerpts from 7 and 8). The best, although again partial, translation is the Nederman version cited above.

11. David Knowles, *The Evolution of Medieval Thought*, 2nd ed. (London: Longman, 1988), 126.

12. Ernst H. Kantorowicz, *The King's Two Bodies: A Study in Mediaeval Political Theology* (Princeton, NJ: Princeton University Press, 1957), 96.

13. See also John of Salisbury, *Policraticus*, prologue.

14. John of Salisbury, *Policraticus*, I, 3; V, 1; and Dickinson, "Introduction: Place of the Policraticus," xvii–xviii.

15. Cary J. Nederman, "Duty to Kill: John of Salisbury's Theory of Tyrannicide," *Review of Politics* 50 (1988), 370; see Kantorowicz, *King's Two Bodies*, 199–200. On the body politic at this time more generally, see Michael Camille, "The Image and the Self: Unwriting Late Medieval Bodies," in *Framing Medieval Bodies*, ed. Sarah Kay and Miri Rubin, 62–99 (Manchester, UK: Manchester University Press, 1994), esp. 68–69. See Miri Rubin, *Corpus Christi: The Eucharist in Late Medieval Culture* (Cambridge: Cambridge University Press, 1991); Kenneth R. Stow, "Holy Body, Holy Society: Conflicting Medieval Structural Conceptions," in *Sacred Space: Shrine, City, Land*, ed. Benjamin Z. Kedar and R. J. Zwi Werblowsky, 151–71 (New York: New York University Press, 1998); and Peter Biller and A. J. Minnis, eds., *Medieval Theology and the Natural Body* (York, UK: York Medieval Press, 1997).

16. John of Salisbury, *Policraticus*, V, 2.

17. It was previously taken in good faith, but has been largely discredited. See Nederman, "Editor's Introduction," in John of Salisbury, *Policraticus*, xxi; Lieberschütz, *Mediaeval Humanism*, 24–25, 24n3. The idea, attributed simply to Plutarch, structures Christine de Pizan's early fifteenth-century text *The Book of the Body Politic*, ed. Kate Langdon Forhan (Cambridge: Cambridge University Press, 1994).

18. Dickinson, "Introduction: Place of the Policraticus," xxi.

19. John of Salisbury, *Policraticus*, V, 2.

20. *The Letters of John of Salisbury*, vol. 1, *The Early Letters (1153–1161)*, ed. W. J. Miller, H. E. Butler, and C. N. L. Brooke, Latin-English ed. (Oxford: Clarendon Press, 1986), 181, letter 111. See Cary J. Nederman, "The Physiological Significance of the Organic Metaphor in John of Salisbury's *Policraticus*," *History of Political Thought* 8 (1987): 218. On this theme, see Ewart Lewis, "Organic Tendencies in Medieval Political Thought," *American Political Science Review* 32 (1938): 849–76; and Cary J. Nederman, "Body Politics: The Diversification of Organic Metaphors in the Later Middle Ages," *Pensiero Politico Medievale* 2 (2004): 59–87.

21. See Nederman, "Duty to Kill," 370–71.

22. Otto Gierke, *Political Theories of the Middle Age*, trans. Frederic William Maitland (Cambridge: Cambridge University Press, 1900), 23–24.

23. Lieberschütz, *Mediaeval Humanism*, 45.

24. Walter Ullmann, *The Growth of Papal Government in the Middle Ages: A Study in the Ideological Relation of Clerical to Lay Power*, 2nd ed. (London: Methuen, 1962), 420.

25. *The Letters of John of Salisbury*, vol. 2, *The Later Letters (1163–1180)*, ed. W. J. Miller and C. N. L. Brooke, Latin-English ed. (Oxford: Clarendon Press, 1979), 726, letter 305; see Ullmann, *Growth of Papal Government*, 426.

26. John of Salisbury, *Policraticus*, VIII, 17.

27. Le Goff, *Medieval Civilization, 400–1500*, 264, 265.

28. On the positive attribution, see Nederman, "Physiological Significance of the Organic Metaphor," 211–23.

29. J. H. Waszink, ed., *Timaeus: A Calcidio Translatus Commentarioque Instructus* (*Plato Latinus*, ed. Raymond Klibansky, vol. 4) (London: Warburg Institute and E. J. Brill, 1962).

30. For a discussion, see Stephen Gersh, *Middle Platonism and Neoplatonism in the Latin Tradition*, 2 vols. (Notre Dame, IN: University of Notre Dame Press, 1986), vol. 2, chap. 6; Gretchen J. Reydams-Schils, "Calcidius on the Human and the World Soul and Middle-Platonist Psychology," *Apeiron* 39 (2006): 177–200; and Gretchen J. Reydams-Schils, "Meta-Discourse: Plato's *Timaeus* According to Calcidius," *Phronesis* 52 (2007): 301–27. The question of Calcidius's sources is discussed in J. C. M. van Winden, *Calcidus on Matter: His Doctrine and Sources: A Chapter in the History of Platonism* (Leiden, Netherlands: E. J. Brill, 1959); and John Phillips, "Numenian Psychology in Calcidius?" *Phronesis* 48 (2003): 132–51. For a geographical reading, see Keith D. Lilley, "Mapping Cosmopolis: Moral Topographies of the Medieval City," *Environment and Planning D: Society and Space* 22 (2004): 681–98.

31. Paul Edward Dutton, "*Illustre civiatis et populi exemplum*: Plato's *Timaeus* and the Transformation from Calcidus to the End of the Twelfth Century of a Tripartite Scheme of Society," *Mediaeval Studies* 45 (1985): 85; and, more generally, see his "Medieval Approaches to Calcidius," in *Plato's* Timaeus *as Cultural Icon*, ed. Gretchen J. Reydams-Schils, 183–205 (Notre Dame, IN: University of Notre Dame Press, 2003).

32. Dutton, "*Illustre civiatis et populi exemplum*," 108; and John of Salisbury, *Metalogicon*, IV, 18.

33. Dutton, "*Illustre civiatis et populi exemplum*," 110–11.

34. John of Salisbury, *Policraticus*, V, 7.

35. John of Salisbury, *Policraticus*, VI, 25.

36. John of Salisbury, *Policraticus*, VI, 6.

37. Kantorowicz, *King's Two Bodies*; see Paul E. Sigmund, introduction to *St. Thomas Aquinas on Politics and Ethics*, ed. Paul E. Sigmund, xii–xxvii (New York: W. W. Norton, 1988), xiv.

38. For a useful survey, see Nederman, "Duty to Kill." As Carlyle and Carlyle, *History of Mediæval Political Theory*, 3:126–27, suggest, this key theme of the distinction between a king and tyrant is found throughout the Middle Ages, but John of Salisbury is its privileged site. They trace its influence back to Isidore of Seville, and through him to Cicero, rather than directly to Aristotle.

39. This fits with the analysis of Cary J. Nederman and Catherine Campbell, in "Priests, Kings, and Tyrants: Spiritual and Temporal Power in John of Salisbury's *Policraticus*," *Speculum* 66 (July 1991): 572–90. They argue that there is no clear hierarchy established between secular and spiritual power, but that this does not mean an inconsistency on John's part.

40. Forham, "Salisburian Stakes," 403; and Nederman, "Physiological Significance of the Organic Metaphor," 222.

41. Forham, "Salisburian Stakes," 406.

42. See Janet Martin, "Uses of Tradition: Gellius, Petronius and John of Salisbury," *Viator* 10 (1979): 62.

43. John Dickinson, preface to *Statesman's Book of John of Salisbury*, xi. More generally on its reception, see Walter Ullmann, "John of Salisbury's Policraticus in the Later Middle Ages," in *Geschichtsschreibung und geistiges Leben im Mittelalter: Festschrift für Heinz Löwe zur 65. Geburtstag*, ed. Karl Hauck and Hubert Mordek, 519–45 (Cologne: Böhlau, 1978).

44. These texts were unknown to John. But the use he made of the *Organon* in the *Metalogicon* shows the likely impact they would have had on his thought. For a discussion of his library, see Webb, *John of Salisbury*, chap. 6.

45. J. A. Watt, "Spiritual and Temporal Powers," in *The Cambridge History of Medieval Political Thought, c. 350–c. 1450*, ed. J. H. Burns, 367–423 (Cambridge: Cambridge University Press, 1988), 367.

46. *The Bible: Authorized King James Version*, ed. Robert Carroll and Stephen Prickett (Oxford: Oxford University Press, 2008). On this version, see Charles C. Butterworth, *The Literary Lineage of the King James Bible, 1340–1611* (Philadelphia: University of Pennsylvania Press, 1941).

47. The Latin in Matthew uses the possessive: *"Converte gladium tuum in locum suum"*; but the Greek is vaguer: *"Apostrepson son ten machairan eis ton topon autes."*

48. Le Goff, *Medieval Civilization, 400–1500*, 273. For a discussion, see Gerard E. Caspary, *Politics and Exegesis: Origen and the Two Swords* (Berkeley: University of California Press, 1979), esp. the introduction.

49. Joseph Canning, *A History of Medieval Political Thought, 300–1450* (London: Routledge, 1996), 49–50. We can see it in that sense, for instance, in John of Salisbury, *Policraticus*, IV, 3.

50. Bernard of Clairvaux, *De consideratione ad Eugenium Papam* (Paris: Mequignon Juniorem, 1828), trans. John D. Anderson and Elizabeth T. Kennan as *Five Books on Consideration: Advice to a Pope* (Kalamazoo: Cistercian Publications, 1976), IV, III.7.

51. Gerhart B. Ladner, "The Concepts of 'Ecclesia' and 'Christianitas' and Their Relation to the Idea of Papal 'Plenitudo Potestatis' from Gregory VII to Boniface VIII," *Miscellanea Historiae Pontificiae* 18 (1954): 57–58.

52. See his letter to Emperor Anastasius in 494, in J. H. Robinson, *Readings in European History*, 2 vols. (Boston: Ginn, 1905), 1:72–73.

53. Walter Ullmann, *Medieval Political Thought* (Harmondsworth, UK: Peregrine, 1975), 41. For a more general discussion in the context of the early medieval church, see Walter Ullmann, "Leo I and the Theme of Papal Primacy," *Journal of Theological Studies*, n.s., 11 (1960): 25–51.

54. Larry Scanlan, *Narrative, Authority, and Power: The Medieval Exemplum and the Chaucerian Tradition* (Cambridge: Cambridge University Press, 1994), 5. See also 37–54 on the *auctoritas/potestas* distinction.

55. "Encyclica Invitatoria ad Episcopos Teutonicos," in *Monumenta Germaniae Historica: Constitutiones et acta publica imperatorum et regum*, ed. Ludwig Weiland (1893; repr., Hanover, Germany: Hahnsche Buchhandlung, 1999), 1:253–54 (citations refer to the 1999 edition).

56. Arthur Monaghan, *Consent, Coercion, and Limit: The Medieval Origins of*

Parliamentary Democracy (Kingston and Montreal: McGill-Queen's University Press, 1987), 18–19.

57. Discussed in Ullmann, *Medieval Political Thought*, 54.

58. Francis de Meyronnes, *De principatu regni Siciliae*, 1, ed. Pierre de Lapparent, in "L'œuvre politique de François de Meyronnes, ses rapports avec celle de Dante," *Archives d'Histoire Doctrinale et Littéraire du Moyen Age* 15–17 (1940–42): 96–97; discussed in Michael Wilks, *The Problem of Sovereignty in the Later Middle Ages: The Papal Monarchy with Augustinus Triumphus and the Publicists* (Cambridge: Cambridge University Press, 1964), 138–39.

59. Gregory VII, *Dictatus Papae*, in *Quellen zur Geschichte des Papsttums und des Römanischen Katholizismus*, ed. D. Carl Mirbt, 4th ed. (Tübingen, Germany: J. C. B. Mohr, 1924), 146–47, translated in Ewart Lewis, *Medieval Political Ideas*, 2 vols. (London: Routledge and Kegan Paul, 1954), 2:380–81.

60. See Gregory VII, *Dictatus Papae*, 8; and Gregory VII's letter to Bishop Herman of Metz, March 1081, in Brian Tierney, *The Crisis of Church and State, 1050–1300* (Toronto: University of Toronto Press, 1988), 67.

61. Paulus Hinschius, ed., *Decreteles Pseudo-Isidorianae et Capitula Angilramni*, 2 vols. (Leipzig, Germany: Bernardi Tauchnitz, 1863).

62. Harald Kleinschmidt, *Understanding the Middle Ages: The Transformation of Ideas and Attitudes in the Medieval World* (Woodbridge, UK: Boydell Press, 2000), 324. Part of the dispute was with Emperor Henry IV. On this earlier dispute, see Carlyle and Carlyle, *History of Mediævel Political Theory*, vol. 4, pt. 3, chap. 1.

63. Bernard of Clairvaux, "Letter to Pope Eugenius III," in Cary J. Nederman and Kate Langham Fordan, *Medieval Political Theory—A Reader: The Quest for the Body Politic, 1100–1400* (London: Routledge, 1993), 22–23.

64. See Nederman and Fordan's introduction to Bernard of Clairvaux, "Letter to Pope Eugenius III," in *Medieval Political Theory*, 21.

65. Thomas Aquinas, *Scriptum super libros Sententiarum Magistri Petri Lombardi Episcopi Parisiensis* (Paris: Lethielleux, 1929–47), IV, d. 37 q. 2. a. 2, exp. text.

66. Thomas Aquinas, *Summa Theologiæ*, Latin-English ed., trans. various, 60 vols. (London: Blackfriars and Eyre and Spottiswoode, 1963–76), IaIIæ 40.

67. Bernard of Clairvaux, *De consideratione*, II, VI.13.

68. Carlyle and Carlyle, *History of Mediævel Political Theory*, 4:335.

69. Carlyle and Carlyle, *History of Mediævel Political Theory*, 4:211.

70. Carlyle and Carlyle, *History of Mediævel Political Theory*, 4:395.

71. See Daniel Waley, *The Papal State in the Thirteenth Century* (London: Macmillan, 1961), on Innocent III's direct rule over the lands of the church; and Sidney R. Packard, *Europe and the Church under Innocent III*, rev. ed. (New York: Russell and Russell, 1968), chap. 1. Also helpful is Kenneth Pennington, "The Legal Education of Pope Innocent III," *Bulletin of Medieval Canon Law*, n.s., 4 (1974): 70–77.

72. See the letters excerpted in Tierney, *Crisis of Church and State*, 132, 133.

73. See Carlyle and Carlyle, *History of Mediævel Political Theory*, 5:152–53; and Charles Howard McIlwain, *The Growth of Political Thought in the West: From the Greeks to the End of the Middle Ages* (London: Macmillan, 1932), 231–33.

74. Michael Wilks, *The Problem of Sovereignty in the Later Middle Ages: The Papal*

Monarchy with Augustinus Triumphus and the Publicists (Cambridge: Cambridge University Press, 1964), 398–99. The emergence of this designation in the eleventh century is discussed by Agostino Paravicini-Bagliani, *The Pope's Body*, trans. David S. Peterson (Chicago: University of Chicago Press, 2000), 58–59.

75. Takashi Shogimen, *Ockham and Political Discourse in the Late Middle Ages* (Cambridge: Cambridge University Press, 2007), 226–27. On Aegidius, see chap. 6.

76. Kantorowicz, *King's Two Bodies*, 89.

77. Innocent III, "Per Venerabilem," in *Quellen zur Geschichte des Papsttums und des Römanischen Katholizismus*, ed. Mirbt, 176; translated in Tierney, *Crisis of Church and State*, 137; see McIlwain, *Growth of Political Thought*, 231.

78. Innocent III, "Per Venerabilem," 176, 136.

79. J. P. Canning, "Law, Sovereignty and Corporation Theory, 1300–1450," in *The Cambridge History of Medieval Political Thought, c. 350–c. 1450*, ed. J. H. Burns, 454–76 (Cambridge: Cambridge University Press, 1988), 469.

80. John A. Scott, *Dante's Political Purgatory* (Philadelphia: University of Pennsylvania Press, 1996), 25.

81. Kimble, *Geography in the Middle Ages*, 75–76.

82. On the intellectual rebirth of this period, see Charles Homer Haskins, *The Renaissance of the Twelfth Century* (Cambridge, MA: Harvard University Press, 1927); Christopher Brooke, *The Twelfth Century Renaissance* (London: Thames and Hudson, 1969); and R. N. Swanson, *The Twelfth-Century Renaissance* (Manchester, UK: Manchester University Press, 1998).

83. Thomas Gilby, *Principality and Polity: Aquinas and the Rise of State Theory in the West* (London: Longmans, Green, 1958), 80.

84. Menso Folkerts, "Arabic Mathematics in the West," in *The Development of Mathematics in Medieval Europe: The Arabs, Euclid, Regiomontanus* (Aldershot, UK: Ashgate Variorium, 2006), 1; and Swanson, *Twelfth-Century Renaissance*, 51–52.

85. On the broader topic of Arab geography, see Ralph W. Brauer, *Boundaries and Frontiers in Medieval Muslim Geography* (Philadelphia: American Philosophical Society, 1995).

86. Folkerts, "Arabic Mathematics in the West."

87. Kimble, *Geography in the Middle Ages*, 63.

88. Shlomo Pines, "Philosophy, Mathematics and the Concepts of Space in the Middle Ages," in *The Interaction between Science and Philosophy*, ed. Y. Elkana, 75–91 (Atlantic Highlands, NJ: Humanities Press, 1974). For another valuable discussion, see Edward Grant, "Place and Space in Medieval Physical Thought," in *Motion and Time, Space and Matter: Interrelations in the History of Philosophy and Science*, ed. Peter K. Machamer and Robert G. Turnbull, 137–67 (Columbus: Ohio State University Press, 1976). This essentially suggests that the medieval period provided the basis for the "infinite container for the motions of bodies" described by Newton (161).

89. Folkerts, "Arabic Mathematics in the West," 11.

90. Folkerts, "Euclid in Medieval Europe," in *Development of Mathematics in Medieval Europe*, 2; and Howard Rollin Patch, *The Tradition of Boethius: A Study of His Importance in Medieval Culture* (1935; repr., New York: Russell and Russell, 1970), 4 (citations refer to the 1970 edition).

91. Folkerts, "Arabic Mathematics in the West," 11; and Folkerts, "Euclid in Medieval Europe," 8.

92. R. W. Southern, *The Making of the Middle Ages* (London: Hutchinson, 1953), 65.

93. Nicholas, *Evolution of the Medieval World*, 341.

94. Alfred Ivry, "Averroes," in *Medieval Philosophy*, ed. John Marenton, 49–64 (London: Routledge, 1998), 58.

95. Ivry, "Averroes," 52.

96. Wilks, *Problem of Sovereignty in the Later Middle Ages*, 85.

97. Nederman, *Worlds of Difference*, 21.

98. Brown, *Augustine of Hippo*, 91.

99. David Bradshaw, *Aristotle East and West: Metaphysics and the Division of Christendom* (Cambridge: Cambridge University Press, 2004), 73.

100. Plotinus, *Enneads*, trans. A. H. Armstrong, Greek-English ed., 6 vols. (London: William Heinemann, 1966–88). For a discussion, see Lloyd Gerson, ed., *The Cambridge Companion to Plotinus* (Cambridge: Cambridge University Press, 2006).

101. Brown, *Augustine of Hippo*, 91.

102. Brown, *Augustine of Hippo*, 91.

103. Klibansky, *Continuity of Platonic Translation*, 36. On this see Stephen Gersh and Maarten J. F. M. Hoenen, eds., *The Platonic Tradition in the Middle Ages* (Berlin: Walter de Gruyter, 2002). See John Dillon, *The Heirs of Plato: A Study of the Old Academy (347–274 BC)* (Oxford: Clarendon Press, 2003); and Gersh, *Middle Platonism and Neoplatonism in the Latin Tradition*.

104. Klibansky, *Continuity of Platonic Translation*, 28.

105. Klibansky, *Continuity of Platonic Translation*, 13, 35.

106. Ernst H. Kantorowicz, "Plato in the Middle Ages," *Selected Studies*, 184–93 (Locust Valley, NY: J. J. Augustin, 1965), 184.

107. Gilby, *Principality and Polity*, 78; see Copleston, *History of Medieval Philosophy*, 153–54; and Knowles, *Evolution of Medieval Thought*, 186.

108. Albertus was an important scholar in his own right. Of particular interest was his taking forward of Aristotle's notion of place from the *Physics*, in an interpretation that proved fundamental for Scholasticism. See Sister Jean Paul Tilmann, *An Appraisal of the Geographical Works of Albertus Magnus and His Contribution to Geographical Thought* (Ann Arbor: Department of Geography, University of Michigan, 1971) (this contains a translation of *De Natura Locorum*, 25–145; the Latin text is found as "De Natura Loci," in Alberti Magni, *Opera Omnia*, V.2 [Aschendorff: Monasterri Westfalorum, 1980], 1–44); J. A. Weishiepl, *Albertus Magnus and the Sciences: Commemorative Essays* (Toronto: Pontifical Institute of Medieval Studies, 1980); and Odd Langholm, *Economics in the Medieval Schools: Wealth, Exchange, Value, Money and Usury According to the Paris Theological Tradition, 1200–1350* (Leiden, Netherlands: E. J. Brill, 1992), chap. 7. More generally, see Helen S. Lang, *Aristotle's* Physics *and Its Medieval Varieties* (Albany: SUNY Press, 1992).

109. See G. H. Luquet, "Hermann l'Allemand," *Revue de l'histoire des religions* 44 (1901): 407–22; and Judson Boyce Allen, "Hermann the German's Averroistic Aristotle and Medieval Poetic Theory," *Mosaic* 9 (1976): 67–81.

110. On William, see Bernard G. Dod, "Aristoteles Latinus," in *The Cambridge His-*

tory of Later Medieval Philosophy, ed. Norman Kretzmann, Anthony Kenny, and Jan Pinborg, 45–79 (Cambridge: Cambridge University Press, 1982), 62–64.

111. Ullmann, *Medieval Political Thought*, 171.

112. Copleston, *History of Medieval Philosophy*, 155.

113. "Decreta magistri Petri de Corbolio," in *Chartularium Universitatis Parisiensis*, ed. Henricus Denifle and Armilio Chatelain, 2 vols. (Paris: Fratrum Delalain, 1889–91), 1:70.

114. Knowles, *Evolution of Medieval Thought*, 206.

115. Gregory IX, letter of April 13, 1231, in *Chartularium Universitatis Parisiensis*, ed. Denifle and Chatelain, 1:138. Both indications to the *Chartularium* are made by Knowles, *Evolution of Medieval Thought*, 206–7.

116. Gilby, *Principality and Polity*, 79.

117. Knowles, *Evolution of Medieval Thought*, 207–8; generally on this period, see Fernard van Steenberghen, *Aristotle in the West: The Origins of Latin Aristotelianism*, trans. Leonard Johnston (Louvain, Belgium: E. Nauwelaerts, 1955), esp. chap. 4.

118. Ullmann, *Medieval Political Thought*, 167; and Robert Goodwin, "General Introduction," in *Selected Writings of St. Thomas Aquinas*, trans. Robert Goodwin (Indianapolis: Library of Liberal Arts, 1965), xiii.

119. See Dod, "Aristoteles Latinus," 59–60.

120. Ralph Lerner and Muhsin Mehdi, eds., *Medieval Political Philosophy: A Sourcebook* (New York: Free Press, 1963), 273; and Dod, "Aristoteles Latinus," 77–78.

121. Dod, "Aristoteles Latinus," 77–78; see William F. Boggess, "Hermannus Alemannus's Rhetorical Translations," *Viator* 2 (1972): 227–50.

122. Dod, "Aristoteles Latinus," 63.

123. Ernst Kantorowicz, *Frederick the Second, 1194–1250* (London: Constable, 1931), 340–41. More generally, see Knowles, *Evolution of Medieval Thought*, 171–73.

124. John B. Morrall, *Political Thought in Medieval Times*, 3rd ed. (London: Hutchinson University Library, 1971), 69.

125. Janet Coleman, *A History of Political Thought: From the Middle Ages to the Renaissance* (Oxford: Blackwell, 2000), 61–62.

126. Ullmann, *Medieval Political Thought*, 159. Conor Martin, "Some Medieval Commentaries on Aristotle's *Politics*," *History* 36 (1951): 29–44.

127. Anthony Black, *Political Thought in Europe, 1250–1450* (Cambridge: Cambridge University Press, 1992), 154.

128. See Tierney, *Crisis of Church and State*, 2; and Mario Grignashchi, "La definition du 'civis' dans la scolastique," in *Gouvernés et Gouvernants*, Recueils de la société Jean Bodin, 24 (Brussels: Éditions de la Librarie Encyclopédique, 1966), 71–88.

129. See Cary J. Nederman, "Aristotle as Authority: Alternative Aristotelian Sources of Late Medieval Political Theory," *History of European Ideas* 4 (1987): 31–44.

130. Gilby, *Principality and Polity*, 79.

131. It is suggested, for example, by Marcia L. Colish, *Medieval Foundations of the Western Intellectual Tradition, 400–1400* (New Haven, CT: Yale University Press, 1997), 295; but compellingly disputed by Jean-Pierre Torrell, *Initiation à saint Thomas d'Aquin: Sa personne et son oeuvre* (Fribourg, Switzerland: Éditions Universitaires de Fribourg, 1993), 255–58.

132. Dod, "Aristoteles Latinus," 68; see Jean Dunbabin, "The Reception and Interpretation of Aristotle's *Politics*, in *The Cambridge History of Later Medieval Philosophy*, ed. Norman Kretzmann, Anthony Kenny, and Jan Pinborg, 723–37 (Cambridge: Cambridge University Press, 1982), 723.

133. Joseph Canning, *The Political Thought of Baldus de Ubaldis* (Cambridge: Cambridge University Press, 1987), 161. For a very helpful discussion, see Nicolai Rubinstein, "The History of the Word *Politicus* in Early-Modern Europe," in *The Languages of Political Theory in Early Modern Europe*, ed. Antony Pagden, 41–56 (Cambridge: Cambridge University Press, 1987), esp. 41–48.

134. Canning, *Political Thought of Baldus de Ubaldis*, 161.

135. Ullmann, *Medieval Political Thought*, 171, 175.

136. R. W. Dyson, introduction to Saint Thomas Aquinas, *Political Writings*, ed. and trans. R. W. Dyson (Cambridge: Cambridge University Press, 2002), xxiv.

137. Hay, *Medieval Centuries*, 155.

138. Laurence Paul Hemming, *Worship as a Revelation: The Past, Present and Future of Catholic Liturgy* (London: Burns and Oates, 2008), 61.

139. Ullmann, *Medieval Political Thought*, 174. The description apparently comes from the work of M. Grabmann, but Ullmann does not provide a more exact reference.

140. Knowles, *Evolution of Medieval Thought*, 79.

141. Morrall, *Political Thought in Medieval Times*, 70; and van Steenberghen, *Aristotle in the West*, 183.

142. Gilby, *Principality and Polity*, 95–96, 249, 265–71.

143. Jean Dunbabin, "Aristotle in the Schools," in *Trends in Medieval Political Thought*, ed. Beryl Smalley, 65–85 (Oxford: Basil Blackwell, 1965). Generally, see Jürgen Miethke, "Practical Intentions of Scholasticism: The Case of Political Theory," in *Universities and Schooling in Medieval Society*, ed. William J. Courtney and Jürgen Miethke, 211–28 (Leiden, Netherlands: Brill, 2000).

144. Saint Thomas Aquinas, *In Libros Politicorum Aristotelis Expositio*, ed. Raymundi M. Spiazzi (Turin: Marietti, 1951); Aquinas, *Commentary on Aristotle's Politics*, trans. Richard J. Regan (Indianapolis: Hackett Publishing, 2007); Aquinas, *In Decem Libros Ethicorum Aristotelis ad Nicomachum Expositio*, ed. Raymundi M. Spiazzi (Turin: Marietti, 1949); Aquinas, *Commentary on the Ethics of Aristotle*, trans. Charles I. Litzinger (1964; repr., Notre Dame, IN: Dumb Ox, 1993) (citations refer to the 1993 edition).

145. Jeremy Catto, "Ideas and Experience in the Political Thought of Aquinas," *Past and Present* 71 (1976): 8–9; see Mark D. Jordan, *Rewritten Theology: Aquinas after His Readers* (Oxford: Blackwell, 2006), 77.

146. Aquinas, *In Libros Politicorum Aristotelis Expositio*, prologue. This perhaps explains the lack of treatment of his political thought in many of the standard introductions. For valuable discussions, see R. J. Henle, ed., *The Treatise on Law: Saint Thomas Aquinas, Summa Theologiae, I–II: qq. 90–97* (Notre Dame, IN: University of Notre Dame Press, 1993); Paul E. Sigmund, "Law and Politics," in *The Cambridge Companion to Aquinas*, ed. Norman Kretzmann and Eleonore Stump, 217–31 (Cambridge: Cambridge University Press, 1993). The best study is John Finnis, *Aquinas: Moral, Political, and Legal Theory* (Oxford: Oxford University Press, 1998), esp. chap. 7.

147. Aquinas, *In Libros Politicorum Aristotelis Expositio*, prologue.

148. Aquinas, *In Decem Libros Ethicorum Aristotelis*, I, 1.

149. Aquinas, *In Libros Politicorum Aristotelis Expositio*, prologue.

150. Aristotle, *Politics*, 1253a7–18; Aristotle, *Nicomachean Ethics*, 1098a3–5, 1139a5–6; and Aristotle, *Eudemian Ethics*, 1242a23–24.

151. Aquinas, *In Libros Politicorum Aristotelis Expositio*, I, 1.

152. Aquinas, *In Libros Politicorum Aristotelis Expositio*, I, 1.

153. See, for example, Aquinas, *Summa Theologiæ*, Ia 96.4; and Aquinas, *In Decem Libros Ethicorum Aristotelis*, I, 1.

154. Ullmann, *Medieval Political Thought*, 175–76.

155. Remigio Dei Girolami, "De Bono Communi," in *La "teologia politica communal" di Remigio de' Girolami*, ed. Maria Consiglia De Matteis (Bologna: Patron Editore, 1977), 1–52, 18 [fol. 100r]. See L. Minio-Paluello, "Remigio Girolami's *De Bono Communoi: Florence at the Time of Dante's Banishment and the Philosopher's Answer to the Crisis*," *Italian Studies* 11 (1956): 56–71; Charles T. Davis, *Dante's Italy and Other Essays* (Philadelphia: University of Pennsylvania Press, 1984), chap. 8; and M. Michèle Mulchahey, "Education in Dante's Florence Revisited: Remigio de' Girolami and the Schools of Santa Maria Novella," in *Medieval Education*, ed. Ronald B. Begley and Joseph W. Koterski, S.J., 143–81 (New York: Fordham University Press, 2005).

156. A. P. d'Entrèves, introduction to Saint Thomas Aquinas, *Selected Political Writings* (Oxford: Blackwell, 1981), xv.

157. Morrall, *Political Thought in Medieval Times*, 19; see Alexander Passerin d'Entrèves, *The Medieval Contribution to Political Thought: Thomas Aquinas, Marsilius of Padua, Richard Hooker* (New York: Humanities Press, 1959), 20.

158. d'Entrèves, *Medieval Contribution to Political Thought*, 23–25.

159. Saint Thomas Aquinas, "De Virtutibus in Communi," in *Quaestiones Disputatae*, 2 vols. (Rome: Marietti, 1953), 2:701–827; and Aquinas, *Disputed Questions on the Virtues*, ed. E. M. Atkins and Thomas Williams, trans. E. M. Atkins (Cambridge: Cambridge University Press, 2005), "On the Virtues in General," IX.

160. Gilby, *Principality and Polity*, 312.

161. Aquinas, *De Regimine Principum*, I, 4; see Ptolemy of Lucca, *De Regimine Principum*, IV, 1. Thomae Aquinatis, *Opuscula Omnia necnon Opera Minora: Tome Primus Opuscula Philosophica* (Paris: Lethielleux, 1949), 221–426, which contains the non-Thomist parts in an appendix; and Ptolemy of Lucca, *On the Government of Rulers: De Regimine Principum, with Portions Attributed to Thomas Aquinas*, trans. James M. Blythe (Philadelphia: University of Pennsylvania Press, 1997).

162. Aquinas, *Summa Theologiæ*, IaIIæ 90.2, 90.3.

163. Aquinas, *In Libros Politicorum Aristotelis Expositio*, I, 1.

164. Aquinas, *Summa Theologiæ*, IaIIæ 90.2.

165. Aquinas, *In Libros Politicorum Aristotelis Expositio*, prologue.

166. Aquinas, *Summa Theologiæ*, IaIIæ 90.3.

167. Gilby, *Principality and Polity*, 261.

168. Dino Bigongiari, introduction to *The Political Ideas of St. Thomas Aquinas: Representative Selections*, ed. Dino Bigongiari, vi–xxxvii (New York: Hafner Publishing Company, 1957), x; and Gilby, *Principality and Polity*, 262.

169. d'Entrèves, introduction to Saint Thomas Aquinas, *Selected Political Writings*, xxv.

170. Aquinas, *In Libros Politicorum Aristotelis Expositio*, III, 2.

171. Aquinas, *In Libros Politicorum Aristotelis Expositio*, III, 5.

172. Aquinas, *Summa Theologiæ*, IaIIæ 95.4.

173. d'Entrèves, *Medieval Contribution to Political Thought*, 32.

174. Thomas Aquinas, *Summa contra Gentiles* (Rome: Commissio Leonina, 1894); *The Summa Contra Gentiles*, trans. English Dominican Fathers, 4 vols. (London: Burns Oates and Washbourne, 1924–28), 4:76; and *Summa Theologiæ*, IIIa 8.1–4. See d'Entrèves, introduction to Saint Thomas Aquinas, *Selected Political Writings*, xxv. On this notion, see Henri de Lubac, *Corpus Mysticum: The Eucharist and the Church in the Middle Ages*, trans. Gemma Simmonds with Richard Price, ed. Laurence Paul Hemming and Susan Frank Parsons (London: SCM Press, 2006), esp. chap. 5; and for a helpful discussion, Laurence Paul Hemming, "Henri de Lubac: Reading *Corpus Mysticum*," *New Blackfriars* 90 (2009): 519–34. For de Lubac, as Hemming makes clear, the *Corpus Mysticum* is at once the Eucharist, the body of believers, and Christ's body.

175. Ullmann, *Medieval Political Thought*, 180.

176. Finnis, *Aquinas*, 328.

177. Aquinas, *Summa contra Gentiles*, IV, 76.

178. Thomas Aquinas, *Contra errores Graecorum* (Rome: Commissio Leonina, 1968), II, 32.

179. Bigongiari, introduction to *Political Ideas of St. Thomas Aquinas*, xxxv.

180. Thomas Aquinas, *Commentum in Matthaeum et Joannem Evangelistas*, in *Opera Omnia* (Parma: Typus Petri Fiaccadori, 1860), vol. 10, XVIII (p. 171); discussed in Shogimen, *Ockham and Political Discourse*, 224.

181. Aquinas, *Scriptum super libros Sententiarum*, II d. 44 q. 2 a. 3 exp. text.

182. Aquinas, *Summa Theologiæ*, IIaIIæ 60.6.

183. Shogimen, *Ockham and Political Discourse*, 224.

184. Aquinas, *Summa Theologiæ*, IIaIIæ 87.4.

185. Aquinas, *Summa Theologiæ*, IIaIIæ 87.3.

186. Aquinas, *Summa Theologiæ*, IaIIæ 105.3.

187. See, for example, Aquinas, *Summa Theologiæ*, IaIIæ 89.6.

188. Thomas Aquinas, *In VIII Libros Physicorum Aristotelis*, in *Opera Omnia* (Rome: Commissio Leoninae, 1884), vol. 2; Aquinas, *Commentary on Aristotle's Physics*, trans. Richard J. Blackwell, Richard J. Spath, and W. Edmund Thirlkel (London: Routledge and Kegan Paul, 1963), IV, 3; see also IV, 6, where Aquinas treats *spatium* as effectively the *extensio* within a containing body, following Aristotle in showing that this is not place.

189. Aquinas, *Summa Theologiæ*, Ia 46.1.

190. For discussions, see James A. Weisheipl, *Friar Thomas D'Aquino: His Life, Thought and Work* (Garden City, NY: Doubleday, 1974), 189–95; Torrell, *Initiation à saint Thomas d'Aquin*, 247–49; Holly Hamilton Bleakley, "The Art of Ruling in Aquinas' *De regimine principum*," *History of Political Thought* 20 (1999): 575–602; Catto, "Ideas and Experience in the Political Thought of Aquinas"; and McIlwain, *Growth of Political Thought*, 235–36.

191. Blythe, preface to Ptolemy of Lucca, *On the Government of Rulers*, vii.

192. Aquinas, *De Regimine Principum*, I, 1 (2 in Latin). *Sociabile* appears as *sociale* and *socialis* in different manuscripts.

193. Aquinas, *De Regimine Principum*, I, 15.

194. Aquinas, *De Regimine Principum*, I, 15.

195. Aquinas, *De Regimine Principum*, I, 2.

196. Aquinas, *De Regimine Principum*, II, 1.

197. Aquinas, *De Regimine Principum*, II, 2.

198. Aquinas / Ptolemy of Lucca, *De Regimine Principum*, II, 3–7.

199. Ptolemy of Lucca, *De Regimine Principum*, II, 7.

200. Ptolemy of Lucca, *Determinatio Compendiosa de Jurisdictione Imperii*, ed. Marius Krammer (Hanover, Germany: Impensis Bibliopolii Hahniani, 1909), XV.

CHAPTER 6

1. On Philip see Edgard Boutaric, *Le France sous Philippe le Bel: Etude sur les institutions politiques et administratives du Moyen Age* (1861; repr., Brionne, France: Le Portulan, 1971) (citations refer to the 1971 edition); Jean Favier, *Philippe le Bel* (Paris: Fayard, 1978); Joseph R. Strayer, *The Reign of Philip the Fair* (Princeton, NJ: Princeton University Press, 1980), esp. chap. 4; Franklin J. Pegues, *The Lawyers of the Last Capetians* (Princeton, NJ: Princeton University Press, 1962), chap. 1; and Joseph R. Strayer, *Medieval Statecraft and the Perspectives of History: Essays by Joseph R. Strayer* (Princeton, NJ: Princeton University Press, 1971), 195–212. Charles T. Wood, ed., *Philip the Fair and Boniface VIII: State vs. Papacy* (New York: Holt, Rinehart and Winston, 1967), provides a selection of sources.

2. For a discussion, see John R. Eastman, *Papal Abdication in Later Medieval Thought* (Lewiston, NY: Edwin Mellon Press, 1990). An old but still very valuable biographical study is T. S. R. Boase, *Boniface VIII* (London: Constable, 1933); on the dispute see also Jean Coste, *Boniface VIII en process: Articles d'accusation et depositions des témoins (1303–1311)* (Rome: "L'Erma" di Bretschneider, 1995).

3. Antonio Gramsci, *Prison Notebooks*, vol. 2, ed. and trans. Joseph A. Buttigieg (New York: Columbia University Press, 1996), 368.

4. John A. Scott, *Dante's Political Purgatory* (Philadelphia: University of Pennsylvania Press, 1996), 26, 138–39; and see Peter Armour, "Dante and Popular Sovereignty," in *Dante and Governance*, ed. John Woodhouse, 27–45 (Oxford: Clarendon Press, 1997), 32.

5. Hostiensis, *Summa*, IV, 17, 13, in Wilhelm Georg Grewe, ed., *Fontes Historiae Iuris Gentium: Quellen Zur Geschichte Des Volkerrechts* (Berlin: Walter de Gruyter, 1995), 298. See Brian Tierney, "A Conciliar Theory of the Thirteenth Century," *Catholic Historical Review* 36 (1951): 415–40.

6. R. W. Dyson, ed., *Quaestio de Potestate Papae (Rex Pacificus) / An Enquiry into the Power of the Pope: A Critical Edition and Translation* (Lewiston, NY: Edwin Mellon, 1999), 38, 88. See Walter Ullmann, "A Medieval Document on Papal Theories of Government," *English Historical Review* 61 (1946): 180–201.

7. *Quaestio de Potestate Papae*, 24, 76–77.

8. *Acta inter Bonifacium VIII, Benedictum XI, Clementem V et Philippum Pul-*

chrum Regem Christian, 1614, fol. 164v, discussed by Ullmann, "Medieval Document on Papal Theories of Government," 181.

9. The text of this can be found in Grewe, *Fontes Historiae Iuris Gentium,* 304–5; excerpts translated in Brian Tierney, *The Crisis of Church and State, 1050–1300* (Toronto: University of Toronto Press, 1988), 205–6. See Boase, *Boniface VIII,* 304. Walter Ullmann, *The Growth of Papal Government in the Middle Ages: A Study in the Ideological Relation of Clerical to Lay Power,* 2nd ed. (London: Methuen, 1962), 22, notes that "Son" was a long-standing way of popes addressing emperors.

10. Excerpts from this, and a putative reply, can be found in Pierre Dupuy, ed., *Histoire du différend d'entre le pape Boniface VIII et Philippe le Bel, Roi de France* (Paris: Cramoisy, 1655), 44; and Boase, *Boniface VIII,* 304–5. See Scott, *Dante's Political Purgatory,* 26; and J. A. Watt, introduction to John of Paris, *On Royal and Papal Power,* trans. J. A. Watt (Toronto: Pontifical Institute of Medieval Studies, 1971), 23–24.

11. Reported in R. W. Dyson, introduction to James of Viterbo, *On Christian Government: De Regimine Christiano,* ed. and trans. R. W. Dyson (Woodbridge, UK: Boydell Press, 1995), xiv. Dupuy, *Histoire du différend,* 102, has this as a "wild animal" (*animal brutum*).

12. The Latin text can be found in D. Carl Mirbt, ed., *Quellen zur Geschichte des Papsttums und des Römanischen Katholizismus,* 4th ed. (Tübingen, Germany: J. C. B. Mohr, 1924), 210–11; the English translation is widely available, such as Tierney, *Crisis of Church and State,* 188–89. For a discussion, see Boase, *Boniface VIII,* chap. 12, 315–37; and Ernst H. Kantorowicz, *The King's Two Bodies: A Study in Mediaeval Political Theology* (Princeton, NJ: Princeton University Press, 1957), 194–96.

13. James of Viterbo, *De regime christiano,* esp. pt. 1, is a detailed explication of these terms. For the Latin text, see *Le plus ancient traité de l'Église: Jacques de Viterbe—De regimine christiano (1301–1302),* ed. H.-X. Arquillière (Paris: Gabriel Beauchesne, 1926); the translation by Dyson is cited above. As Wilks suggests, "The triad 'one society, one head, one power' beats like a refrain through the work of the publicists." Michael Wilks, *The Problem of Sovereignty in the Later Middle Ages: The Papal Monarchy with Augustinus Triumphus and the Publicists* (Cambridge: Cambridge University Press, 1964), 67–68.

14. Ullmann, *Growth of Papal Government,* 456.

15. R. W. Carlyle and A. J. Carlyle, *A History of Mediæval Political Theory in the West,* 6 vols. (Edinburgh: William Blackwell and Sons, 1903–36), 5:374; and F. W. Powicke, *The Christian Life in the Middle Ages* (Oxford: Clarendon Press, 1935), 73. Ernst Kantorowicz, *Frederick the Second, 1194–1250* (London: Constable, 1931), 39, 441, sees him as the last in the line from Gregory VII through Innocent III.

16. *Quaestio de Potestate Papae,* 37, 87.

17. "A Dispute between a Priest and a Knight," Latin and English text, ed. and trans. Norma N. Erickson, *Proceedings of the American Philosophical Society* 111 (1967): 288–309. It also appears in R. W. Dyson, ed., *Three Royalist Tracts, 1296–1302: Antequam essent clerici; Disputatio inter Clericum et Militem; Quaestio in utramque partem* (Bristol, UK: University of Durham / Thoemmes Press, 1999). The text is anonymous, although it used to be attributed to William of Ockham.

18. "Dispute between a Priest and a Knight," 300, 308–9.

19. "Dispute between a Priest and a Knight," 288.

20. Boase, *Boniface VIII*, 323. For general discussions of John, and his later influence, see Janet Coleman, "Medieval Discussions of Property: *Ratio* and *Dominium* according to John of Paris and Marsilius of Padua," *History of Political Thought* 4 (Summer 1983): 209–28; and Janet Coleman, "*Dominium* in the Thirteenth and Fourteenth-Century Political Thought and Its Seventeenth-Century Heirs: John of Paris and Locke," *Political Studies* 33 (1985): 73–100.

21. Janet Coleman, "Property and Poverty," in *The Cambridge History of Medieval Political Thought, c. 350–c. 1450*, ed. J. H. Burns, 607–48 (Cambridge: Cambridge University Press, 1988), 640.

22. The text may have had relatively little impact at the time. See Arthur Monaghan, "General Introduction: Origins of the Church-State Problem," in John of Paris, *On Royal and Papal Power*, trans. Arthur Monaghan (New York: Columbia University Press, 1974), xlvi.

23. John of Paris, *De regia potestate et papali*, 1. Johannes Quidort von Paris, *Über königliche und päpstliche Gewalt (De regia potestate et papali)*, ed. Fritz Bleienstein, Latin-German ed. (Stuttgart: Ernst Klett Verlag, 1969); and John of Paris, *On Royal and Papal Power*.

24. Brian Tierney, *Religion, Law and the Growth of Constitutional Thought, 1150–1650* (Cambridge: Cambridge University Press, 1982), 35.

25. John of Paris, *De regia potestate et papali*, 1.

26. John of Paris, *De regia potestate et papali*, 1.

27. John of Paris, *De regia potestate et papali*, 3; see Arthur Monaghan, *Consent, Coercion, and Limit: The Medieval Origins of Parliamentary Democracy* (Kingston and Montreal: McGill-Queen's University Press, 1987).

28. Monaghan, *Consent, Coercion, and Limit*, 204.

29. Saint Thomas Aquinas, *Summa Theologiæ*, Latin-English ed., trans. various, 60 vols. (London: Blackfriars and Eyre and Spottiswoode, 1963–76), IaIIæ 95.3, endorsing Isidore's claim that "law will be honorable, fair, possible, according to nature and the custom of the country, befitting place and time"; see IaIIæ 97; and Monaghan, *Consent, Coercion, and Limit*, 204–5.

30. John of Paris, *De regia potestate et papali*, 3, referencing Augustine, *De Civitate Dei*, IV, 15. See Engelbert of Admont, "De Ortu et fine Rom. Imperij," in *Politica Imperialia*, ed. Melchior Goldast, 754–812 (Frankfurt: Iohannis Bringeri, 1614), chap. 16.

31. John of Paris, *De regia potestate et papali*, 15.

32. John of Paris, *De regia potestate et papali*, 18 [30].

33. John of Paris, *De regia potestate et papali*, 18 [30].

34. Coleman, "Property and Poverty," 638–39. In "The Dominican Political Theory of John of Paris in Its Context," in *The Church and Sovereignty, c590–1918: Essays in Honour of Michael Wilks*, ed. Diana Wood, 187–223 (Oxford: Basil Blackwell, 1991), Coleman contests the standard understanding that it is a single tract, suggesting instead that it was composed over time (ca. 1297–98) and was a general canonist opinion rather than simply direct advice to Philip in his struggle with Boniface.

35. John of Paris, *De regia potestate et papali*, 8.

36. Tierney, *Crisis of Church and State*, 196.

37. John of Paris, *De regia potestate et papali*, 10; see 21.

38. John of Paris, *De regia potestate et papali*, 21; see Carlyle and Carlyle, *History of Mediævel Political Theory*, 5:432.

39. *Quaestio in utramque partem*, II. The text appears in Dyson, *Three Royalist Tracts*.

40. *Quaestio in utramque partem*, II.

41. *Quaestio in utramque partem*, III. As Wilks, *Problem of Sovereignty in the Later Middle Ages*, 261, notes, the analogy is inadequate for papal purposes and could easily be turned against the popes, because "two swords immediately suggests two distinct powers, temporal and spiritual," the point the church was attempting to combat.

42. Given its importance, remarkably no modern, let alone critical, edition of this text exists, although there are plans for a "best" edition with a facing-page Italian translation in the *Aegidii Romani Opera Omnia*. The most recent editions are reprints: of Aegidii Romani, *De regimine principum Libri III* (Rome: Antonium Baldum, 1556; repr., Frankfurt: Unveränderter Nachdruck, 1968); and Aegidii Columnae Romani, *De regimine principum Lib. III* (Rome: Bartholomaum Zannettum, 1607; repr., Aalen, Germany: Scientia Verlag, 1967). On the link to Aquinas and Ptolemy, see Carlyle and Carlyle, *History of Mediævel Political Theory*, 5:64, 5:380. For a general discussion of this work, see M. S. Kempshall, *The Common Good in Late Medieval Thought* (Oxford: Clarendon Press, 1999), chap. 5.

43. For a discussion of this literature, see István P. Bejczy and Cary J. Nederman, eds., *Princely Virtues in the Middle Ages, 1200–1500* (Turnhout, Belgium: Brepols, 2007).

44. R. W. Dyson, introduction to *Giles of Rome's On Ecclesiastical Power: A Medieval Theory of World Government*, ed. and trans. R. W. Dyson, Latin-English ed., xi–xxxiv (New York: Columbia University Press, 2004), xi.

45. Walther I. Brandt, introduction to Pierre Dubois, *The Recovery of the Holy Land*, trans. Walther I. Brandt, 3–65 (New York: Columbia University Press, 1956), 46.

46. Charles S. Briggs, *Giles of Rome's De Regimine Principum: Reading and Writing Politics at Court and University, c. 1275–c.1525* (Cambridge: Cambridge University Press, 1999); David C. Fowler, Charles F. Briggs, and Paul G. Remley, eds., *The Governance of Kings and Princes: John Trevisa's Middle English Translation of the De Regimine Principum of Aegidius Romanus* (New York: Garland, 1997). See the preface, x–xi, on dating and dissemination. For its ownership (in Latin) by King Richard III, along with Geoffrey of Monmouth's *History of the Kings of Britain*, some Chaucer, a Middle English summary of the Old Testament, and the Wycliff translation of the New Testament, see Anne F. Sutton and Livia Visser-Fuchs, *Richard III's Books: Ideals and Reality in the Life and Library of a Medieval Prince* (Stroud, UK: Sutton, 1997), 283–85.

47. Giorgio Agamben, *The Kingdom and the Glory: For a Theological Genealogy of Economy and Government*, trans. Lorenzo Chiesa with Matteo Mandarini (Stanford, CA: Stanford University Press, 2011), 100.

48. John A. Watt, *Theory of the Papal Monarchy in the Thirteenth Century: The Contribution of the Canonists* (New York: Fordham University Press, 1966), 143. On his work generally, see M. Anthony Hewson, *Giles of Rome and the Medieval Theory of Conception: A Study of the De formatione corporis humani in utero* (London: Athlone,

1975); and Cecilia Trifolgi, "Giles of Rome on Natural Motion in the Void," *Mediaeval Studies* 54 (1992): 136–61.

49. Aegidius Romanus, "De renunciatione papae," in Juan Tomás de Rocaberti, *Bibliotheca Maxima Pontificia*, 21 vols. (Rome: Buagni, 1698), 2:1–64. There is a modern edition, Aegidius Romanus, *De Renuncitatione pape*, ed. John R. Eastman (Lewiston, NY: Edwin Mellon Press, 1992), with critical apparatus in German, and an English summary on 363–83. See also Eastman, *Papal Abdication in Later Medieval Thought*; and Monaghan, *Consent, Coercion, and Limit*, 159, 190.

50. For a discussion, see David Luscombe, "The 'Lex Divinitatis' in the Bull 'Unam sanctum' of Pope Boniface VIII," in *Church and Government in the Middle Ages: Essays Presented to C. R. Cheney on his 70th Birthday*, ed. C. N. L. Brooke, D. E. Luscombe, G. H. Martin, and Dorothy Owen, 205–21 (Cambridge: Cambridge University Press, 1976).

51. Aegidius Romanus, *De Ecclesiastica Potestate*, I, v. The edition used is *Giles of Rome's On Ecclesiastical Power*, cited above.

52. Aegidius Romanus, *De Ecclesiastica Potestate*, II, iv.

53. Henry of Cremona, "De Potestate Papae," in Richard Scholz, *Die Publizistik zur Zeit Philipps des Schönen und Bonifaz VIII: Ein Beitrag zur Geschichte der politischen Anschauungen des Mittelalters* (Stuttgart: Ferdinand Enke, 1903), 459–71; see Carlyle and Carlyle, *History of Mediævel Political Theory*, 5:398–402. Henry agrees with the anonymous pamphlet that Scholz also provides (pp. 471–84, which was possibly written by him) and with Ptolemy of Lucca on this point. See Carlyle and Carlyle, *History of Mediævel Political Theory*, 5:402; for the possible attribution to Henry, see 5:395.

54. Hostiensis, *Lectura or Commentum super decretalibus*, III, xxx, 26; and *Decretales Gregorii Papæ IX, Suæ integritati* (Lyon: Sumptibus Ioannis Pillehotte, 1613), III, xxx, 26. See Watt, *Theory of the Papal Monarchy*, 105.

55. Aegidius Romanus, *De Ecclesiastica Potestate*, II, vii.

56. Aegidius Romanus, *De Ecclesiastica Potestate*, II, xii.

57. Aegidius Romanus, *De Ecclesiastica Potestate*, I, iv.

58. Aegidius Romanus, *De Ecclesiastica Potestate*, II, vii; see II, x; II, xii; III, xi.

59. Aegidius Romanus, *De Ecclesiastica Potestate*, III, 1; III, 2. See Monaghan, *Consent, Coercion, and Limit*, 191.

60. Aegidius Romanus, *De Ecclesiastica Potestate*, III, ix; III, x; III, xi.

61. Aegidius Romanus, *De Ecclesiastica Potestate*, I, v.

62. Aegidius Romanus, *De Ecclesiastica Potestate*, III, ix. See also the views of Augustinius, discussed in Wilks, *Problem of Sovereignty in the Later Middle Ages*, 263–65.

63. Aegidius Romanus, *De Ecclesiastica Potestate*, I, v.

64. Aegidius Romanus, *De Ecclesiastica Potestate*, I, iv, and passim.

65. Aegidius Romanus, *De Ecclesiastica Potestate*, II, v; see II, xv, where he tries to reconcile this account (where the drawn sword is the spiritual) with Bernard's account (where the drawn one, which should be returned to the scabbard, is the material one). His response is a straightforward fudge: "So also in our proposition: the drawn sword can sometimes be taken to represent the material sword and sometimes the spiritual, and the undrawn sword can sometimes be considered as representing the one sword and sometimes the other" (II, xv).

66. Aegidius Romanus, *De Ecclesiastica Potestate*, III, i.

67. *Ordo Romanus*, XIV, cited in James of Viterbo, *De regime christiano*, II, 8. The text of *Ordo Romanus* can be found in *Patrologiæ Cursus Completus, Series Latina*, vol. 78, col. 1242.

68. Aegidius Romanus, *De Ecclesiastica Potestate*, I, iv.

69. Aegidius Romanus, *De Ecclesiastica Potestate*, I, iv.

70. See James of Viterbo, *De regime christiano*, II, 3; and Dyson, introduction to James of Viterbo, *On Christian Government*, xxv.

71. Innocent III, "Novit Ille," in *Patrologiæ Cursus Completus, Series Latina*, vol. 215, cols. 325–28; translated in Bennett D. Hill, *Church and State in the Middle Ages* (New York: John Wiley and Sons, 1970), 150–52. See Dyson, introduction to James of Viterbo, *On Christian Government*, viii, n. 11; and Watt, *Theory of the Papal Monarchy*, 133.

72. James of Viterbo, *De regime christiano*, I, 5. See Dyson's note on 32n17.

73. James of Viterbo, *De regime christiano*, II, 7.

74. James of Viterbo, *De regime christiano*, I, prologue.

75. James of Viterbo, *De regime christiano*, I, 1.

76. James of Viterbo, *De regime christiano*, I, 3.

77. Dante Aligheri, *The Divine Comedy*, trans. with a commentary by Charles S. Singleton, Italian-English ed., 3 vols. (Princeton, NJ: Princeton University Press, 1970–75). Reference will be made to the three parts separately: *Inferno*, *Purgatorio*, and *Paradiso*, and to canto and line of the Italian text. I have also consulted the verse translation by John Ciardi: Dante Aligheri, *The Divine Comedy* (New York: W. W. Norton, 1970).

78. A. d'Entrèves, *Dante as a Political Thinker* (Oxford: Clarendon, 1952), 63.

79. Dante Aligheri, *Monarchia*, trans. Federico Sanguineti, Latin-Italian ed. (Milan: Garzani, 1985); trans. Donald Nicholl in *Monarchy and Three Political Letters* (London: Weidenfeld and Nicholson, 1954); and in *Monarchy*, trans. and ed. Prue Shaw (Cambridge: Cambridge University Press, 1996). There are helpful notes in Richard Kay, ed., *Dante's* Monarchia (Toronto: Pontifical Institute of Medieval Studies, 1998).

80. d'Entrèves, *Dante as a Political Thinker*, 1.

81. Joan M. Ferente, "Dante and Politics," in *Dante: Contemporary Perspectives*, ed. Amilcare A. Iannucci, 181–94 (Toronto: University of Toronto Press, 1997), 184.

82. John M. Najemy, "Dante and Florence," in *The Cambridge Companion to Dante*, ed. Rachel Jacoff, 67–79, 80–99 (Cambridge: Cambridge University Press, 1993), 80.

83. Joan M. Ferente, *The Political Vision of the* Divine Comedy (Princeton, NJ: Princeton University Press, 1984), 44.

84. Ferente, *Political Vision of the* Divine Comedy, 7; see Canning, *History of Medieval Political Thought*, 153. A more general, though less useful, account is Stewart Farnell, *The Political Ideas of the Divine Comedy: An Introduction* (Lanham, MD: University Press of America, 1985).

85. Ferente, *Political Vision of the* Divine Comedy, 46; see John A. Scott, *Understanding Dante* (Notre Dame, IN: University of Notre Dame Press, 2004), 144. Of course, the *Commedia* cannot be reduced to any one reading.

86. Dante, *Purgatorio*, VI, 77, 89.

87. Ferente, *Political Vision of the* Divine Comedy, 46–47.

88. Ferente, *Political Vision of the* Divine Comedy, 47.

89. Ferente, *Political Vision of the* Divine Comedy, 48.

90. Dante, *Monarchia*, III, xiii; see d'Entrèves, *Dante as a Political Thinker*, 57.

91. Dante, *Monarchia*, I, v; see d'Entrèves, *Dante as a Political Thinker*, 19, 48.

92. d'Entrèves, *Dante as a Political Thinker*, 20; see Charles T. Davis, *Dante and the Idea of Rome* (Oxford: Clarendon, 1957), 187.

93. Ovid, *Fasti*, trans. James George Frazer, Latin-English ed., 2nd ed. (Cambridge, MA: Harvard University Press, 1989), II, 684. See chap. 2 for a fuller discussion of this phrase. As Wilks, *Problem of Sovereignty in the Later Middle Ages*, 401, notes, some papal apologists saw the pope, as bishop of Rome, as holding the city that led to rulership of the earth, again equating *urbis et orbis*.

94. Ferente, *Political Vision of the* Divine Comedy, 75; see n. 40 for the reference to Ovid.

95. d'Entrèves, *Dante as a Political Thinker*, 7. For a fuller discussion, see Nancy Lenekeith, *Dante and the Legend of Rome* (London: Warburg Institute, University of London, 1952); Davis, *Dante and the Idea of Rome*; and J. K. Hyde, *Society and Politics in Medieval Italy: The Evolution of the Civil Life, 1000–1350* (London: Macmillan, 1973), esp. chap. 5, "Politics in the Age of Dante," 124–52.

96. Dante, *Monarchia*, III, xvi (xv in some editions); see Engelbert of Admont, "De Ortu et fine Rom. Imperij." See Ferente, *Political Vision of the* Divine Comedy, 35; and Donna Mancusi-Ungaro, *Dante and the Empire* (New York: Peter Lang, 1987), 45. On Engelbert see George Bingham Fowler, *Intellectual Interests of Engelbert of Admont* (New York: AMS Press, 1967); and his "Engelbert of Admont's *Tractatus de Officiis et Abusionibus Eorum*," in *Essays in Medieval Life and Thought: Presented in Honor of Austin Patterson Evans*, ed. John M. Mundy, Richard W. Emery, and Benjamin N. Nelson, 109–22 (New York: Columbia University Press, 1955).

97. Dante, *Monarchia*, I, ii. Andreas Osiander, *Before the State: Systematic Political Change in the West from the Greeks to the French Revolution* (Oxford: Oxford University Press, 2007), 315, contends that Dante is the first to promulgate this view, but this is surely misjudged.

98. Canning, *History of Medieval Political Thought*, 152.

99. Dante, *Monarchia*, I, v. See George Holmes, "*Monarchia* and Dante's Attitude to the Popes," in *Dante and Governance*, ed. John Woodhouse, 46–57 (Oxford: Clarendon Press, 1997), 53.

100. Dante, *Purgatorio*, VI, 105; and Charles Till Davis, "Dante and the Empire," in *Cambridge Companion to Dante*, ed. Jacoff, 68. For a more general discussion, see Mancusi-Ungaro, *Dante and the Empire*; and the title essay of Charles T. Davis, *Dante's Italy and Other Essays* (Philadelphia: University of Pennsylvania Press, 1984).

101. See Kantorowicz, *Frederick the Second*, 668.

102. Dante, *Monarchia*, II, iii.

103. Scott, *Dante's Political Purgatory*, 29.

104. Dante, *Monarchia*, II, i; II, vi; see Davis, "Dante and the Empire," 71. He had earlier made this claim in *Convivio* (ed. Giorgio Inglese [Milan: RCS Rizeoli Libri,

1993]; trans. Philip H. Wickstead [London: J. M. Dent, 1924]), IV, iv; drawing on Virgil, *Aeneid*, I, 278–79 (in *Works*, trans. H. Rushton Fairclough, revised by G. P. Goold, Latin-English ed., 2 vols. [Cambridge, MA: Harvard University Press, 1999–2000]): "I am imposing no bounds on his realm, no temporal limits / I have given *imperium* without end." See Scott, *Dante's Political Purgatory*, 38; and chap. 2.

105. Scott, *Dante's Political Purgatory*, 104.

106. Davis, "Dante and the Empire," 72; Scott, *Dante's Political Purgatory*, 93; and Holmes, "*Monarchia* and Dante's Attitude to the Popes," 53.

107. Dante, *Monarchia*, III, ix.

108. Dante, *Purgatorio*, XVI, 106–8. See Scott, *Dante's Political Purgatory*, 122, 155; Ferente, *Political Vision of the* Divine Comedy, 230; and Ernst H. Kantorowicz, "Dante's Two Suns," in *Selected Studies*, 325–38 (Locust Valley, NY: J. J. Augustin, 1965).

109. Dante, *Monarchia*, III, iv.

110. Francesco Maiolo, *Medieval Sovereignty: Marsilius of Padua and Bartolus of Saxoferrato* (Delft: Eburon, 2007), 252; drawing on Cecil N. Sidney Woolf, *Bartolus of Sassoferrato: His Position in the History of Medieval Thought* (Cambridge: Cambridge University Press, 1913), 306.

111. Dante, *Monarchia*, III, iv. One of the most crucial words in this passage is disputed. Ricci has *regnum*, whereas Shaw has *regimen*. If *regimen*, then the word should be translated as "government" or "power"; though Shaw renders it as "realm," thus equivocating on the sense. See *Monarchia*, ed. Pier Giorgio Ricci (Milan: Edizione Nationale, 1965), vol. 5; *Monarchia*, trans. and ed. Prue Shaw (Cambridge: Cambridge University Press, 1995). Kay, *Dante's* Monarchia, 230n23, notes that the variant exists but gives *regnum* in his text and translates it as "rule." (Kay references *Dantis Alagherii De monarchia libri III*, ed. Ludwig Bertalot, 2nd ed. [Geneva: Olschki, 1920]; and *Monarchia*, trans. with notes by Bruno Nardi, in Dante, *Opera Minori* [Milan: Ricciardi, 1979]. I am grateful to Arielle Saiber for advice on this point.)

112. Dante, *Monarchia*, III, iv, vii, viii, ix, etc. The extensive use of Aristotle is noted by Theodore M. Sumberg, *Political Literature of Europe: Before and after Machiavelli* (Lanham, MD: University Press of America, 1993), 140; and discussed in Larry Peterman, "Dante's *Monarchia* and Aristotle's Political Thought," *Studies in Medieval and Renaissance History* 10 (1973): 3–40. More generally, see Mary Elizabeth Sullivan, "Justice, Temptation, and the Limits of Princely Virtue in Dante's Conception of the Monarch," in *Princely Virtues in the Middle Ages, 1200–1500*, ed. Bejczy and Nederman, 123–38.

113. Scott, *Understanding Dante*, 160. See Holmes, "*Monarchia* and Dante's Attitude to the Popes," 55, who notes that his rejection was fairly common at the time.

114. Ferente, "Dante and Politics," 186, drawing upon Ovidio Capitani, "Riferimento stovico a pubblicistica rel commento di Bruno Nardi alla *Monarchia* Dantesca," *Letture Classica* 9–10 (1982): 217–45.

115. Dante, *Monarchia*, III, x. As Nick Havely, *Dante and the Franciscans: Poverty and the Papacy in the "Commedia"* (Cambridge: Cambridge University Press, 2004), 154–59, shows, this is because the third book of *Monarchia* has the question of poverty as a key theme.

116. Dante, *Monarchia*, I, xvi; III, x. As Dante surely must have known, Boniface used this image in *Unam sanctum* to talk of the church's indivisible nature.

117. Dante, *Paradiso*, VI, 1–2. For a discussion, see John Woodhouse, "Dante and Governance: Contexts and Contents," in *Dante and Governance*, ed. Woodhouse, 10.

118. Dante, *Purgatorio*, XXXII, 129.

119. Dante, *Inferno*, XIX, 117. On these last two passages, see Scott, *Dante's Political Purgatory*, 199. More generally see Peter Armour, "Gold, Silver and True Treasure: Economic Imagery in Dante," *Romance Studies* 3 (1994): 7–30. Peter Armour, *Dante's Griffin and the History of the World: A Study of the Earthly Paradise* (Purgatorio, *cantos xxix–xxxiiii)* (Oxford: Clarendon, 1989), esp. chap. 4 on the *Monarchia*, offers an intriguing discussion.

120. Havely, *Dante and the Franciscans*, esp. 44, 113.

121. Michael Caesar, introduction to *Dante: The Critical Heritage 1314 (?)–1870*, ed. Michael Caesar, 1–88 (London: Routledge, 1988), 3. More generally on its reception, see Simon A. Gilson, *Dante and Renaissance Florence* (Cambridge: Cambridge University Press, 2005).

122. Guido Vernani, "The Refutation of the *Monarchia* Composed by Dante," in Anthony K. Cassell, *The* Monarchia *Controversy: An Historical Study with Accompanying Translations of Dante Aligheri's* Monarchia, Guido Vernani's *Refutation of the "Monarchia" Composed by Dante, and Pope John XXII's Bull* Si fratrum (Washington, DC: Catholic University of America Press, 2004), 174–97. On the idea of a single monarch, see Dante, *Monarchia*, I, xv.

123. Caesar, introduction to *Dante*, 3.

124. C. W. Previté-Orton, "Marsilius of Padua," *Proceedings of the British Academy* 21 (1935): 137.

125. Alan Gewirth, *Marsilius of Padua, The Defender of Peace*, vol. 1, *Marsilius of Padua and Medieval Political Philosophy* (New York: Columbia University Press, 1951), 3, and n. 1; and Gewirth, introduction to vol. 2, *The Defensor Pacis*, trans. Alan Gewirth (New York: Columbia University Press, 1956), xix.

126. Cary J. Nederman, *Worlds of Difference: European Discourses of Toleration, c. 1100–c. 1550* (University Park: Pennsylvania State University Press, 2000), 121. As Ladner, "Concepts of 'Ecclesia' and 'Christianitas,'" notes, *plenitudo potestatis* did originally apply to spiritual, not temporal, power. For a discussion, see William D. McCready, "Papal Plenitudo Potestatis and the Source of Temporal Authority in Late Medieval Papal Hierocratic Theory," *Speculum* 48 (1973): 654–74.

127. The Latin text appears in *The Defensor Pacis of Marsilius of Padua*, ed. C. W. Previté-Orton (Cambridge: Cambridge University Press, 1928); and is translated by Annabel Brett as Marsilius of Padua, *The Defender of the Peace* (Cambridge: Cambridge University Press, 2005). The earlier translation by Gewirth (vol. 2, *Defensor Pacis*) has also been consulted.

128. Both texts appear in Latin and French in Marsile de Padoue, *Oeuvres Mineures*, ed. Colette Jeudy and Jeannine Quillet (Paris: Éditions du CNRS, 1979); and in English in Marsiglio of Padua, *Writings on the Empire: Defensor Minor and De translatione Imperii*, ed. Cary J. Nederman (Cambridge: Cambridge University Press, 1993).

129. Previté-Orton, "Marsilius of Padua," 142.

130. Previté-Orton, "Marsilius of Padua," 137–38.

131. See Gerson Moreno-Riaño, "Marsilius of Padua's Forgotten Discourse," *History of Political Thought* 29 (2008): 441–59.

132. Gewirth, *Marsilius of Padua*, 1:12.

133. Cary J. Nederman, "Editor's Introduction," in Marsiglio of Padua, *Writings on the Empire*, xi.

134. Gewirth, *Marsilius of Padua*, 1:94.

135. See, for instance, Paul E. Sigmund Jr., "The Influence of Marsilius of Padua on XVth-Century Conciliarism," *Journal of the History of Ideas* 23 (1962): 392–402; and Brian Tierney, "Marsilius on Rights," *Journal of the History of Ideas* 52 (1991): 3–17.

136. Previté-Orton, "Marsilius of Padua," 143.

137. Previté-Orton, "Marsilius of Padua," 153.

138. Maiolo, *Medieval Sovereignty*, 199n79.

139. This is one of the arguments of George Garnett, *Marsilius of Padua and "the Truth of History"* (Oxford: Oxford University Press, 2006).

140. C. J. Nederman, "From *Defensa Pacis* to *Defensa Minor*: The Problem of Empire in Marsiglio of Padua," *History of Political Thought* 16 (1995): 322. Lupold of Bebenberg, *Tractatus de iuribus regni et imperii romani*, in *Politische Schriften des Lupold von Bebenburg*, ed. Jürgen Miethke and Christoph Flüeler (Hanover, Germany: Hahnsche, Monumenta Germanicae Historica, 2004), 233–409; two brief excerpts are found in Ewart Lewis, *Medieval Political Ideas*, 2 vols. (London: Routledge and Kegan Paul, 1954), 1:310–12, 1:500–502. On Lupold, see Hanns Gross, "Lupold of Bebenberg: National Monarchy and Representative Government in Germany," *Il pensiero politico* 7 (1974): 3–14.

141. Nederman, "From *Defensa Pacis* to *Defensa Minor*," 315, 329.

142. Maiolo, *Medieval Sovereignty*, 195.

143. Garnett, *Marsilius of Padua*, 113.

144. Marsilius of Padua, *Defensor Pacis*, I, 19.8; II, 11.8, 18.7, 26.3. In the second and fourth of these, he (like Ockham) notes Bernard of Clairvaux's argument in *De consideratione*, IV, III.6, that says that without any known riches, Peter was able to feed the sheep.

145. Marsilius of Padua, *De translatione imperii*, 2.

146. Nederman, "Editor's Introduction," in Marsiglio of Padua, *Writings on the Empire*, xii; Gewirth, *Marsilius of Padua*, 1:84.

147. Marsilius of Padua, *De translatione imperii*, 9.

148. Marsilius of Padua, *De translatione imperii*, 1.

149. Marsilius of Padua, *De translatione imperii*, 1; see Nederman, "Editor's Introduction," in Marsiglio of Padua, *Writings on the Empire*, xiii; Garnett, *Marsilius of Padua*, 5. The reference is to Radulphi de Columna (Landolph of Colonna), "De translatione Imperii," in *Monarchia S. Romani Imperii*, ed. Melchoir Goldast, 3 vols. (Hanover, Germany: Biermanni, 1611–14), 2:88–95.

150. Previté-Orton, "Marsilius of Padua," 147. On the use of Aristotle, see C. J. Nederman, "Character and Community in the *Defensa Pacis*: Marsiglio of Padua's Adaptation of Aristotelian Moral Psychology," *History of Political Thought* 13 (1992): 377–90; developed in his *Community and Consent: The Secular Political Theory of*

Marsilius of Padua's Defensor Pacis (Lanham, MD: Rowman and Littlefield, 1995); and the more extensive treatment in Jeannine Quillet, *La philosophie politique de Marsile de Padoue* (Paris: J. Vrin, 1970), esp. chap. 5.

151. Nederman, "Editor's Introduction," in Marsiglio of Padua, *Writings on the Empire*, xxi.

152. Marsilius of Padua, *Defensor Pacis*, I, 4.3.

153. Marsilius of Padua, *Defensor Pacis*, II, 2.4.

154. Marsilius of Padua, *Defensor Pacis*, II, 2.4.

155. Marsilius of Padua, *Defensor Pacis*, II, 2.5.

156. Marsilius of Padua, *Defensor Pacis*, II, 13, 14; see Shogimen, *Ockham and Political Discourse*, 50. For a discussion, see Conal Condren, "Rhetoric, Historiography and Political Theory: Some Aspects of the Poverty Controversy Reconsidered," *Journal of Religious History* 13 (1984): 15–34.

157. Coleman, "Property and Poverty," 643–44.

158. Marsilius of Padua, *Defensor Pacis*, II, 12.33.

159. Marsilius of Padua, *Defensor Pacis*, II, 13.5.

160. Marsilius of Padua, *Defensor Pacis*, II, 13.6; and see Brett's note to her translation, 267n11, which also refers to Ludwig's Appeal of Sachsenshausen from May 1324. See also William of Ockham, *A Short Discourse on the Tyrannical Government*, ed. Arthur Stephen McGrade, trans. John Kilcullen (Cambridge: Cambridge University Press, 1992), III, 9.

161. Bernard of Clairvaux, *De consideratione*, II, VIII.16. On Bernard, see also Steven Botterill, *Dante and the Mystical Tradition: Bernard of Clairvaux in the* Commedia (Cambridge: Cambridge University Press, 1994).

162. Tierney, *Religion, Law and the Growth of Constitutional Thought*, 23, 85. There was vagueness over the terminology, especially concerning the relation and distinction between *auctoritas* and *potestas*. See Brian Tierney, *Foundations of the Conciliar Theory: The Contribution of the Medieval Canonists from Gratian to the Great Schism*, enlarged ed. (Leiden, Netherlands: Brill, 1998), esp. 131; and also Gaines Post, *"Plena Potestas* and Consent in Medieval Assemblies," in *Studies in Medieval Legal Thought: Public Law and the State, 1100–1322* (Princeton, NJ: Princeton University Press, 1964), 91–162; Monaghan, *Consent, Coercion, and Limit*, 121–26; and Kenneth Pennington, *Pope and Bishops: The Papal Monarchy in the Twelfth and Thirteenth Centuries* (Philadelphia: University of Pennsylvania Press, 1984), chap. 2.

163. Monaghan, *Consent, Coercion, and Limit*, 20.

164. For Aegidius, see the references earlier in this chapter. Augustinus Triumphus, *Summa de Ecclesiastica Potestate* (Rome, 1473); "Tractatus brevis du duplici potestate prelatorum et laicorum, qualiter se habeant," in Scholz, *Die Publizistik zur Zeit Philipps des Schönen und Bonifaz VIII*, 486–501; an excerpt appears in *Medieval Political Ideas*, ed. Lewis, 2:384–91. On Augustinus, see Wilks, *Problem of Sovereignty in the Later Middle Ages*. A similar case is mounted in James of Viterbo, *De regime christiano*.

165. Joseph Canning, "The Role of Power in the Political Thought of Marsilius of Padua," *History of Political Thought* 20 (Spring 1999): 26. On this see Marsilius of Padua, *Defensor Pacis*, II, 1.5, 23.

166. Marsilius of Padua, *Defensor Pacis*, II, 4.2.

167. Marsilius of Padua, *Defensor Pacis*, II, 4.3.

168. Marsilius of Padua, *Defensor Pacis*, II, 4.4.

169. For his characterization of the standard papal reading, see Marsilius of Padua, *Defensor Pacis*, II, 3.6, 27.12.

170. Marsilius of Padua, *Defensor Pacis*, II, 28.24. Brett notes that this is from the "ordinary gloss" on Luke 22:38, but not one attributed to Ambrose.

171. Marsilius of Padua, *Defensor Pacis*, II, 28.24.

172. Gewirth, *Marsilius of Padua*, 1:126.

173. Gewirth, *Marsilius of Padua*, 1:127.

174. John B. Morrall, *Political Thought in Medieval Times*, 3rd ed. (London: Hutchinson University Library, 1971), 107, 113. See Gewirth, *Marsilius of Padua*, 1:236n2, who notes that the transition from "political" to "republican" is in part down to Cicero, who uses *res publica* in the general sense of the *politeia*, but not in the specific popular sense. Aquinas uses it in both ways.

175. Gewirth, introduction to vol. 2, *Defensor Pacis*, lxvii.

176. Marsilius of Padua, *Defensor Pacis*, I, 2.2.

177. Marsilius of Padua, *Defensor Pacis*, I, 2.2; see Marsilius of Padua, *Defensor Minor*, 2.5.

178. Marsilius of Padua, *Defensor Pacis*, I, 2.2; see Aristotle, *Politics*, 1279a34–b10, 1307b30. Aristotle's point is that of the three kinds of rule—by one, the few, or the many—there are good and bad versions: tyranny (*tyrannis*) and kingship (*basileias*) for monarchy; oligarchy and aristocracy; democracy and constitutional government (*politeias*). On Marsilius's reading of this, see Nederman, *Community and Consent*, 20–21.

179. Annabel Brett, "Notes on the Translation," in Marsilius of Padua, *Defender of the Peace*, xlix.

180. Marsilius of Padua, *Defensor Pacis*, I, 19.11; see Gewirth, *Marsilius of Padua*, 1:128

181. Otto Gierke, *Political Theories of the Middle Age*, trans. Frederic William Maitland (Cambridge: Cambridge University Press, 1900), 97. In *Defensor Pacis*, II, 24.12, Marsilius uses the corporeal metaphor to think about the body of the church. On this see Renée Baernstein, "Corporatism and Organicism in Discourse I of Marsilius of Padua's *Defensor Pacis*," *Journal of Medieval and Early Modern Studies* 26 (1996): 113–38.

182. Gierke, *Political Theories of the Middle Age*, 97.

183. See Nicolai Rubinstein, "Marsilius of Padua and Italian Political Thought of His Time," in *Europe in the Late Middle Ages*, ed. J. R. Hale, J. R. L. Highfield, and B. Smalley, 44–75 (Evanston, IL: Northwestern University Press, 1965), 74.

184. d'Entrèves, *Medieval Contribution to Political Thought*, 82.

185. Garnett, *Marsilius of Padua*, 6; and Gewirth, introduction to vol. 2, *Defensor Pacis*, lxxvii.

186. See Garnett, *Marsilius of Padua*, 165. See Walter Ullmann, "Personality and Territoriality in the 'Defensor Pacis,'" reprinted in Ullmann, *Law and Jurisdiction in the Middle Ages*, ed. George Garnett, 397–410 (London: Variorum, 1988), 401.

187. Previté-Orton, "Marsilius of Padua," 152; the reference is to *Defensor Pacis*, I, 17.3.

188. As d'Entrèves, *Medieval Contribution to Political Thought*, 82, suggests.

189. Marjorie Reeves, "Marsiglio of Padua and Dante Aligheri," in *Trends in Medieval Political Thought*, ed. Beryl Smalley, 86–104 (Oxford: Basil Blackwell, 1965), 99; Gewirth, *Marsilius of Padua*, 1:128; and Nederman, "From *Defensa Pacis* to *Defensa Minor*," 316.

190. Gewirth, *Marsilius of Padua*, 1:256–57. On this generally, with a specific focus on coercion, see Canning, "Role of Power in the Political Thought of Marsilius of Padua"; and his "Power and Powerless in the Political Thought of Marsilius of Padua," in *The World of Marsilius of Padua*, ed. Gerson Moreno-Riaño, 211–25 (Turnhout, Belgium: Brepols, 2006).

191. Marsilius of Padua, *Defensor Pacis*, II, 2.4.

192. Aristotle, *Physics* IV, 221a–b.

193. Arthur Stephen McGrade, *The Political Thought of William of Ockham: Personal and Institutional Principles* (Cambridge: Cambridge University Press, 1974), 230–31.

194. McGrade, introduction to William of Ockham, *Short Discourse on the Tyrannical Government*, xvi.

195. William of Ockham, *Opera Philosophica et Theologica: Opera Theologica*, vol. 9, *Quodlibera Septem*, ed. Joseph C. Wey (St. Bonaventure, NY: St. Bonaventure University, 1980); William of Ockham, *Quodlibetal Questions*, trans. Alfred J. Freddoso and Francis E. Kelly, 2 vols. (New Haven, CT: Yale University Press, 1991).

196. McGrade, *Political Thought of William of Ockham*, 3; see William J. Courtney, "The Academic and Intellectual Worlds of Ockham," in *The Cambridge Companion to Ockham*, ed. Paul Vincent Spada, 17–30 (Cambridge: Cambridge University Press, 1999), 27.

197. Marilyn McCord Adams, *William Ockham*, 2 vols. (continuous pagination) (Notre Dame, IN: University of Notre Dame, 1987). See also Gordon Leff, *William of Ockham: The Metamorphosis of Scholastic Discourse* (Manchester, UK: Manchester University Press, 1975), which devotes only pp. 614–43 to his "polemical or so-called political writings" (614).

198. McGrade, *Political Thought of William of Ockham*, 173.

199. Janet Coleman, "Ockham's Right Reason and the Genesis of the Political as 'Absolutist,'" *History of Political Thought* 20 (Spring 1999): 37–38.

200. McGrade, introduction to William of Ockham, *Short Discourse on the Tyrannical Government*, xvi; John Kilcullen, "The Political Writings," in *Cambridge Companion to Ockham*, ed. Spada, 319.

201. See David Burr, *The Spiritual Franciscans: From Protest to Persecution in the Century after Saint Francis* (University Park: Pennsylvania State University Press, 2001). Many of these documents are available at Papal Encyclicals Online (http://www.papalencyclicals.net/all.htm) and The Franciscan Archive (http://www.franciscan-archive.org/index2.html) (last accessed November 2012).

202. The Latin is available at http://www.papalencyclicals.net/Nichol03/exiit-l

.htm, the English at http://www.papalencyclicals.net/Nichol03/exiit-e.htm (last accessed November 2012).

203. Kilcullen, "Political Writings," 306.

204. *Regula bullata*, VI, in *Opuscula sancti patris Francisci Assisiensis*, ed. Caietanus Esser (Grottaferrata, Italy: Collegii S. Bonaventurae Ad Claras Aquas, 1978).

205. Malcolm D. Lambert, *Franciscan Poverty: The Doctrine of the Absolute Poverty of Christ and the Apostles in the Franciscan Order, 1210–1323*, 2nd ed. (St. Bonaventure, NY: Franciscan Institute, St. Bonaventure University, 1998), 54. On the historical background more generally, see Lester K. Little, *Religious Poverty and the Profit Economy in Medieval Europe* (London: Paul Elek, 1978); Virpi Mäkinen, *Property Rights in the Late Medieval Discussion on Franciscan Poverty* (Leuven, Belgium: Peeters, 2001); and Bronislaw Geremek, *Poverty: A History*, trans. Agnieszka Kolakowski (Oxford: Blackwell, 1994), esp. chap. 1.

206. Coleman, "Property and Poverty," 612–15. For a more general background, see Carlyle and Carlyle, *History of Mediæval Political Theory*, vol. 1, chap. 12; vol. 2, chap. 6.

207. Coleman, "Property and Poverty," 621.

208. San Bonaventura, *Apologia Pauperum*, in *Apologia dei poveri contro il calunniatore*, trans. E. Piacenti, Latin-Italian ed. (Vicenza, Italy: LIEF, 1988), chap. 7, 3. On the pre-Bonaventura period, see Rosalind B. Brooke, *Early Franciscan Government: Elias to Bonaventure* (Cambridge: Cambridge University Press, 1959).

209. Annabel S. Brett, *Liberty, Right and Nature: Individual Rights in Later Scholastic Thought* (Cambridge: Cambridge University Press, 1997), 18. Brett cites Bonaventura, *Apologia Pauperum*, XI, in *Opera Omnia*, 10 vols. (Florence: Quaracchi, 1897), 8:312. See also Coleman, "Property and Poverty," 635–36.

210. Gregory IX, "Quo Elongati," http://www.documentacatholicaomnia.eu/04z/z_1230-09-28__SS_Gregorius_VIIII__Quo_Elongati__ES.doc.html (last accessed November 2012). Lambert, *Franciscan Poverty*, 141, claims that there is not a distinction between *dominium* and *proprietas* in these texts.

211. Coleman, "Property and Poverty," 636.

212. Brett, *Liberty, Right and Nature*, 19. For a discussion, see Brian Tierney, *Origins of Papal Infallibility, 1150–1350: A Study on the Concepts of Infallibility, Sovereignty and Tradition in the Middle Ages* (Leiden, Netherlands: E. J. Brill, 1972); and Arthur Stephen McGrade, "Ockham and the Birth of Individual Rights," in *Authority and Power: Studies on Medieval Law and Government Presented to Walter Ullmann on his Seventieth Birthday*, ed. Brian Tierney and Peter Linehan, 149–65 (Cambridge: Cambridge University Press, 1980). On the emergence of natural rights discourse generally, see Francis Oakley, *Natural Law, Laws of Nature, Natural Rights: Continuity and Discontinuity in the History of Ideas* (London: Continuum, 2005); and Richard Tuck, *Natural Rights Theories: Their Origin and Development* (Cambridge: Cambridge University Press, 1979).

213. Lambert, *Franciscan Poverty*, 150, 153; see Sean Kinsella, "The Poverty of Christ in the Medieval Debates between the Papacy and the Franciscans," *Laurentianum* 36 (1995): 492–98.

214. Brian Tierney, *The Idea of Natural Rights: Studies on Natural Rights, Natural*

Law and Church Law, 1150–1625 (Grand Rapids, MI: William B. Eerdmans, 1997), 95; and Tierney, *Origins of Papal Infallibility*, 173.

215. Lambert, *Franciscan Poverty*, 221. Previous popes, including notably Celestine V, had been more sympathetic to Franciscan ideals. See Tierney, *Foundations of the Conciliar Theory*, 144–45. John XXII was not as intractable as he might assume, and later writers often made use of a problematic historical record, or deliberately reformulated things to serve their purposes. For a revisionist view, see Patrick Nold, *Pope John XXII and His Franciscan Cardinal: Bertrand de la Tour and the Apostolic Poverty Controversy* (Oxford: Clarendon, 2003); and Gabrielle Gonzales, "'The King of the Locusts Who Destroyed the Poverty of Christ': Pope John XXII, Marsilius of Padua, and the Franciscan Question," in *World of Marsilius of Padua*, ed. Moreno-Riaño, 65–88.

216. John XXII, "Quorundam exigit," in *Bullarium Franciscanum*, ed. J. H. Sbaralea and F. Eubel, 8 vols. (Rome: Typis Sacrae Congregationis de Propaganda Fide, 1759–1904), 5:130.

217. Coleman, "Property and Poverty," 641; on Aquinas on property, see 622–25.

218. On this chronology, see Lambert, *Franciscan Poverty*, chap. 10, esp. 243, and for the point about the omissions in the final bull, 261.

219. On Ubertino, see Lambert, *Franciscan Poverty*, chap. 8.

220. John XXII, "Quia vir reprobus," in *Bullarium Franciscanum*, 5:408–49. See Coleman, "Property and Poverty," 642.

221. As Shogimen, *Ockham and Political Discourse*, 49, notes, "The early fourteenth-century poverty controversy was not essentially doctrinal. It was a legalistic dispute over competing concepts of papal authority."

222. The *Sentences* of Peter Lombard were a major work of medieval theology, on which many later theologians wrote commentaries. See Peter Lombard, *The Sentences*, trans. Giulio Silano, 4 vols. (projected) (Toronto: Pontifical Institute of Mediaeval Studies, Toronto, 2007–); and Philipp W. Rosemann, *The Story of a Great Medieval Book: Peter Lombard's Sentences* (Peterborough, Ontario: Broadview Press, 2007).

223. Adams, *William Ockham*, xv–xvi; and "Principal Dates in Ockham's Life," in William of Ockham, *A Letter to the Friars Minor and Other Writings*, ed. Arthur Stephen McGrade and John Kilcullen, trans. John Kilcullen (Cambridge: Cambridge University Press, 1995), xxxv.

224. See H. S. Offler, "Introduction to Tractatus Contra Benedictum," in Guillelmi de Ockham, *Opera Politica*, ed. H. S. Offler et al., 4 vols., 3:158–64 (Manchester, UK: Manchester University Press / Oxford: British Academy and Oxford University Press, 1940–1997), 3:161.

225. Brian Tierney, "Natural Law and Canon Law in Ockham's *Dialogus*," in *Aspects of Late Medieval Government: Essays Presented to J. R. Lander*, ed. J. G. Rowe, 3–24 (Toronto: University of Toronto Press, 1986), 6.

226. See Ockham, "A Letter to the Friars Minor," in *Opera Politica*, 1:6–17; *A Letter to the Friars Minor*, 3–15; and Shogimen, *Ockham and Political Discourse*, 2. On these issues, see Gordon Leff, *Heresy in the Later Middle Ages: The Relation of Heterodoxy to Dissent, c. 1250–c. 1450*, 2 vols. (Manchester, UK: Manchester University Press, 1967), esp. pt. 1.

227. Ockham, "Letter to the Friars Minor."

228. Kilcullen, "Political Writings," 302.

229. Reported in Nederman and Fordan, introduction to *Medieval Political Theory*, 207. On the dispute between John XXII and Ludwig, see H. S. Offler, "Empire and Papacy: The Last Struggle," in *Transactions of the Royal Historical Society*, ser. 5, 6 (1956): 21–47. Nold, *Pope John XXII and His Franciscan Cardinal*, 176, cautions against conflating the apostolic poverty controversy with the Ludwig debate.

230. Shogimen, *Ockham and Political Discourse*, 74.

231. The Latin text of this and Ockham's other key political texts, with the exception of the *Dialogus*, are available in Ockham, *Opera Politica*. The *Opus* appears in 1:292–368 and 2:375–858; and translated selections in Ockham, *Letter to the Friars Minor*, 19–115.

232. Coleman, "Property and Poverty," 642; and Lambert, *Franciscan Poverty*, 264. The discussion of the bread is in *Opus*, 2–32; see McGrade, *Political Thought of William of Ockham*, 3.

233. There is no reliable printed edition of the Latin text. For an online critical edition, with translations of parts, see William of Ockham, *Dialogus*, ed. John Kilcullen, George Knysh, Volker Leppin, John Scott, and Jan Ballweg, http://www.britac.ac.uk/pubS/dialogus/ockdial.html (accessed November 2012). Selections from pt. 3 appear in translation in Ockham, *Letter to the Friars Minor*, 119–298. Pt. 2 seems never to have been written, and texts ordinarily included appear to have a different provenance.

234. Ockham, *Tractatus Contra Benedictum* and *Tractatus Contra Ioannem*, in *Opera Politica*, 3:29–56, 3:165–322. These texts are not available in English translation.

235. Ockham, *Opera Politica*, 1:15–217; and selections in *Letter to the Friars Minor*, 301–33. On these see H. S. Offler, "The Origin of Ockham's *Octo Quaestiones*," *English Historical Review* 8223 (1967): 323–32. Bayley has described them as a kind of *Summa Politica*. Charles C. Bayley, "Pivotal Concepts in the Political Philosophy of William of Ockham," *Journal of the History of Ideas* 10 (1949): 206.

236. Ockham, *Opera Politica*, 4:97–260; and *Short Discourse on the Tyrannical Government*.

237. William of Ockham, *De Imperatorum et pontificum potestate*, in *Opera Politica*, 4:279–355; and William of Ockham, *On the Power of Emperors and Popes*, trans. and ed. Annabel S. Brett (Bristol: University of Durham Press / Thoemmes Press, 1998); see Brett's introduction, 21. As with most of his writings, this was not Ockham's own title for the work.

238. Ockham, *Opus Nonagintan Dierum*, 93

239. Coleman, "Property and Poverty," 647.

240. Coleman, "Property and Poverty," 647.

241. Ockham, *Opus Nonagintan Dierum*, 93; and Ockham, *Short Discourse on the Tyrannical Government*, II, 12. The reference is to Bernard of Clairvaux, *De consideratione*, IV, III.6; see II, VI.10. Exactly the same argument is made by Marsilius of Padua, *Defensor Pacis*, discourse 2, 11.8. For a useful discussion of Ockham's thought on property, see Tierney, *Idea of Natural Rights*, 157–94.

242. See Ockham, *Dialogus*, pt. 3, tract 1; and McGrade, *Political Thought of William of Ockham*, 24. Ockham had written many commentaries on Aristotle's logical works in his Oxford years.

243. Takashi Shogimen, "William of Ockham and Guido Terreni," *History of Political Thought* 19 (Winter 1998): 517–30.

244. Tierney, *Idea of Natural Rights*, 103.

245. Kilcullen, "Political Writings," 314; and Brett, introduction to Ockham, *On the Power of Emperors and Popes*, 40, 48–49.

246. Shogimen, *Ockham and Political Discourse*, 38, 72.

247. Lambert, *Franciscan Poverty*, 241.

248. Shogimen, *Ockham and Political Discourse*, 103; and, more generally, Tierney, *Origins of Papal Infallibility*, chap. 7.

249. William of Ockham, *An Princeps*, in Ockham, *Opera Politica*, 1:228–67; translated in *Medieval Political Thought in Early Fourteenth-Century England: Treatises by Walter of Milemete, William of Pagula, and William of Ockham*, ed. Cary Nederman, 153–97 (Turnhout, Belgium: Brepols, 2002), chap. 11. For a discussion, see Tierney, *Origins of Papal Infallibility*, chap. 6.

250. See, for example, Ockham, *On the Power of Emperors and Popes*, chap. 14; and H. S. Offler, "Introduction to *Epistola ad Fratres Minores*," in Ockham, *Opera Politica*, III, 3.

251. Ockham, *On the Power of Emperors and Popes*, prologue; see Ockham, *Short Discourse on the Tyrannical Government*, prologue.

252. Ockham, "Letter to the Friars Minor"; Ockham, *On the Power of Emperors and Popes*, XXVIII; see Offler, "Introduction to *Epistola ad Fratres Minores*," 4.

253. Offler, "Introduction to Tractatus Contra Benedictum," in Ockham, *Opera Politica*, III, 159.

254. Shogimen, *Ockham and Political Discourse*, 157.

255. See, for example, Ockham, *An Princeps*, chaps. 1–6; and Ockham, *Tractatus Contra Benedictum*, IV, 12.

256. Shogimen, *Ockham and Political Discourse*, 170–71.

257. Ockham, *Short Discourse on the Tyrannical Government*, IV, chap. 1; Ockham, *Tractatus Contra Benedictum*, VI; and Kilcullen, "Political Writings," 323n65.

258. See McGrade's note to Ockham, *Short Discourse on the Tyrannical Government*, 71n1.

259. Ockham, *An Princeps*, chap. 9.

260. Ockham, *Short Discourse on the Tyrannical Government*, V, 5.

261. Innocent III, "Vergentis in senium," in *Patrologiæ Cursus Completus, Series Latina*, vol. 214, col. 539. It is alluded to in Ockham, *Short Discourse on the Tyrannical Government*, II, 10; cited in Ockham, *An Princeps*, 2; and quoted without attribution in Ockham, *Dialogus*, III, tract 1, 1, 9. McGrade provides a translation of this key passage in his edition of *Short Discourse*, 36n41. See Coleman, *"Dominium."*

262. Ockham, *On the Power of Emperors and Popes*, XVII, XX.

263. Innocent III, "Per Venerabilem," in *Quellen zur Geschichte des Papsttums und des Römanischen Katholizismus*, ed. Mirbt, 175–77; translated in Tierney, *Crisis of Church and State*, 136–38; see Ockham, *Short Discourse on the Tyrannical Government*, VI, 1.

264. Ockham, *On the Power of Emperors and Popes*, XIX. Brett notes that Ockham's challenge is confined to the different origins of spiritual and temporal power,

rather than a refutation of the idea of the *translatio imperii* (Ockham, *On the Power of Emperors and Popes*, 117n1). In *Short Discourse on the Tyrannical Government*, VI, 3, Ockham quotes extensively from the "Donation." He notes that some think these words are "apocryphal and of no authority" (VI, 4); his tactic is rather to contest its standard interpretation. Yet the only extant manuscript breaks off partway through chap. 5, so it is impossible to know how his argument would have continued.

265. Ockham, *Dialogus*, pt. 3, tract 1, II, 30. It should be noted that the very nature of the *Dialogus*—a dialogue between opposing points of view—makes it very difficult to ascribe views there to Ockham himself.

266. Shogimen, *Ockham and Political Discourse*, 249.

267. Tierney, *Idea of Natural Rights*, 29; see Michel Villey, "Le genèse du droit subjectif chez Guillaume d'Occam," *Archives de philosophie du droit* 9 (1964): 97–127, esp. 109. See also Brett, *Liberty, Right and Nature*, 4–5, 51. Tuck, *Natural Rights Theories*, 22–23, contests this idea, suggesting that the *ius-potestas* relation is not the central issue, but it is the conflation of *ius* and *dominium*, which he suggests came rather earlier than Ockham.

268. Shogimen, *Ockham and Political Discourse*, 262.

269. Philotheus Bochus, "Editor's Introduction," in William of Ockham, *Philosophical Writings: A Selection* (Indianapolis: Hackett, 1990), ix–li, 1.

270. Bochus, "Editor's Introduction," in Ockham, *Philosophical Writings*, li.

271. Kilcullen, "Political Writings," 311, 322n343.

272. Coleman, "Ockham's Right Reason," 39.

273. Coleman, "Ockham's Right Reason," 46.

274. McGrade, *Political Thought of William of Ockham*, 225–26.

275. Ockham, *Dialogus*, pt. 3, tract 1, II, 13.

276. See also Ockham, *Opus Nonagintan Dierum*, 93, where it is clear that the rule of the first does not extend to nonbelievers, drawing on Corinthians 5:12.

277. Bochus, "Editor's Introduction," in Ockham, *Philosophical Writings*, 1. The polemical nature of his engagements is a key theme of Shogimen, *Ockham and Political Discourse*.

278. Tierney, *Idea of Natural Rights*, 93.

279. Antonio Gramsci, *Prison Notebooks*, vol. 3, ed. and trans. Joseph A. Buttigieg (New York: Columbia University Press, 2007), 70.

280. Gramsci, *Prison Notebooks*, vol. 2, 338.

281. Antonio Gramsci, *Selections from the Cultural Writings*, ed. David Forgacs and Geoffrey Nowell-Smith, trans. William Boelhower (London: Lawrence and Wishart, 1985), 187–88. See Dante, *De vulgar eloquentia*, ed. and trans. Steven Botterill, Latin-English ed. (Cambridge: Cambridge University Press, 1996); and for a discussion, Marianne Shapiro, *De Vulgari Eloquentia: Dante's Book of Exile* (Lincoln: University of Nebraska Press, 1990).

282. W. H. V. Reade, "Introduction: The Political Theory of Dante," in Dante, *De Monarchia*, ed. E. Moore, v–xxxi (Oxford: Clarendon, 1916), xxiii.

283. J. A. Watt, "Spiritual and Temporal Powers," in *The Cambridge History of Medieval Political Thought, c. 350–c. 1450*, ed. J. H. Burns, 367–423 (Cambridge: Cambridge University Press, 1988), 416–17.

284. See Arthur Monaghan, *From Personal Duties towards Personal Rights: Late Medieval and Early Modern Political Thought, 1300–1600* (Montreal and Kingston: McGill-Queen's University Press, 1994), 20.

285. Caesar, introduction to *Dante*, 3.

286. See Ephraim Emerton, *Humanism and Tyranny: Studies in the Italian Trecento* (Gloucester, MA: Peter Smith, 1964), 18–19, who notes that it has been described as "the abdication document of the mediæval empire." For the text, see Lorenz Weinrich, ed., *Quellen zur Verfassungsgeschichte des Römisch-Deutschen Reiches im Spätmittelalter (1250–1500): Diplomata et acta publica*, 315–77 (Darmstadt, Germany: Wissenschaftliche Buchgesellschaft, 1983); a translation can be found in Oliver J. Thatcher and Edgar H. McNeal, *A Sourcebook for Medieval History: Selected Documents Illustrating the History of Europe in the Middle Age* (New York: AMS Press, 1905), 283–308.

287. Annabel S. Brett, *Changes of State: Nature and the Limits of the City in Early Modern International Law* (Princeton, NJ: Princeton University Press, 2011), 170.

288. Knowles, *Evolution of Medieval Thought*, 305.

CHAPTER 7

1. George Mousourakis, *A Legal History of Rome* (London: Routledge, 2007), ix; and Barry Nicholas, *An Introduction to Roman Law* (Oxford: Oxford University Press, 1962), 1. On this law in its mature state, see Jill Harries, *Law and Empire in Late Antiquity* (Cambridge: Cambridge University Press, 1999).

2. O. F. Robinson, *The Sources of Roman Law: Problems and Methods for Ancient Historians* (London: Routledge, 1997), 2. The references are to Livy, *Ab urbe condita*, III, 34; and Cicero, *De Legibus*, II, 9.

3. George Mousourakis, *The Historical and Institutional Context of Roman Law* (Aldershot, UK: Ashgate, 2003), 418; see Mousourakis, *Legal History of Rome*, 181–82.

4. Mousourakis, *Legal History of Rome*, 182–83; and Robinson, *Sources of Roman Law*, 18–19.

5. *Codex Theodosianus*, ed. Theodore Mommsen (with Paul Meyer on vol. 3), 3 vols. (Berlin: Weidmann, 1957); trans. Clyde Pharr as *The Theodosian Code and Novels and the Sirmodian Constitutions* (New York: Greenwood Publishers, 1982). See Jill Harries and Ian Wood, eds., *The Theodosian Code: Studies in the Imperial Law of Late Antiquity* (London: Duckworth, 1993).

6. On the wider context of the time, see John W. Barker, *Justinian and the Later Roman Empire* (Madison: University of Wisconsin Press, 1966); J. A. S. Evans, *The Age of Justinian: The Circumstances of Imperial Power* (London: Routledge, 1996); and William Rosen, *Justinian's Flea: Plague, Empire and the Birth of Europe* (London: Jonathan Cape, 2007).

7. These are sometimes known as constitutions, from the Latin *constitutones*. There is an English translation as "The Code of Justinian," in *The Civil Law*, ed. S. P. Scott, 17 vols. (Cincinnati: Central Trust, 1932), vols. 12–14.

8. On why it was revised, see Caroline Humfress, "Law and Legal Practice in the

Age of Justinian," in *The Cambridge Companion to the Age of Justinian*, ed. Michael Mass, 161–84 (Cambridge: Cambridge University Press, 2005), 164–65.

9. *The Digest of Justinian*, ed. Theodor Mommsen with Paul Krueger, translation ed. Alan Watson, Latin-English ed., 4 vols. (Philadelphia: University of Pennsylvania Press, 1985).

10. Justinian, *Institutes*, trans. Peter Birks and Grant McLeod, with the Latin text of Paul Kruger, Latin-English ed. (London: Duckworth, 1987).

11. On these see P. N. Ure, *Justinian and His Age* (Harmondsworth, UK: Penguin, 1951), chap. 8. As Robinson, *Sources of Roman Law*, 21, 60, notes, they had little impact on Western Europe, but were of course foundational for Byzantine law.

12. Mousourakis, *Legal History of Rome*, 190–91. For a good general overview, see Quirinus Breen, "Justinian's Corpus Juris Civilis," *Oregon Law Review* 23 (1944): 219–48.

13. See Tony Honoré, *Tribonian* (London: Duckworth, 1978), chap. 5.

14. Birks and McLeod, introduction to Justinian, *Institutes*, 10; J. H. A. Lokin, "The Novels of Leo and the Decisions of Justinian," in *Analecta Atheniensia ad ius byzantinum spectantia*, ed. Spyros Troianos, 131–40 (Athens: Ant. N. Sakkoulas, 1997).

15. Peter Stein, *Roman Law in European History* (Cambridge: Cambridge University Press, 1999), 33.

16. For a discussion of this, see Birks and McLeod, introduction to Justinian, *Institutes*, 11.

17. J. G. A. Pocock, *The Ancient Constitution and the Feudal Law: English Historical Thought in the Seventeenth Century* (New York: W. W. Norton, 1967), 12–13.

18. Evans, *Age of Justinian*, 205; and Alexander Passerin d'Entrèves, *Natural Law: An Introduction to Legal Philosophy*, 3rd ed. (New Brunswick, NJ: Transaction, 1994), 22.

19. Mousourakis, *Legal History of Rome*, 155.

20. Walter Ullmann, *A Short History of the Papacy in the Middle Ages*, 2nd ed. (London: Routledge, 2003), 13.

21. For a full discussion, see Charles M. Radding and Antonio Ciaralli, *The Corpus Iuris Civilis in the Middle Ages: Manuscripts and Transmission from the Sixth Century to the Juristic Revival* (Leiden, Netherlands: Brill, 2007).

22. Mousourakis, *Historical and Institutional Context of Roman Law*, 423. See Susan Reynolds, "Medieval Law," in *The Medieval World*, ed. Peter Linehan and Janet L. Nelson, 485–502 (London: Routledge, 2001), 490.

23. Mousourakis, *Legal History of Rome*, 193; and Mousourakis, *Historical and Institutional Context of Roman Law*, 423.

24. Evans, *Age of Justinian*, 207.

25. Michael Ott, "Irnerius," in *The Catholic Encyclopedia* (New York: Robert Appleton Company, 1910), vol. 8. For a helpful discussion see Quirinus Breen, "The Twelfth-Century Revival of the Roman Law," *Oregon Law Review* 24 (1944–45): 244–87.

26. The key work here is Hermann Kantorowicz, ed., *Studies in the Glossators of the Roman Law: Newly Discovered Writings of the Twelfth Century* (1938; repr., Darmstadt, Germany: Scientia Verlag Aalen, 1969) (citations refer to the 1969 edition).

See also his "The Quaestiones Disputatae of the Glossators," *Revue d'Histoire du Droit (Tijdschrift voor Rechtsgeschiednis)* 16 (1939): 1–67.

27. R. N. Swanson, *The Twelfth-Century Renaissance* (Manchester, UK: Manchester University Press, 1998), 35.

28. Mousourakis, *Historical and Institutional Context of Roman Law*, 425. When Aquinas quotes Justinian, he refers to him as "the jurist."

29. Mousourakis, *Legal History of Rome*, 194.

30. François Rabelais, *Pantagruel*, in *Oeuvres de François Rabelais*, ed. Abel Lefranc, 6 vols. (Paris: Éduoard Champion, 1913–55), 3:58; and *The Histories of Gargantua and Pantagruel*, trans. J. M. Cohen (Harmondsworth, UK: Penguin, 1955), 183.

31. Walter Ullmann, *The Medieval Idea of Law as Represented by Lucas de Penna: A Study in Fourteenth-Century Legal Scholarship* (1946; repr., New York: Barnes and Noble, 1969), 1–2 (citations refer to the 1969 edition).

32. James Tully, "The Pen Is a Mighty Sword," in James Tully, ed., *Meaning and Context: Quentin Skinner and His Critics*, 7–25 (Princeton, NJ: Princeton University Press, 1988), 11; Quentin Skinner, *The Foundations of Modern Political Thought*, vol. 1, *The Renaissance* (Cambridge: Cambridge University Press, 1978), 9.

33. Bartolus, on *Codex*, VII.45.13, in *Opera quae nunc extant omnia*, 11 vols. (Basel, Switzerland: Ex Officina Episcopiana, 1588–89), 8:195.

34. Alan Harding, "The Origins of the Concept of the State," *History of Political Thought* 15 (1994): 63.

35. Ullmann, *Medieval Idea of Law*, 163; and Cecil N. Sidney Woolf, *Bartolus of Sassoferrato: His Position in the History of Medieval Thought* (Cambridge: Cambridge University Press, 1913), 4.

36. On his life and work, see Woolf, *Bartolus of Sassoferrato*; Leo Mucha Mladen, "Bartolus the Man," *Annals of the New York Academy of Sciences* 314 (1978): 311–48; and Augusto Miceli, "Bartolus of Sassoferrato," *Louisiana Law Review* 37 (1976–77): 1027–36. On the context he worked within, see Anna T. Sheedy, *Bartolus on Social Conditions in the Fourteenth Century* (New York: Columbia University Press, 1942), esp. chap. 1. An excellent multilingual collection of essays on his work is Università Degli Studi di Perugia, *Bartolo da Sassoferrato: Studi e Documenti per il VI centenario*, 2 vols. (Milan: Giuffrè, 1962). See also Kenneth Pennington, *The Prince and the Law, 1200–1600: Sovereignty and Rights in the Western Legal Tradition* (Berkeley: University of California Press, 1993), 196–201; and James Franklin, *The Science of Conjecture: Evidence and Probability before Pascal* (Baltimore: Johns Hopkins University Press, 2001), 24–27.

37. I have used the Basel edition of 1588–89; see n. 33 above. The eleventh volume is an index. There is no modern critical edition, and there are variants between the editions of the fifteenth to sixteenth centuries, based on different manuscripts. Crucially, the *Consilia* are not numbered the same in the different editions.

38. These works are collected in vol. 10 of *Opera*, entitled *Consilia, Quaestiones, et Tractatus*. Three have appeared in a modern critical edition in Diego Quaglioni, *Politica e diritto nel trecento Italiano: Il "De tyranno" di Bartolo da Sassoferratto (1314–1357) con l'edizione critica dei trattati "De Guelphis et Gebellinis," "De regimine*

civitas" e "De tyranno" (Firenze, Italy: Leo S. Olschki, 1983). "De tyranno" and "De Guelphis et Gebellinis" are translated in Ephraim Emerton, *Humanism and Tyranny: Studies in the Italian Trecento* (Gloucester, MA: Peter Smith, 1964), 126–54, 273–84.

39. Magnus Ryan, "Rules and Justice, 1200–1500," in *Medieval World*, ed. Linehan and Nelson, 513.

40. Stein, *Roman Law in European History*, 71.

41. Skinner, *Foundations of Modern Political Thought*, 1:9.

42. Oldradus de Ponte, [*Consilia et Quaestiones*] (Venetiis, 1490). See Brendan McManus, "The *Consilia* and *Quaestiones* of Oldradus de Ponte," *Bulletin of Medieval Canon Law* 23 (1999): 85–113; see Norman Zacour, *Jews and Saracens in the Consilia of Oldradus de Ponte* (Toronto: Pontifical Institute of Medieval Studies, 1990).

43. On *consilia*, see Peter Riesenberg, "The Consilia Literature: A Prospectus," *Manuscripta* 6 (1962): 3–22; and Julius Kirschner, "*Consilia* as Authority in Late Medieval Italy," in *Legal Consulting in the Civil Law Tradition*, ed. Mario Ascheri, Ingrid Baumgärtner, and Julius Kirschner, 107–40 (Berkeley: Robbins Collection, 1999); and Julius Kirschner, "Some Problems in the Interpretation of Legal Texts Regarding the Italian City States," *Archiv für Begriffsgeschichte* 19 (1975): 16–27.

44. The text can be found in Emil Friedberg, ed., *Corpus Iuris Canonici*, 2 vols. (Graz, Austria: Akademische Druck U. Verlagsanstalt, 1959), vol. 2, cols. 1151–53. On this, see Georges Lizerand, "Les constitutions *Romani principes* et *Pastoralis cura* et leurs sources," *Nouvelle revue historique de droit français et étranger* 37 (1913): 725–57; Walter Ullmann, "The Development of the Medieval Idea of Sovereignty," *English Historical Review* 64 (1949): 1–33; and William M. Bowsky, *Henry VII in Italy: The Conflict of Empire and City-State, 1310–1313* (Westport, CT: Greenwood, 1960).

45. See Sophia Menache, *Clement V* (Cambridge: Cambridge University Press, 1998).

46. On the Avignon papacy, see G. Mollat, *Les papes d'Avignon (1305–1378)* (Paris: Letouzey and Ané, 1949).

47. J. Canning, "Law, Sovereignty and Corporation Theory, 1300–1450," in *The Cambridge History of Medieval Political Thought, c. 350–c. 1450*, ed. J. H. Burns, 454–76 (Cambridge: Cambridge University Press, 1988), 469; see Joseph Canning, *The Political Thought of Baldus de Ubaldis* (Cambridge: Cambridge University Press, 1987), 47; and Woolf, *Bartolus of Sassoferrato*, 76–77.

48. Canning, *Political Thought of Baldus de Ubaldis*, 22.

49. Walter Ullmann, *Medieval Political Thought* (Harmondsworth, UK: Peregrine, 1975), 198. See his *Law and Politics in the Middle Ages: An Introduction to the Sources of Medieval Political Ideas* (Cambridge: Cambridge University Press, 1975), 186, 287; and *Short History of the Papacy*, 282.

50. Bartolus, on *Digest*, XLVIII.17.1, in *Opera*, 6:528. Woolf, *Bartolus of Sassoferrato*, 78, is confusing here, since he translates *terris Imperii* as "territory of the Empire."

51. Woolf, *Bartolus of Sassoferrato*, 113.

52. Engelbert of Admont, "De Ortu et fine Rom. Imperij," in *Politica Imperialia*, ed. Melchior Goldast (Frankfurt: Iohannis Bringeri, 1614), 754–812, XII. See Woolf, *Bartolus of Sassoferrato*, 280–81n2. The list in Aristotle is dubious; Augustine's model is found in *De Civitate Dei*, XIX, 7.

53. Bartolus, "De tyranno," II. For a discussion of his work on tyranny, see Jérémie

Barthas, "Formes de gouvernement ou modalités de la prevue? Eclaircissements sur la *Traité du Tyran* de Bartole de Sassoferrato," in *Della tirannia: Machiavelli con Bartolo*, ed. Jérémie Barthas, 47–73 (Firenze, Italy: Leo S. Olschki, 2007); and for its later impact, see Julius Kirschner, "Bartolo of Sassoferrato's *De Tyranno* and Sallustio Buongug-lielmi's *Consilium* on Niccolò Fortebracci's Tyranny in Città di Castello," *Mediaeval Studies* 68 (2006): 303–31.

 54. Woolf, *Bartolus of Sassoferrato*, 267.

 55. Skinner, *Foundations of Modern Political Thought*, 1:51. Skinner makes it clear that this challenges the interpretations of Ullmann and Woolf. The latter claims there are only two occasions where Bartolus even mentions Aristotle. Woolf, *Bartolus of Sassoferrato*, 385 and n. 3.

 56. Bartolus, "Istri nostril Nucerini [Cons I. 102]," ed. Susanne Lepsius, in *A Renaissance of Conflicts: Visions and Revisions of Law and Society in Italy and Spain*, ed. John A. Marino and Thomas Kuehn, 161–70 (Toronto: Centre for Reformation and Renaissance Studies, 2004). And, in the same volume, see Susanne Lepsius, "Public Responsibility for Failure to Prosecute Crime? An Inquiry into an Umbrian Case by Bartolo da Sassoferrato," 131–60.

 57. Bartolus, on *Digest*, II.1.1, in *Opera*, 1:160.

 58. Bartolus, Consilia, II, 29, in *Opera*, 10:162–63; see Peter Riesenberg, "Civism and Roman Law in Fourteenth-Century Italian Society," in *Explorations in Economic History* 7, no. 2 (1969): 243. His reference is to Bartolus, *Consilia* (Venice, 1576), vol. 2, *consilium* 29, but I have been unable to locate this edition or to find the *consilium* in another edition. More generally on citizenship, see Riesenberg, "Citizenship at Law in Late Medieval Italy," *Viator: Medieval and Renaissance Studies* 5 (1974): 333–46; and Julius Kirschner, "*Civitas Sibi Faciat Civem*: Bartolus of Sassoferrato's Doctrine on the Making of a Citizen," *Speculum* 48 (1973): 694–713. On the English case, see Keechang Kim, *Aliens in Medieval Law: The Origins of Modern Citizenship* (Cambridge: Cambridge University Press, 2000).

 59. Bartolus, on *Codex*, I.1, §13, in *Opera*, 7:6–20, 7:9. A modern edition of the Latin text as "Commentariae in Codicem (*Lib. 1, tit. 1, de Summâ Trinitate*)" can be found in Friedrich Carl von Savigny, *A Treatise on the Conflict of the Laws and the Limits of Their Operation in Respect of Place and Time*, trans. William Guthrie (Edinburgh: T. & T. Clark, 1880), 433–52. It was translated into English as *On the Conflict of the Laws*, trans. Joseph Henry Beale (Cambridge, MA: Harvard University Press, 1914); and J. A. Clarence Smith, "Bartolo on the Conflict of Laws," *American Journal of Legal History* 14 (1970): 157–83 and 247–75.

 60. Bartolus, on *Codex*, I.1, §46.

 61. Bartolus, on *Codex*, I.1, §45.

 62. D. L. Carey Miller, "Property," in *A Companion to Justinian's Institutes*, ed. Ernest Metzger, 42–79 (London: Duckworth, 1998), 44; see also the glossary in that volume, 253.

 63. Miller, "Property," 44. For a helpful discussion, see Andrew Lintott, *Judicial Reform and Land Reform in the Roman Republic: A New Edition, with Translation and Commentary, of the Laws from Urbino* (Cambridge: Cambridge University Press, 1992), chap. 3.

64. Robinson, *Sources of Roman Law*, 118.

65. Andrew Borkowski and Paul du Plessis, *Textbook on Roman Law*, 3rd ed. (Oxford: Oxford University Press, 2005), 156–57. On *res mancipi*, see also Ditlev Tamm, *Roman Law and European Legal History* (Copenhagen: Djøf Publishing, 1997), 75–80.

66. Bartolus, on *Codex*, I.1, §46.

67. See Stuart Elden, *Terror and Territory: The Spatial Extent of Sovereignty* (Minneapolis: University of Minnesota Press, 2009), xxviii–xxx.

68. Ulpian in *Digest*, XXXXVIII.22.7.19.

69. *Digest*, L.16.239. If the text of Frontinus on land disputes, discussed in chap. 2, is indeed genuinely attributed, then this is clearly an echo.

70. Bartolus, on *Digest*, L.16.239, in *Opera*, 8:699.

71. For what little is known of Sextus Pomponius, see Henry John Roby, *An Introduction to the Study of Justinian's Digest* (Cambridge: Cambridge University Press, 1884), clxxi–clxxii; H. F. Jolowicz, *Historical Introduction to the Study of Roman Law* (Cambridge: Cambridge University Press, 1954), 395; Mousourakis, *Legal History of Rome*, 116–17; and Harries, *Law and Empire in Late Antiquity*, 15–17.

72. Ulpian has 2,464 extracts; Paulus, 2,081; Papanian, 601; Pomponius, 578. See Ure, *Justinian and His Age*, 141; and C. F. Kolbert, "General Introduction," in Justinian, *The Digest of Roman Law: Theft, Rapine, Damage and Insult*, trans. C. F. Kolbert (Harmondsworth, UK: Penguin, 1979), 39.

73. *Digest*, I.2.2–53. For a discussion, see Tamm, *Roman Law and European Legal History*, 19–24.

74. On this generally, see J. G. Collier, *Conflict of Laws*, 3rd ed. (Cambridge: Cambridge University Press, 2001), chap. 1; and Hessel E. Yntema, "The Historic Bases of Private International Law," *American Journal of Comparative Law* 2 (1953): 297–317. See Peter Stein, "Bartolus, the Conflict of Laws and the Roman Law," in *Multum non Multa: Festschrift für Kurt Lipstein aus Anlass seines 70. Geburtstages*, ed. Peter Feuerstein and Cliver Parry (Heidelberg, Germany: C. F. Müller Juristischer Verlag, 1980), 251–58. Only eighteen years after Bartolus's death, a work that Ullmann has described as "the first systematic treatise on international law" was written. See Giovanni de Legnano, *Tractatus de bello, de repressaliis et de duello*, ed. Thomas Erskine Holland (1365; repr., Oxford: Oxford University Press, 1917) (citations refer to the 1917 edition); Walter Ullmann, *The Origins of the Great Schism: A Study in Fourteenth-Century Ecclesiastical History*, new ed. (Hamden, CT: Archon Books, 1972), 148; see Ullmann, *Medieval Idea of Law*, 173ff. On the territory/jurisdiction relation more recently, see Richard T. Ford, "Law's Territory (A History of Jurisdiction)," *Michigan Law Review* 97 (1999): 843–930.

75. Francesco Maiolo, *Medieval Sovereignty: Marsilius of Padua and Bartolus of Saxoferrato* (Delft, Netherlands: Eburon, 2007), 234; drawing upon Guido Fassò, *Storia della filosofia del diritto—I—Antichità e medioevo* (1966; repr., Roma-Bari: Laterza, 2001), 227–28 (citations refer to the 2001 edition).

76. Annabel S. Brett, *Changes of State: Nature and the Limits of the City in Early Modern International Law* (Princeton, NJ: Princeton University Press, 2011), 170, draws quite a sharp distinction between territorial jurisdiction and territorial sovereignty;

but as chapter 9 below shows, the key theorists of the latter drew extensively on their Italian predecessors.

77. See Maiolo, *Medieval Sovereignty*, 240–41. For a conceptual discussion, see Bartolus, "*De iurisdictione*," in *Opera*, 10:392–96. Generally, on the context, see Julius Kirschner, ed., *Origins of the State in Italy, 1300–1600* (Chicago: University of Chicago Press, 1995).

78. Woolf, *Bartolus of Sassoferrato*, 388–89.

79. See Floriano Jonas Cesar, "Popular Autonomy and Imperial Power in Bartolus of Saxoferrato: An Intrinsic Connection," *Journal of the History of Ideas* 65 (2005): 369–81.

80. Bartolus, on *Digest*, II.1.3, in *Opera*, 1:163–64.

81. Bartolus, on *Digest*, XLIX.15.24, in *Opera*, 6:637. *Digest*, XLIX.15.24, from Ulpian, is on enemies of Rome and prisoners of war.

82. Bartolus, on *Digest*, VI.1.1, in *Opera*, 1:552–53. A discussion of the term *dominus mundi* is one of the themes of Christian Fasolt, *The Limits of History* (Chicago: University of Chicago Press, 2004), esp. chap. 4. For a discussion of Fasolt's wider historiographical claims, see Ian Hunter, "The State of History and the Empire of Metaphysics," *History and Theory* 44 (2005): 289–303.

83. Bartolus, on *Digest*, XLIX.15.24, in *Opera*, 6:637.

84. Bartolus, on *Codex*, I.3.31, in *Opera*, 7:75.

85. This is a distinction that Clement V also made in *Pastoralis cura*. See Joseph Canning, "A State Like Any Other? The Fourteenth-Century Papal Patrimony through the Eyes of Roman Law Jurists," in *The Church and Sovereignty, c590–1918: Essays in Honour of Michael Wilks*, ed. Diana Wood, 245–60 (Oxford: Basil Blackwell, 1991), 251; and Woolf, *Bartolus of Sassoferrato*, 75–76.

86. Woolf, *Bartolus of Sassoferrato*, 78.

87. Magnus Ryan, "Bartolus of Sassoferrato and Free Cities," *Transactions of the Royal Historical Society*, 6th ser., 10 (2000): 76; and Woolf, *Bartolus of Sassoferrato*, 79.

88. Canning, "State Like Any Other?," 251.

89. Woolf, *Bartolus of Sassoferrato*, 83.

90. Woolf, *Bartolus of Sassoferrato*, 99.

91. Ullmann, *Medieval Idea of Law*, 90.

92. Ullmann, *Medieval Idea of Law*, 90.

93. Magnus Ryan, "Freedom, Law, and the Medieval State," in *States and Citizens: History, Theory, Prospects*, ed. Quentin Skinner and Bo Stråth, 51–62 (Cambridge: Cambridge University Press, 2003), 60; Woolf, *Bartolus of Sassoferrato*, 154, 197, 379; and Marcel David, "Le contenu de l'hegemonie imperiale dans la doctrine de Bartole," in Università Degli Studi di Perugia, *Bartolo da Sassoferrato*, 2:201–16.

94. Bartolus, on *Digest*, XLVII.1.7, in *Opera*, 6:422.

95. Emerton, *Humanism and Tyranny*, 64.

96. Bartolus, *De tyranno*, VI.

97. Bartolus, on *Digest*, IV.4.3, in *Opera*, 1:430; see Bartolus, on *Digest*, XLIX.1.1, 6:580–81.

98. Anthony Black, *Political Thought in Europe, 1250–1450* (Cambridge: Cambridge University Press, 1992), 116.

99. Bartolus, *De Guelphis et gebellinis*, III. For a discussion, see Ryan, "Bartolus of Sassoferrato and Free Cities," 84–85.

100. Quentin Skinner, *The Foundations of Modern Political Thought*, vol. 2, *The Age of Reformation* (Cambridge: Cambridge University Press, 1978), 131; see 181.

101. Skinner, *Foundations of Modern Political Thought*, 1:65.

102. Ernst H. Kantorowicz, *The King's Two Bodies: A Study in Mediaeval Political Theology* (Princeton, NJ: Princeton University Press, 1957), 298.

103. Woolf, *Bartolus of Sassoferrato*, 109.

104. Canning, *Political Thought of Baldus de Ubaldis*, 96. For a discussion of the territorial aspect, see Pietro Vaccari, "'Utrum jurisdictio cohaereat territorio': La dottrina di Bartolo," in Università Degli Studi di Perugia, *Bartolo da Sassoferrato*, 2:737–53.

105. Joseph Canning, "A Fourteenth-Century Contribution to the Theory of Citizenship: Political Man and the Problem of Created Citizenship in the Thought of Baldus de Ubaldis," in *Authority and Power: Studies on Medieval Law and Government Presented to Walter Ullmann on His Seventieth Birthday*, ed. Brian Tierney and Peter Linehan (Cambridge: Cambridge University Press, 1980), 198–99.

106. Bartolus, on *Digest*, II.1.1, in *Opera*, 1:160.

107. Bartolus, on *Digest*, XXXIII.9.4; and *Digest*, L.16.2, in *Opera*, 4:249, 6:684–86.

108. Bartolus, "De fluminibus," II ["De Insula"], 1. This text appears in *Opera*, 10:363–92; and as *Tractatus de fluminibus*, ed. Hercule Buttrigario (Bononiae: Ioannem Roscium, 1576); which also exists in facsimile (Turin: Bottega d'Erasmo, 1964). The first part, "De alluvione," is available in a modern critical edition, ed. Osvaldo Cavallar, in *Renaissance of Conflicts*, ed. Marino and Kuehn, 83–116.

109. Miceli, "Bartolus of Sassoferrato," 1033–34.

110. Bartolus, on *Codex*, I.3.31, in *Opera*, 7:75. On the distinct territories under the same lord, see Woolf, *Bartolus of Sassoferrato*, 91–94.

111. M. H. Keen, "The Political Thought of the Fourteenth-Century Civilians," in *Trends in Medieval Political Thought*, ed. Beryl Smalley, 105–26 (Oxford: Basil Blackwell, 1965), 119.

112. Bartolus, "Et primo super Constitutione Extravaganti, Ad reprimendum," in *Opera*, 10:262.

113. Maiolo, *Medieval Sovereignty*, 259.

114. Harold Dexter Hazeltine, "The Legal and Political Ideas of the Post-Glossators," in Ullmann, *Medieval Idea of Law*, xxxiii; and Maiolo, *Medieval Sovereignty*, 260.

115. Maiolo, *Medieval Sovereignty*, 261.

116. See Woolf, *Bartolus of Sassoferrato*, 94–98; and Ryan, "Bartolus of Sassoferrato and Free Cities," 75–76.

117. Lorenzo Valla, "Elegantiae linguae Latinae," in *Opera Omnia* (Basel, Switzerland: Henri, 1465), 633.

118. Valla, "Elegantiae linguae Latinae," 635.

119. Valla, "Elegantiae linguae Latinae," 635–36, 643. On Valla's criticisms, see Skinner, *Foundations of Modern Political Thought*, 1:105–6; and Donald R. Kelley, "Legal Humanism and the Sense of History," *Studies in the Humanities* 13 (1966): 184–99.

120. Emerton, *Humanism and Tyranny*, 120. Salutati wrote a treatise entitled "De Tyranno" (The tyrant), which Emerton juxtaposes to Bartolus's earlier text "De Tyrannia" (On tyranny). Salutati's text appears in Emerton, *Humanism and Tyranny*, 70–116.

121. Osvaldo Cavallar, Susanne Degenring, and Julius Kirshner, *A Grammar of Signs: Bartolo da Sassoferrato's Tract on Insignia and Coats of Arms* (Berkeley: Robbins Collection, University of California at Berkeley, 1994).

122. Valentine Groebner, *Defaced: The Visual Culture of Violence in the Late Middle Ages*, trans. Pamela Selwyn (New York: Zone Books, 2008), 54–56.

123. These questions are the classic issues raised in Justinian, *Institutes*, II.1, 20–24; and in the *Digest*, XLI.1.7, and indeed, some of Bartolus's text is a commentary on these. For a discussion, see Nicholas, *Introduction to Roman Law*, 132–33; and Borkowski and du Plessis, *Textbook on Roman Law*, 187–88.

124. Bartolus, "De fluminibus."

125. Bartolus, "De fluminibus," prologue.

126. The figures of the first part appear in *Renaissance of Conflicts*, ed. Marino and Kuehn, 117–29. These are the figures from Bartolus's own manuscript; later editions have them redrawn, becoming both more precise and more ornate.

127. Osvaldo Cavallar, "River of Law: Bartolus's Tiberiadis (De alluvione)," in *Renaissance of Conflicts*, ed. Marino and Kuehn, 34–35; see 59 ("a remarkable novelty"). On this generally, see Carla Frova, "Le traité de fluminibus de Bartolo da Sassoferrato (1355)," *Médiévales* 186 (1999): 81–89; Sheedy, *Bartolus on Social Conditions in the Fourteenth Century*, 37, 44–45; and Jan van Maanen, "Teaching Geometry to 11 Year Old 'Medieval Lawyers,'" *Mathematical Gazette* 76 (1992): 37–45. The first part (37–42) provides a discussion of the text and the geometrical problems; 43–45 is a classroom exercise.

128. Franklin, *Science of Conjecture*, 353.

129. Cavallar, "River of Law," 32, 72.

130. Cavallar, "River of Law," 32, 48, 54–55.

131. See Bartolus, "De fluminibus," I (pp. 18–19).

132. The most comprehensive study is Canning, *Political Thought of Baldus de Ubaldis*. See also J. A. Wahl, "Immortality and Inalienability: Baldus de Ubaldis," *Medieval Studies* 32 (1970): 308–28. For a discussion of his work in context, see also Ullmann, *Origins of the Great Schism*, 143–60; and Pennington, *Prince and the Law*, chap. 6.

133. Franklin, *Science of Conjecture*, 30.

134. Canning, *Political Thought of Baldus de Ubaldis*, 7.

135. Kenneth Pennington, "Baldus de Ubaldis," *Rivista internazionale di diritto commune* 8 (1997): 35–61; and Hermann Lange, *Die Consilien des Baldus de Ubaldis (†1400)* (Mainz, Germany: Akademie der Wissenschaften und der Literatur, 1974). The literature on specific opinions is extensive. Generally, see Kenneth Pennington, "The Consilia of Baldus de Ubaldis," *Tijdschrift voor Rechtsgeschiedenis* 56 (1988): 85–92.

136. Pennington, *Prince and the Law*, 204.

137. Baldus, on *Digest*, I.14.3, *Commentaria in Digestum vetus nunc fidelis sime restituta* (Lugduni, 1562), fols. 54r–55v.

138. Canning, "State Like Any Other?," 259.

139. Baldus, on *Codex*, I.2.12, in *Iurisconsulti Omnium suae tempestatis celeber-rimi*, 9 vols. (1577), vol. 5, fol. 20v. See Canning, "State Like Any Other?," 260.

140. Canning, *Political Thought of Baldus de Ubaldis*, 21.

141. Baldus, *Super decretalibus* (Lugduni, 1551), fol. 189v.

142. Canning, *Political Thought of Baldus de Ubaldis*, 45.

143. Canning, *Political Thought of Baldus de Ubaldis*, 17.

144. J. A. Wahl, "Baldus de Ubaldis and the Foundations of the Nation-State," *Manuscripta* 21 (1977): 84.

145. "Key to Technical Terms," in Tacitus, *Annals of Imperial Rome*, trans. Michael Grant, rev. ed. (Harmondsworth, UK: Penguin, 1975), 402.

146. *Digest*, XLIX.15.24.

147. Justinian, *Institutes*, I, 12.

148. Baldus, on *Codex*, VI.24.1, in *Iurisconsulti Omnium*, vol. 7, fol. 70v.

149. Baldus Perusini, "De allodiis," 7, in *In usus feudorum* (Lugduni, 1550), fol. 100r.

150. Joseph Canning, "The Role of Power in the Political Thought of Marsilius of Padua," *History of Political Thought* 20 (Spring 1999): 34.

151. Joseph Canning, "Ideas of the State in Thirteenth and Fourteenth-Century Commentators on the Roman Law," *Transactions of the Royal Historical Society*, 5th ser., 33 (1983): 16.

152. Baldus, *Super decretalibus*, fol. 143r; see Canning, "Ideas of the State," 13.

153. Baldus, *Super decretalibus*, fol. 28v.

154. Canning, *Political Thought of Baldus de Ubaldis*, 127; see 131.

155. Baldus, *Consilia*, 5 vols. (Lyon: Vincent de Portonaris, 1543), vol. 3, *consilium* 159, fol. 34r.

156. Gaius, *Digest*, I.1.9.

157. Justinian, *Institutes*, I, 1–2. See glossary, in *Companion to Justinian's* Institutes, ed. Metzger, 250; and "Vocabulary" in Justinian, *Institutes*, 153, 156.

158. Baldus, on *Digest*, I.1.9, *Commentaria*, fol. 13r.

159. See, for example, Baldus, on *Codex*, VI.24.1, and IX.1.11, *Iurisconsulti Omnium*, vol. 7, fol. 70v; vol. 8, fols. 203v–204r. On extraterritorial jurisdiction, see Baldus, on *Digest*, II.12, *Commentaria*, fol. 55v.

160. See Bartolus, on *Digest*, II.1.1, in *Opera*, 1:160; and the discussion above.

161. Baldus, on *Digest*, I.1.9, *Commentaria*, fol. 13r.

162. See J. P. Canning, "The Corporation in the Political Thought of the Italian Jurists of the Thirteenth and Fourteenth Centuries," *History of Political Thought* 1 (1980): 24n51.

163. Baldus, on *Codex*, VII.53.5, *Iurisconsulti Omnium*, vol. 8, fol. 73r. See Otto Gierke, *Das Deutsche Genossenschaftsrecht*, 4 vols. (Berlin: Weidmannsche Buchhandlung, 1868–1913), 3:432; Kantorowicz, *King's Two Bodies*, 299–300; Canning, *Political Thought of Baldus de Ubaldis*, 187; and Engin Isin, *Cities without Citizens: Modernity of the City as a Corporation* (Montréal: Black Rose Books, 1992), 27–28. For the phrase *unum corpus mysticum* in Aquinas, see chap. 5.

164. Canning, "Corporation in the Political Thought of the Italian Jurists." For

a discussion, see also Walter Ullmann, "The Delictal Responsibility of Medieval Corporations," *Law Quarterly Review* 64 (1948): 77–96, which stresses the importance of Bartolus in developing this notion. I am grateful to Laurence Paul Hemming for an instructive discussion on this point.

165. Philip Pettit, *Made with Words: Hobbes on Language, Mind and Politics* (Princeton, NJ: Princeton University Press, 2008), 76–77.

166. Canning, *Political Thought of Baldus de Ubaldis*, 207–8; Canning, "Corporation in the Political Thought of the Italian Jurists," 31; and Pennington, *Prince and the Law*, 208.

167. Canning, "Ideas of the State," 25.

168. Canning, *Political Thought of Baldus de Ubaldis*, 208.

169. Canning, *Political Thought of Baldus de Ubaldis*, 66, 227.

170. Ullmann, *Origins of the Great Schism*, 5; and Mollat, *Les papes d'Avignon*.

171. Michael Wilks, *The Problem of Sovereignty in the Later Middle Ages: The Papal Monarchy with Augustinus Triumphus and the Publicists* (Cambridge: Cambridge University Press, 1964), 407.

172. Ullmann, *Origins of the Great Schism*, v.

173. Stein, *Roman Law in European History*, 91. Stein notes that Irnerius may have had a role in this adoption.

174. Ullmann, *Short History of the Papacy*, 185.

175. Nicholas, *Introduction to Roman Law*, 49. See *Digest*, XIV.2.9, where Antonius declares, "I am the Lord of the world [*dominus mundi*]."

176. *Digest*, I.4.1.

177. Wim Blockmans and Peter Hoppenbrouwers, *Introduction to Medieval Europe, 300–1550*, trans. Isola van den Hoven (London: Routledge, 2007), 16.

178. R. W. Carlyle and A. J. Carlyle, *A History of Mediæval Political Theory in the West*, 6 vols. (Edinburgh: William Blackwell and Sons, 1903–36), 6:29.

179. Canning, "Ideas of the State," 23, 26.

180. For an analysis that similarly finds the state emerging from medieval law, but in more everyday procedures and institutions than in Roman and canon law, see Alan Harding, *Medieval Law and the Foundations of the State* (Oxford: Oxford University Press, 2002).

181. Skinner, *Foundations of Modern Political Thought*, 1:11; and Peter Riesenberg, *Inalienability of Sovereignty in Medieval Political Thought* (New York: Columbia University Press, 1956), 82–83.

182. See Walter Ullmann, *Medieval Papalism: The Political Theories of the Medieval Canonists* (London: Methuen, 1949), 83. For a general discussion of these issues, see Gaines Post, *Studies in Medieval Legal Thought: Public Law and the State, 1100–1322* (Princeton, NJ: Princeton University Press, 1964).

183. "Acta Regni Karoli IV," in *Monumenta Germaniae Historica: Constitutiones et Acta Publica Imperorum et Regnum*, vol. 8, *1345–1348*, ed. Karolus Zeumer and Ricardus Salomon (Hanover, Germany: Impensis Bibliopolii Hahniani, 1910–26), 151; discussed in Harding, *Medieval Law and the Foundations of the State*, 254.

184. Woolf, *Bartolus of Sassoferrato*, 380. Oldradus, *Consilia et Quaestiones*, CCXXXI (no pagination); Woolf's reference is to CCXXX and the 1472 edition.

185. Armin Wolf, "Die Gesetzgebung der entstehenden Territorialstaaten," in *Handbuch der Quellen und Literatur der neueren europaischen Privatrechtsge-schichte: Erster Band Mittelalter (1100–1500)*, ed. Helmut Coing, 517–800 (Munich: C. H. Beck'sche, 1973), 529; and Armin Wolf, *Gesetzgebung in Europa, 1100–1500, Zur Enstehung der Territorialstaaten* (Munich: C. H. Beck'sche Verlagsbuchhandlung, 1996), 18.

186. *The Letters of John of Salisbury*, vol. 2, *The Later Letters (1163–1180)*, ed. W. J. Miller and C. N. L. Brooke, Latin-English ed. (Oxford: Clarendon Press, 1979), 581, letter 275.

187. Henry de Bracton, *De legibus et consuetudinibus Angliæ / On the Laws and Customs of England*, trans. Samuel E. Thorne, Latin-English ed., 4 vols. (Cambridge, MA: Belknap Press, 1968), 2:33.

188. John B. Morrall, *Political Thought in Medieval Times*, 3rd ed. (London: Hutchinson University Library, 1971), 94.

189. Barbara H. Rosenwein, *Negotiating Space: Power, Restraint, and Privileges of Immunity in Early Medieval Europe* (Ithaca, NY: Cornell University Press, 1999), chap. 9.

190. Woolf, *Bartolus of Sassoferrato*, 381.

191. Woolf, *Bartolus of Sassoferrato*, 382.

192. Ullmann, *Medieval Political Thought*, 29; and Gaines Post, "Two Notes on Nationalism in the Middle Ages," *Traditio* 9 (1953): 296, 312.

193. Wilks, *Problem of Sovereignty in the Later Middle Ages*, 434; see 442.

194. Wilks, *Problem of Sovereignty in the Later Middle Ages*, 437.

195. Brian Tierney, "Some Recent Works on the Political Theorists of the Medieval Canonists," *Traditio* 10 (1954): 615; and Canning, "Law, Sovereignty and Corporation Theory," 464n40.

196. Kantorowicz, *King's Two Bodies*, 97; see Kantorowicz, *Frederick the Second*.

197. Tierney, "Some Recent Works on the Political Theorists of the Medieval Canonists," 619.

198. As Skinner notes (*Foundations of Modern Political Thought*, 2:351), "It has even been argued by one recent authority that Bartolus and Baldus together constructed the entire 'legal foundations' on which 'the modern theory of the State rests.'" The reference is to Wahl, "Baldus de Ubaldis," 80.

199. See Baldus, on *Codex*, VII.50.3, in *Iurisconsulti Omnium*, vol. 7, fol. 65r: "*Princeps enim legitime electus est in terris Deus* & maxime Papa." This is cited in Wahl, "Baldus de Ubaldis," 91, but without the link to the pope, making it appear that it is the *princeps*, rather than the pope as *princeps*, who has this right. For further caution, see Kenneth Pennington, "Was Baldus an Absolutist? The Evidence of His *Consilia*," in *VI Centenario della morte di Baldo degli Ubaldi, 1400–2000*, ed. Carlo Frova, Maria Grazia, Nico Ottaviani, and Stefania Zucchini, 1–16 (Perugia, Italy: Università degli Studi, 2005).

200. See Kleinschmidt, *Understanding the Middle Ages*, 50.

201. Mousourakis, *Historical and Institutional Context of Roman Law*, 421.

202. Benjamin Arnold, *Princes and Territories in Medieval Germany* (Cambridge: Cambridge University Press, 1991), 284.

203. See Robert Folz, *The Concept of Empire in Western Europe from the Fifth to the Fourteenth Century*, trans. Sheila Ann Ogilvie (London: Edward Arnold, 1969), 157.

204. Nicholas of Cusa, *De concordantia catholica libri tres*, in *Opera Omnia*, ed. Gerhardus Kallen (Hamburg: Felicis Meiner, 1959–68), vol. 14; and *The Catholic Concordance*, ed. and trans. Paul E. Sigmund (Cambridge: Cambridge University Press, 1991). See also Nicholas of Cusa, *Writings on Church and Reform*, trans. Thomas M. Izbicki, Latin-English ed. (Cambridge, MA: Harvard University Press, 2008).

205. Charles Howard McIlwain, *The Growth of Political Thought in the West: From the Greeks to the End of the Middle Ages* (London: Macmillan, 1932), 348–49.

206. Paul E. Sigmund Jr., "The Influence of Marsilius of Padua on XVth-Century Conciliarism," *Journal of the History of Ideas* 23 (1962): 395. Paul E. Sigmund, *Nicholas of Cusa and Medieval Political Thought* (Cambridge, MA: Harvard University Press, 1963), is the major study in English; see also Morimichi Watanabe, *The Political Ideas of Nicholas of Cusa with Special Reference to his* De concordantia catholica (Geneva: Librarie Droz, 1963).

207. Cusa, *De concordantia catholica*, preface, 3. He returns to this body imagery in III, xli, 588.

208. Cusa, *De concordantia catholica*, III, ii, 294.

209. Karl Jaspers, *Anselm and Nicholas of Cusa*, ed. Hannah Arendt, trans. Ralph Manheim (New York: Harvest, 1966), 128.

210. Cusa, *De concordantia catholica*, II, xxix, 224.

211. Woolf, *Bartolus of Sassoferrato*, 306.

212. Cusa, *De concordantia catholica*, III, ii, 299.

213. Cusa, *De concordantia catholica*, III, iii, 313.

214. Cusa, *De concordantia catholica*, II, xiii, 116.

215. Cusa, *De concordantia catholica*, III, vi, 343. The reference to *Ego Ludovicus* is to *Decretum Gratiani*, I, lxiii, xxx, in *Patrologiæ Cursus Completus, Series Latina*, vol. 187, cols. 339–40. The word used by Louis is *territoriis*.

216. Cusa, *De concordantia catholica*, III, vi, 344.

217. Jaspers, *Anselm and Nicholas of Cusa*, 107.

218. See also Lauro Martines, *Lawyers and Statecraft in Renaissance Florence* (Princeton, NJ: Princeton University Press, 1968).

219. Francisco de Vitoria, "De Indus," in *Vorlesungen: Völkerrecht Politik Kirche*, ed. Urich Horst, Heinz-Gerhard Justenhoven, and Joachim Stüben, Latin-German ed., 2 vols. (Stuttgart: W. Kohlhammer, 1997), 2:370–541; Francisco de Vitoria, "On the American Indians," in *Political Writings*, ed. Anthony Pagden and Jeremy Lawrance, 231–92 (Cambridge: Cambridge University Press, 1991). On this aspect of his work, see James Muldoon, *Popes, Lawyers and Infidels: The Church and the Non-Christian World, 1250–1550* (Liverpool, UK: Liverpool University Press, 1979), 143–52. More generally, see Bernice Hamilton, *Political Thought in Sixteenth-Century Spain: A Study of the Political Ideas of Vitoria, De Soto, Suárez, and Molina* (Oxford: Clarendon Press, 1963); and Antony Anghie, *Imperialism, Sovereignty and the Making of International Law* (Cambridge: Cambridge University Press, 2005), chap. 1.

220. Vitoria, "De potestate civili," 8, *Vorlesungen*, 1:134; and Vitoria, "On Civil Power," *Political Writings*, 16.

221. Vitoria, "De potestate ecclesiae I," I, ii, *Vorlesungen*, 1:170; and Vitoria, "I On the Power of the Church," *Political Writings*, 50.

222. Vitoria, "De potestate ecclesiae I," V, *Vorlesungen*, 1:228–62; and Vitoria, "I On the Power of the Church," *Political Writings*, 82–101; see Hamilton, *Political Thought in Sixteenth-Century Spain*, 83–87.

223. Vitoria, "De potestate ecclesiae I," I, x, *Vorlesungen*, 1:174; and Vitoria, "I On the Power of the Church," *Political Writings*, 52.

224. J. Neville Figgis, "Bartolus and the Development of European Political Ideas," *Transactions of Royal Historical Society*, n.s., 19 (1905): 147; see Mladen, "Bartolus the Man," 323. Figgis also mentions Albericus Gentilis.

225. Canning, *Political Thought of Baldus de Ubaldis*, 229. Canning mentions Bodin, Francisco Suárez, and Grotius.

226. On this work see Richard Tuck, *Natural Rights Theories: Their Origin and Development* (Cambridge: Cambridge University Press, 1979); and Richard Tuck, *The Rights of War and Peace: Political Thought and the International Order from Grotius to Kant* (Oxford: Oxford University Press, 1999), chap. 3. On the broader context, see Olaf Asbach and Peter Schröder, eds., *War, the State and International Law in Seventeenth-Century Europe* (Farnham, UK: Ashgate, 2010).

227. Hugo Grotius, *De jure belli ac pacis* (1625; repr., Amsterdam: Gasparis Fritsch, 1735) (citations refer to the 1735 edition); Hugo Grotius, *The Rights of War and Peace*, ed. Richard Tuck, trans. John Morrice (Indianapolis: Liberty Fund, 2005). While Bartolus and Baldus are noted here, they are more frequently referenced in earlier works such as Grotius, *De Jure Praedae Commentarius*, Latin-English ed., 2 vols. (Oxford: Oxford University Press, 1950).

228. Michael Roberts, *Essays in Swedish History* (London: Weidenfeld and Nicolson, 1967), 216. See Bruce D. Porter, *War and the Rise of the State: The Military Foundations of Modern Politics* (New York: Free Press, 1994).

229. Roberts, *Essays in Swedish History*, 216. For a helpful discussion, see Robert Kolb, "The Origin of the Twin Terms jus ad bellum / jus in bello," *International Review of the Red Cross* 320 (1997): 553–62.

230. Grotius, *De jure belli ac pacis*, II, iii, iv.1.

231. Grotius, *De jure belli ac pacis*, III, vi, iv.2.

232. Grotius, *De jure belli ac pacis*, II, vi, vii.

233. Grotius, *De jure belli ac pacis*, II, viii, ix.1.

234. Grotius, *De jure belli ac pacis*, II, ii, ii.5. The last point references Servius Maurus Honoratus, *Servii Grammatici qui feruntur in Vergilii carmina commentarii*, ed. Georg Thilo and Hermann Hagen, 3 vols. (Leipzig, 1881–1902), on Virgil, *Aeneid*, IV, 58.

235. Grotius, *De jure belli ac pacis*, II, iii, vii.

236. Grotius, *De jure belli ac pacis*, II, iii, xvi.1.

237. Hugo Grotius, *Mare Liberum: De jure quod Batavis competit ad indicana commercio* (1609; repr., Leiden, Netherlands: Elzeviriana, 1618) (citations refer to the 1618 edition).

238. John Selden, *Mare Clausum seu de dominio maris* (1618; repr., London: Will. Stanebeii, 1636) (citations refer to the 1636 edition); and John Selden, *Mare Clausum:*

The Right and Dominion of the Sea, trans. J. H. Gent (London: Andrew Kembe and Edward Thomas, 1663). The explicit challenge to Grotius is in I, xxvi; for references to the Post-Glossators, see, for example, I, xvi. For a valuable contemporary genealogy of some of these issues, see Daniel Heller-Roazen, *The Enemy of All: Piracy and the Law of Nations* (New York: Zone Books, 2009).

239. Selden, *Mare Clausum*, I, v.

240. On this "battle of the books," see Philip E. Steinberg, "Lines of Division, Lines of Connection: Stewardship in the World Ocean," *Geographical Review* 89 (1999): 254–64; and Philip E. Steinberg, *The Social Construction of the Ocean* (Cambridge: Cambridge University Press, 2001), esp. 89–98.

241. Selden, *Mare Clausum*, I, xvi.

242. Grotius, *De jure belli ac pacis*, II, iii, xiii.2; see also II, ii, ii.4, where he discusses a speculative origin of property, developing from claims over movable things to immovable things.

CHAPTER 8

1. See Kirsten A. Seaver, *The Frozen Echo: Greenland and the Exploration of North America, ca. A.D. 1000–1500* (Stanford, CA: Stanford University Press, 1996).

2. Seaver, *Frozen Echo*, chap. 8.

3. Treaty of Tordesillas, June 7, 1494, Spanish-English ed., in Frances Gardiner Davenport, *European Treaties Bearing on the History of the United States and Its Dependencies to 1648*, 4 vols. (Washington, DC: Carnegie Institution, 1917), 1:86–100, clause 1.

4. David Storey, *Territory: The Claiming of Space* (Harlow, UK: Prentice Hall, 2001), 16–17; see Robert D. Sack, *Human Territoriality: Its Theory and History* (Cambridge: Cambridge University Press, 1986), 131–32. On its impact on the ocean, see Philip E. Steinberg, "Lines of Division, Lines of Connection: Stewardship in the World Ocean," *Geographical Review* 89 (1999): 254–64; and Steinberg, *The Social Construction of the Ocean* (Cambridge: Cambridge University Press, 2001), 75–79.

5. On cartography in the wake of Tordesillas, see James R. Akerman, "The Structuring of Political Territory in Early Printed Atlases," *Imago Mundi* 47 (1995): 138–54; Jerry Brotton, *Trading Territories: Mapping the Early Modern World* (London: Reaktion Books, 1997); and Frank Lestringant, *Mapping the Renaissance World: The Geographical Imagination in the Age of Discovery*, trans. David Fausett (Berkeley: University of California Press, 1994). The original French title of the latter suggests the more evocative "workshop of the cosmographer." See Frank Lestringant, *L'Atelier du cosmographe, ou l'image du monde à la Renaissance* (Paris: Albin Michel, 1991).

6. Treaty of Tordesillas, clause 1.

7. Treaty of Tordesillas, clause 3.

8. Patricia Seed, *Ceremonies of Possession in Europe's Conquest of the New World, 1492–1640* (Cambridge: Cambridge University Press, 1995), 102. For a more nuanced analysis of the use of law to acquire property in land in the New World, see Lauren Benton, *A Search for Sovereignty: Law and Geography in European Empires, 1400–1900* (Cambridge: Cambridge University Press, 2010); and Laura Benton and Benjamin Strau-

mann, "Acquiring Empire by Law: From Roman Doctrine to Early Modern European Practice," *Law and History Review* 28 (2010): 1–38.

9. Denys Hay, *The Medieval Centuries* (London: Methuen, 1953), 164.

10. For a discussion, see Emilie Savage-Smith, "Celestial Mapping," in *The History of Cartography*, vol. 2, bk. 1, *Cartography in the Traditional Islamic and South Asian Societies*, ed. J. B. Harley and David Woodward, 12–70 (Chicago: University of Chicago Press, 1992).

11. Geoffrey Chaucer, *The Canterbury Tales*, ed. Jill Mann (London: Penguin, 2005), lines 3208–12.

12. See David Eugene Smith, *History of Mathematics* (1925; repr., New York: Dover, 1958), 2:188n3 (citations refer to the 1958 edition), who suggests that augrim is "algorism, from al-Khowârizmi." See L. C. Karpinski, "Augrim Stones," *Modern Language Notes* 27 (1912): 206–9.

13. Geoffrey Chaucer, *The Treatise on the Astrolabe*, ed. Andrew Edmund Brae (London: John Russell Smith, 1870); see J. D. North, *Chaucer's Universe* (Oxford: Clarendon Press, 1988).

14. Leonarde Digges, *A Boke named Tectonicon* (1556; repr., London: Thomas Marshe, 1570) (citations refer to the 1570 edition). For a discussion see G. L'E. Turner, "Mathematical Instrument-Making in London in the Sixteenth Century," in *English Map-Making, 1500–1650: Historical Essays*, ed. Sarah Tyacke, 93–106 (London: British Library, 1983). More generally, see A. W. Richeson, *English Land Measuring to 1800: Instruments and Practices* (Cambridge, MA: Society for the History of Technology / MIT Press, 1966); and John Roche, "The Cross-Staff as a Surveying Instrument in England, 1500–1640," in *English Map-Making*, ed. Tyacke, 107–11.

15. John Dee, "Mathematicall Preface," in *The Elements of Geometrie of the most auncient Philosopher Euclide of Megara*, trans. H. Billingsley (London: John Daye, 1570), n. (p. 3).

16. Dee, "Mathematicall Preface," 11.

17. Dee, "Mathematicall Preface," 5.

18. Denis Cosgrove, "The Geometry of Landscape: Practical and Speculative Arts in Sixteenth-Century Venetian Land Territories," in *The Iconography of Landscape: Essays on the Symbolic Representation, Design and Use of Past Environments*, ed. Denis Cosgrove and Stephen Daniels, 254–76 (Cambridge: Cambridge University Press, 1988), 259.

19. Cosgrove, "Geometry of Landscape," 271.

20. See Seed, *Ceremonies of Possession in Europe's Conquest of the New World*, 115.

21. Matthew H. Edney, "The Irony of Imperial Mapping," in *The Imperial Map: Cartography and the Mastery of Empire*, ed. James R. Akerman, 11–45 (Chicago: University of Chicago Press, 2009), 45. The literature on imperial practices is extensive. For helpful surveys of the way that the law and techniques were used, see also Jess Edwards, *Writing, Geometry, and Space in Seventeenth Century England and America: Circles in the Sand* (London: Routledge, 2006); Ken MacMillan, *Sovereignty and Possession in the English New World: The Legal Foundations of Empire, 1576–1640* (Cambridge:

Cambridge University Press, 2006); and James R. Akerman, ed., *The Imperial Map: Cartography and the Mastery of Empire* (Chicago: University of Chicago Press, 2009).

22. On the impact on knowledge more generally, see Stephen Greenblatt, *Marvelous Possessions: The Wonder of the New World* (Oxford: Clarendon Press, 1991); Peter Grafton, *New Worlds, Ancient Texts: The Power of Tradition and the Shock of Discovery* (Cambridge, MA: Belknap Press, 1992); and Tzvetan Todorov, *The Conquest of America: The Question of the Other*, trans. Richard Howard (New York: Harper and Row, 1999). An exemplary study is Walter Mignolo, *The Darker Side of the Renaissance: Literacy, Territoriality, and Colonization* (Ann Arbor: University of Michigan Press, 1995).

23. Bertrand Badie, *The Imported State: The Westernization of the Political Order*, trans. Claudia Royal (Stanford, CA: Stanford University Press, 2000).

24. Seed, *Ceremonies of Possession in Europe's Conquest of the New World*, 126.

25. Jerry Brotton, *The Renaissance Bazaar: From the Silk Road to Michelangelo* (Oxford: Oxford University Press, 2003), 21.

26. Brotton, *Renaissance Bazaar*, 23. Jacob Burckhardt, *The Civilisation of the Renaissance in Italy*, trans. S. G. C. Middlemore (London: Penguin, 1990). See Claire Farago, *Reframing the Renaissance: Visual Culture in Europe and Latin America, 1450–1650* (New Haven, CT: Yale University Press, 1995); and Francesca Fiorani, *The Marvel of Maps, Art, Cartography, and Politics in Renaissance Italy* (New Haven, CT: Yale University Press, 2005).

27. On the exchange between East and West in the Renaissance and before, see Lisa Jardine and Jerry Brotton, *Global Interests: Renaissance Art between East and West* (Ithaca, NY: Cornell University Press, 2000); Brotton, *Renaissance Bazaar*; and Suzanne Conklin Akbari, *Idols in the East: European Representations of Islam and the Orient, 1100–1450* (Ithaca, NY: Cornell University Press, 2009).

28. Samuel Y. Edgerton Jr., *The Renaissance Rediscovery of Linear Perspective* (New York: Harper and Row, 1975); and Samuel Y. Edgerton, *The Mirror, the Window and the Telescope: How Renaissance Linear Perspective Changed Our Vision of the Universe* (Ithaca, NY: Cornell University Press, 2009). See also Gunnar Olsson, *Abysmal: A Critique of Cartographic Reason* (Chicago: University of Chicago Press, 2007), 131–33.

29. Evelyn Edson, *Mapping Time and Space: How Medieval Mapmakers Viewed Their World* (London: British Library, 1997), 165.

30. Christian Jacob, *The Sovereign Map: Theoretical Approaches in Cartography throughout History*, ed. Edward H. Dahl, trans. Tom Conley (Chicago: University of Chicago Press, 2006), 62.

31. George H. T. Kimble, *Geography in the Middle Ages* (London: Methuen, 1938), 215. See also Denis Cosgrove, *Apollo's Eye: A Cartographic Genealogy of the Earth in the Western Imagination* (Baltimore: Johns Hopkins University Press, 2001), 102–5; and Evelyn Edson, *The World Map, 1300–1492: The Persistence of Tradition and Transformation* (Baltimore: Johns Hopkins University Press, 2007), chap. 5.

32. The standard collection of his writings is Niccolò Machiavelli, *Opere*, ed. Sergio Bertelli and Franco Gaeta, 8 vols. (Milan: Feltrinelli, 1960–65). There is a comprehen-

sive selection in English in *The Chief Works and Others*, trans. Allan Gilbert (Durham, NC: Duke University Press, 1965).

33. Machiavelli, *Il Principe*, in *Opere*, 1:3–105; Machiavelli, *The Prince*, in *Selected Political Writings*, ed. and trans. David Wootton (Indianapolis: Hackett, 1994), 5–80. I have also consulted Machiavelli, *The Prince*, ed. Quentin Skinner and Russell Price (Cambridge: Cambridge University Press, 1988).

34. Machiavelli, *Discorsi sopra la prima deca di Tito Livio*, in *Opere*, 1:109–506; and Machiavelli, *The Discourses*, ed. Bernard Crick, trans. Leslie J. Walker, S.J. (Harmondsworth, UK: Penguin, 1970). Where available in *Selected Political Writings*, 81–217, I have generally used this.

35. Machiavelli, *Il Principe*, 14.

36. Machiavelli, letter to Francesco Vettori, December 10, 1513, in *Opere*, 6:304; and Machiavelli, *Selected Political Writings*, 3.

37. For a discussion, see David Wootton, introduction to Machiavelli, *Selected Political Writings*, xvii–xx.

38. Machiavelli, *Discorsi*, I, preface.

39. J. G. A. Pocock, "Custom and Grace, Form and Matter: An Approach to Machiavelli's Concept of Innovation," in *Machiavelli and the Nature of Political Thought*, ed. Martin Fleisher, 153–74 (New York: Atheneum, 1972), 159.

40. Janet Coleman, *A History of Political Thought: From the Middle Ages to the Renaissance* (Oxford: Blackwell, 2000), 250; Sebastian de Grazia, *Machiavelli in Hell* (New York: Harvester Wheatsheaf, 1989), 277–79. For a full analysis, see Allan H. Gilbert, *Machiavelli's* Prince *and Its Forerunners*: The Prince *as a Typical Book* de Regimine Principum (New York: Barnes and Noble, 1938).

41. Claude de Seyssel, *La monarchie de France et deux autres fragments politiques*, ed. Jacque Poujol (1518; repr., Paris: Librairie d'Argences, 1961) (citations refer to the 1961 edition); Erasmus, *Institutio principis christiani*, in *Opera Omnia Desiderii Erasmi Roterodami*, vol. 4-1 (1516; repr., Amsterdam: North-Holland Publishing, 1974) (citations refer to the 1974 edition); trans. Neil M. Cheshire and Michael J. Heath as *The Education of a Christian Prince*, in *Collected Works of Erasmus*, ed. A. H. T. Levi, vol. 27 (Toronto: University of Toronto Press, 1986).

42. Wootton, introduction to Machiavelli, *Selected Political Writings*, xix.

43. Michel Foucault, *Sécurité, Territoire, Population: Cours au Collège de France (1977–1978)*, ed. Michel Senellart (Paris: Seuil/Gallimard, 2004), 99.

44. Michel Foucault, *Politics, Philosophy, Culture: Interviews and Other Writings, 1977–84*, ed. Lawrence D. Kritzman (London: Routledge, 1990), 76.

45. Foucault, *Sécurité, Territoire, Population*, 67.

46. Russell Price, "Appendix B: Notes on the Vocabulary of *The Prince*," in *The Prince*, ed. Skinner and Price, 102.

47. Maurizio Viroli, *From Politics to Reason of State: The Acquisition and Transformation of the Language of Politics, 1250–1600* (Cambridge: Cambridge University Press, 1992), 130. See Michel Senellart, *Les arts de gouverner: Du* regimen *médiéval au concept de gouvernement* (Paris: Éditions de Seuil, 1995), 212.

48. Quentin Skinner, *Visions of Politics*, vol. 2, *Renaissance Virtues* (Cambridge: Cambridge University Press, 2002), 376.

49. Machiavelli, *Il Principe*, 1.

50. The analysis of this sentence is indebted to de Grazia, *Machiavelli in Hell*, 158. See also Romain Descendre, *L'État du monde: Giovanni Botero entre raison d'État et géopolitique* (Geneva: Librarie Droz, 2009), 93. On this term more generally, see J. H. Hexter, *The Vision of Politics on the Eve of the Reformation: More, Machiavelli and Seyssel* (New York: Basic Books, 1973), chap. 3; and Nicolai Rubinstein, "Notes on the Word *Stato* in Florence before Machiavelli," in *Florilegium Historiale: Essays Presented to Wallace K. Ferguson*, ed. J. G. Rowe and W. H. Stockdale, 313–26 (Toronto: University of Toronto Press, 1971).

51. Hexter, *Vision of Politics*, 133.

52. Machiavelli, *Il Principe*, 10.

53. Machiavelli, *Il Principe*, 11.

54. Machiavelli, *Il Principe*, 24. These three examples are rendered by Price as "sufficient territory and power," "more territory and power," and "not much power and territory." See "Appendix B," 102. In his translation (Oxford: Oxford University Press, 2005), Peter Bondanella suggests the context of the second of these "obviously demands 'more territory,' not 'more state'" (103).

55. Machiavelli, *Discorsi*, II, 19.

56. Machiavelli, *Discorsi*, II, 19.

57. Machiavelli, *Discorsi*, II, 19.

58. Machiavelli, *Il Principe*, 3.

59. Machiavelli, *Discorsi*, II, 21.

60. Machiavelli, *Discorsi*, II, 19.

61. Machiavelli, *Discorsi*, II, 1.

62. Machiavelli, *Discorsi*, I, 26. See also II, preface.

63. Machiavelli, *Discorsi*, I, 1.

64. Machiavelli, *Discorsi*, I, 6.

65. Machiavelli, *Discorsi*, I, 6.

66. Machiavelli, *Discorsi*, II, 19.

67. Machiavelli, *Il Principe*, 10. For a note on *contado*, see M. E. Bratchell, *Medieval Lucca and the Evolution of the Renaissance State* (Oxford: Oxford University Press, 2008), 205.

68. Machiavelli, *Il Principe*, 3.

69. Machiavelli, *Il Principe*, 5.

70. Machiavelli, *Il Principe*, 3. For a discussion, see J. G. A. Pocock, *The Machiavellian Moment: Florentine Political Thought and the Atlantic Republican Tradition* (Princeton, NJ: Princeton University Press, 1975), 163–64.

71. Machiavelli, *Il Principe*, 3; see 5.

72. Machiavelli, *Il Principe*, 3.

73. Machiavelli, *Il Principe*, 20.

74. Machiavelli, *Il Principe*, 7.

75. Machiavelli, *Il Principe*, 7.

76. Machiavelli, *Discorsi*, I, 37; see II, 7.

77. Machiavelli, *Discorsi*, I, 55.

78. Crick, introduction to *Discourses*, 41–42.

79. Machiavelli, *Il Principe*, 14. See Machiavelli, *Discorsi*, III, 39. His inspiration here is Xenophon, *Cyropaedia*, I, ii, 9–11; VIII, I, 38–39 (Greek-English ed., 2 vols. [London: William Heinemann, 1914]). See also Machiavelli, *Il Principe*, dedication, where he talks of political observation in terms of cartographic surveying.

80. Machiavelli, *Dell'arte della guerra*, in *Opere*, 2:309–520; and Machiavelli, *The Art of War*, in *Chief Works and Others*, 2:561–726.

81. Sydney Anglo, *Machiavelli: A Discussion* (London: Victor Gollancz, 1969), 157.

82. See esp. Machiavelli, *Dell'arte della guerra*, 426–27; and Machiavelli, *Art of War*, 649–50.

83. Machiavelli, *Dell'arte della guerra*, 512; and Machiavelli, *Art of War*, 719. On the multitude, see Machiavelli, *Discorsi*, I, 59: "Though in one sense there is nothing more formidable than a disorganized and headless multitude [*moltitudine*], in another sense there is nothing more weak."

84. Machiavelli, *Istorie fiorentine*, in *Opere*, vol. 7; in *Chief Works and Others*, 3:1025–1435. For minor exceptions, see V, 24, 28, on the terrain of Verona; and V, 30, on Marradi.

85. De Grazia, *Machiavelli in Hell*, 159; and Giuseppe Bonghi, "Biografia di Niccolò Machiavelli," 1996, http://www.classicitaliani.it/bonghiG/bonghi_bio_Machiavelli.htm.

86. Wootton, introduction to Machiavelli, *Selected Political Writings*, xii.

87. Francesco Guicciardini, *Dialogo e discorsi del reggimento di Firenze*, ed. Roberto Palmarocchi (Bari, Italy: Gius. Laterza and Figli, 1932), 72–73; and Francesco Guicciardini, *Dialogue on the Government of Florence*, ed. and trans. Alison Brown (Cambridge: Cambridge University Press, 1994), 70. On the relationship, see Felix Gilbert, *Machiavelli and Guicciardini: Politics and Writing in Sixteenth-Century Florence* (Princeton, NJ: Princeton University Press, 1965); Pocock, *Machiavellian Moment*, pt. 2; and Viroli, *From Politics to Reason of State*, chap. 4.

88. Guicciardini, *Dialogo e discorsi del reggimento di Firenze*, 162, 158.

89. Quentin Skinner, "Political Philosophy," in *The Cambridge History of Renaissance Philosophy*, ed. Quentin Skinner and Eckhard Kessler, 389–452 (Cambridge: Cambridge University Press, 1988), 443.

90. On Erasmus, see James D. Tracy, *The Politics of Erasmus: A Pacifist Intellectual and His Political Milieu* (Toronto: University of Toronto Press, 1978); and Lisa Jardine, *Erasmus, Man of Letters: The Construction of Charisma in Print* (Princeton, NJ: Princeton University Press, 1993).

91. Erasmus, *Institutio principis christiani*, 1, 164, 233.

92. Erasmus, *Institutio principis christiani*, 1, 167, 237.

93. Erasmus, *Institutio principis christiani*, 1, 164, 233.

94. Erasmus, *Institutio principis christiani*, 9, 208, 277.

95. Peter Ackroyd, *The Life of Thomas More* (London: Vintage, 1999), 41.

96. On More, see J. H. Hexter, *More's* Utopia: *The Biography of an Idea* (New York: Harper and Row, 1952); Hexter, *Vision of Politics*, chap. 2; and Stephen Greenblatt, *Renaissance Self-Fashioning: From More to Shakespeare* (Chicago: University of Chicago Press, 1980), chap. 1. On his vocabulary, see Ladislaus J. Bolchazy, ed., *A Concordance*

to the Utopia *of St. Thomas More and a Frequency Word List* (Hildesheim, Germany: Georg Olms, 1978).

97. Thomas More, *Utopia*, I, in *The Complete Works of St. Thomas More*, ed. Edward Surtz, S.J., and J. H. Hexter, Latin-English ed. (New Haven, CT: Yale University Press, 1965), vol. 4, 50, 51.

98. More, *Utopia*, II, 110–16, 111–17. For a discussion, see Brian R. Goodey, "Mapping 'Utopia': A Comment on the Geography of Sir Thomas More," *Geographical Review* 60 (1970): 15–30.

99. More, *Utopia*, II, 112, 113.

100. More, *Utopia*, I, 66, 67. See Ackroyd, *Life of Thomas More*, 173: "wholesale attack upon the policy of land enclosure for the rearing of sheep, which had led to the removal of fields for cultivation, the destruction of houses and the eviction of tenants." For a recent helpful discussion, see Alvaro Sevilla-Buitrago, "Territory and the Governmentalisation of Social Reproduction: Parliamentary Enclosure and Spatial Rationalities in the Transition from Feudalism to Capitalism," *Journal of Historical Geography* 38, no. 3 (July 2012): 209–19.

101. William Tyndale, *The Obedience of a Christian Man*, ed. David Daniell (London: Penguin, 2000).

102. David Daniell, *William Tyndale: A Biography* (New Haven, CT: Yale University Press, 2001), 246; and J. F. Mozley, *William Tyndale* (London: Society for Promoting Christian Knowledge, 1937), 143.

103. A revised edition was published in 1537. I have consulted the London: Thomas Bertheleti, 1546, ed.; but used Donald W. Rude, *A Critical Edition of Sir Thomas Elyot's The Boke named the Governour* (New York: Garland, 1992). Page numbers refer to the latter edition. Works supporting Henry's decisions were common. See, for instance, Thomas Starkey, *An Exhortation to the People Instructing Them to Unity and Obedience*, ed. James M. Pictor (1534; repr., New York: Garland, 1988) (citations refer to the 1988 edition); and his slightly more critical *A Dialogue between Pole and Lupset*, ed. T. F. Mayer (ca. 1533–35; repr., London: Offices of the Royal Historical Society, 1989) (citations refer to the 1989 edition).

104. Robin Headlam Wells, *Shakespeare, Politics and the State* (Houndmills, UK: Macmillan, 1986), 36.

105. For a summary life of Elyot, see the entry in *The Oxford Dictionary of National Biography*, at http://www.oxforddnb.com/view/article/8782.

106. See Rude, introduction to *Critical Edition of Sir Thomas Elyot's The Boke named the Governour*.

107. Elyot, *Boke named the Governour*, I, i (p. 15).

108. Elyot, *Boke named the Governour*, I, xi (p. 50). On maps at the time more generally, see J. B. Harley, "Meaning and Ambiguity in Tudor Cartography," in *English Map-Making, 1500–1650: Historical Essays*, ed. Sarah Tyacke, 22–45 (London: British Library, 1983).

109. Note in Thomas Elyot, *The Dictionary of syr Thomas Eliot knyght* (London: Thomæ Bertheleti, 1531; reproduced by Menston: Scholar Press, 1970).

110. Elyot, *Dictionary of syr Thomas Eliot knyght*. See also the later definition of a

forest by John Manwood: "A Forrest is a certen Territorie of wooddy grounds & fruitfull pastures," in *A Treatise and discourse of the lawes of the forrest* (London: T. Wight and B. Norton, 1598), fol. 1.

111. See, for example, Greenblatt, *Renaissance Self-Fashioning*, 17–21; Lisa Jardine, *Worldly Goods* (London: Macmillan, 1996), 425–36; and Ernest B. Gilman, *The Curious Perspective: Literary and Pictorial Wit in the Seventeenth Century* (New Haven, CT: Yale University Press, 1978), 98–104. The key study is Mary F. S. Hervey, *Holbein's "Ambassadors": The Picture and the Men—A Historical Study* (London: George Bells and Sons, 1900).

112. On the developments of the time, see Turner, "Mathematical Instrument-Making in London in the Sixteenth Century."

113. Petrus Apianus, *Eyn newe unnd wolgegründte underweysung aller Kauffmanss Rechnung* (Ingolstadt, Germany: G. Apianum, 1527).

114. Michael Levey, *The German School* (London: National Gallery, 1959), 53.

115. For a list, see Levey, *German School*, 50; and Hervey, *Holbein's "Ambassadors,"* 212–18.

116. Brotton, *Renaissance Bazaar*, 6–16; see Jardine and Brotton, *Global Interests*, 49–54. Levey, *German School*, 47–54, provides a detailed account.

117. On Luther see Roland H. Bainton, *Here I Stand: A Life of Martin Luther* (London: Hodder and Stoughton, 1951); Heiko A. Oberman, *Luther: Man between God and the Devil*, trans. Eileen Walliser-Schwarzbart (London: Fontana, 1993); and Michael A. Mullett, *Martin Luther* (London: Routledge, 2004).

118. Le Goff, *Medieval Civilization, 400–1500*, 116.

119. Bernd Moeller, *Imperial Cities and the Reformation*, trans. H. C. Erik Midelfort and Mark U. Edwards (Philadelphia: Fortress Press, 1972), 3.

120. R. W. Carlyle and A. J. Carlyle, *A History of Mediævel Political Theory in the West*, 6 vols. (Edinburgh: William Blackwell and Sons, 1903–36), 6:273.

121. Luther, "Vom kriege wider die Türcken," *Martin Luthers Werke: Kritische Gesamtausgabe*, 70 vols., 30.2:107–48 (Weimar, Germany: Hermann Böhlaus Nachfolger, 1883–), 30.2:112; and Luther, "Ob kriegsleutte auch ynn seligem stande seyn künden," in *Werke*, 19:625.

122. Gordon Ruff, "Luther and Government," in *Luther: A Profile*, ed. H. G. Koenigsberger, 125–49 (London: Macmillan, 1973), 127. See W. D. J. Cargill Thompson, *The Political Thought of Martin Luther* (Brighton, UK: Harvester Press, 1984); Sheldon S. Wolin, *Politics and Vision: Continuity and Innovation in Western Political Thought* (Boston: Little, Brown, 1960), chap. 5; Jan Herman Brinks, "Luther and the German State," *Heythrop Journal* 39 (January 1998): 1–17; and James D. Tracy, ed., *Luther and the Modern State in Germany* (Kirksville, MO: Sixteenth Century Essays and Studies, 1986).

123. J. W. Allen, *A History of Political Thought in the Sixteenth Century* (London: Methuen, 1928), 29.

124. Luther, "An den Christlichen Adel deutscher Nation von des Christlichen standes besserung," in *Werke*, 6:404–69; Luther, "To the Christian Nobility of the German Nation," *Luther's Works*, ed. Jaroslav Pelikan and Helmut T. Lehmann, 55 vols. (Philadelphia: Muhlenberg Press, 1955–86), 44:123–217; "Von dem Bapstum zu Rome widder den hochberumpten Romanisten zu Leipzck," in *Werke*, 6:285–324; and "On

the Papacy in Rome, against the Most Celebrated Romanist in Leipzig," in *Works*, 39:55–104.

125. Luther, "Von Weltlicher Oberkeytt, wie weyt man yhr gehorsam schuldig sey," in *Werke*, 11:245–80; and Luther, "On Temporal Authority: To What Extent It Should Be Obeyed," in *Works*, 45:81–129.

126. Harro Höpfl, introduction to *Luther and Calvin on Secular Authority*, vii–xxiii (Cambridge: Cambridge University Press, 1991), xii.

127. Luther, "6043," in *Martin Luthers Werke: Kritische Gesamtausgabe: Tischreden*, 6 vols. (Weimar, Germany: Hermann Böhlaus Nachfolger, 1912–21), 5:456; see "6462," 5:675. Both passages are cited in Black, "Donation of Constantine," 76.

128. See Quentin Skinner, *The Foundations of Modern Political Thought*, vol. 2, *The Age of Reformation* (Cambridge: Cambridge University Press, 1978), 14–15, 138.

129. Martin Luther, "Bredigt in der Schloßkirche zu Weimar" (1522), in *Werke*, 10.3:380.

130. John Neville Figgis, *Political Thought from Gerson to Grotius: 1414–1625: Seven Studies* (1960; repr., Kitchener, Ontario: Batoche Books, 1999), 84 (citations refer to the 1999 edition).

131. Luther, "Von Weltlicher Oberkeytt," 247; and Luther, "On Temporal Authority," 85.

132. Luther, "Genesisvorlesung," in *Werke*, 44:530–31; and Luther, "Lectures on Genesis," in *Works*, 7:312.

133. Luther, "Genesisvorlesung," in *Werke*, 43:74–75; and Luther, "Lectures on Genesis," in *Works*, 3:279.

134. Cargill Thompson, *Political Thought of Martin Luther*, 62.

135. *Passional Christi und Antichristi* (Wittenberg, 1521).

136. Melanchthon wrote a short text in praise of Irnerius and Bartolus: Philipp Melanchthon, "De Irnerio et Bartolo iurisconsultis oratio recitata a D. Sebaldo Munster," in Guido Kisch, *Melanchthons Rechts- und Soziallehre* (Berlin: Walter de Gruyter, 1967), 214–20. For a discussion, see Kisch's text, 136–38.

137. Michael Gaismair, "Michael Gaismairs erste Landesordnung (14.5.1525)," and "Michael Gaismairs zweite Landsordnung (Januar–März 1526)," in Jürgen Bückling, *Michael Gaismair: Reformer—Sozialrebell—Revolutionär: Seine Rolle im Tiroler "Bauenkreig" (1525/32)* (Stuttgart, Germany: Klett-Cotta, 1978); and "Michael Gaismair's Territorial Constitution for the Tyrol," in *The German Peasants' War: A History in Documents*, ed. and trans. Tom Scott and Bob Scribner (Atlantic Highlands, NJ: Humanities Press, 1991), 265–69.

138. Skinner, *Visions of Politics*, 2:250–51. For a more radical German, see Thomas Müntzer, *Thomas-Müntzer-Ausgabe: Schriften und Fragmente*, 3 vols. (Leipzig: Evangelische Verlagsanstalt, 2010), vol. 1; and *Sermon to the Princes*, presented by Wu Ming, trans. Michael G. Baylor (London: Verso, 2010). For a discussion of his impact within Marxism, see Alberto Toscano, *Fanaticism: On the Uses of an Idea* (London: Verso, 2010), 68–92.

139. Jean Calvin, *Institutio Christianae Religionis* (1536; repr., Geneva: Oliua Roberti Stephani, 1559) (citations refer to the 1559 edition). Some of this is in John Calvin, *Selections from His Writings*, ed. John Dillenberger (Missoula, MT: Scholars Press,

1975). On Calvin's politics, see Wolin, *Politics and Vision*, chap. 6. Harro Höpfl, *The Christian Polity of John Calvin* (Cambridge: Cambridge University Press, 1982).

140. Skinner, *Foundations of Modern Political Thought*, 2:62–63.

141. Henry VIII, *State Papers Published under the Authority of Her Majesty's Commission* (London: George E. Eyre and William Spottiswoode, 1849), 7:261–62; see Walter Ullmann, "'This Realm of England Is an Empire,'" *Journal of Ecclesiastical History* 30 (1979): 188.

142. Charles de Grassaille [Carolo Degrassalio], *Regulia Franciae libri duo, jura omnia et dignitates Christianiss Galliae Regum, continentes* (Paris, 1545), I, i.

143. See Elisabeth Labrousse, *La révocation de l'édit de Nantes. Une foi, une loi, un roi?* (Paris: Éditions Labor et Fides, 1985). Skinner, *Foundations of Modern Political Thought*, 2:250, attributes this to Michel de l'Hôpital (1507–73). It is later satirized by Gottfried Leibniz, "Mars Christianissimus" (1683), in *Sämliche Schriften und Briefe*, ed. Leibniz-Archiv Hannover (Darmstadt, Germany: Otto Reichl, 1923–), ser. 4, II, 453; and Gottfried Leibniz, *Political Writings*, ed. and trans. Patrick Riley (Cambridge: Cambridge University Press, 1988), 123.

144. Otto Gierke, *Political Theories of the Middle Age*, trans. Frederic William Maitland (Cambridge: Cambridge University Press, 1900).

145. These are some of the issues addressed in the remarkable novel by an Italian collective under the name of Luther Blissett, *Q*, trans. Shaun Whiteside (London: Arrow Books, 2004).

146. Foucault, *Sécurité, Territoire, Population*, 99.

147. Guillaume de La Perrière, *Le miroir politique, Oeuvre non moins utile que necessaire à tous Monarches, Roys, Princes, Seigneurs, Magistrats, & autres surintendans & gouverneurs de Republicques* (Lyon: Macé Bonhomme, 1555), 15. There is an English translation, but this is pretty loose. Guillaume de La Perrière, *The Mirrour of Policie. A worke nolesse profitable than necessarie, for all Magistrates, and Governours of Estates and Commonweales* (London: Adam Islip, 1598).

148. La Perrière, *Le miroir politique*, 15.

149. See, for example, Mark Neocleous, *The Fabrication of Social Order: A Critical Theory of Police Power* (London: Pluto Press, 2000).

150. Allen, *History of Political Thought in the Sixteenth Century*, 284n2.

151. Michel Foucault, *"Il faut défendre la société": Cours au Collège de France (1975–1976)*, ed. Maurio Bertani and Alessandro Fontata (Paris: Seuil/Gallimard, 1997), 103–7. On Hotman, see Étienne Blocaille, *Étude sur François Hotman: La Franco-Gallia* (1902; repr., Geneva: Slatkine Reprints, 1970) (citations refer to the 1970 edition); and Donald R. Kelley, *François Hotman: A Revolutionary's Ordeal* (Princeton, NJ: Princeton University Press, 1973).

152. Beatrice Reynolds, *Proponents of Limited Monarchy in Sixteenth-Century France: Francis Hotman and Jean Bodin* (New York: Columbia University, 1931), 41.

153. François Hotman, *Francogallia*, Latin text ed. Ralph E. Giesey, trans. J. H. M. Salmon, Latin-English ed. (Cambridge: Cambridge University Press, 1972). Giesey and Salmon, "Editors' Introduction," 99, are good on the differences between the editions.

154. Hotman, *Francogallia*, 458, 459.

155. Reynolds, *Proponents of Limited Monarchy*, 52.

156. François Hotman, *Antribonian ou Discours d'un grand et renomme Iuris-consulte de nostre temps* (1567; repr., Paris: Ieremie Perier, 1603) (citations refer to the 1603 edition). For a discussion, see Jean-Louis Ferrary, "À propos d'un texte de François Hotman: Les jurists humanistes et l'édition du *Corpus iuris ciuilis* glosé," in *A Ennio Cortese*, ed. Domenico Maffei, Italo Birocchi, Mario Caravale, Emanuele Conte, and Ugo Petronio, 3 vols. (Rome: Il Cigno, 2002), 2:86–104. See also Giesey and Salmon, "Editors' Introduction," in Hotman, *Francogallia*, 33.

157. J. G. A. Pocock, *The Ancient Constitution and the Feudal Law: English Historical Thought in the Seventeenth Century* (New York: W. W. Norton, 1967), 22–24.

158. Hotman, *Francogallia*, 150, 151.

159. Peter Stein, *Roman Law in European History* (Cambridge: Cambridge University Press, 1999), 79.

160. J. U. Lewis, "Jean Bodin's 'Logic of Sovereignty,'" *Political Studies* 16 (1968): 208.

161. M. J. Tooley, introduction to Jean Bodin, *Six Books of the Commonwealth*, abridged and trans. M. J. Tooley, vii–xlii (Oxford: Basil Blackwell, 1955), xiv.

162. Black, *Monarchy and Community*, 80; Wilks, *Problem of Sovereignty in the Later Middle Ages*, 151; and, more generally, Marcel David, *La souveraineté et les limites juridiques du pouvoir monarchique du IXe au XVe siècle* (Paris: Librarie Dalloz, 1954). The last is good but very vague in the use of territorial aspects.

163. Skinner, *Foundations of Modern Political Thought*. See R. Mousnier, "The Exponents and Critics of Absolutism," in *The New Cambridge Modern History*, vol. 4, *The Decline of Spain and the Thirty Years War*, ed. J. P. Cooper (Cambridge: Cambridge University Press, 1970), 104; and Jean Gottmann, *The Significance of Territory* (Charlottesville: University Press of Virginia, 1973), 43.

164. Franklin, introduction to Jean Bodin, *On Sovereignty: Four Chapters from The Six Books of the Commonwealth*, ed. and trans. Julian H. Franklin, ix–xxvi (Cambridge: Cambridge University Press, 1992), xv.

165. David Parker, "Law, Society and the State in the Thought of Jean Bodin," *History of Political Thought* 11 (1981): 284–85.

166. Jean Bodin, *Les six livres de la république*, ed. Christiane Frémont, Marie-Dominique Couzinet, and Henri Rochais, 6 vols. (Paris: Fayard, 1986). This is based on the 10th French ed., 1593.

167. Jean Bodin, *De republica libri sex* (Paris: J. de Puys, 1586).

168. On the differences, see Kenneth Douglas McRae, introduction to Jean Bodin, *The Six Bookes of a Commonweale: A Facsimile Reprint of the English Translation of 1606*, trans. Richard Knolles, ed. Kenneth Douglas McRae (Cambridge, MA: Harvard University Press, 1962), A 28–38.

169. Bodin, *Six Bookes*.

170. McRae, preface to Bodin, *Six Bookes*, A ix.

171. Bodin, *Six Books of the Commonwealth*.

172. Bodin, *On Sovereignty*. This includes I, viii; I, x; II, i; and II, v.

173. See Franklin, "Note on the Text," in Bodin, *On Sovereignty*, xxxv–xxxviii.

174. The key edition that does this is the Italian, based on the 1583 Paris edition, reprinted in 1961. Jean Bodin, *I sei libri dello Stato di Jean Bodin*, ed. Margherita Isnardi

Parente and Diego Quaglioni, 3 vols. (Turin: Unione Tipografico-Editrice Torinese, 1964–97).

175. Ralph E. Giesey, "Medieval Jurisprudence in Bodin's Concept of Sovereignty," in *Jean Bodin: Verhandlungen der internationalen Bodin Tagung in München*, ed. Horst Denzer, 151–66 (Munich: C. H. Beck, 1973), 168n3.

176. Jean Bodin, *Methodus ad facilem historiarum cognitionem* (Amsterdam, 1650; reproduction, Aalen, Germany: Scientia Verlag, 1967); trans. Beatrice Reynolds as *Method for the Easy Comprehension of History* (New York: W. W. Norton, 1945).

177. Jean Bodin, *De la démonomanie des sorciers* (Paris: Jacques du Puys, 1580); and Jean Bodin, *On the Demon-Mania of Witches*, trans. Randy A. Scott (Toronto: Centre for Reformation and Renaissance Studies, 1995).

178. Alan Harding, "The Origins of the Concept of the State," *History of Political Thought* 15 (1994): 68. On Bodin generally, see also his *Medieval Law and the Foundations of the State* (Oxford: Oxford University Press, 2002), 316–21.

179. Skinner, *Foundations of Modern Political Thought*, 2:284.

180. Julian H. Franklin, *Jean Bodin and the Rise of Absolutist Theory* (Cambridge: Cambridge University Press, 1973), 102.

181. Bodin, *Les six livres*, I, x.

182. Bodin, *Les six livres*, I, ix. The Knolles translation includes the gloss "which hold all their territories of others" after "feudal kings," a phrase not in the French or Latin.

183. Bodin, *Les six livres*, I, viii.

184. Bodin, *Les six livres*, I, i.

185. Bodin, *Les six livres*, II, i.

186. Bodin, *Les six livres*, II, i.

187. Bodin, *De republica*, I, viii.

188. Bodin, *De republica*, I, viii.

189. Skinner, *Foundations of Modern Political Thought*, 2:288.

190. Bodin, *Les six livres*, I, i.

191. Bodin, *Les six livres*, I, ii.

192. Bodin, *Les six livres*, I, vi.

193. Bodin, *Six Bookes*, 264.

194. Bodin, *Les six livres*, III, i; see IV, i, for the same phrasing.

195. Bodin, *Les six livres*, II, vi. See Bodin, *Six Bookes*, 239.

196. Bodin, *Les six livres*, I, vii.

197. Bodin, *Les six livres*, VI, ii.

198. Bodin, *Les six livres*, VI, ii.

199. Bodin, *Les six livres*, I, vii. Many other instances could be given.

200. Bodin, *Six Bookes*, III, vi. This is not in the French and not really in the Latin. There are several other minor instances in this chapter.

201. Bodin, *Les six livres*, III, v.

202. Bodin, *Les six livres*, III, vi.

203. Bodin, *De republica*, III, v.

204. Bodin, *Les six livres*, III, vi.

205. Bodin, *De republica*, I, vii.

206. Bodin, *De republica*, III, vi.

207. Bodin, *Six Bookes*, 355.

208. Bodin, *Les six livres*, III, vi.

209. Bodin, *De republica*, III, vi. Not all this passage is in the French.

210. Bodin, *Les six livres*, III, vi.

211. Bodin, *Les six livres*, III, vi.

212. Bodin, *Les six livres*, III, vi.

213. Bodin, *Les six livres*, II, i.

214. Tierney, *Religion, Law and the Growth of Constitutional Thought*, 84.

215. Bodin, *Les six livres*, VI, vi, passim.

216. Bodin, *Les six livres*, IV, ii.

217. Bodin, *Les six livres*, VI, vi.

218. Bodin, *Les six livres*, II, i.

219. Bodin, *Les six livres*, VI, vi. For a discussion of this generally, see Franklin, *Jean Bodin and the Rise of Absolutist Theory*.

220. Bodin, *Methodus ad facilem historiarum cognitionem*, preface.

221. Perry Anderson, *Lineages of the Absolutist State* (London: NLB, 1974), 50.

222. Giesey, "Medieval Jurisprudence," 174.

223. Bodin, *Les six livres*, III, vii; see I, Epistola.

224. Harold Dexter Hazeltine, "The Legal and Political Ideas of the Post-Glossators," in *The Medieval Idea of Law as Represented by Lucas de Penna: A Study in Fourteenth-Century Legal Scholarship*, ed. Walter Ullmann, xv–xxxix (1946; repr., New York: Barnes and Noble, 1969), xxv, xxxvi–xxxvii (citations refer to the 1969 edition).

225. Giesey, "Medieval Jurisprudence," 186.

226. Michel de Montaigne, *Les essais*, ed. Pierre Villey, 3 vols. (Paris: PUF, 1965); and Michel de Montaigne, *The Complete Essays*, trans. and ed. M. A. Screech (London: Penguin, 2003), II, v.

227. Elizabeth Hodges, *Urban Poetics in the French Renaissance* (Aldershot, UK: Ashgate, 2008), 126.

228. Quentin Skinner, *The Foundations of Modern Political Thought*, vol. 1, *The Renaissance* (Cambridge: Cambridge University Press, 1978), 253; and Robert J. Collins, "Montaigne's Rejection of Reason of State in 'De l'Utile et de l'honneste,'" *Sixteenth Century Journal* 23 (1992): 71–94.

229. Vom Thumshirn, representative for Saxe-Altenburg, November 28, 1648, quoted in Geoffrey Parker, *The Thirty Years' War* (London: Routledge and Kegan Paul, 1984), 219.

230. Friedrich Meinecke, *Machiavellism: The Doctrine of* Raison d'État *and Its Place in Modern History*, trans. Douglas Scott (Boulder, CO: Westview, 1984), 1.

231. Guicciardini, *Dialogo e discorsi del reggimento di Firenze*, 163; and Guicciardini, *Dialogue on the Government of Florence*, 159.

232. Hexter, *Vision of Politics*, 168

233. Skinner, "Political Philosophy," 442. On Botero, see Viroli, *From Politics to*

Reason of State, chap. 6; Kenneth C. Schellhase, "Botero, Reason of State, and Tacitus," in *Botero e la "Ragion di stato": Atti del convegno in memoria di Luigi Firpo (Torino, 8–10 marzo 1990)* (Firenze, Italy: Leo S. Olschki, 1992); and Descendre, *L'État du monde.*

234. Giovanni Botero, *Della ragione di stato* (Milan: Nella Stamparia del q. Pacifico Pontio, 1596); and Giovanni Botero, *The Reason of State,* trans. P. J. Waley and D. P. Waley (London: Routledge and Kegan Paul, 1956), I, 1.

235. See, for example, Rome: Vincenzo Pellagallo, 1590.

236. Descendre, *L'État du monde,* 94.

237. Descendre, *L'État du monde,* 79.

238. Foucault, *Sécurité, Territoire, Population,* 243.

239. Giovanni Botero, *Della cause della grandezza delle citta* (Milan: Nella Stamparia del q. Pacifico Pontio, 1596), I, 1; Giovanni Botero, *A Treatise Concerning the Causes of the Magnificency and Greatness of Cities,* trans. Robert Peterson (London: Richard Ockould and Henry Tomes, 1606); reprinted with modernized spelling in Botero, *Reason of State.* There is another early translation: Giovanni Botero, *The Cause of the Greatness of Cities,* trans. Sir T.H. (London: Henry Seile, 1635).

240. Botero, *Della cause della grandezza delle citta,* I, 3.

241. "The territory [*territorio*] of Bourges" (I, 9; not in 1596 edition, in 1606); "whether the territory [*territorio*] be fruitful" (I, 9).

242. Botero, *Della cause della grandezza delle citta,* I, 8, 9.

243. Giovanni Botero, *Le relationi universali* (Venetia, Italy: Dusinelli, 1595). The fifth book was finally published in Carlo Gioda, *La vita e le opere di Giovanni Botero,* 3 vols. (Milan: Ulrico Hoepli, 1894–95), vol. 3. There is a selection in a more recent edition: Aldo Albònico, *Il mondo Americano di Giovanni Botero: Con una selezione dale* Epistolae *e dale* Relationi Universali (Rome: Bulzoni, 1990). There is quite a loose English translation, which incorporates materials not by Botero: *The Worlde; or, A Historicall Description of the most famous kingdoms and common-weales therein* (London: John Iaggard, 1601); and later as *Relations of the Most Famous Kingdoms and Commonweales through the World* (London: John Iaggard, 1616).

244. Robert Bireley, *The Counter-Reformation Prince: Anti-Machiavellianism or Catholic Statecraft in Early Modern Europe* (Chapel Hill: University of North Carolina Press, 1990), 21.

245. Bireley, *Counter-Reformation Prince,* 48.

246. Sebastian Münster, *Cosmographia, Das ist Beschreibung der gantzen Welt,* 1544, facsimile edition of the 1628 Basel edition, 2 vols. (Lindau: Antiqua-Verlag, 1984). On this, see Matthew McLean, *The* Cosmographia *of Sebastian Münster* (Aldershot, UK: Ashgate, 2007).

247. On this map, see Bernhard Klein, *Maps and the Writing of Space in Early Modern England and Ireland* (Houndmills, UK: Palgrave, 2001), 36–38.

248. The latter was translated into English in 1606. On this period generally, see David Woodward, ed., *The History of Cartography,* vol. 3, *Cartography in the European Renaissance,* 2 pts. (Chicago: University of Chicago Press, 2007).

249. Botero, *Worlde,* 1.

250. Botero, *Della ragione di stato,* III, 3.

251. Botero, *Della ragione di stato,* III, 4.

252. Botero, *Della ragione di stato*, I, 5. This phrase is not in either the 1590 or 1596 edition and first appears in the 1606 one.

253. Botero, *Della ragione di stato*, III, 4.

254. Descendre, *L'État du monde*, 169.

255. Descendre, *L'État du monde*, 14.

256. Descendre, *L'État du monde*, 14. Chap. 6 is entitled "Territorialisation of politics," and, of this, section 2 is entitled "Territory in *Ragione di Stato*."

257. Descendre, *L'État du monde*, 242.

258. Descendre, *L'État du monde*, 339.

259. Martin van Gelderen, *The Political Thought of the Dutch Revolt, 1555–1590* (Cambridge: Cambridge University Press, 1992), 186.

260. Justus Lipsius, *Politica: Six Books of Politics or Political Instruction*, ed. and trans. Jan Waszink, Latin-English ed. (Assen, Netherlands: Van Gorcum, 2004).

261. Ronald Mellor, *Tacitus: The Classical Heritage* (New York: Garland, 1995), 41.

262. Justus Lipsius, *De constantia libri duo* (Francofurdi: Ioannem Wechelum, 1590); and Justus Lipsius, *On Constancy*, trans. Sir John Stradling (1595), ed. John Sellars (Exeter: Bristol Phoenix Press, 2006).

263. Sellars, introduction to *On Constancy*, 4; and Waszink, introduction to Lipsius, *Politica*, 148.

264. Lipsius, *De constantia*, preliminary matter.

265. Lipsius, *Politica*, I, 1, notae. The *Notae* were published in 1589 as *Ad libros Politicorum breves notae*, and expanded in 1596; they are included in the Waszink edition, although only in Latin.

266. Lipsius, *Monita et exempla politica, libro duo: Qui virtutes et vitia principium spectant* (Antwerp, 1605); in *Opera Omnia*, 4 vols. (Vesaliae: Andreae ab Hoogenhuysen, 1645), vol. 4.

267. Lipsius, *Adversus Dialogistam liber De Una Religione* (Frankfurt, Germany: Ioannem Wechelum, 1591).

268. Lipsius, *De Militia Romana libri quinque: Commentarius ad Polybium* (Antwerp, 1614); in *Opera Omnia*, vol. 3.

269. Peter H. Wilson, *Europe's Tragedy: A New History of the Thirty Years War* (London: Penguin, 2009), 140.

270. See Gerhard Oestreich, *Neostoicism and the Early Modern State*, ed. Brigitta Oestreich and H. G. Koenigsberger, trans. David McLintock (Cambridge: Cambridge University Press, 1982).

271. Lipsius, *De constantia*, preliminary matter.

272. Lipsius, *De constantia*, preliminary matter.

273. Lipsius, *De constantia*, preliminary matter.

274. Waszink, introduction to Lipsius, *Politica*, 92, citing a letter.

275. Senellart, *Les arts de gouverner*, 233.

276. Lipsius, *De constantia*, II, 11.

277. Lipsius, *Politica*, IV, xiv.

278. Plutarch, *Moralia*, 190e5 (trans. Frank Cole Babbitt, Greek-Latin ed., 14 vols. [London: William Heinemann, 1931]). Spartan general Lysander: "When the Argives seemed to make out a better case than the Spartans to a land [*khoras*] in dispute, he

drew his sword and said, 'He that is master of this can best talk about bounds [*kraton beltista peri ges horon dialegetai*] of countries." See also *Moralia*, 229d6; and *Lysander*, 445d, in *Lives*, trans. Bernadotte Perrin, Greek-English ed., 11 vols. (Cambridge, MA: Harvard University Press, 1914–26), vol. 4.

279. Kleinschmidt, *Understanding the Middle Ages*, 60.

280. Lipsius, *De constantia*, I, 11.

281. Quentin Skinner, *Hobbes and Republican Liberty* (Cambridge: Cambridge University Press, 2008), 47.

282. Stephanus Junius Brutus, the Celt, *Vindiciae, contra Tyrannos: Sive, De Principis in Populum, Populique in Principem, legitima potestate* (Edimburgi, 1579) (the place was false and the publisher was actually Thomas Guérin in Basle); *Vindiciae, contra Tyrannos; or, Concerning the Legitimate Power of a Prince over the People, and of the People over a Prince*, ed. and trans. George Garnett (Cambridge: Cambridge University Press, 1994).

283. Oestreich, *Neostoicism and the Early Modern State*, 144.

284. *Vindiciae, contra Tyrannos*, 222, 176.

285. *Vindiciae, contra Tyrannos*, 233–34, 183. Garnett adds a reference to *Digest* L.17.36, from Pomponius, *Sabinus*, 27, which reads, "It is culpable to involve oneself in an affair with which one has no concern [*culpa est immiscere se rei ad se non pertinenti*]."

286. *Henry VI, Part 2*, III, i.

287. *King Lear*, I, i. At least, that is how it is in the composite text that is usually used of *King Lear*, which builds on the later folio, but usually incorporates material that was originally in the earlier quarto. I have used the Arden edition, ed. Kenneth Muir (London: Routledge, 1972).

288. *King Lear*, I, i. A variant noted by Muir substitutes "spacious" for "precious."

289. Henry S. Turner, "*King Lear* Without: The Heath," *Renaissance Drama* 28 (1997): 158.

290. *King Lear*, I, i. "With champains rich'd, / With plenteous rivers and" is not in the quarto.

291. *King Lear*, I, i. These lines are not in the quarto.

292. Turner, "*King Lear* Without," 178. See Edward Grant, *Much Ado about Nothing: Theories of Space and Vacuum from the Middle Ages to the Scientific Revolution* (Cambridge: Cambridge University Press, 1981).

293. *King Lear*, I, i.

294. These points are greatly expanded in my "The Geopolitics of *King Lear*: Territory, Land, Earth," *Law and Literature* 35 (2013).

295. Harry V. Jaffa, "The Limits of Politics: *King Lear*, Act I, Scene i," in Allan Bloom with Harry V. Jaffa, *Shakespeare's Politics*, 113–45 (New York: Basic Books, 1964), 113.

296. John Gillies, *Shakespeare and the Geography of Difference* (Cambridge: Cambridge University Press, 1994), chap. 2; and his "The Scene of Cartography in *King Lear*," in *Literature, Mapping, and the Politics of Space in Early Modern Britain*, ed. Andrew Gordon and Bernhard Klein, 109–37 (Cambridge: Cambridge University Press, 2001), 116.

297. Jaffa, "Limits of Politics," 127.

298. *King Lear*, I, i.

299. *King Lear*, I, i.

300. Jaffa, "Limits of Politics," 122.

301. Jaffa, "Limits of Politics," 123–24.

302. *King Lear*, I, i.

303. Emily W. Leider, "Plainness of Style in *King Lear*," *Shakespeare Quarterly* 21 (1970): 46.

CHAPTER 9

1. W. J. Torrance Kirby, *Richard Hooker, Reformer and Platonist* (Aldershot, UK: Ashgate, 2005), ix; see W. D. J. Cargill-Thompson, "The Philosopher of the 'Politic Society': Richard Hooker as a Political Thinker," in his *Studies in the Reformation: Luther to Hooker*, ed. C. W. Dugmore, 131–91 (London: Athlone Press, 1980).

2. The standard modern edition is found in the *Folger Library Edition of the Works of Richard Hooker*, ed. W. Speed Hill, 5 vols. (Cambridge, MA: Belknap Press, 1977–90). There is a more convenient edition of the preface and books 1 and 8 in Richard Hooker, *Of the Laws of Ecclesiastical Polity*, ed. Arthur Stephen McGrade (Cambridge: Cambridge University Press, 1989).

3. Stanley Archer, *Richard Hooker* (Boston: Twayne Publishers, 1983), 99.

4. Hooker, *Of the Laws*, III, 1.14.

5. Robert Eccleshall, *Order and Reason in Politics: Theories of Absolute and Limited Monarchy in Early Modern England* (Oxford: Oxford University Press, 1978), 127.

6. Hooker, *Of the Laws*, I, 8.10.

7. Hooker, *Of the Laws*, VII, 8.

8. Hooker, *Of the Laws*, VII, 8.1.

9. Hooker, *Of the Laws*, VII, 8.2.

10. Hooker, *Of the Laws*, V, 80.11.

11. Hooker, *Of the Laws*, VIII, 3.5; see VIII, 6.8.

12. Robert K. Faulkner, *Richard Hooker and the Politics of a Christian England* (Berkeley: University of California Press, 1981), 13.

13. Hooker, *Of the Laws*, VII, 2.1.

14. Hooker, *Of the Laws*, VIII, 6.3.

15. Hooker, *Of the Laws*, VIII, 8.4.

16. Faulkner, *Richard Hooker*, 100.

17. Johannes Althusius, *Politica methodice digesta atq; exemplis sacris & profanes illustrata* (Groningae: Iohannes Radaeus, 1610; 3rd ed., 1614). There is an abridged modern edition: Johannes Althusius, *Politica Methodice Digesta of Johannes Althusius (Althaus)*, ed. Carl Joachim Friedrich (Cambridge, MA: Harvard University Press, 1932); Johannes Althusius, *The Politics of Johannes Althusius: An Abridged Translation of the Third Edition of* Politica Methodice Digesta atque exemplis sacris et profanes illustrata, *and including the Prefaces to the First and Third Editions*, trans. Frederick S. Carney (London: Eyre and Spottiswoode, 1964; repr., Indianapolis: Liberty Fund, 1995) (citations refer to the 1995 edition). A comprehensive study, which was fundamental in his rediscovery, is Otto Gierke, *The Development of Political Theory*, trans. Bernard

Freyd (London: George Allen and Unwin, 1939). See also Hanns Gross, *Empire and Sovereignty: A History of the Public Law Literature in the Holy Roman Empire, 1599–1804* (Chicago: University of Chicago Press, 1973), chap. 3; and Thomas O. Hueglin, *Early Modern Concepts for a Late Modern World: Althusius on Community and Federalism* (Waterloo, Ontario: Wilfrid Laurier University Press, 1999).

18. Quentin Skinner, *The Foundations of Modern Political Thought*, vol. 2, *The Age of Reformation* (Cambridge: Cambridge University Press, 1978), 341. On this, see Jesse Chupp and Cary J. Nederman, "The Calvinist Background to Johannes Althusius's Idea of Religious Toleration," in *Jurisprudenz, Politische Theorie und Politische Theologie*, ed. Frederick S. Carney, Heinz Schilling, and Dieter Wyduckel, 243–60 (Berlin: Duncker and Humblot, 2004); and John Witte Jr., *The Reformation of Rights: Law, Religion, and Human Rights in Early Modern Calvinism* (Cambridge: Cambridge University Press, 2007), chap. 3.

19. Friedrich, introduction to Althusius, *Politica Methodice Digesta*, lxxxvi.

20. See, for example, Althusius, *Politica*, IX, 21.

21. Witte Jr., *Reformation of Rights*, 161.

22. Witte Jr., *The Reformation of Rights*, 169.

23. Althusius, *Politica*, preface to the 3rd ed.

24. Althusius, *Politica*, I, 1–2.

25. Friedrich, introduction to Althusius, *Politica Methodice Digesta*, lxxxiv.

26. Althusius, *Politica*, preface to the 3rd ed.

27. Althusius, *Politica*, V, 8.

28. Althusius, *Politica*, V, 28–29.

29. Althusius, *Politica*, VI, 1–6.

30. Althusius, *Politica*, VI, 39–41.

31. Althusius, *Politica*, VIII, 51.

32. Althusius, *Politica*, VI, 42–43.

33. Althusius, *Politica*, IX, 33–34; see XXXVIII, 61.

34. Althusius, *Politica*, XVIII, 106.

35. Althusius, *Politica*, VII, 1–2. This chapter is not in earlier editions.

36. Althusius, *Politica*, VIII, 53–54. The ellipses do not omit text, but extensive references to jurists.

37. Althusius, *Politica*, IX, 12–13. Other references to laws working within the "territory of the realm [*territorio regni*]," *in territorio*, or *extra territorium*: in XI, 5, 16; XIII, 3–5; XVIII, 112; XXVIII, 2; etc.

38. Althusius, *Politica*, IX, 14.

39. Udalricus Zasius, *Opera Omnia*, ed. Johann Ulrich Zasius and Joachim Münsinger von Frundeck, 7 vols. (Lyon, 1550; repr., Aalen, Germany: Scientia Verlag, 1966) (citations refer to the 1966 edition).

40. Zasius, *Opera Omnia*, 1:24 (col. 43).

41. Zasius, *Opera Omnia*, 1:210 (col. 415); and 6:256 (col. 508).

42. Zasius, *Opera Omnia*, consilia XVI, in 6:256 (col. 508).

43. Matthias Stephani, *Tractatus de jurisdictione, Liber II* (Frankfurt, Germany: Petrum Kopffium, 1610), I, 7.

44. Stephani, *Tractatus de jurisdictione Liber II*, I, 7.

45. Stephani, *Tractatus de jurisdictione Liber II*, I, 7.

46. Stephani, *Tractatus de jurisdictione Liber II*, I, 7.

47. Andreas Knichen, *De sublimi et regio territorii iure*, in *Opera* (Hanover, Germany: Iohan Aubrii, 1613), 1–92.

48. Gross, *Empire and Sovereignty*, 142.

49. D. Willoweit, "Territorium," in *Handwörterbuch zur deutschen Rechtsgeschichte*, ed. Adalbert Erler and Ekkehard Kaufmann, vol. 5/33 (Berlin: Erich Schmidt, 1991), cols. 149–51.

50. Almut Höfert, "States, Cities, Citizens in the Later Middle Ages," in *States and Citizens: History, Theory, Prospects*, ed. Quentin Skinner and Bo Stråth, 63–75 (Cambridge: Cambridge University Press, 2003), 64.

51. Knichen, *De sublimi et regio territorii iure*, "Tobiæ Lagi Responsatio."

52. Knichen, *De sublimi et regio territorii iure*, I, 202.

53. Knichen, *De sublimi et regio territorii iure*, "Epistola liminaris."

54. See Michael Stolleis, *Geschichte des öffentlichen Rechts in Deutschland Erster Band: Reichspublizistik und Policeywissenschaft, 1600–1800* (Munich: C. H. Beck, 1988), 147.

55. Knichen, *De sublimi et regio territorii iure*, III, 1–6.

56. Knichen, *De sublimi et regio territorii iure*, I, 195.

57. Knichen, *De sublimi et regio territorii iure*, esp. IV, 409–22.

58. On some of the nuances, see Werner Köster, *Die Rede über den "Raum": Zur semantischen Karriere eines deutschen Konzepts* (Heidelberg, Germany: Synchron, 2002), 54.

59. Knichen, *De sublimi et regio territorii iure*, I, 1.

60. Gross, *Empire and Sovereignty*, 141.

61. Knichen, *De sublimi et regio territorii iure*, IV, 35.

62. Knichen, *De sublimi et regio territorii iure*, I, 262; see I, 276.

63. Knichen, *De sublimi et regio territorii iure*, I, 334–54.

64. Knichen, *De sublimi et regio territorii iure*, V, 133–35.

65. See Gross, *Empire and Sovereignty*, 141–46; Kleinschmidt, *Understanding the Middle Ages*, 58–60; and Stolleis, *Geschichte des öffentlichen Rechts in Deutschland*, 147, 185.

66. D. Willoweit, "Landesobrigkeit," in *Handwörterbuch zur deutschen Rechtsgeschichte*, vol. 1, 1404–5, A. 3, 173; see Stolleis, *Geschichte des öffentlichen Rechts in Deutschland*, 185.

67. Francis Bacon, *The Works of Francis Bacon*, ed. James Spedding, Robert Leslie Ellis, and Douglas Denon Heath, 7 vols. (London: Longmans, 1862–72), 6:587.

68. Bacon, *Works*, 6:587.

69. Bacon, *Works*, 6:445.

Bacon, *Works*, 6:445.

70. Bacon, *Works*, 1:793–94, 5:80.

71. Bacon, *Works*, 7:47–64.

72. Of James's own writings, which provide a strong defense of the divine right of kings, see King James VI and I, *Political Writings*, ed. Johann Somerville (Cambridge: Cambridge University Press, 1994).

73. Bacon, *Works*, 7:47.

74. Bacon, *Works*, 7:48.

75. Bacon, *Works*, 7:48–49.

76. Bacon, *Works*, 7:51.

77. See Markku Peltonen, "Politics and Science: Francis Bacon and the True Greatness of States," *Historical Journal* 35 (1992): 299; and Markku Peltonen, "Bacon's Political Philosophy," in *The Cambridge Companion to Bacon*, ed. Markku Peltonen, 283–310 (Cambridge: Cambridge University Press, 1996).

78. On the attributions of spurious texts to Ralegh, see Anna Beer, "Textual Politics: The Execution of Sir Walter Ralegh," *Modern Philology* 94 (1996): 19–38; and generally, Stephen J. Greenblatt, *Sir Walter Ralegh: The Renaissance Man and His Roles* (New Haven, CT: Yale University Press, 1973).

79. Kenneth Olwig, *Landscape, Nature and the Body Politic: From Britain's Renaissance to America's New World* (Madison: University of Wisconsin Press, 2002), 249n13.

80. Walter Ralegh, "Maxims of State," in *The Works of Walter Ralegh*, 8 vols. (Oxford: Oxford University Press, 1829), 8:1; see "The Cabinet-Council," in *Works*, 8:37.

81. Jean Bodin, *The Six Bookes of a Commonweale: A Facsimile Reprint of the English Translation of 1606*, trans. Richard Knolles, ed. Kenneth Douglas McRae (Cambridge, MA: Harvard University Press, 1962), I, ix.

82. Ralegh, "A Discourse of the Original and Fundamental Cause of Natural, Arbitrary, Necessary, and Unnatural War," in *Works*, 8:270.

83. Ralegh, "Maxims of State," in *Works*, 21.

84. Stephen Toulmin, *Cosmopolis: The Hidden Agenda of Modernity* (Chicago: University of Chicago Press, 1990), 46.

85. Toulmin, *Cosmopolis*, 57.

86. Geoffrey Parker, *The Thirty Years' War* (London: Routledge and Kegan Paul, 1984), 61.

87. Emile Boutroux, "Descartes and Cartesianism," in *The Cambridge Modern History*, vol. 4, *The Thirty Years War* (New York: Macmillan, 1906), 778.

88. Boutroux, "Descartes and Cartesianism," 787.

89. See John Rockford Vrooman, *René Descartes: A Biography* (New York: G. P. Putnam's Sons, 1970), 244.

90. In a slightly different register, see Antonio Negri, *Political Descartes: Reason, Ideology and the Bourgeois Project*, trans. Matteo Mandarini and Alberto Toscano (London: Verso, 2006).

91. The libretto was believed lost, but a version attributed to Descartes appeared as "Un ballet de Descartes," by Albert Thibaudet and Johan Nordström, *Revue de Genève* 1 (1920): 163–85; and in the 2nd ed. of *Oeuvres de Descartes*, ed. Charles Adam and Paul Tannery, 13 vols. (Paris: Vrin, 1964–74) (hereafter AT), 5:616–27. The standard case against is Richard A. Watson, "René Descartes n'est pas l'auteur de 'La naissance de la paix,'" *Archives de Philosophie* 53 (September 1990): 389–401. The only "proof" is Descartes's mention of the ballet in a letter (AT, 5:455–61), but this merely says he is enclosing a copy of the libretto, *not* that he wrote it. See Stephen Gaukroger, *Descartes: An Intellectual Biography* (Oxford: Clarendon Press, 1995), 415.

92. Descartes, *Discours de la méthode*, AT, 6:11–13. English translations of most works are largely based on those in *The Philosophical Writings of Descartes*, trans. John Cottingham, Robert Stoothoff, and Dugald Murdoch, 2 vols. (Cambridge: Cambridge University Press, 1985). This text, as with most translations, has the AT pagination in the margins.

93. Descartes, "Objectiones Septimæ," AT, 7:536–61.

94. Descartes, *Meditationes de Prima Philosophia*, AT, 7:44.

95. Descartes, *Meditationes de Prima Philosophia*, AT, 7:24.

96. Descartes, *Meditationes de Prima Philosophia*, AT, 7:31.

97. Descartes, *Principia Philosophiae*, AT, vol. VIII-1, 41; see 10; and *Regulae ad directionem ingenii*, AT, 10:442.

98. Descartes, *Principia Philosophiae*, AT, vol. VIII-1, 47–48.

99. Descartes, *Principia Philosophiae*, AT, vol. VIII-1, 48.

100. Descartes, letter to Mersenne, December 1637, AT, 1:478.

101. Descartes, *Meditationes de Prima Philosophia*, AT, 7:69.

102. Descartes, *La geometrie*, AT, 6:369.

103. Descartes, *La geometrie*, AT, 6:475.

104. Jacob Klein, *Greek Mathematical Thought and the Origin of Algebra* (New York: Dover, 1992), 210–11.

105. See Vincenzo de Risi, *Geometry and Monadology: Leibniz's Analysis Situs and Philosophy of Space* (Basel, Switzerland: Birkhäuser, 2007), 6.

106. H. G. Alexander, introduction to *The Leibniz-Clarke Correspondence*, ed. H. G. Alexander, ix–lvi (Manchester, UK: Manchester University Press, 1956), xxxiii.

107. See Klein, *Greek Mathematical Thought*, 256n209.

108. Shlomo Pines, "Philosophy, Mathematics and the Concepts of Space in the Middle Ages," in *The Interaction between Science and Philosophy*, ed. Y. Elkana, 75–91 (Atlantic Highlands, NJ: Humanities Press, 1974), 86–87; and Edward A. Maziarz and Thomas Greenwood, *Greek Mathematical Philosophy* (New York: Frederick Ungar, 1968), 256.

109. Descartes to Mersenne, July 27, 1638, AT, 2:268.

110. Descartes, *Discours de la méthode*, AT, 6:36.

111. Descartes, *Discours de la méthode*, AT, 6:36.

112. This discussion draws extensively on the analysis in Stuart Elden, *Speaking against Number: Heidegger, Language and the Politics of Calculation* (Edinburgh: Edinburgh University Press, 2006), 131–37.

113. See Dorothy Waley Singer, *Giordano Bruno: His Life and Thought with Annotated Translation of His Work on the Infinite Universe and Worlds* (London: Constable and Company, 1950); and Frances Yates, *Giordano Bruno and the Hermetic Tradition* (London: Routledge and Kegan Paul, 1964).

114. In an extensive literature, see Alexandre Koyré, *Metaphysics and Measurement: Essays in Scientific Revolution* (London: Chapman and Hall, 1958); and Eileen Reeves, *Galileo's Glassworks: The Telescope and the Mirror* (Cambridge, MA: Harvard University Press, 2008).

115. Galileo Galilei, *Le operazioni del compasso geometrico et militare* (Padua, Italy: Pietro Marinelli, 1606).

116. Alexandre Koyré, *From the Closed World to the Infinite Universe* (New York: Harper, 1958), 47.

117. Cusa, *De staticis experimentis*, in *Opera Omnia*, ed. Ludovici Baur (Hamburg: Felicis Meiner, 1959–68), vol. 5.

118. Cusa, *De docta ignorantia*, in *Opera Omnia*, ed. Ernestus Hoffmann and Raymundus Klibansky, 1:111 (bk. 2, 13); Cusa, *De staticis experimentis*, in *Opera Omnia*, ed. Baur, 5:222. See Ernst Cassirer, *The Individual and the Cosmos in Renaissance Philosophy*, trans. Mario Domandi (Oxford: Basil Blackwell, 1963), 52–53. Though, as Keith Thomas, *Religion and the Decline of Magic: Studies in Popular Beliefs in Sixteenth- and Seventeenth-Century England* (Harmondsworth, UK: Penguin, 1971), 430–31, points out, calculation was often seen as magic by religious authorities, and mathematics as a black art.

119. See, in particular, Edmund Husserl, *The Crisis of European Sciences and Transcendental Phenomenology*, trans. David Carr (Evanston, IL: Northwestern University Press, 1970).

120. Ernst Cassirer, *The Myth of the State* (1946; repr., New York: Doubleday, 1955), 205 (citations refer to the 1955 edition). On the context generally, see Stephen Gaukroger, *The Emergence of a Scientific Culture: Science and the Shaping of Modernity, 1210–1685* (Oxford: Oxford University Press, 2006).

121. These are collected in AT, vol. 7. One exception is the discussion by Pierre Gassendi, "Objectiones Quintæ," AT, 7:271–74.

122. Antoine Arnauld, "Objectiones Quartæ," AT, 7:204.

123. Thomas Hobbes, *Leviathan*, ed. Richard Tuck (Cambridge: Cambridge University Press, 1996), I, 4 [15]. References are given to book, chapter, and pagination of the first edition, often found in the margins of modern editions.

124. Tuck, introduction to Hobbes, *Leviathan*.

125. Quentin Skinner, *Visions of Politics*, vol. 3, *Hobbes and Civil Science* (Cambridge: Cambridge University Press, 2002), 318.

126. Michael Allen Gillespie, *The Theological Origins of Modernity* (Chicago: University of Chicago Press, 2008), 233. See also Frithiof Brandt, *Thomas Hobbes' Mechanical Conception of Nature* (Copenhagen: Levin and Munksgaard, 1928), chap. 4. On his attitude to science more generally, see Steven Shapin and Simon Schaffer, *Leviathan and the Air-pump: Hobbes, Boyle, and the Experimental Life* (Princeton, NJ: Princeton University Press, 1985); and Gaukroger, *Emergence of a Scientific Culture*, 282–89. Hobbes wrote a preface to Marin Mersenne, *Ballistica et acontismologia* (Paris: A. Bertier, 1644). For a helpful general discussion, see J. A. Bennett, "The Mechanics' Philosophy and the Mechanical Philosophy," *History of Science* 24 (1986): 1–28.

127. Thomas Hobbes, *The Correspondence*, ed. Noel Malcolm, 2 vols. (Oxford: Clarendon Press, 1994), 1:51; see Malcolm's note to 2:825.

128. Hobbes, *Leviathan*, I, 5 [18].

129. In Hobbes, *Correspondence*, 2:530, 2:532.

130. Spinoza, *Ethica*, in *Opera*, ed. Carl Gebhardt, 4 vols. (Heidelberg, Germany: Universitætsbuchhandlung, 1925), vol. 3; Spinoza, *Ethics*, ed. and trans. G. H. R. Parkin-

son (Oxford: Oxford University Press, 2000). The latter is actually a translation of the edition in *Benedicti de Spinoza Opera*, ed. J. Van Vloten and J. P. N. Land, 4 vols. (The Hague: Martinus Nijhoff, 1914), vol. 1.

131. See, for example, Gilles Deleuze, "Spinoza and the Three 'Ethics,'" in *The New Spinoza*, ed. Warren Montag and Ted Stolze (Minneapolis: University of Minnesota Press, 1997), 33n2, which suggests Desargues's "projective optical geometry," rather than that of Descartes or Hobbes, is the model. Deleuze's source is Yvonne Toros, "Spinoza et l'espace projectif" (thesis, University of Paris-VIII, St.-Denis, 1990).

132. Warren Montag, *Bodies, Masses, Power: Spinoza and His Contemporaries* (London: Verso, 1999), 2.

133. Koyré, *From the Closed World*, 155–56.

134. Matthew Stewart, *The Courtier and the Heretic: Leibniz, Spinoza, and the Fate of God in the Modern World* (New Haven, CT: Yale University Press, 2005), 327.

135. For a sampling, see Montag and Stolze, *New Spinoza*. See also Etienne Balibar, *Spinoza et la politique* (Paris: PUF, 1985); and Pierre Macherey, *Hegel ou Spinoza* (Paris: Maspéro, 1977).

136. There is no discussion, for instance, in his *Tractatus Politicus*. See *Opera*, vol. 3; and *The Political Works: The Tractatus Theologico-Politicus in Part and the Tractatus Politicus in Full*, ed. and trans. A. G. Wernham, Latin-English ed. (Oxford: Clarendon Press, 1958).

137. Spinoza, *Tractatus Theologico-Politicus*, VIII, 3 [119]. The Wernham edition omits much of the theological discussion. I have therefore used the Latin text in *Oeuvres III*, ed. Fokke Akkerman, Latin-French ed. (Paris: PUF, 1999); and Spinoza, *Theological-Political Treatise*, ed. Jonathan Israel, trans. Michael Silverthorne and Jonathan Israel (Cambridge: Cambridge University Press, 2007). The references give the chapter and section, and the Gebhardt edition page numbers in brackets; these appear in the margins of most modern editions.

138. Spinoza, *Tractatus Theologico-Politicus*, XVII, 12 [208]; XVII, 14 [210].

139. For the influence of Spinoza, see Henry de Boulainviller, *Oeuvres Philosophiques*, ed. Renée Simon (The Hague: Martinus Nijhoff, 1973). His historical writings can be seen in *Etat de la France*, 3 vols. (London: T. Wood and S. Palmer, 1727); part of the third volume was translated as *An Historical Account of the Ancient Parliaments of France, or States-General of the Kingdom*, trans. Charles Forman, 2 vols. (London: J. Brindley, 1739). On his work generally, see Renée Simon, *Henry de Boulainviller: Historien, Politique, Philosophe, Astrologue, 1658–1722* (Paris: Boivie, 1941); and on the political purposes of his histories, see Michel Foucault, *"Il faut défendre la société": Cours au Collège de France (1975–1976)*, ed. Maurio Bertani and Alessandro Fontata (Paris: Seuil/Gallimard, 1997). For a valuable discussion, see Peter Gratton, *The State of Sovereignty: Lessons from the Political Fictions of Modernity* (Albany: State University of New York Press, 2011), introduction.

140. Gottfried Leibniz, "De usu geometriae," *Sämliche Schriften und Briefe*, ed. Leibniz-Archiv Hannover (Darmstadt, Germany: Otto Reichl, 1923–), ser. 6, 3:449.

141. Leibniz, *The Labyrinth of the Continuum: Writings on the Continuum Problem, 1672–1686*, trans. and ed. Richard T. W. Arthur, Latin-English ed. (New Haven,

CT: Yale University Press, 2001). This collection draws on the *Sämliche Schriften und Briefe*, ser. 6, vols. 2 and 3.

142. Alexander, *Leibniz-Clarke Correspondence*. For discussions, see Alexandre Koyré, "Leibniz and Newton," in *Leibniz: A Collection of Critical Essays*, ed. Harry G. Frankfurt, 239–79 (Notre Dame, IN: University of Notre Dame Press, 1976); A. Rupert Hall, *Philosophers at War: The Quarrel between Newton and Leibniz* (Cambridge: Cambridge University Press, 1980); E. J. Aiton, *Leibniz: A Biography* (Bristol, UK: Adam Hilger, 1985); and Domenico Bertoloni Meli, *Equivalence and Priority: Newton versus Leibniz* (Oxford: Clarendon Press, 1993).

143. Leibniz to Newton, March 7–17, 1692–93, in *Sämliche Schriften und Briefe*, ser. 3, 5:512; also in Isaac Newton, *Philosophical Writings*, ed. Andrew Janiak (Cambridge: Cambridge University Press, 2004), 106.

144. Isaac Newton, *Principia Mathematica*, Definitio VIII, scholium. I have used Isaac Newton, *Philosophiae Naturalis Principia Mathematica*, ed. Alexandré Koyre and I. Bernard Cohen, 2 vols. (Cambridge: Cambridge University Press, 1972); and the selections in *Philosophical Writings*. See Daniel Garber, "Leibniz: Physics and Philosophy," in *The Cambridge Companion to Leibniz*, ed. Nicholas Jolley, 270–352 (Cambridge: Cambridge University Press, 1995), 301–5. On Newton, Lisa Jardine, *Ingenious Pursuits: Building the Scientific Revolution* (London: Little, Brown, 1999), is helpful.

145. Janiak, introduction to Newton, *Philosophical Writings*, xix.

146. Newton, "De Gravitatione et aequipondio fluidorum," in *Unpublished Scientific Writings of Isaac Newton*, ed. A. Rupert Hall and Marie Boas Hall (Cambridge: Cambridge University Press, 1962), 103; and Newton, *Philosophical Writings*, 25.

147. Newton, "De Gravitatione," 91; Newton, *Philosophical Writings*, 12. See also the suggestion that "time, space, place, and motion are very familiar to everyone," in Newton, *Principia Mathematica*, Definitio VIII, scholium.

148. Newton, "De Gravitatione," 91; Newton, *Philosophical Writings*, 13; see Newton, *Principia Mathematica*, Definitio VIII, scholium.

149. Newton, *Principia Mathematica*, Definitio VIII, scholium.

150. Leibniz, "Phoranomus seu de potential et legibus naturæ" (1689), *Opuscules et fragments inédits de Leibniz* (Paris: Alcan, 1903), 590; Leibniz, "On Copernicanism and the Relativity of Motion," in Leibniz, *Philosophical Essays*, ed. Roger Ariew and Daniel Garber (Indianapolis: Hackett, 1989), 91.

151. Leibniz, "Spatium et motus revera relations" (1677?), in *Sämliche Schriften und Briefe*, ser. 6, vol. 4, 1968.

152. Leibniz, "Remarques sur les Objections de M. Foucher" (1695), in *Die Philosophischen Schriften von Gottfried Wilhelm Leibniz*, ed. C. J. Gerhardt, 7 vols. (Hildesheim, Germany: Georg Olms, 1961), 4:491; and Leibniz, *Philosophical Essays*, 146.

153. Leibniz, "Specimen Dynamicum" (1695), in *Mathematische Schriften*, ed. C. I. Gerhardt (Hildesheim, Germany: Georg Olms, 1962), 6:247; and Leibniz, *Philosophical Texts*, trans. and ed. R. S. Woolhouse and Richard Francks (Oxford: Oxford University Press, 1998), 168.

154. Maria Roza Antognazza, *Leibniz: An Intellectual Biography* (Cambridge: Cambridge University Press, 2009), 104.

155. Leibniz, "Fünftes Schreiben" (1716), in *Philosophischen Schriften*, 7:406; and Leibniz, "Fifth Paper," in Alexander, *Leibniz-Clarke Correspondence*, 77.

156. John Sallis, *Force of Imagination: The Sense of the Elemental* (Bloomington: Indiana University Press, 2000), 134. Sallis references Leibniz, "Lettre au Père Bouvet à Paris, 1697," in *Opera Philosophica quae extant Latina, Gallica, Germanica Omnia*, ed. J. E. Erdmann (Aalen, Germany: Scientia, 1959), 146.

157. According to Antognazza, *Leibniz*, 52–53, this apparent choice precipitated an intellectual crisis for the young Leibniz.

158. Leibniz, "Extrait d'une lettre" (1691), in *Philosophischen Schriften*, 4:467; see Sallis, *Force of Imagination*, 134.

159. Garber, "Leibniz: Physics and Philosophy," 284.

160. Leibniz, "Initia rerum mathematicarum metaphysica" (1715), in *Mathematische Schriften*, ed. C. I. Gerhardt, 7:17–29 (Hildesheim, Germany: Georg Olms, 1962), 7:18; and Leibniz, "Metaphysical Foundations of Mathematics," *Selections*, ed. Philip P. Wiener (New York: Charles Scribner's Sons, 1951), 202.

161. Leibniz, "Initia rerum mathematicarum metaphysica," in *Mathematische Schriften*, 7:18; and Leibniz, *Selections*, 202.

162. Leibniz, "Initia rerum mathematicarum metaphysica," in *Mathematische Schriften*, 7:20–21; and Leibniz, *Selections*, 206.

163. Denis E. Cosgrove, "Images of Renaissance Cosmography, 1450–1650," in *The History of Cartography*, vol. 3, *Cartography in the European Renaissance*, 2 pts., ed. David Woodward, pt. 1, 55–98 (Chicago: University of Chicago Press, 2007), 55.

164. Much of this is influenced by Gilles Deleuze, *Le pli: Leibniz et le baroque* (Paris: Les Éditions de Minuit, 1984).

165. Quentin Skinner, "The State," in *Political Innovation and Conceptual Change*, ed. Terence Ball, James Farr, and Russell L. Hanson, 90–131 (Cambridge: Cambridge University Press, 1989), 126.

166. Skinner, *Hobbes and Republican Liberty*, 48.

167. On his reading generally, see Skinner, *Visions of Politics*, vol. 3, chap. 2.

168. Hobbes, *Leviathan*, III, 39. See J. A. Watt, "Spiritual and Temporal Powers," in *The Cambridge History of Medieval Political Thought, c. 350–c. 1450*, ed. J. H. Burns, 367–423 (Cambridge: Cambridge University Press, 1988), 419.

169. Hobbes, *Leviathan*, III, 42 [315].

170. Hobbes, *Leviathan*, IV, 44 [336].

171. Hobbes, *Leviathan*, II, 29 [171–72].

172. This point is stressed by Skinner, *Visions of Politics*, 3:19. His references are to the opening letter of, and the conclusion to, *Leviathan*.

173. Tuck, introduction to Hobbes, *Leviathan*, x.

174. Hobbes, *Leviathan*, IV, 47 [386].

175. Hobbes, *De Cive*, VI, 8. Critical editions of seventeenth-century Latin and English texts appear as *De Cive: Latin Version*, ed. Howard Warrender (Oxford: Clarendon Press, 1983); and *De Cive: English Version*, ed. Howard Warrender (Oxford: Clarendon Press, 1983); see *On the Citizen*, ed. Richard Tuck and Michael Silverthorne (Cambridge: Cambridge University Press, 1998).

176. Hobbes, *Leviathan*, IV, 47 [386].

177. On this relation, see Patricia Springborg, "Thomas Hobbes and Cardinal Bel-larmine: Leviathan and the 'Ghost of the Roman Empire,'" *History of Political Thought* 16 (1995): 503–31.

178. Roberto Bellarmino, *Disputationes* (Ingolstadt, Germany: Dauidis Sartorii, 1590).

179. Hobbes, *Leviathan*, III, 33 [206].

180. Hobbes, *De Cive*, XVII, 22.

181. Hobbes, *Leviathan*, III, 42 [302]. See also I, 13 [60]: "the territories of other Christian princes."

182. Hobbes, *Leviathan*, III, 42 [314].

183. Hobbes, *Leviathan*, III, 42 [312].

184. Hobbes, *Leviathan*, III, 42 [297].

185. Hobbes, *Leviathan*, III, 42 [302].

186. Hobbes, *Leviathan*, I, 17 [87].

187. See Patricia Springborg, "Hobbes's Biblical Beasts: Leviathan and Behemoth," *Political Theory* 23 (1995): 353–75.

188. See, for example, A. P. Martinich, *The Two Gods of Leviathan* (Cambridge: Cambridge University Press, 1992), 362–67; and Quentin Skinner, *Hobbes and Republi-can Liberty* (Cambridge: Cambridge University Press, 2008), 190–96.

189. See Gunnar Olsson, *Abysmal: A Critique of Cartographic Reason* (Chicago: University of Chicago Press, 2007), 233–35.

190. This can be found in the Tuck edition of *Leviathan*, 2. The better-known one is on xciii. For a brief discussion of the variants, and debates concerning the artist, see Arnold A. Rogow, *Thomas Hobbes: Radical in the Service of Reaction* (New York: W. W. Norton, 1986), 156–60; and in more detail, Keith Brown, "The Artist of the Levia-than Title Page," *British Library Journal* 4 (1978): 24–36.

191. Martinich, *Two Gods of Leviathan*, 362–63; and Brown, "Artist of the Levia-than Title Page."

192. Hobbes, *Leviathan*, I, 20 [103].

193. On the "trophy," see Skinner, *Hobbes and Republican Liberty*, 194.

194. Hobbes, *Leviathan*, III, 42 [279]; see Skinner, *Hobbes and Republican Liberty*, 193–94.

195. See Rogow, *Thomas Hobbes*, 159.

196. George Lawson, *Politica Sacra et Civilis*, ed. Conal Condren (Cambridge: Cambridge University Press, 1992). On his work, see Conal Condren, *George Lawson's Politica and the English Revolution* (Cambridge: Cambridge University Press, 1989).

197. On Filmer, see James Daly, *Sir Robert Filmer and English Political Thought* (Toronto: University of Toronto Press, 1979).

198. Robert Filmer, "The Necessity of the Absolute Power of all Kings: and in par-ticular, of the King of England," in *Patriarcha and Other Writings*, ed. Johann Somer-ville (Cambridge: Cambridge University Press, 1991), 172–83. On this relation, see Daly, *Sir Robert Filmer and English Political Thought*, 21–23.

199. Filmer, "Observations upon Aristotles Politiques," in *Patriarcha and Other Writings*, 235–86.

200. Filmer, *Patriarcha*, III, 13.

201. Johann Somerville, "The Authorship and Dating of Some Works," in Filmer, *Patriarcha and Other Writings*, xxxii–xxxvii, xxxxii.

202. Filmer, *Patriarcha*, I, 10.

203. Filmer, *Patriarcha*, I, 10.

204. Edward Coke, *The First Part of the Institutes of the Laws of England* (London: E. and R. Brooke, 1794), chap. 9, section 73 (fol. 58b).

205. Filmer, *The Free-holders Grand Inquest*, in *Patriarcha and Other Writings*, 81.

206. Filmer, *Patriarcha*, I, 7.

207. John Locke, *First Treatise*, in *Two Treatises of Government*, ed. Peter Laslett (1960; repr., Cambridge: Cambridge University Press, 1988) (citations refer to the 1988 edition), IV, 41; VII, 75.

208. Locke, *First Treatise*, X, 136.

209. Laslett, introduction to Locke, *Two Treatises of Government*, 34.

210. Laslett, introduction to Locke, *Two Treatises of Government*, 47.

211. Laslett, introduction to Locke, *Two Treatises of Government*, 61.

212. Skinner, *Foundations of Modern Political Thought*, 2:239.

213. On this theme, see J. W. Gough, *John Locke's Political Philosophy: Eight Studies* (Oxford: Clarendon Press, 1950), chap. 4; and C. B. Macpherson, *The Political Theory of Possessive Individualism: Hobbes to Locke* (Oxford: Clarendon Press, 1962), chap. 5.

214. Locke, *Second Treatise*, in *Two Treatises of Government*, V, 26.

215. Locke, *Second Treatise*, in *Two Treatises of Government*, V, 26.

216. Locke, *Second Treatise*, in *Two Treatises of Government*, V, 27.

217. Locke, *Second Treatise*, in *Two Treatises of Government*, V, 28.

218. Locke, *Second Treatise*, in *Two Treatises of Government*, V, 28.

219. Macpherson, *Political Theory of Possessive Individualism*, 215.

220. Locke, *Second Treatise*, V, 32.

221. Locke, *Second Treatise*, in *Two Treatises of Government*, V, 32.

222. Locke, *Second Treatise*, in *Two Treatises of Government*, V, 33.

223. Locke, *Second Treatise*, in *Two Treatises of Government*, V, 38.

224. Locke, *Second Treatise*, in *Two Treatises of Government*, V, 49.

225. Locke, *Second Treatise*, in *Two Treatises of Government*, V, 41.

226. Macpherson, *Political Theory of Possessive Individualism*, 212.

227. Locke, *Second Treatise*, IX, 124.

228. Locke, *Second Treatise*, in *Two Treatises of Government*, V, 38; see V, 45.

229. Locke, *Second Treatise*, in *Two Treatises of Government*, VI, 73.

230. Locke, *Second Treatise*, in *Two Treatises of Government*, VIII, 119; see VIII, 122.

231. See the anonymous text, probably by James Tyrell, *Patriarcha non Monarcha: The Patriarch Unmonarch'd* (London: Richard Janeway, 1681); and Henry Neville, *Plato Redivivus; or, a Dialogue Concerning Government*, in *Two English Republican Tracts*, ed. Caroline Robbins, 61–200 (Cambridge: Cambridge University Press, 1969).

232. Quentin Skinner, *Liberty before Liberalism* (Cambridge: Cambridge University Press, 1997), 15.

233. James Harrington, *The Commonwealth of Oceana* and *A System of Politics*, ed. J. G. A. Pocock (Cambridge: Cambridge University Press, 1992), 43–44, 48–49. See Pocock, *Machiavellian Moment*, 390–92, chap. 12.

234. Harrington, *Commonwealth of Oceana*, 86, 107, 234.

235. Harrington, *Commonwealth of Oceana*, 11.

236. See also "A System of Politics," in Harrington, *Commonwealth of Oceana*, esp. 271–72.

237. On this theme, see George Gale, "John Locke on Territoriality: An Unnoticed Aspect of the Second Treatise," *Political Theory* 1 (November 1973): 472–85. Gale suggests that what he, rather than Locke, calls "territoriality" is implicit rather than stated as a theme. For an attempt to develop Locke's arguments, see Cara Nine, "A Lockean Theory of Territory," *Political Studies* 56 (2008): 148–65.

238. Both these essays are in Locke, *Political Essays*, ed. Mark Goldie (Cambridge: Cambridge University Press, 1997).

239. John Locke, "The Fundamental Constitutions of Carolina," in *Political Essays*, 161.

240. Theodoro Reinkingk, *Tractatus de regimine saeculari et ecclesiastico* (1619; repr., Marpurgi: Nicolai Hampelii and Johan Jacobi Genathi, 1632), with revised editions up to a sixth in 1659 (citations refer to the 1632 edition). See Gross, *Empire and Sovereignty*, 200–201.

241. Reinkingk, *Tractatus de regimine saeculari et ecclesiastico*, I, v, i; see I, ii, ix.

242. Reinkingk, *Tractatus de regimine saeculari et ecclesiastico*, I, v, iii–iv.

243. Reinkingk, *Tractatus de regimine saeculari et ecclesiastico*, I, v, vi.

244. Reinkingk, *Tractatus de regimine saeculari et ecclesiastico*, I, v, vi.

245. S. H. Steinberg, *The Thirty Years' War and the Conflict for European Hegemony, 1600–1660* (New York: W. W. Norton, 1966).

246. See Robert Ergang, *The Myth of the All-Destructive Fury of the Thirty Years' War* (Pocono Pines, PA: Craftsman, 1956); and Theodore K. Rabb, "The Effects of the Thirty Years' War on the German Economy," *Journal of Modern History* 34 (March 1962): 40–51. As J. V. Polišenský, *War and Society in Europe, 1618–1648* (Cambridge: Cambridge University Press, 1978), 199, notes, "Generalisations about this 'pre-statistical' age are always risky."

247. Peter H. Wilson, *Europe's Tragedy: A New History of the Thirty Years War* (London: Penguin, 2009), 9; and Steinberg, *Thirty Years' War*, 2.

248. Stewart, *Courtier and the Heretic*, 40.

249. See, for example, Leo Gross, "The Peace of Westphalia, 1648–1948," *American Journal of International Law* 42 (1948): 20–41; Martin Wight, *Systems of States*, ed. Hedley Bull (Leicester, UK: Leicester University Press, 1977); and Gianfranco Poggi, *The Development of the Modern State: A Sociological Introduction* (Stanford, CA: Stanford University Press, 1978). For a survey of debates, see James A. Caporaso, "Changes in the Westphalian Order: Territory, Public Authority, and Sovereignty," *International Studies Review* 2 (2000): 1–28; and the papers in that issue generally.

250. See, for example, Stephen D. Krasner, "Westphalia and All That," in *Ideas and Foreign Policy: Beliefs, Institutions and Political Change*, ed. Judith Goldstein and Robert O. Keohane (Ithaca, NY: Cornell University Press, 1993), 235–36; and his "Com-

promising Westphalia," *International Security* 20 (1995–96): 115–51. See also Stéphane Bealuac, *The Power of Language in the Making of International Law: The Word* Sovereignty *in Bodin and Vattel and the Myth of* Westphalia (Leiden, Netherlands: Martinus Nijhoff, 2004), chap. 5.

251. Wilson, *Europe's Tragedy*, 751, 754.

252. Benno Teschke, *The Myth of 1648: Class, Geopolitics and the Making of Modern International Relations* (London: Verso, 2003), 11.

253. Teschke, *Myth of 1648*, 151, 245, for instance.

254. Teschke, *Myth of 1648*, 245.

255. Teschke, *Myth of 1648*, 230.

256. Teschke, *Myth of 1648*, 230–31, for example.

257. Teschke, *Myth of 1648*, 264–65.

258. See also Benno Teschke, "'The Metamorphoses of European Territoriality: A Historical Reconstruction,'" in *State Territoriality and European Integration*, ed. Michael Burgess and Hans Vollaard, 37–67 (London: Routledge, 2006).

259. Hippolithus a Lapide [Bogislaw Philipp von Chemnitz], *Dissertatio de ratione status in imperio nostro Romano-Germanico* (no place: no publisher, 1640).

260. See Foucault, *Sécurité, Territoire, Population*, 245.

261. See Heinz H. F. Eulau, "Theories of Federalism under the Holy Roman Empire," *American Political Science Review* 35 (1941): 643–64.

262. Friedrich Meinecke, *Machiavellism: The Doctrine of* Raison d'État *and Its Place in Modern History*, trans. Douglas Scott (Boulder, CO: Westview, 1984), 134.

263. Rudolf Hoke, "Hippolithus a Lapide," in *Staatsdenker im 17. und 18. Jahrhundert: Reichspublizistik, Politik, Naturrecht*, ed. Notker Hammerstein et al., 118–28 (Frankfurt am Main: Metzner, 1977), 119; and R. Mousnier, "The Exponents and Critics of Absolutism," in *The New Cambridge Modern History*, vol. 4, *The Decline of Spain and the Thirty Years War*, ed. J. P. Cooper (Cambridge: Cambridge University Press, 1970), 110–11.

264. Foucault, *Sécurité, Territoire, Population*, 246, 296.

265. "The Religious Peace of Augsburg, 1555," §§15–24, in Gerhard Benecke, *Germany in the Thirty Years War* (London: Edward Arnold, 1978), 8–9.

266. Wilson, *Europe's Tragedy*, 41.

267. John Ruggie, "Territoriality and Beyond: Problematizing Modernity in International Relations," *International Organization* 47 (1993): 157.

268. Wilson, *Europe's Tragedy*, 42.

269. Wilson, *Europe's Tragedy*, 201, 264.

270. Andreas Osiander, "Sovereignty, International Relations, and the Westphalian Myth," *International Organization* 55 (2001): 270–72; see Osiander, *The States System of Europe, 1640–1990: Peacemaking and the Conditions of European Stability* (Oxford: Clarendon Press, 1994), 12, 40.

271. Gagliardo, *Reich and Nation*, 16, citing Conrad Bornhak, *Deutsche Verfassungsgeschichte vom westfälischen Frieden an* (Stuttgart, Germany: Ferdinand Enke, 1934), 46–54.

272. Wilson, *Europe's Tragedy*, 758.

273. Wilson, *Europe's Tragedy*, 754.

274. Innocent X, "Zelo domus Dei," in *Quellen zur Geschichte des Papsttums und des Römanischen Katholizismus*, ed. D. Carl Mirbt, 4th ed. (Tübingen, Germany: J. C. B. Mohr, 1924), 382–83. See Moritz Borsch, "Papal Policy 1590–1648," in *The Cambridge Modern History*, vol. 4, *The Thirty Years War* (New York: Macmillan, 1906), 688.

275. Wilson, *Europe's Tragedy*, 755.

276. Treaty of Osnabrück, Article VIII, §1, see 4; and Treaty of Münster, §64, see 67. The treaties can be found in *Instrumentum Pacis Westphalicae / Die Westfälischen Friedensverträge 1648*, Latin-German ed. (Bern, Switzerland: Herbert Lang, 1949); and Clive Parry, ed., *The Consolidated Treaty Series*, vol. 1, *1648–49* (Dobbs Ferry, NY: Oceana, 1969) (although the referencing is different in the latter).

277. Treaty of Osnabrück, Article V, 12 (§30).

278. Osiander, "Sovereignty, International Relations, and the Westphalian Myth," 272.

279. Gagliardo, *Reich and Nation*, 4.

280. Gagliardo, *Reich and Nation*, 13, 227, 296; see Teschke, *Myth of 1648*, 242.

281. Gerhard Oestreich, *Neostoicism and the Early Modern State*, ed. Brigitta Oestreich and H. G. Koenigsberger, trans. David McLintock (Cambridge: Cambridge University Press, 1982), 245.

282. George Pagès, *La guerre de trente ans, 1618–1648* (Paris: Payot, 1939), 244; see Fritz Dickmann, *Der Wastfälische Friede*, 3rd ed. (Münster, Germany: Aschendorff, 1972), 129, 133.

283. Treaty of Osnabrück, Article VIII, §1; and Treaty of Münster, §64.

284. Johann Jacob Moser, *Grund-riss de heutigen Staats-Verfassung des Teutschen Reichs*, 5th ed. (Tübingen, Germany: Cotta, 1745), 492; discussed by Osiander, "Sovereignty, International Relations, and the Westphalian Myth," 272.

285. Osiander, "Sovereignty, International Relations, and the Westphalian Myth," 283.

286. Gross, *Empire and Sovereignty*, 293–94, with a reference to Albrecht Randelzhofer, *Völkerrechtliche Aspekte des Heilige Römischen Reiches nach 1648* (Berlin: Duncker and Humblot, 1967), 166.

287. Gagliardo, *Reich and Nation*, viii.

288. Parker, *Thirty Years' War*, 174.

289. Treaty of Münster, Article 65.

290. Wilson, *Europe's Tragedy*, 776.

291. For a discussion of how this developed, see Janice E. Thomson, *Mercenaries, Pirates, and Sovereigns: State-Building and Extraterritorial Violence in Early Modern Europe* (Princeton, NJ: Princeton University Press, 1994).

292. E. A. Beller, "The Thirty Years' War," in *The New Cambridge Modern History*, ed. J. P. Cooper (Cambridge: Cambridge University Press, 1970), 355.

293. For brief exceptions, see Samuel Pufendorf, *Elementorum jurisprudentiae universalis*, Latin-English ed., 2 vols. (Oxford: Clarendon Press, 1931), I, def. 5, on the exercise of eminent domain over *territoria*; and Samuel Pufendorf, *De jure naturae et gentium*, in *Gesammelte Werke*, vol. 4, ed. Frank Böhling (Berlin: Akademie Verlag, 1998), VII, 2, 20.

294. Severinus de Monzambano [Samuel Pufendorf], *Moderni status imperii Roma-*

no-Germanici (Utopiæ: Udonem Neminem, 1668); trans. Edmund Bohun, ed. Michael J. Seidler, *The Present State of Germany* (Indianapolis: Liberty Fund, 2007). On Pufendorf generally, see David Boucher, "Resurrecting Pufendorf and Capturing the Westphalian Moment," *Review of International Studies* 27 (2001): 557–77; and Tuck, *Rights of War and Peace*, chap. 5.

295. Samuel von Pufendorf, *Der statu imperii Germanici*, in *Die Verfassung des deutschen Reiches*, ed. Horst Denzer, Latin-German ed. (Stuttgart: Insel Verlag, 1994), VI, 9. For a discussion, see Peter Schröder, "The Constitution of the Holy Roman Empire after 1648: Samuel Pufendorf's Assessment in His *Monzambano*," *Historical Journal* 42 (1999): 961–83; and Peter H. Wilson, "Still a Monstrosity? Some Reflections on Early Modern German Statehood," *Historical Journal* 49 (2006): 565–76.

296. Pufendorf, *Jus feciale divinum* (1695), in *Gesammelte Werke*, vol. 9, ed. Detlef Döring (2004), XIII, 24. This text was only published posthumously.

297. Wight, *Systems of States*, 21. See also Hendrik Spruyt, *The Sovereign State and Its Competitors: An Analysis of Systems Change* (Princeton, NJ: Princeton University Press, 1995), 205.

298. Terry Nardin, *Law, Morality and the Relations of States* (Princeton, NJ: Princeton University Press, 1983), 57–58.

299. See, for example, Leibniz, "In Severinum de Monzambano" (1668–72), *Schriften und Briefe*, ser. 4, 1:500–502; "Monita Quædam as Samuelis Puffendorfii principia," in *Opera Omnia*, ed. L. Dutens (1768, repr. Hildesheim, Germany: Georg Olms, 1989), vol. 6, pt. 3, 275–84 (citations refer to the 1989 edition); and "Opinion on the Principles of Pufendorf" (1706), in *Political Writings*, 65–75. The former is a commentary on *De statu imperii germanici*; the latter on *De officio hominis et civis*.

300. On the broader context of his political thought, see John Hostler, *Leibniz's Moral Philosophy* (London: Duckworth, 1975); Gregory Brown, "Leibniz's Moral Philosophy," in *Cambridge Companion to Leibniz*, ed. Jolley, 411–41; Patrick Riley, *Leibniz' Universal Jurisprudence: Justice as the Charity of the Wise* (Cambridge, MA: Harvard University Press, 1996). On the phrase "justice as the charity of the wise," see also Antognazza, *Leibniz*, 259–60.

301. Nicholas Jolley, introduction to *Cambridge Companion to Leibniz*, ed. Nicholas Jolley, 1–17 (Cambridge: Cambridge University Press, 1995), 3–4; and Roger Ariew, "G. W. Leibniz, Life and Works," in *Cambridge Companion to Leibniz*, ed. Jolley, 19–20.

302. Leibniz to Hobbes, July 13/23, 1670, in Hobbes, *Correspondence*, 2:713, 2:716–17; Leibniz to Hobbes, ca. 1674, in vol. 2, 731, 733.

303. Antognazza, *Leibniz*, 41, notes that when he had to sell his father's library, Althusius's juridical encyclopedia was one of a few books that he retained.

304. Leibniz, "Specimen difficultatis in jure, seu Dissertatio Casibus perplexis," in *Opera Omnia*, ed. Ludovici Dutens, 6 vols. (Geneva: Fratres de Tournes, 1768), vol. 4, pt. 3, 45–158.

305. See, in particular, Gottfried Leibniz, "Codex Iuris Gentium: Praefatio" (1693), in *Sämliche Schriften und Briefe*, ser. 4, 5:48–79; excerpts in *Political Writings*, 165–76. This was a compilation of legal texts, stemming from his researches into the history of the Brunswick family for the Duke of Hanover. For a discussion, see Roger Berkowitz, *The Gift of Science: Leibniz and the Modern Legal Tradition* (Cambridge, MA: Harvard

University Press, 2005), chaps. 1–3; and Werner Conze, *Leibniz als Historiker* (Berlin: Walter de Gruyter, 1951).

306. Leibniz, "Préceptes pour advancer les sciences," in *Philosophischen Schriften*, 7:167; see Ian Hacking, *The Taming of Chance* (Cambridge: Cambridge University Press, 1990), 86.

307. Hacking, *Taming of Chance*, 185.

308. Hacking, *Taming of Chance*, 18–19. See Leibniz, "Entwurff gewißer Staats-Tafeln" (1680), in *Sämliche Schriften und Briefe*, ser. 4, 3:341–49; §3.C of this volume generally; and §7 of ser. 4, vol. 4. For a discussion, see Jean-Marc Rohrbasser and Jacques Véron, "Leibniz et la mortalité mesure des 'apparences' et calcul de la vie moyenne," *Population* 53, no. 2 (1998): 29–44; Matthew L. Jones, *The Good Life in the Scientific Revolution: Descartes, Pascal, Leibniz and the Cultivation of Virtue* (Chicago: University of Chicago Press, 2006), chaps. 5 and 6.

309. Leibniz to Duke of Hanover, October 1671, in *Sämliche Schriften und Briefe*, ser. 2, 1:263; see Stewart, *Courtier and the Heretic*, 90.

310. Leibniz, *Protogaea*, trans. and ed. Claudine Cohen and Andre Wakefield, Latin-English ed. (Chicago: University of Chicago Press, 2008).

311. Chenxi Tang, *The Geographical Imagination of Modernity: Geography, Literature, and Philosophy in German Romanticism* (Stanford, CA: Stanford University Press, 2008), 140.

312. Carl J. Friedrich, "Philosophical Reflections of Leibniz on Law, Politics, and the State," in *Leibniz*, ed. Frankfurt, 48.

313. Riley, preface to the 2nd ed., in Leibniz, *Political Writings*, ix.

314. Foucault, *Sécurité, Territoire, Population*, 304. See Pierre Costabel, *Leibniz and Dynamics: The Texts of 1692*, trans. R. E. W. Maddison (Paris: Hermann, 1993); and André Robinet, *G. W. Leibniz: Le Meilleur des Mondes par la Balance de l'Europe* (Paris: PUF, 1973).

315. Leibniz, "Vom Naturrecht," in G. F. Guhrauer, *Deutsche Schriften*, 2 vols. (Berlin, 1838–40), 1:414–16; and Leibniz, "On Natural Law," in *Political Writings*, 78.

316. See Riley's note, *Political Writings*, 79n1.

317. Leibniz, "Vom Naturrecht," in *Deutsche Schriften*, 1:416; and Leibniz, *Political Writings*, 79.

318. Aiton, *Leibniz*, 74; Ariew, "G. W. Leibniz, Life and Works," 28; and Antognazza, *Leibniz*, 7–8, 116–17.

319. Antognazza, *Leibniz*, 204.

320. See Ian Hunter, "Conflicting Obligations: Pufendorf, Leibniz and Barbeyrac on Civil Authority," *History of Political Thought* 25 (2004): 683.

321. Caesarinus Fürstenerius [Leibniz], "De Jure Suprematus ac Legationis Principum Germaniae" (1677–88), in *Sämliche Schriften und Briefe*, ser. 4, 2:15–270. Some brief excerpts can be found in *Political Writings*, 111–20.

322. Aiton, *Leibniz*, 75.

323. Leibniz, "Entrétiens de Philarete et d'Eugène sur la question du temps agitée à Nimwègue touchant le droit d'ambassade des électeurs et princes de l'empire" (1677), in *Sämliche Schriften und Briefe*, ser. 4, 2:289–338.

324. Aiton, *Leibniz*, 75; and Antognazza, *Leibniz*, 205.

325. Antognazza, *Leibniz*, 206. The Akademie *Sämliche Schriften und Briefe* edition provides a critical version of the text.

326. Leibniz, "De Jure Suprematus," IX; and Leibniz, *Political Writings*, 113.

327. Leibniz, "De Jure Suprematus," IX; and Leibniz, *Political Writings*, 113.

328. Leibniz, "De Jure Suprematus," IX; and Leibniz, *Political Writings*, 114.

329. Leibniz, "De Jure Suprematus," X; and Leibniz, *Political Writings*, 114.

330. Leibniz, "De Jure Suprematus," X; and Leibniz, *Political Writings*, 115.

331. Leibniz, "De Jure Suprematus," X; and Leibniz, *Political Writings*, 115.

332. Leibniz, "De Jure Suprematus," X; and Leibniz, *Political Writings*, 115–16. See I: "quando Galli vocant *la souveraineté*, mihi Suprematum dicere fas sit"; "Entrétiens de Philarete et d'Eugène," 305; and "De Libero Territorio" (1682), *Sämliche Schriften und Briefe*, ser. 4, 2:394–401.

333. Leibniz, "De Jure Suprematus," XII.

334. Leibniz, "De Jure Suprematus," IX; and Leibniz, *Political Writings*, 114. See also "Codex Iuris Gentium: Praefatio," in *Sämliche Schriften und Briefe*, ser. 4, 5:74; and Leibniz, *Political Writings*, 175.

335. Friedrich, "Philosophical Reflections of Leibniz on Law, Politics, and the State," 62.

336. Leibniz, "Extrait d'une Lettre a l'Auteur du *Journal des Sçavans*" (1678), *Sämliche Schriften und Briefe*, ser. 4, 2:360.

337. Leibniz, "Entrétiens de Philarete et d'Eugène," 305. I do not think that we should extrapolate from this claim, in the mouth of one of the characters in the dialogue, that Leibniz makes a three-way distinction between majesty, supremacy, and territorial superiority, effectively equating the first two and marginalizing the third. See Gross, *Empire and Sovereignty*, 343; and Gierke, *Development of Political Theory*, 220n166.

338. See Riley, introduction to Leibniz, *Political Writings*, 26–28; and John H. Herz, "Rise and Demise of the Territorial State," *World Politics* 9 (1957): 473–93.

339. Leibniz, "De Jure Suprematus," XIII.

340. Riley, *Leibniz' Universal Jurisprudence*, 227.

341. Leibniz, "De Jure Suprematus," preface; and Leibniz, "Codex Iuris Gentium: Praefatio," XV.

342. Friedrich Hertz, "Leibniz as a Political Thinker," in *Festschrift für Heinrich Benedikt überreucht zum 70 Geburtstag*, ed. Hugo Hantsch and Alexander Novotny, 26–38 (Vienna: Notring de wissenschaftlichen Verbände Österreichs, 1957), 28.

343. Stewart, *Courtier and the Heretic*, 80.

344. Leibniz, "Mars Christianissimus" (1683), in *Sämliche Schriften und Briefe*, ser. 4, 2:451–502; and Leibniz, *Political Writings*, 121–45. He used the same title for a work written in 1688–89.

345. Antognazza, *Leibniz*, 222–23.

346. Stewart, *Courtier and the Heretic*, 268–70.

347. Riley, *Leibniz' Universal Jurisprudence*, 2.

348. Leibniz, "Preface de la seconde edition des entretiens sur le droit d'Ambassade des princes d'Allemagne" (1682), in *Sämliche Schriften und Briefe*, ser. 4, 2:291.

349. Leibniz, "De Jure Suprematus," XI; and Leibniz, *Political Writings*, 117.

350. Riley, introduction to Leibniz, *Political Writings*, 27.

351. Leibniz, "Extrait d'une Lettre a l'Auteur du *Journal des Sçavans*," 2:360.

352. Leibniz, "Preface de la seconde edition," 291. See Herz, "Rise and Demise of the Territorial State," 478.

353. Leibniz, "Extrait d'une Lettre a l'Auteur du *Journal des Sçavans*," 2:360.

354. Some of these claims are developed in more systematic form in the writings of Johannes Hertius. See Ioannis Nicolai Hertii, "Dissertatio de superioritate territoriali" (1682), and "Dissertatio de subjectione territoriali" (1698), in *Commentationum atque Opusculorum de selectis et rarioribus*, 2 vols. (Frankfurt am Main: Ioh. Beniam. Andreae, 1737), vol. 1, bk. 2, 127–257, 257–79.

CODA

1. Mary Fulbrook, "Introduction: States, Nations and the Development of Europe," in *National Histories and European History*, ed. Mary Fulbrook, 1–17 (London: University College London Press, 1993), 3.

2. John Breuilly, "Sovereignty and Boundaries: Modern State Formation and National Identity in Germany," in *National Histories and European History*, ed. Fulbrook, 108.

3. See the special issue of the *Scandinavian Journal of History* 10 (1985). More generally, see Samuel Clark, *State and Status: The Rise of the State and Aristocratic Power in Western Europe* (Cardiff: University of Wales Press, 1995); and Charles Tilly, *Coercion, Capital and European States, AD 990–1992* (Oxford: Blackwell, 1992).

4. See Marc Raeff, *The Well-Ordered Police State: Social and Institutional Change through Law in the Germanies and Russia, 1600–1800* (New Haven, CT: Yale University Press, 1983).

5. Breuilly, "Sovereignty and Boundaries," 113.

6. Breuilly, "Sovereignty and Boundaries," 123–24.

7. Breuilly, "Sovereignty and Boundaries," 132.

8. Marcelo Escolar, "Exploration, Cartography and Modernization," in *State/Space: A Reader*, ed. Neil Brenner, Bob Jessop, Martin Jones, and Gordon MacLeod, 29–52 (Oxford: Blackwell, 2002), 33.

9. J. B. Harley, *The New Nature of Maps: Essays in the History of Cartography*, ed. Paul Laxton (Baltimore: Johns Hopkins University Press, 2001), 59.

10. Roger J. P. Kain and Elizabeth Bagnet, *The Cadastral Map in the Service of the State: A History of Property Mapping* (Chicago: University of Chicago Press, 1992), 343.

11. Christian Jacob, *The Sovereign Map: Theoretical Approaches in Cartography throughout History*, ed. Edward H. Dahl, trans. Tom Conley (Chicago: University of Chicago Press, 2006), xviii.

12. J. B. Harley, "Silences and Secrecy: The Hidden Agenda of Cartography in Early Modern Europe," *Imago Mundi* 40 (1988): 59. See David Buisseret, ed., *Monarchs, Ministers and Maps: The Emergence of Cartography as a Tool of Government in Early Modern Europe* (Chicago: University of Chicago Press, 1992).

13. Charles W. J. Withers, *Placing the Enlightenment: Thinking Geographically about the Age of Reason* (Chicago: University of Chicago Press, 2007), 97; and Denis

Cosgrove, *Apollo's Eye: A Cartographic Genealogy of the Earth in the Western Imagination* (Baltimore: Johns Hopkins University Press, 2001), 184.

14. James Scott, *Seeing Like a State: How Certain Schemes to Improve the Human Condition Have Failed* (New Haven, CT: Yale University Press, 1998), 2.

15. Jeppe Strandsbjerg, *Territory, Globalization and International Relations: The Cartographic Reality of Space* (London: Palgrave Macmillan, 2010).

16. Matthew H. Edney, *Mapping an Empire: The Geographical Construction of British India, 1765–1843* (Chicago: University of Chicago Press, 1997); and Sanjay Chaturvedi, "The Excess of Geopolitics: Partition of 'British India,'" in *Partitions: Reshaping States and Minds,* ed. Stefano Bianchini, Sanjay Chaturvedi, Rada Iveković, and Ranabir Samaddar, 126–60 (Abingdon, UK: Frank Cass, 2005).

17. Raymond B. Craib, *Cartographic Mexico: A History of State Fixations and Fugitive Landscapes* (Durham, NC: Duke University Press, 2004).

18. Josef W. Konvitz, *Cartography in France, 1660–1848: Science, Engineering and Statecraft* (Chicago: University of Chicago Press, 1987); Tom Conley, *The Self-Made Map: Cartographic Writing in Early Modern France* (Minneapolis: University of Minnesota Press, 1996); and Chandra Mukerji, *Territorial Ambitions and the Garden of Versailles* (Cambridge: Cambridge University Press, 1997).

19. Friedrich Kratochwil, "Of Systems, Boundaries and Territoriality: An Inquiry into the Formation of the State System," *World Politics* 39 (October 1986): 33.

20. See Anne Marie Claire Godlewska, *Geography Unbound: French Geographic Science from Cassini to Humboldt* (Chicago: University of Chicago Press, 1999), esp. 27–28, 66–86; Withers, *Placing the Enlightenment,* 102–4; David N. Livingstone, *Putting Science in Its Place: Geographies of Scientific Knowledge* (Chicago: University of Chicago Press, 2003), 124ff; and Cosgrove, *Apollo's Eye,* 184, 202.

21. Chenxi Tang, *The Geographical Imagination of Modernity: Geography, Literature, and Philosophy in German Romanticism* (Stanford, CA: Stanford University Press, 2008), 134.

22. Withers, *Placing the Enlightenment,* 29.

23. Breuilly, "Sovereignty and Boundaries," 109.

24. Henri Lefebvre, *La production de l'espace* (Paris: Anthropos, 1974), 325; and Henri Lefebvre, *The Production of Space,* trans. Donald Nicholson-Smith (Oxford: Blackwell, 1991), 285.

25. Yves Lacoste, *La Géographie ça sert d'abord à faire la guerre* (Paris: Maspero, 1976).

26. Jean Baudrillard, *Simulacres et simulation* (Paris: Galilée, 1981), 10; and James Corner, "The Agency of Mapping: Speculation, Critique and Invention," in *Mappings,* ed. Denis Cosgrove, 213–52 (London: Reaktion, 1999), 222.

27. John Pickles, *A History of Spaces: Cartographic Reason, Mapping and the Geo-Coded World* (London: Routledge, 2004), 31; and Geoff King, *Mapping Reality: An Exploration of Cultural Cartographies* (London: Palgrave, 1996), 16–17.

28. See, for example, Andrew Barry, *Political Machines: Governing a Technological Society* (London: Athlone, 2001); Timothy Mitchell, *Rule of Experts: Egypt, Techno-Politics, Modernity* (Berkeley: University of California Press, 2002); and Joe Painter, "Rethinking Territory," *Antipode* 42 (2010): 1090–1118.

29. See Bertrand Badie, *The Imported State: The Westernization of the Political Order*, trans. Claudia Royal (Stanford, CA: Stanford University Press, 2000).

30. Bouda Etemad, *Possessing the World: Taking the Measurements of Colonisation from the 18th to the 20th Century*, trans. Andrene Everson (New York: Berghahn, 2007); see also José Rabasa, *Writing Violence on the Northern Frontier: The Historiography of Sixteenth-Century New Mexico and Florida and the Legacy of Conquest* (Durham, NC: Duke University Press, 2000).

31. See, in an extensive literature, Denis Cosgrove, "The Measures of America," in *Taking Measures across the American Landscape*, ed. James Corner and Alex S. MacLean, 3–19 (New Haven, CT: Yale University Press, 1996); Edwin Danson, *Drawing the Line: How Mason and Dixon Surveyed the Most Famous Border in America* (New York: John Wiley and Sons, 2001); William D. Pattison, *Beginnings of the American Rectangular Land Survey System, 1784–1800* (Chicago: University of Chicago Press, 1957); and more generally, Matthew Hannah, *Governmentality and the Mastery of Territory in Nineteenth Century America* (Cambridge: Cambridge University Press, 2000). On Jefferson's land-surveying projects, see Scott, *Seeing Like a State*, 49–51; Denis Cosgrove, *Geography & Vision: Seeing, Imagining and Representing the World* (London: I. B. Tauris, 2008), chap. 5; Edward S. Casey, *Representing Place: Landscape Painting and Maps* (Minneapolis: University of Minnesota Press, 2002); and Denis Cosgrove, *Social Formation and Symbolic Landscape*, with a new introduction (Madison: University of Wisconsin Press, 1998), chap. 6.

32. Edmund Gunter, *Use of the Sector, Crosse-Staffe, and Other Instruments* (1622; repr., London: Walter J. Johnson, 1971) (citations refer to the 1971 edition).

33. On Gunter, see Andro Linklater, *Measuring America: How the United States was Shaped by the Greatest Land Sale in History* (New York: Plume, 2003); and A. W. Richeson, *English Land Measuring to 1800: Instruments and Practices* (Cambridge, MA: Society for the History of Technology / MIT Press, 1966).

34. Paul Alliès, *L'invention du territoire* (Grenoble: Presses Universitaires de Grenoble, 1980), 147, 152. Lefebvre's notion of "the production of space" has been similarly related to territory by Rhys Jones, *Peoples/States/Territories: The Political Geographies of British State Transformation* (Oxford: Blackwell, 2007), 33–34.

35. Alliès, *L'invention du territoire*, 184.

36. Pierre Dockés, *L'espace dans la pensée économique du XVIe au XVIIIe siècle* (Paris: Flammarion, 1969); Frank Swetz, *Capitalism and Arithmetic: The New Math of the 15th Century* (La Salle, IL: Open Court, 1987); and Richard W. Hadden, *On the Shoulders of Merchants: Exchange and the Mathematical Conception of Nature in Early Modern Europe* (Albany: State University of New York Press, 1994).

37. Karl Marx, *Grundrisse: Foundations of the Critique of Political Economy (Rough Draft)*, trans. Martin Nicolaus (Harmondsworth, UK: Penguin, 1973), 524, 539.

38. For a discussion, see Quentin Skinner, *Visions of Politics*, vol. 3, *Hobbes and Civil Science* (Cambridge: Cambridge University Press, 2002), 318–20.

39. William Petty, "Preface to Political Arithmetic," in *The Economic Writings of Sir William Petty*, ed. Charles Henry Hull, 2 vols. (Cambridge: Cambridge University Press, 1899), 1, 244. On Petty, see also Hacking, *Taming of Chance*, chap. 12; Juri Myk-

känen, "'To Methodize and Regulate Them': William Petty's Governmental Science of Statistics," *History of the Human Sciences* 7 (1994): 65–88; David N. Livingstone, *The Geographical Tradition: Episodes in the History of a Contested Enterprise* (Oxford: Blackwell, 1992), 90–94; Mary Poovey, *A History of the Modern Fact: Problems of Knowledge in the Sciences of Wealth and Society* (Chicago: University of Chicago Press, 1998); and Patrick Carroll, *Science, Culture and Modern State Formation* (Berkeley: University of California Press, 2006), chap. 3.

40. This text is included in *Economic Writings of Sir William Petty*, vol. 2. See Withers, *Placing the Enlightenment*, 198. On political arithmetic more generally, see Peter Buck, "Seventeenth-Century Political Arithmetic: Civil Strife and Vital Statistics," *Isis* 68 (1977): 67–84; and Andrea A. Rusnock, *Vital Accounts: Quantifying Health and Population in Eighteenth-Century England and France* (Cambridge: Cambridge University Press, 2002).

41. Lorraine Daston, *Classical Probability in the Enlightenment* (Princeton, NJ: Princeton University Press, 1988); and her "Enlightenment Calculations," *Critical Inquiry* 21 (1994): 182–202; Alain Desrosières, *The Politics of Large Numbers: A History of Statistical Reasoning*, trans. Camille Naish (Cambridge, MA: Harvard University Press, 1998); Tore Frängsmyr, J. L. Heilbron, and Robin E. Rider, eds., *The Quantifying Spirit in the 18th Century* (Berkeley: University of California Press, 1990); M. Norton Wise, ed., *The Values of Precision* (Princeton, NJ: Princeton University Press, 1995); and William Clark, Jan Godlinski, and Simon Schaffer, eds., *The Sciences in Enlightened Europe* (Chicago: University of Chicago Press, 1999).

42. Max Weber, "Politik als Beruf," in *Gesammelte Politische Schriften*, ed. Johannes Winckelmann (Tübingen, Germany: Mohr, 1988), 510–11.

43. For some important exceptions, see the collection of essays in Neil Brenner, Bob Jessop, Martin Jones, and Gordon MacLeod, eds., *State/Space: A Reader* (Oxford: Blackwell, 2002).

44. See Hendrik Spruyt, *The Sovereign State and Its Competitors: An Analysis of Systems Change* (Princeton, NJ: Princeton University Press, 1995).

45. Quentin Skinner, "Language and Political Change," in *Political Innovation and Conceptual Change*, ed. Terence Ball, James Farr, and Russell L. Hanson, 6–23 (Cambridge: Cambridge University Press, 1989).

46. Rousseau, *Du contract social*, II, 10.1, *Oeuvres complètes*, vol. 3; and Rousseau, *The Social Contract and Other Later Political Writings*, ed. and trans. Victor Gourevitch (Cambridge: Cambridge University Press, 1997).

47. Rousseau, *Sur l'économie politique*, 35, *Oeuvres complètes*, vol. 3; "Discourse on Political Economy" is in *Social Contract*.

48. Rousseau, *Du contract social*, II, 9.5.

49. Rousseau, *L'État de guerre*, 57. The French text is in *Oeuvres complètes*, vol. 3. The English text "The State of War" is in *Social Contract*.

50. Rousseau, *Du contract social*, I, 9.6–7.

51. Rousseau, *Du contract social*, I, 9.5.

52. Rousseau, *Du contract social*, I, 9.4. For a discussion, see William E. Connolly, *The Ethos of Pluralization* (Minneapolis: University of Minnesota Press, 1995), 166–67.

53. Rousseau, *Du contract social*, IV, 3.6.

54. David Hume, *Political Essays*, ed. Knud Haakonssen (Cambridge: Cambridge University Press, 1994).

55. Of late eighteenth-century theorists, perhaps the most detailed discussion of these issues is Daniel Nettelbladt, *Erörterungen einiger einzelner Lehren des teuschen Staatsrechtes*, in Christian Wolff, *Gesammelte Werke Materialien und Dokumente*, ed. Bernhard Martin Scherl, vol. 49 (Hildesheim, Germany: Georg Olms, 1998), 258–81. Nettelbladt argued that territorial supremacy was equal to the rights of other states in Europe, as long as it did not violate the laws of the empire. On this, see Gross, *Empire and Sovereignty*, 422–26.

CPSIA information can be obtained
at www.ICGtesting.com
Printed in the USA
LVHW090921060319
609643LV00036B/371/P